Chemometrics and Authenticity of Foods of Plant Origin

One of the challenges facing the world is feeding the ever-increasing population, with food security being a growing 21st-century problem. This stresses the need for coordinated international systems to prevent and mitigate food fraud in global food supply chains. Food fraud, which is usually financially motivated, has significant consequences including unfair competition, major damage to markets and organizations, loss of consumer confidence, and it raises food safety issues.

A shift toward a more plant-based diet can be endorsed to promote sustainability but also to improve public health and minimize animal suffering. The aim of the book is to deal with issues related to authenticity and chemometrics of the most important food products of plant origin, such as cereals, nuts, legumes, table olives and olive oil, coffee, tea, fruits and vegetables, fruit juices, spices, mushrooms, beers and wines, and honey, using state-of-the-art analytical techniques and instrumentation coupled with available chemometric tools.

Chemometrics and Authenticity of Foods of Plant Origin

Sofia Agriopoulou
Maria Tarapoulouzi
Theodoros Varzakas

CRC Press
Taylor & Francis Group
Boca Raton London New York

CRC Press is an imprint of the
Taylor & Francis Group, an **informa** business

First edition published 2023
by CRC Press
6000 Broken Sound Parkway NW, Suite 300, Boca Raton, FL 33487–2742

and by CRC Press
4 Park Square, Milton Park, Abingdon, Oxon, OX14 4RN

CRC Press is an imprint of Taylor & Francis Group, LLC

ISBN: 978-1-032-19945-0 (hbk)
ISBN: 978-1-032-20972-2 (pbk)
ISBN: 978-1-003-26616-7 (ebk)

DOI: 10.1201/9781003266167

Typeset in Palatino
by Apex CoVantage, LLC

Dedications

Lecturer Dr Sofia Agriopoulou

Dedicated to the people I met along the way who significantly influenced my way of thinking.

Post-Doctoral Researcher Dr Maria Tarapoulouzi

Dedicated to my husband Dimitris Koukounis and to my parents Stamatis and Eleni Tarapoulouzi for their continuous support. Moreover, it is also dedicated to my daughters Lena and Tonia Koukouni.

Professor Dr Theodoros Varzakas

Dedicated to God, Jesus Christ, the Holy Spirit and Virgin Mary. Also dedicated to my family and to the memory of my father and the memory of Ioannis Arvanitoyannis, my mentor.

Contents

Contributors

Rubén Agregán
Centro Tecnológico de la Carne de Galicia
Parque Tecnológico de Galicia
San Cibrao das Viñas, Ourense, Spain

Sofia Agriopoulou
Department of Food Science and Technology
University of the Peloponnese
Antikalamos, Kalamata, Greece

Silvana Mariela Azcarate
Facultad de Ciencias Exactas y Naturales
Universidad Nacional de La Pampa
Santa Rosa, La Pampa, Argentina

Davide Ballabio
Department of Earth and Environmental Sciences
University of Milano-Bicocca
20126 Milano, Italy

José Manuel Camiña
Facultad de Ciencias Exactas y Naturales
Universidad Nacional de La Pampa
Santa Rosa, La Pampa, Argentina

Viviana Consonni
Department of Earth and Environmental Sciences
University of Milano-Bicocca
Milano, Italy

Marilena E. Dasenaki
Department of Chemistry
National and Kapodistrian University of Athens, Panepistimiopolis
Athens, Greece

Rubén Domínguez
Centro Tecnológico de la Carne de Galicia
Parque Tecnológico de Galicia
San Cibrao das Viñas, Ourense, Spain

Noemí Echegaray
Centro Tecnológico de la Carne de Galicia
Parque Tecnológico de Galicia
San Cibrao das Viñas, Ourense, Spain

Hesham el Enshashy
Institute of Bioproduct Development (IBD)
Universiti Teknologi Malaysia (UTM)
Johor Bahru, Malaysia

Ángeles Fernández-Recamales
Faculty of Experimental Sciences
University of Huelva
Huelva, Spain

Adriana S. França
Departamento de Engenharia Mecânica
Universidade Federal de Minas Gerais
Belo Horizonte, Brasil

Antonia Garrido Frenich
Department of Chemistry and Physics
University of Almeria
Almeria, Spain

Raúl González-Domínguez
Faculty of Experimental Sciences
University of Huelva
Huelva, Spain

Fabio Gosetti
Department of Earth and Environmental Sciences
University of Milano-Bicocca
Milano, Italy

Fayyaz Salih Hussain
National Centre of Excellence in Analytical Chemistry
University of Sindh
Jamshoro, Sindh-Pakistan

Anastasios Koidis
Institute for Global Food Security
Queen's University
Belfast, Northern Ireland, United Kingdom

Marios G. Kostakis
Department of Chemistry
National and Kapodistrian University of Athens, Panepistimiopolis
Athens, Greece

José M. Lorenzo
Centro Tecnológico de la Carne de Galicia
Parque Tecnológico de Galicia
San Cibrao das Viñas, Ourense, Spain

Ioannis C. Martakos
Department of Chemistry
National and Kapodistrian University of Athens, Panepistimiopolis
Athens, Greece

Tahir Mehmood
Centre for Applied Molecular Biology (CAMB)
University of the Punjab
Punjab, Pakistan

Najma Memon
National Centre of Excellence in Analytical Chemistry
University of Sindh
Jamshoro, Sindh-Pakistan

Antonella Evelyn Montemerlo
Instituto de Ciencias de la Tierra y Ambientales de La Pampa (INCITAP-CONICET)
Santa Rosa, La Pampa, Argentina

Paulo E. S. Munekata
Centro Tecnológico de la Carne de Galicia
Parque Tecnológico de Galicia
San Cibrao das Viñas, Ourense, Spain

Dimitrios D. Ntakoulas
General Chemical Lab of Research and Analysis
Lamia, Greece

Leandro S. Oliveira
Departamento de Engenharia Mecânica
Universidade Federal de Minas Gerais
Belo Horizonte, Brasil

Ioannis N. Pasias
General Chemical Lab of Research and Analysis
Lamia, Greece

Konstantina Pasvanka
Department of Chemistry
National and Kapodistrian University of Athens
Athens, Greece

Mirian Pateiro
Centro Tecnológico de la Carne de Galicia
Parque Tecnológico de Galicia
San Cibrao das Viñas, Ourense, Spain

Charalampos Proestos
Department of Chemistry
National and Kapodistrian University of Athens, Panepistimiopolis
Athens, Greece

David Psomiadis
Imprint Analytics GmbH
Neutal, Austria

Kalomoira G. Raptopoulou
General Chemical Lab of Research and Analysis
Lamia, Greece

Araceli Rivera-Pérez
Department of Chemistry and Physics
University of Almeria
Almeria, Spain

Jose Manuel Lorenzo Rodriguez
Centro Tecnológico de la Carne de Galicia
Parque Tecnológico de Galicia
San Cibrao das Viñas, Ourense, Spain

Roberto Romero-González
Department of Chemistry and Physics
University of Almeria
Almeria, Spain

Sevastianos Roussos
Aix Marseille Univ, Avignon Université, CNRS, IRD, IMBEMarseille, France

María Santos-Martín
Faculty of Experimental Sciences
University of Huelva
Huelva, Spain

Ana Sayago
Faculty of Experimental Sciences
University of Huelva
Huelva, Spain

Xenofontas Stampelos
Regional Unit of Chios
Region of North Aegean
Chios, Greece

Maria Tarapoulouzi
Department of Chemistry
University of Cyprus
Nicosia, Cyprus

Nikolaos S. Thomaidis
Department of Chemistry
National and Kapodistrian University of Athens, Panepistimiopolis
Athens, Greece

Roberto Todeschini
Department of Earth and Environmental Sciences
University of Milano-Bicocca
Milano, Italy

Theodoros Varzakas
Department of Food Science and Technology
University of the Peloponnese
Antikalamos, Kalamata, Greece

Institute of Bioproduct Development (IBD)
Universiti Teknologi Malaysia (UTM)
Johor Bahru, Malaysia

Marcelo Wagner
Instituto de Ciencias de la Tierra y Ambientales de La Pampa (INCITAP-CONICET)
Santa Rosa, La Pampa, Argentina

Philipp Weller
Competency Center for Chemometrics
Mannheim University of Applied Sciences
Mannheim, Germany

Fadim Yemiş
Department of Chemical Processing Technologies
Manisa Celal Bayar University
Akhisar, Manisa, Turkey

Nilgün Yenil
Chemistry Department
Celal Bayar University
Muradiye, Manisa, Turkey

Jorgelina Zaldarriaga Heredia
Facultad de Ciencias Exactas y Naturales
Universidad Nacional de La Pampa
Santa Rosa, La Pampa, Argentina

Nikoleta Zisi
Imprint Analytics GmbH
Neutal, Austria

1 Introduction

Sofia Agriopoulou, Maria Tarapoulouzi, and Theodoros Varzakas

CONTENTS

1 INTRODUCTION

The food industry is one of the most critical and rapidly growing industries in the world due to the huge increase in the human population and the increased interest of consumers in consuming high-quality products. Food adulteration and fraud are by no means modern phenomena and may be as old as food production (Hong et al., 2017). However, food authenticity research has flourished over the last two decades due to the efforts of food producers to develop high-quality products of specific territorial origin and/or processing, and authentic products have a higher market price (Louppis et al., 2017; Van der Veer et al., 2011). Food adulteration is described as the act of deliberately reducing the quality of food either by adding or exchanging low-quality materials or by eliminating various important ingredients. When the cheapest and lowest quality items are added to an original product, they threaten the health of the consumer, and the product is considered and declared counterfeit (Fakhlaei et al., 2020; Ríos-Reina et al., 2016).

Food adulteration for economic gain is becoming a major issue in many countries as a result of growing world trade. Numerous food fraud scandals have occurred around the world leading to international consequences. One of the possible sources of chemical contamination of food is any intentional adulteration with low-quality (sometimes unsafe) ingredients for economic purposes. Food fraud is not only an illegal activity but can also cause health problems for consumers. For this reason, the authenticity of food is regulated by national and international legislation, such as the Codex Alimentarius standards (Ferreiro-González et al., 2018).

Human health is directly affected by food trends, thus acute or chronic exposure to adulterated products may increase the risk of causing adverse health effects. Even major health risks, including liver, vision, skin and stomach disorders, are directly related to adulterated food intake. Foods such as vegetables, fruits, fish or meat adulterated with formaldehyde have been found to be responsible for asthma and cancer (Bansal et al., 2017). There is also evidence that the addition of dyes is responsible for genotoxicity, hypersensitivity and carcinogenesis (Shomaji, et al., 2021). Various cases of vegetables and other foods have been adulterated with harmful chemicals. In most cases, cheap industrial-grade dyes are used instead of food coloring to maximize profit. Copper sulfate, malachite green and Sudan red are some of the dyes not used in commercial food preparation, but unreliable entities use these substances to color vegetables in the food supply chain to make them look fresh and vibrant (Shomaji et al., 2021).

Recently, a variety of analytical methodologies have been developed for the certification and traceability of food, and a wide variety of applications of such methodologies have been reported in the scientific literature (Biancolillo et al., 2020). Ensuring food safety and quality has become a major challenge mainly due to the globalization of food supply and demand for minimally processed food products (Pigott, 1984). Controlling adulterated foods often requires the application of long-term methodologies and the use of precise analytical instruments used by highly trained personnel. Therefore, there is still a constant demand for new analytical methods capable of detecting small concentrations of chemicals in complex matrices such as food cheaply, sensitively and quickly, without requiring sample pretreatment and the use of non-destructive techniques. In addition, it is very interesting that they can be used directly in the sector or on production lines by people with minimum scientific training or even with non-scientific backgrounds (Revenga-Parra et al., 2020).

DOI: 10.1201/9781003266167-1

1.2 AUTHENTICITY OF FOODS AND BEVERAGES

1.2.1 Some Basic Terms

The automation and computerization of analytical laboratories have led to many changes. One of them is the acquisition of large volumes of data, creating a new scientific science known as the "big data science", which has had a strong impact on many disciplines. In chemistry, the term "big data" refers to large and complex data sets that contain useful and invisible information related to chemistry and should be extracted using complex data analysis tools (Parastar and Tauler, 2018). However, having a large amount of data does not mean that sufficient answers can be given unless the right data processing tools are implemented. Data collection is not synonymous with having information. The data must be treated and interpreted in order to be transformed into useful information for the user or the analyst. This issue, the proper use of big data and how it could meet ISO/IEC 17025 requirements in laboratory accreditation have already been described (Ghernaout et al., 2018).

The nomenclature used to refer to such tools depends on the field of study. Analytical chemistry is the area with the greatest variability of terms. Some authors use the terms "pattern recognition methods" or "multivariate analysis methods", but the most commonly used term to refer to methods used to treat chemistry-related data is "chemometric tools". Chemometrics was originally defined as "an approach to the science of analysis and measurement that uses mathematical, statistical, and other methods of standard logic to determine (often indirectly) the properties of substances that would otherwise be very difficult to measure directly" (Lavine, 2000). Currently, the International Union of Pure and Applied Chemistry (IUPAC) considers chemometrics to be the "science of relating measurements made to a chemical system or process to the state of the system by applying mathematical or statistical methods" (Hibbert, 2016).

1.2.2 Chemometric Techniques

Chemometrics as mentioned above is a well-known discipline based on extracting relevant information from large data sets with multiple variants. In this regard, many studies can be found in the scientific literature published in recent years on food certification, which focus on the development of distinctive models/classifications or quantitative models of fingerprint instruments provided by different analytical platforms. Multivariate classification methods are defined as chemometric techniques designed to find mathematical models that can identify the class in which each object/sample belongs to the base of a particular data set. For example, these methods have been used to classify samples according to their geographical origin or type of variety or to identify adulterants in foods such as olive oil, wine, milk, cheese and many more (Jiménez-Carvelo et al., 2021).

The main purpose of multivariate pattern recognition methods is to perform the most appropriate data processing in order to model and characterize a set of objects (samples) that exhibit a particular characteristic or behavior (Kramer and Twigg, 1970). For this purpose, important and invisible information is extracted to establish relationships between the objects/samples of the set or between the set of objects/samples and one or more characteristics, depending on the similarity of their chemical measurements taken through spectra, chromatograms, elementary analyses, images and so on. These tools must also be able to classify new samples into a specific group and reliably predict the value of a particular property in a fast and objective manner (Brereton, 2015).

Pattern recognition methods are divided into two main groups: non-supervised methods, whose main tools are principal component analysis (PCA) and hierarchical cluster analysis (HCA); and supervised methods such as k-nearest neighbors (kNN; Steinbach and Tan, 2009), partial least squares–discriminant analysis (PLS-DA; Ballabio and Consonni, 2013), and soft independent modeling of class analogy (SIMCA), among others (Kavakiotis et al., 2017). Figure 1.1 shows a simple flowchart of conventional pattern recognition methods.

An exploratory analysis is generally used to detect the data structure and to determine if there are trends in the data set. Key analysis (PCA) is a valuable statistical tool, the goal of which is to maximize data fluctuation information and display it visually in as few data as possible. It is mainly used to obtain a general overview of the data, to provide information about physical groups of objects/samples and to reduce the number of variables needed to represent the system by providing a new set of latent variables known as "main components" (Bro and Smilde, 2014). However, PCA has sometimes been misused and also used in some studies as a classification method to develop and validate classification models (Chung et al., 2015; Hakki Boyaci et al., 2014; Sun et al., 2015); this is a serious mistake that unfortunately still happens. Complex analysis is based on the inherent similarity between groups of objects/samples. The results of the HCA are

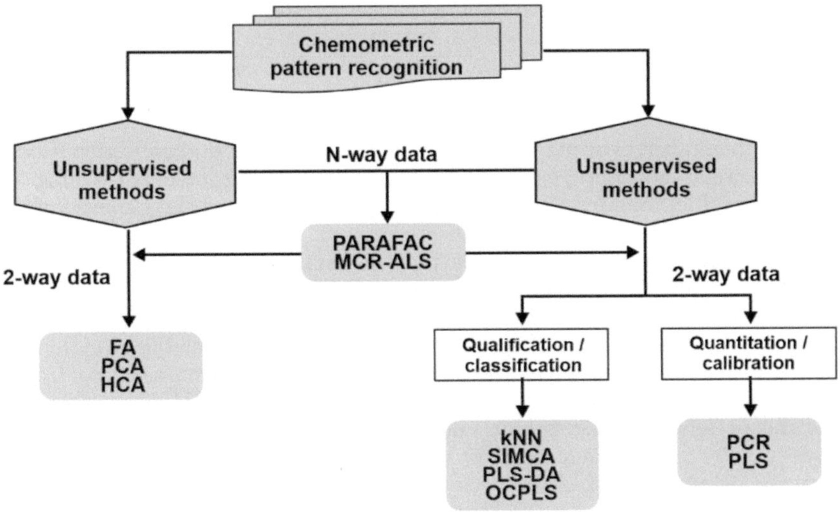

Figure 1.1 Overview of conventional methods for chemometric pattern recognition.
Source: Adapted from Jiménez-Carvelo et al. (2019).

presented as a tree diagram where the objects/samples are distributed in a branched tree where the data are organized into categories and subcategories (branches) and the nodes represent the clusters according to their similarity (Drab and Daszykowski, 2014).

The supervised analysis methods are divided into two groups: (1) classification or certification methods and (2) calibration or quantification methods. Multivariate classification/certification methods are defined as chemometric techniques designed to find mathematical models that can identify in which class each object/sample belongs to, based on a specific data set, by involving the use of various chemometric algorithms based on two main statistical backgrounds related to class discrimination and modeling approaches (Belvilacqua et al., 2014; Gacula and Singh, 1984; Gould, 1989; Gould and Gould,1988; Gromski et al., 2015; Snedecor and Cochran, 1980).

Multivariate calibration/quantification methods are in fact multivariable regression methods that aim to determine the functional relationships between the analytical signal obtained from a set of samples and a characteristic feature of those samples such as their composition (Bender et al., 1982; Hubbard, 1990). The most widely used algorithm is partial least squares (PLS) regression (Mehmood and Ahmed, 2016). It should be noted that, although the classification is intrinsically a qualitative process, the assignment of the objects or samples to a specific class can have a qualitative basis (as by kNN or SIMCA) or a quantitative basis (as by PLS-DA). Indeed, the PLS-DA method involves performing a multivariate regression and placing a numeric value on each object/sample first, and then classifying them into a specific class (Brereton and Lloyd, 2014). In addition, there are other kinds of multivariate methods that are applied in the interpretation of second-order data. That means that a matrix of data is obtained for each sample rather than a vector of data (first-order data). In this case, the most common methods are parallel factor analysis (PARAFAC) and multivariate curve resolution–alternating least squares (MCR-ALS).

The development of a pattern recognition supervised model involves two stages. The first stage is to build the model using a set of objects or samples whose class or particular features are known (i.e., training set or calibration set). In this stage, an internal validation or cross-validation could be applied in order to assess the goodness of fit of the model from the objects/samples of the training set. However, cross-validation by its own designs purpose is never able to achieve all the necessary objectives of a proper validation alone (Esbensen and Geladi, 2010). The second stage is to evaluate and externally validate the performance of the model built in the previous stage; this is done using additional objects or samples (i.e., test set or validation set) that fulfill the same requirements but were not part of the original training set (Szymańska et al., 2015; Westad and Marini, 2015). In these methods, it is assumed that there are enough reference objects/samples that act as analytical standards because the outcomes of interest (i.e., the qualitative class or the value of one or more quantitative features) are formerly known or have been accurately measured (Ballabio et al., 2018;

Cuadros Rodríguez et al., 2016). There are not definitive rules on the minimum number of objects/samples that are necessary for model development, as this depends on the particular problem; however, it would be desirable to devote 40%–50% of the reference object/sample for the validation set (Alewijn et al., 2016; Olivieri, 2014; Olivieri et al., 2006).

As regards the effective use of classification methods, some authors argue that it is better to use class modeling methods such as SIMCA to perform an adequate food authentication (Rodionova et al., 2016). This is because class modeling methods operate, in the training stage, by defining a well-delimited acceptance region that contains all the objects/samples of the target class; consequently, only new objects/samples located in the acceptance region are assigned as belonging to the target class.

As stated above, the supervised multivariate methods are split in two groups: (1) qualification or classification methods and (2) quantification methods. In turn, classification methods are conventionally divided into discriminant analysis methods and class modeling methods depending on how the model is built. Discriminant analysis, as PLS-DA, works by establishing the boundaries between the different classes defined by the training objects, while the class modeling methods, as SIMCA, define successive enclosed space domain that contain the objects of each class. Nevertheless, the models generated by decision tree methods (DT), such as classification and regression tree (CART) or random forest (RF), do not establish the separation of data in different classes as way above but the samples are divided into subsets (or classes) based on the value of certain variables, and this process is repeated on each derived subset of samples (Ai et al., 2014; Cutler et al., 2007). Consequently, the classifications are based on a set of concatenated decisions, similar to artificial neural networks (ANN) (Marini, 2009). Figure 1.2a shows a flowchart of the most

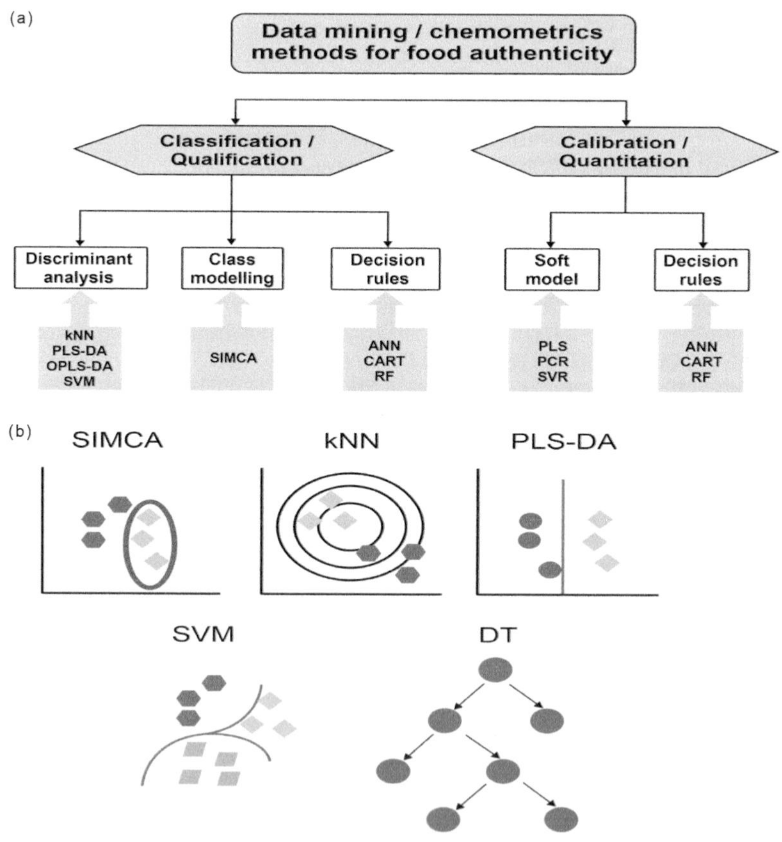

Figure 1.2 (a) Diagram showing the most usual data mining chemometric methods used in food quality and authenticity. (b) Simple graphical description of how some of the most usual data mining/chemometric methods carry out the classification.

Source: Jiménez-Carvelo et al. (2019).

common data mining/chemometrics methods used for the analytical evaluation of food quality and authenticity, and Figure 1.2b shows schematically how these methods operate in general.

1.2.3 Data Fusion

The combination of two or more spectral techniques—that is, different data from different techniques that help to provide more accurate knowledge of a sample and consequently classifications with lower error rates and predictions with less uncertainty than the application of a single technique—is called data fusion. The scientific community has focused on measurements on multiple spectral printing platforms, which provide a larger number increasing the likelihood of better understanding (Tarapoulouzi and Theocharis, 2019, 2021; Tarapoulouzi et al., 2020). The challenge is to combine not only multiple variables but also preprocessing them by combining different data blocks, selecting the most appropriate variables and of course, validating the models at the end. In addition, data merging consists not only of the direct joining of the panels but also of special preprocessing of both separate data and data selected in the process for use in chemometrics. Multiblock orthogonal component analysis (MOCA) has recently become available and is a new method used to interpret fused data (Borràs et al., 2015; Maléchaux et al., 2020; Mishra et al., 2021; Tarapoulouzi and Theocharis, 2021). The MOCA method is applied for fast and accurate analysis of multiple data blocks and extracts two sets of components: the joint components and the unique components. Data fusion strategies have yielded more efficient classification models than models constructed using the spectral information of a single technique (Ríos-Reina et al., 2021).

1.2.4 Chapter Summaries

Chapter 2 presents the basic methods of exploratory analysis and classification of data in the context of food authenticity through chemometrics.

Chapter 3 presents the evaluation of the authenticity of olive oil in terms of geographical origin, varietal origin, cultivation and/or processing parameters. Techniques such as LC and GC with different detectors have been used to investigate the authenticity of olive oil, mainly due to their ease of use and the ability to identify different compounds with different properties. Spectroscopic techniques also provide reliable and robust results, with NMR appearing to be the technique with the highest potential, as it can detect a large number of compounds.

In Chapter 4, the applications of different analytical techniques and chemometric approaches for the investigation of olive authenticity are presented. Varietal and/or geographical origin and cultivation system (organic or conventional) are the two main issues of authenticity in table olives. Liquid chromatography and gas chromatography with the use of mass spectrometry, in combination with advanced chemometrics offer researchers excellent tools to check the authenticity of table olives.

In Chapter 5, the topic of analytical techniques and associated chemometric methods used for certification of authenticity of coffee and tea is discussed. Authenticity of both products can be related to their botanical species and genotypes, geographic provenance, adulteration, processing practices and organoleptic attributes.

In Chapter 6, a summary of the most common analytical and chemometric tools used for fruit and vegetable authentication purposes is provided. Recent literature on the application of chromatographic-based techniques, multi-elemental and isotopic analysis, as well as non-targeted fingerprinting approaches (e.g., metabolomics, spectroscopic techniques), in combination with multivariate statistics, for the authentication of fruits and vegetables in terms of variety, geographical origin, and agricultural practices are also discussed.

Chapter 7 deals with the assessment of the authenticity of fruit juices, through detailed authenticity checks, which is practically a combination of multiple procedures, techniques and methods aimed at detecting and preventing fraudulent practices, especially fraud that may increase and may be associated with health effects such as undeclared sugar addition or the use of synthetic ingredients.

Chapter 8 offers the most novel information about cereals quality analysis, including a wide spectrum of instrumental techniques (e.g., NIR, MIP-OES, ICP-OES, ICP-MS, HPLC, NMR) in conjunction with chemometric methods.

In Chapter 9, different modern analytical techniques used for the detection of adulteration and authenticity of honey are presented. The different chemometric methods used to provide safe information about the adulteration and authenticity of honey are also discussed.

In Chapter 10, a comprehensive overview of analytical-chemometric tools applied during the last decade (2012–2022) to authenticate seasonings is presented, including adulterant detection, quality control, geographical origin traceability, discrimination according to post-harvest processing, and differentiation of cultivars within the same plant species, among other relevant applications related to their genuineness.

In Chapter 11, an overview of the use of approaches, such as proteomic, especially metabolomic, elemental, isotopic, and that based on vibrational spectroscopy-based analysis in conjunction with chemometrics, that help to identify and order legume samples, is presented.

In Chapter 12, the analytical methods for the rapid detection of adulteration in mushrooms in combination with various chemometric approaches are discussed.

In Chapter 13 various sections on techniques have been reported to measure marker compounds of nuts for the identification of geographical origin and the statistical protocols to find out if underlying patterns are provided.

In Chapter 14, wine and beer authenticity assessment combined with chemometrics is discussed. Advanced techniques for wines and beer authentication and successful classification of various wine and beer products according to their variety, geographical origin, and certain aspects of the making process are presented.

In Chapter 15, legislation, and food authenticity issues are discussed. Issues related to food legislation are described, which regulate the entire food production chain, setting controls and regulations, from animal feed, farm controls, pre-processing, to food handling and commercial transactions ensuring that food products are not exposed to any fraudulent practices aimed at deceiving the consumer.

REFERENCES

Ai, F. F., Bin, J., Zhang, Z. M., Huang, J. H., Wang, J. B., Liang, Y. Z., and Yu, L., 2014. Application of random forests to select premium quality vegetable oils by their fatty acid composition. *Food Chemistry, 143*, 472–478.

Alewijn, M., Van der Voet, H., and Van Ruth, S., 2016. Validation of multivariate classification methods using analytical fingerprints—Concept and case study on organic feed for laying hens. *Journal of Food Composition and Analysis, 51*, 15–23.

Ballabio, D., and Consonni, V., 2013. Classification tools in chemistry. Part 1: Linear models. PLS-DA. *Analytical Methods, 5*, 3790–3798.

Ballabio, D., Grisoni, F., and Todeschini, R., 2018. Multivariate comparison of classification performance measures. *Chemometrics and Intelligent Laboratory Systems, 174*, 33–44.

Bansal, S., Singh, A., Mangal, M., Mangal, A. K., and Kumar, S., 2017. Food adulteration: Sources, health risks, and detection methods. *Critical Reviews in Food Science and Nutrition, 57*, 1174–1189.

Biancolillo, A., Marini, F., Ruckebusch, C., and Vitale, R., 2020. Chemometric strategies for spectroscopy-based food authentication. *Applied Sciences, 10*, 6544.

Bender, F. E., Douglas, L. W., and Kramer, A., 1982. *Statistics for Food and Agriculture*, Avi Publish. Co. Inc. Westport, Connecticut.

Belvilacqua, M., Nescatelli, R., Bucci, R., Magrì, A. D., Magrì, A. L., and Marini, F., 2014. Chemometric classification techniques as a tool for solving problems in analytical chemistry. *Journal of AOAC International, 97*, 19–28.

Borràs, E., Ferré, J., Boqué, R., Mestres, M., Aceña, L., and Busto, O., 2015. Data fusion methodologies for food and beverage authentication and quality assessment—A review. *Analytica Chimica Acta, 891*, 1–14.

Brereton, R. G., 2015. Pattern recognition in chemometrics. *Chemometrics and Intelligent Laboratory Systems, 149*, 90–96.

Brereton, R. G., and Lloyd, G. R., 2014. Partial least squares discriminant analysis: Taking the magic away. *Journal of Chemometrics, 28*, 213–225.

Bro, R., and Smilde, A., 2014. Principal component analysis. *Analytical Methods, 6*, 2812–2831.

Chung, I. M., Kim, J. K., Lee, J. K., and Kim, S. H., 2015. Discrimination of geographical origin of rice (Oryza sativa L.) by multielement analysis using inductively coupled plasma atomic emission spectroscopy and multivariate analysis. *Journal of Cereal Science, 65*, 252–259

Cuadros Rodríguez, L., Pérez Castaño, E., and Ruiz Samblás, C., 2016. Quality performance metrics in multivariate classification methods for qualitative analysis. *Trends in Analytical Chemistry, 80*, 612–624.

Cutler, D. R., Edwards, T. C., Beard, K. H., Cutler, A., Hess, K. T., Gibson, J., and Lawler, J. J., 2007. Random forest for classification in ecology. *Ecology, 88*, 2783–2792.

Drab, K., and Daszykowski, M., 2014. Clustering in analytical chemistry. *Journal of AOAC International, 97*, 29–38.

Esbensen, K. H., and Geladi, P., 2010. Principles of proper validation: Use and abuse of re- sampling for validation. *Journal of Chemometrics, 24*, 168–187.

Fakhlaei, R., Selamat, J., Khatib, A., Razis, A. F. A., Sukor, R., Ahmad, S., and Babadi, A. A., 2020. The toxic impact of honey adulteration: A review. *Foods, 9*, 1538.

Ferreiro-González, M., Espada-Bellido, E., Guillén-Cueto, L., Palma, M., Barroso, C. G., and Barbero, G. F., 2018. Rapid quantification of honey adulteration by visible-near infrared spectroscopy combined with chemometrics. *Talanta, 188*, 288–292.

Gacula, M. C., and Singh, I., 1984. *Statistical Methods in Food and Consumer Research*, Academic Press, New York.

Ghernaout, D., Aichouni, M., and Alghamin, A., 2018. Overlapping ISO/IEC 17025:2017 into big data: A review and perspectives. *International Journal of Science and Qualitative Analysis, 4*, 83–92.

Gould, W. A., 1989. *Total Quality Management for the Industries*, CTI Publications, Maryland.

Gould, W. A., and Gould, R. W., 1988. *Total Quality Assurance for the Food Industry*, CTI Publ. Inc., Maryland, pp. 53–76.

Gromski, P. S., Muhamadali, H., Ellis, D. I., Xu, Y., Correa, E., Turner, M. L., and Goodacre, R., 2015. A tutorial review: metabolomics and partial least squares-discriminant analysis—A marriage of convenience or a shotgun wedding. *Analytica Chimica Acta, 879*, 10–23.

Hakki Boyaci, I., Tümay Temiz, H., Selin Uysal, R., Murat Velioglu, H., Jafarzedh Yadegari, R., and Mahmoudi Rishkan, M., 2014. A novel method for discrimination of beef and horsemeat using Raman spectroscopy. *Food Chemistry, 148*, 37–41.

Hibbert, D. B., 2016. Vocabulary of concepts and terms in chemometrics (IUPAC Recommendations 2016). *Pure and Applied Chemistry, 88*, 407–443.

Hong, E., Lee, S. Y., Jeong, J. Y., Park, J. M., Kim, B. H., Kwon, K., and Chun, H. S., 2017. Modern analytical methods for the detection of food fraud and adulteration by food category. *Journal of the Science of Food and Agriculture, 97*, 3877–3896.

Hubbard, M. R., 1990. *Statistical Quality Control for the Food Industry*, AVI Van Nostrand Reinhold, New York.

Jiménez-Carvelo, A. M., González-Casado, A., Bagur-González, M. G., and Cuadros-Rodríguez, L., 2019. Alternative data mining/machine learning methods for the analytical evaluation of food quality and authenticity—A review. *Food Research International, 122*, 25–39.

Jiménez-Carvelo, A. M., Martín-Torres, S., Ortega-Gavilán, F., and Camacho, J., 2021. PLS-DA vs sparse PLS-DA in food traceability. A case study: Authentication of avocado samples. *Talanta, 224*, 121904.

Kavakiotis, I., Tsave, O., Salifoglou, A., Maglaveras, N., Vlahavas, I., and Chouvarda, I., 2017. Machine learning and data mining methos in diabetes research. *Computational and Structural Biotechnology Journal, 15*, 104–116.

Kramer, A., and Twigg, B. A., 1970. *Quality Control for the Food Industries*, 3rd Ed. Vol. 1. AVI Publ. Co., Westport, Conn.

Lavine, B. K., 2000. Chemometrics. *Analytical Chemistry, 72*, 91–97.

Louppis, A. P., Karabagias, I. K., Kontakos, S., Kontominas, M. G., and Papastephanou, C., 2017. Botanical discrimination of Greek unifloral honeys based on mineral content in combination with physicochemical parameter analysis, using a validated chemometric approach. *Microchemical Journal, 135*, 180–189.

Maléchaux, A., Le Dréau, Y., Artaud, J., and Dupuy, N., 2020. Control chart and data fusion for varietal origin discrimination: Application to olive oil. *Talanta, 217*, 121115.

Marini, F., 2009. Artificial neural networks in foodstuff analyses: trends and perspectives. A review. *Analytica Chimica Acta, 635*, 121–131.

Mehmood, T., and Ahmed, B., 2016. The diversity in the applications of partial least squares: An overview. *Journal of Chemometrics, 30*, 4–17.

Mishra, P., Roger, J., Jouan-Rimbaud-Bouveresse, D., Biancolillo, A., Marini, F., Nordon, A., and Rutledge, D. N., 2021. Recent trends in multi-block data analysis in chemometrics for multi-source data integration. *TrAC Trends in Analytical Chemistry, 137*, 116206.

Olivieri, A. C., 2014. Analytical figures of merits: from univariate to multiway calibration. *Chemical Reviews, 114*, 5358–5378.

Olivieri, A. C., Faber, N. M., Ferré, J., Boqué, R., Kalivas, J. H., and Mark, H., 2006. Uncertainty estimation and figures of merits: for multiway calibration. *Pure and Applied Chemistry, 78*, 633–661.

Parastar, H., and Tauler, R., 2018. Big (Bio) Chemical Data Mining Using Chemometric Methods: A Need for Chemists. *Angewante Chemistry International Edition*, 10.1002/anie.201801134.

Pigott, J. R. Ed., 1984. *Sensory Analysis of Foods*, Elsevier Applied Science Publishers, London.
Revenga-Parra, M., Robledo, S. N., Martínez-Periñán, E., González-Quirós, M. M., Colina, A., Heras, A., Pariente, F., and Lorenzo, E., 2020. Direct determination of monosaccharides in honey by coupling a sensitive new Schiff base Ni complex electrochemical sensor and chemometric tools. *Sensors and Actuators B: Chemical, 312*, 127848.
Ríos-Reina, R., Camiña, J. M., Callejón, R. and Azcarate, S. M., 2021. Spectralprint techniques for wine and vinegar characterization, authentication and quality control: Advances and projections, *TrAC Trends in Analytical Chemistry*, 116121.
Ríos-Reina, R., García-González, D. L., Callejón, R. M., and Amigo, J. M., 2016. NIR spectroscopy and chemometrics for the typification of Spanish wine vinegars with a protected designation of origin, *Food Control, 89*, 108–116.
Rodionova, O. Y., Titova, A. V., and Pomerantsev, A. L., 2016. Discriminant analysis in an inappropriate method of authentication. *Trends in Analytical Chemistry, 78*, 17–22.
Shomaji, S., Masna, N. V. R., Ariando, D., Deb Paul, S., Horace-Herron, K., Forte, D., Mandal, S., and Bhunia, S., 2021. Detecting dye-contaminated vegetables using low-field NMR relaxometry. *Foods, 10*, 2232.
Snedecor, G. W., and Cochran, W. G., 1980. *Statistical Methods*, 7th Ed. Iowa University, Ames, IA, USA.
Steinbach, M., and Tan, P. N., 2009. kNN: k-Nearest Neighbours, in X. In V. K. Wu (Ed.). *The Top Ten Algorithms in Data Mining* (pp. 151–161). Chapman & Hall/CRC Press, Boca Raton.
Sun, X., Lin, W., Li, X., Shen, Q., and Luo, H., 2015. Detection and quantification of extra virgin olive oil adulteration with edible oils by FT-IR spectroscopy and chemometrics. *Analytical Methods, 7*, 3939–3945.
Szymańska, E., Gerretzen, J., Engel, J., Geurts, B., Blanchet, L., and Buydens, L. M. C., 2015. Chemometrics and qualitative analysis have a vibrant relationship. *Trends in Analytical Chemistry, 69*, 34–51.
Tarapoulouzi, M., Kokkinofta, R., and Theocharis, C. R., 2020. Chemometric analysis combined with FTIR spectroscopy of milk and Halloumi cheese samples according to species' origin. *Food Science and Nutrition, 8*, 3262–3273.
Tarapoulouzi, M., and Theocharis, C. R., 2019. Discrimination of Cheddar and Kefalotyri cheese samples: Analysis by chemometrics of proton-NMR and FTIR spectra. *Journal of Agricultural Science and Technology, 9*, 347–355.
Tarapoulouzi, M., and Theocharis, R. C., 2021. Discrimination of Cheddar, Kefalotyri and Halloumi cheese samples by the chemometric analysis of FTIR and ^{1}H-NMR spectra. *Journal of Food Process Engineering*, 13933.
Van der Veer, G., Van Ruth, S. M., and Akkermans, W., 2011. Guidelines for validation of chemometric models for food authentication. *RIKLT Report* 2011.22.
Westad, F., and Marini, F., 2015. Validation of chemometric models—A tutorial. *Analytica Chimica Acta, 893*, 14–24.

2 Authenticity and Chemometrics Basics

Viviana Consonni, Roberto Todeschini, Fabio Gosetti, and Davide Ballabio

CONTENTS

2.1 INTRODUCTION

2.1.1 Chemometrics at a Glance

Chemometrics is an interdisciplinary field that is based on chemistry, classical statistics, multivariate analysis and mathematics, used in a synergic way to solve complex chemical problems; in other words, chemometrics exploits statistical and mathematical strategies to search for useful information in chemical data and gain a deeper understanding of complex chemical systems.

Chemometrics is a quite recent field of scientific research. On June 10, 1974, Svante Wold, an organic chemist at the University of Umeå (Sweden), and Bruce Kowalski, an analytical chemist at the University of Seattle (Washington, United States), for the first time proposed the word *chemometrics* to refer to the use of mathematical methods for analyzing chemical data. Since then, chemometrics has received more and more attention from the scientific community thanks to the significant technological development of the last few years and the consequent availability of large amounts of data that need to be processed with powerful tools capable of providing reliable solutions in a short time.

The chemometric philosophy is based on some well-defined and shared principles: multivariate analysis of the problem, searching for the relevant information of data, model validation to

DOI: 10.1201/9781003266167-2

produce predictive models, definition of the model applicability domain for reliable predictions and graphical visualization of the relevant information.

Highly complex problems are characterized by the contemporary presence of different elements of complexity:

- Several potential variables describing the studied system;
- Lack of a priori knowledge about the most relevant variables;
- Presence of experimental noise, spurious information and chance correlation;
- Redundancy of information due to the presence of several correlated variables;
- Synergic and/or antagonistic effects;
- Non-linearity in the relationships searched for;
- Holistic effects and macro-properties of the studied system.

Chemometrics approaches the complexity inside the data by using several different methods based on one or more data-driven models (i.e., soft models) that provide answers to specific questions. The problem is usually analyzed by a workflow that starts from the definition of the objective(s) and the formulation of some work hypotheses. If no previous data (i.e., historical data) are available, it is necessary to plan an experimental work, always based on the actual knowledge: this is carried out by the well-defined mathematical strategies known as the design of experiments (DoE). DoE allows to obtain experimental data with high information content closely related to the solution of the problem by minimizing the number of experiments and consequently costs, resources and time. It is important to point out that the experimental data have to be meaningful and representative of the problem under study to have reliable outcomes from the chemometric analysis.

Once the data have been collected or experimentally produced, the next steps involve their elaboration by using the most appropriate statistical/mathematical tools depending on the goal. In the initial stage, data exploration is commonly undertaken to look at the data structure, search for the characteristic patterns and understand how much information is enclosed in the data, the correlation between variables and the presence of clusters and/or outliers.

In particular, principal component analysis (PCA) is among the fundamental methods for multivariate data exploration, along with the methods based on similarity/diversity analysis, such as cluster analysis that identifies groups of similar samples. Data exploration is based on unsupervised learning, since no prior knowledge is used to draw considerations on data patterns and clusters; as a consequence, a certain degree of subjectivity is always present, especially in the case of highly complex data.

Besides data exploration, another important phase in the chemometric workflow is the calculation of supervised classification or regression models. If the samples belong to well-defined classes on the basis of theoretical assumptions (natural classes), a priori knowledge or by partitioning of a quantitative experimental response, then supervised classification models can be calculated. Classification methods exploit the information on the variables, which describe the samples and their classes, to build mathematical models that predict the class of new samples, which are associated with unknown experimental response. When the samples are associated with a quantitative response, then regression methods are used to define the relationship between the independent chemical variables and the response and to obtain quantitative predictions for new samples.

Finally, if the classification/regression model can be generalized—that is, it becomes independent of the specific data used in the calculation process—then the model acquires a more relevant theoretical role and the knowledge of the problem will be improved.

2.1.2 Chemometrics for Food Authenticity

In the last few decades, due to the relevant scientific and technological progress, the food characterization requires more sophisticated approaches of investigation to face and solve complex problems. A recent review paper (Oliveri, 2017) presents and discusses the different applications of chemometric methods in food sciences, which include the understanding of the effects of production factors on processes and finished products, optimization of manufactures, process monitoring, quality control and food authentication. The author points toa noticeable growing trend of chemometric applications in this research field since 2008. In particular, chemometrics is often encountered in authentication studies where exploratory methods, such as PCA, are used for data analysis and visualization before the application of supervised learning that can involve both qualitative (classification) and quantitative (regression) modeling. For authenticity verification, the chemical analysis of food is referred to as *targeted analysis* if the amounts of characteristic substances, used as authenticity markers, are determined and as *non-targeted analysis* if the food samples are characterized by one or more types of instrumental signals that holistically encode the entire chemical composition of the authentic products. In this framework, several analytical techniques are applied, which require a processing of their multivariate outcomes. The most common are (Oliveri, 2017) chromatography (HPLC, GC and derived techniques), electrophoresis, spectroscopic methods (including ICP-OES, MIR, NIR, Raman, UV-Vis, fluorescence, MS and NMR) and hyperspectral imaging (HSI). Chemometrics provides reliable statistical methods for the rapid analysis of the complex outcomes of the modern analytical platform, which allow a better understanding and use of the underlying chemical information within the data.

The verification of food authenticity is fundamentally a classification problem; indeed, samples are labeled according to their authenticity status, that is, authentic (positive) or non-authentic (negative). Future samples with unknown authenticity status are required to be predicted by multivariate classification models, which exploit the experimental analytical measures of the samples. The commonly applied methods in this framework are the discriminant classification methods, such as linear discriminant analysis (LDA), partial least squares–discriminant analysis (PLS-DA) and nonlinear classifiers, such as those based on local similarity (kNN, N3 and BNN), support vector machines (SVMs) and artificial neural networks (ANNs). However, class modeling methods play a fundamental role when dealing with authenticity problems as they better manage the natural inherent variability due to the heterogeneous matrix of food samples, which has to be properly considered in order to avoid misleading conclusions and biased predictions for new samples. Among these approaches, SIMCA is probably the most popular one, but UNEQ, potential functions and one-class SVM are also worthy of mention.

2.2 CHEMOMETRIC WORKFLOW

2.2.1 Data Structure

Chemometric methods are usually applied to data that are collected in a numerical table. A common data table can be represented as a matrix, where the n rows denote the samples (e.g., objects, experiments, observations, molecules, spectra) and the p columns denote the variables (e.g., features, descriptors, attributes, signals). Variables (or factors) are the chemical entities selected to study the problem, which can be measured through experiments or obtained by theoretical calculations. The samples, described by the variables, are the observations available to understand the problem through the verification of the hypotheses. When the samples are described by several variables, then they are referred to as multivariate data.

Variables are usually distinguished into two logical blocks: the block **X** of independent variables (also called predictors when used in modeling techniques) and the block **Y** of dependent variables (usually called responses).

The data matrix **X** is formally defined as:

$$\mathbf{X} = \begin{vmatrix} x_{11} & x_{12} & \cdots & \cdots & x_{1p} \\ x_{21} & x_{22} & & & \\ \cdots & \cdots & \cdots & \cdots & \cdots \\ \cdots & \cdots & \cdots & \cdots & \cdots \\ x_{n1} & & & & x_{np} \end{vmatrix} \tag{2.1}$$

where the single matrix element x_{ij} represents the value of the jth variable (jth column) for the ith sample (i.e., the ith row).

Adopting some predefined criterion or additional information, a class vector **c** can include the information about the belonging of each sample to one of the a priori defined G classes (e.g., categories, groups). This vector comprises the labels denoting the class memberships to the G classes.

2.2.2 Data Pre-Treatment

Chemometric methods usually require a pre-treatment of the raw data in order to be able to extract useful and non-trivial information and discard obvious (i.e., intrinsic in the single variable) and useless information. Before carrying out any type of multivariate processing, a preliminary analysis of the data with the aim of checking and preparing the data for the subsequent processing consists of the following operations:

- Manually check to avoid obvious errors of data transcription;
- Add a distinctive code for missing data (e.g., -999 or NaN), chosen in such a way that the code can never be confused with a value that theoretically a variable can assume (e.g., never use zero as a code for missing data);
- If a variable is constant (i.e., its values are all equal for the considered samples), then exclude this variable from any subsequent processing.

A more general check concerns the verification of the type of variables that describe the data system: the presence of non-continuous variables (e.g., binary variables that take only values 0 and 1), discrete ordinal variables (e.g., integer values 1, 2, 3 and 4), or real but highly degenerate variables, which assume a few different values, can cause some trouble, as not all the methods and algorithms are suitable for variables of this type or for providing useful information in these conditions.

If one or more variables have particular statistical distributions (non-normality and asymmetrical distributions), it is preferable to transform them in order to eliminate or attenuate the unwanted characteristics. However, once the transformations of the single variables have been applied (e.g., logarithmic transform), we are not yet sure that, in a multivariate analysis that considers all the variables at the same time, they are comparable to each other, in accordance with the objectives and methods used to achieve them. In these cases, to make the variables homogeneous, it is often advisable to scale the variables by acting separately with the same procedure on each of the variables.

In general, data pre-treatment is a critical stage in multivariate studies as (1) it influences the results of the subsequent modeling, (2) each transformation or scaling changes in some way the information contained in the data and (3) the undetected presence of anomalous data or errors can lead to deviant and not general models and conclusions.

When looking for information on the relationships among variables and/or samples, the comparability of the variables should be maximized to avoid misleading results. One can believe that if the purpose of the analysis is to observe the data as they are, no scaling must be carried out. However, some fundamental issues have to be taken into consideration. First, if the variables represent quantities of different nature and/or different measurement scales, both the variances and the differences between pairs of samples can vary a lot from one variable to another. Even if the variables represent the same type of quantity, the variances may differ significantly from each other. The variables with large variance (or large differences between pairs of samples) are the most influential in determining the principal components or the distances between the pairs of samples in cluster analysis. Therefore, the results of the multivariate analysis will be biased by a few variables that mask the possible contributions from other variables only due to their numerical representation, high variance or measurement scale.

Most of the exploratory data analysis methods, such as PCA and all the methods based on the calculation of distances (e.g., similarity/diversity analysis, cluster analysis), are not invariant to the variable scaling; thus, they require an a priori decision on whether and which type of scaling to use.

2.2.3 Regression and Classification Modeling

The majority of the supervised modeling strategies are based on regression and classification methods, depending on the nature of the considered response. For continuous responses, the typical regression model is defined as:

$$y = f\left(x_1, x_2, \ldots, x_p\right) \quad (2.2)$$

where y is any quantitative response of the system in analysis; $x_1, \ldots, x_p$ are the p independent variables; and f is a quantitative function representing the relationship between the response and the variables, which is estimated by a number of observations (i.e., the samples). Multivariate regression is a fundamental supervised approach that has been widely applied in different scientific fields. Regression methods are commonly applied in the framework of authenticity studies to calculate supervised models able to predict the adulteration level of food products. Typical examples are the prediction of the percentage of methanol in alcoholic beverages or the percentage of barley in powder roasted coffee (Oliveri, 2017).

For discrete responses, like qualitative properties distinguishing, for instance, between authentic and non-authentic food products, the typical classification model is defined as:

$$C = f\left(x_1, x_2, \ldots, x_p\right) \tag{2.3}$$

where C is the class label predicted by the classification model and f is a function representing the relationship between the class assignment and the variables describing the samples.

Once the model has been calculated and properly validated for its prediction ability, it can be used to predict the response for new samples. Moreover, the model generally improves the knowledge of the system/problem under study through the interpretation of the role and influence of each variable on the response.

Regression models can be linear or nonlinear. Linear models can be calculated on all the original variables or a reduced number of linear combinations of the variables. Latent variables in partial least squares regression (PLS) and principal components in principal component regression (PCR) are examples of linear combinations of the original variables, which enhance the modeling performance on highly correlated variables or data sets with more variables than samples.

The linear regression model is formalized as follows:

$$\hat{\mathbf{y}} = \mathbf{Xb} \tag{2.4}$$

where the independent variables are collected into the data matrix $\mathbf{X}$ ($n \times p$), $\hat{y}$ is the vector containing the estimates of the response $\mathbf{y}$, and $\mathbf{b}$ is the vector of the regression coefficients. Ordinary least squares regression (OLS), also known as multiple linear regression (MLR), is the most common regression technique used to estimate the quantitative relationship between variables and the response (Draper and Smith, 1998). OLS calculates the regression coefficients that minimize the residual sum of squares (RSS), as follows:

$$\mathbf{b}_{OLS} = \left(\mathbf{X}^{\mathrm{T}}\mathbf{X}\right)^{-1}\mathbf{X}^{\mathrm{T}}\mathbf{y} \tag{2.5}$$

OLS is definitely one of the most common and popular regression approaches. However, there are some limitations related to the constraints of the inversion of the information matrix ($\mathbf{X}^{\mathrm{T}}\mathbf{X}$): the number n of samples must be larger than the number p of variables and no collinear variables should be present in the data matrix.

To overcome these limitations, the PCR method (Basilevsky, 1994; Massy, 1965) calculates the regression coefficients replacing the original variables by a reduced number of their linear combinations (i.e., principal components) in the following way:

$$\mathbf{b}_{PCR} = \left(\mathbf{L}\Lambda^{-1}\mathbf{L}^{\mathrm{T}}\right)\mathbf{X}^{\mathrm{T}}\mathbf{y} \tag{2.6}$$

where Λ is the matrix of PCA eigenvalues and $\mathbf{L}$ ($p \times M$) is the loading matrix. The optimal number M of principal components to be used in the regression model can be selected by a cross-validation protocol.

Like PCR, PLS identifies a new set of linear combinations, denoted as latent variables (LVs), which are used in place of the original variables. The latent variables are the linear combinations that describe the largest amount of data variance in $\mathbf{X}$ with maximal correlation to the response (Wold, 1966). The NIPALS algorithm (Andersson, 2009), which is one of the several different algorithms to estimate the PLS coefficients (Marengo and Todeschini, 1991), calculates the regression coefficients as follows:

$$\mathbf{b}_{PLS} = \mathbf{W}(\mathbf{P}^{\mathrm{T}}\mathbf{W})^{-1}\mathbf{Q}^{\mathrm{T}} \tag{2.7}$$

where $\mathbf{P}$ ($p \times M$) and $\mathbf{Q}$ ($n \times M$) are the loadings for $\mathbf{X}$ and $\mathbf{y}$, respectively, and $\mathbf{W}$ ($p \times M$) collects the PLS weights. As for PCR, the optimal number M of latent variables can be estimated in cross-validation.

Regression methods usually estimate linear relationships between the chemical space and the response. However, linear relationships are often not sufficient to properly fit experimental data and obtain reliable models. In these cases, the relationships are better described by nonlinear functions. A general nonlinear model can be defined as (Varmuza and Filzmoser, 2009):

$$y_i = b_0 + b_1 \cdot g_1\left(x_{i1}\right) + b_2 \cdot g_2\left(x_{i2}\right) + \ldots + b_q \cdot g_q\left(x_{iq}\right) \tag{2.8}$$

where $g_1, g_2, \ldots$ are nonlinear functions of the original variables x and q is the total number of considered nonlinear functions (i.e., the *basis functions*). Usually, the number q of basis functions is larger than the number p of variables present in the data matrix $\mathbf{X}(n,p)$. In the simplest cases, when $g_m(x_m) = x_m;\ m = 1,p$, the classical linear model is obtained and $q = p$. If the original variables are substituted by the corresponding squared variables, that is, $g_m(x_m) = x_m^2;\ m = 1,p$, a simple quadratic model is obtained and $q = p$.

In the last few years, regression techniques based on artificial neural networks (ANNs) have been increasingly used (Drgan et al., 2016; Hochreiter and Schmidhuber, 1997; Olivecrona et al., 2017; Segler et al., 2017; Wythoff, 1993; Zupan et al., 1997; Žuvela et al., 2018), as well as ensemble approaches such as random forest (RF; Grisoni et al., 2019; Polishchuk et al., 2009; Rakers et al., 2017; Svetnik et al., 2003) and modeling approaches based on local similarity measures, like the k-nearest neighbor (kNN) method and its variants (e.g., weighted kNN, N3 and BNN).

While regression methods are used to model quantitative responses, classification methods (also denoted as classifiers) find the mathematical relationship between a set of descriptive variables (e.g., chemical measurements) and a qualitative response, that is, the membership to a defined category (i.e., class).

The classification methods can be distinguished on the basis of the mathematical form of the decision boundary between classes, which can be linear or nonlinear. A linear classification method calculates the best linear boundary for class discrimination, while nonlinear classification methods define the best nonlinear boundary between the classes. Moreover, classification methods are defined as probabilistic when they assume specific probability distributions in the data and are distinguished into parametric and non-parametric methods depending on whether the probability distributions are characterized by central tendency and dispersion parameters (e.g., mean, variance, covariance). Classification methods can also be defined as distance based, if they require the calculation of distance metrics between samples or between samples and models.

Another important distinction can be made between pure classification and class modeling methods. This is particularly important when dealing with the definition of authenticity of food products. In fact, pure classification methods divide the chemical hyperspace into as many regions as the number of available classes. Each sample is classified as belonging to the class corresponding to the region of the hyperspace where the sample is located. According to this strategy, all the samples are always assigned to one of the defined classes. For example, if a pure classification method is used to discriminate Italian and French wines on the basis of their chemical spectra, new samples of unknown origin will be always classified as Italian or French, even in the case they are from other countries. Therefore, it is important to keep in mind that the pure classification models can only be used to predict unknown samples that belong to one of the classes used to calculate the model.

The class modeling methods represent a different approach to classification, since they focus on modeling the analogies among the samples of a target class and define a boundary to separate the specific class from the rest of the chemical hyperspace. Each class is modeled separately and the samples fitting the class model are considered members of the class, while the samples that do not fit are recognized as non-members of that class. A portion of the data hyperspace can be enclosed within the boundaries of more than one class or not assigned to any of the classes. When more target classes are modeled at the same time, different outcomes can result from a class modeling method for a given sample: (1) the sample is assigned to one of the defined classes; (2) the sample is assigned to more than one class (fuzzy classification); or (3) the sample is not assigned (it cannot be assigned to any of the defined classes).

Besides the classical discriminant analysis (DA) and Bayesian methods (Hastie et al., 2009; Mardia et al., 1988), other classification methods widely used in food studies are PLS-DA (Barker and Rayens, 2003), SIMCA (Wold and Sjostrom, 1977), support vector machines (SVM; Vapnik, 1995), classification and regression trees (CART; Breiman et al., 1984), artificial neural networks (ANNs) and similarity-based methods, such as kNN (Cover and Hart, 1967), N3 and BNN (Todeschini et al., 2015). These methods will be presented in more detail in the following paragraphs as most of the supervised models for the verification of food authenticity result from the application of classification methods.

2.2.4 Variable Selection

When modeling a particular property and many variables describing the samples are available, as in the case of multivariate instrumental measurements, it is reasonable to assume that not all the variables but only a few of them are responsible for the property. Often, the presence of many redundant and non-relevant variables can lower the predictive ability of models, limit the interpretation of model outcomes, add noise and spurious information to the data and increase the computational time.

If there is no a priori knowledge about which variables are more important for the specific property to be modeled, they are searched for by using different strategies. In the last few years, the scientific community has paid great attention to develop suitable variable subset selection (VSS) methods for the identification of a small number of variables that are relevant for building the mathematical model of interest. These methods usually provide a population of models, which can have a similar predictive performance with different subsets of variables and, often, different model size (i.e., the number of variables included in the model).

A preliminary step that is often carried out before applying a VSS method is the variable reduction. This consists in the unsupervised elimination of variables on the basis of considerations other than the correlation with the property to be modeled. Variable reduction usually relies only on the comparison and analysis of the independent variables. For instance, variables that are constant or near constant do not encode relevant information and therefore can be eliminated. Likewise, the variables with low information content (e.g., the Shannon entropy), which are expected to be less effective in modeling the response of interest, are often eliminated. Finally, when two or more variables have an absolute correlation value higher than a predefined threshold (e.g., >0.95), one can be discarded to avoid redundant information.

Unlike variable reduction, variable selection is the supervised identification of the variables that are relevant for modeling the response of interest. The simplest variable selection method is the all subsets models (ASM) method that evaluates all the possible combinations of the p variables; it guarantees that the best subset of variables is found by maximizing the model accuracy but it is very computationally consuming and thus not feasible for large numbers of variables. To overcome this limitation, several variable selection methods have been proposed in the literature; the most popular methods are briefly described below.

The forward selection (FS) is a basic and simple approach (Efroymson, 1960) that, starting from an empty set, adds variables sequentially one at a time; the criterion to select which variable to include is based on the minimization of an error function or the maximization of the model accuracy. The main drawback of this approach is that the results can be biased by the first selected variable and then the global optimum is often not reached when several variables are involved in the calculation.

Genetic algorithms (Gas) are a popular variable selection approach (Leardi, 1996) that is inspired by the natural selection of the individuals of a population during its evolution. In analogy with biological systems, each gene represents a variable and each chromosome (sequence of genes/variables) is a model. The evolution of the population of chromosomes is determined by two processes: (1) in the crossover process, pairs of chromosomes generate new individuals according to a crossover probability; and (2) in the mutation process, some genes of a chromosome can change according to a mutation probability. The quality of each chromosome is evaluated by a selected fitness function according to which the chromosome survives or is excluded from the population. In this way, the population evolves toward models with optimal variables for the problem under analysis.

The sequential replacement (SR) method, originally proposed by Miller (Miller, 1990), is a selection algorithm that aims to find optimal subsets of variables for a specified model size with an iterative procedure. This algorithm replaces the variables of a given model, one at a time, with all of the available ones to evaluate whether the variable replacement leads to better performance of the model. The algorithm randomly initializes a population of models with constraints on the number of variables (size) for each model. After each variable replacement, the new model is evaluated and compared with the previously generated models. Over the years, different versions of the sequential replacement method have been proposed with the aim to lower the required computational time and avoid sub-optimal solutions (Cassotti et al., 2014).

2.2.5 Validation

Supervised modeling requires validation protocols to evaluate if models are reliable and predictive for unknown future samples. Model validation plays a crucial role in the model development process since it consists of the evaluation of the predictive ability of the model and detecting model pathologies such as overfitting, chance correlation, redundancy and useless model complexity. This is particularly important when many variables are available, which make the relationship between response and variables very complex. In these cases, hard models, which are based on a well-defined theory under some theoretical assumptions, constraints, initial conditions and approximations, cannot be easily applied and are replaced by the so-called soft models, which are empirical data-driven instead of theory-driven models. Hence, the fundamental challenge of soft modeling is a model generalization, that is, the ability of models to capture the relationship between the variables and the response even beyond the specific data used in the model calculation.

The model complexity mainly relies on two factors: (1) the kind of functional relationship f (e.g., linear, quadratic, exponential) explaining the effects of the variables on the response and (2) the number of independent variables included in the model. A too complex model could be characterized by low-quality estimates of the model parameters (i.e., estimates with high uncertainty), while a too simple model could be not able to fit the data properly. In statistics, the first case is called *overfitting*, that is, the model outcomes correspond too closely to a particular set of data and may fail to accurately predict future samples. The second case is called *underfitting*, which occurs when a model cannot adequately catch the underlying structure of the data. An underfitted model is typically a model that lacks some parameters and/or variables or is based on a not appropriate functional relationship (e.g., a linear instead of a nonlinear function). Hence, searching for the optimal complexity is the way to balance the two effects to obtain reliable models.

Model validation can be carried out by including a subset of the available samples in the *training set* and the remaining part of the samples in the *test set* (or *validation set*). Training samples are then used to calibrate models, while test samples are used to evaluate the predictive ability of the models, comparing the test set experimental responses with the predicted ones. In some cases, an additional *external test set* can be used to obtain a further estimate of the predictive ability of the final selected model(s); this is particularly useful when the validation set is used to optimize the model parameters. The external test set provides an unbiased estimate of the prediction ability of the model.

Several validation approaches can be adopted to split the data into appropriate subsets that are used to calculate the model and to estimate its predictive ability. The simplest approach is the so-called training/test splitting, which extracts (randomly or using some deterministic algorithms) from the original data a defined percentage of samples (usually 10%–50%) to be used as the test samples. A unique random training/test split can affect the results and thus it should be avoided, when possible (Lanteri, 1992; Wu et al., 1996).

Cross-validation techniques are the most popular techniques for internal validation. They consist of selecting each sample (i.e., leave-one-out [LOO]) or each group of k samples (i.e., leave-more-out [LMO]) only once in the test set; the remaining samples (i.e., $n - 1$ in the case of LOO and $n - k$ for LMO) are used to calibrate partial models that predict the test samples.

Similar to LMO, the Monte Carlo cross-validation (MCCV) randomly selects the samples into cross-validation groups for a very large number of iterations. Bootstrap validation (Efron, 1982; Efron, 1987) selects the training set hundreds of times with replacement, that is, repeated samples are allowed in the training set, with the constraint that the training set dimension is the same as the data matrix (i.e., the number n of samples). Each time, the test set is composed of the samples that have not been selected in the training set. Boosted leave-many-out (boosted LMO) is a non-parametric method used to create test sets including balanced sampling across the response range and/or classes and maximizing the training set diversity by a predefined criterion (Clark, 2003).

2.2.6 Model Quality Evaluation

To evaluate the model performance, one can use indices derived from two kinds of statistics, called *goodness of fit* and *goodness of prediction*. The former pays more attention to the ability of models to fit the training data, while the latter to their prediction power, which is assessed by the aid of validation techniques.

2.2.6.1 *Regression Diagnostics*

The most common indices to evaluate the performance of regression models are:

- The *measures of dispersion*, like the root-mean-square error (*RMSE*) or the standard error (*SE*), which provide simple and unambiguous estimates of the model prediction error:

$$RMSE = \sqrt{\frac{RSS}{n}} = \sqrt{\frac{\sum_{i=1}^{n} (y_i - \hat{y}_i)^2}{n}} \tag{2.9}$$

$$SE = \sqrt{\frac{RSS}{n - p'}} \tag{2.10}$$

- The *correlation-based metrics*, like the coefficient of determination R^2 and similar measures, which are often preferred, since they are independent of the measurement scale of the response and have an upper bound of 1 allowing to directly perceive how good a model is:

$$R^2 = 1 - \frac{RSS}{TSS} = 1 - \frac{\sum_{i=1}^{n} (y_i - \hat{y}_i)^2}{\sum_{i=1}^{n} (y_i - \overline{y})^2} \tag{2.11}$$

In equations (2.9)–(2.11), *RSS* is the residual sum of squares; *TSS* the total sum of squares; y_i and $\hat{y}_i$ are the experimental and calculated responses for the *i*th sample, respectively; $\overline{y}$ is the average response; n is the number of samples; and p' is equal to p (the number of variables) plus 1. These metrics are used also to assess the cross-validation performance (R^2_{CV} or Q^2, $RMSE_{CV}$); in this case, $\hat{y}_i$ refers to the predicted response when the *i*th sample is temporarily left out of the training set (Consonni et al., 2009; Todeschini et al., 2016).

2.2.6.2 Classification Metrics

Several parameters can be used to evaluate the quality of classification models, both for fitting and validation purposes (Frank and Todeschini, 1994). These parameters rely on the misclassifications (i.e., samples assigned to the wrong classes) and on how severe the different classification errors are perceived according to the aim of the classification problem.

The prediction ability of classification models is evaluated by parameters that differ from those of regression methods, since the modeled response is different by nature, that is, qualitative and not quantitative. These parameters are usually calculated by the same validation protocols as the regression models (i.e., single or iterated training/test splitting, internal cross-validation, bootstrap). However, when dealing with classification models some additional constraints should be applied to have a balanced class distribution in the training and test sets. For instance, consider a data set that includes two classes (A and B) and the cross-validation procedure involves the elimination from the training set of one validation group at a time; if the entire class A is removed from the data set to be used as the test set, the validation fails as the model is calibrated on data that do not account for the class A and therefore cannot recognize samples belonging to that class. In conclusion, a good validation protocol should always retain a representative number of samples of all the considered classes in each training group.

The classification metrics are derived from the *confusion matrix*, which is a square matrix with dimension $G \times G$, where G is the number of classes (Table 2.1). Each entry c_{gk} of the confusion matrix represents the number of samples belonging to class g and assigned to class k. The diagonal elements c_{gg} are the counts of the samples that are correctly classified, while the off-diagonal elements represent the number of classification errors. The confusion matrix is generally asymmetric since the number of samples belonging to class g and assigned to class k is not usually equal to the number of samples belonging to k and assigned to g.

Table 2.1 General Representation of a Confusion Matrix with G Classes

		Assigned or Predicted Classes				
		1	2	. . .	G	
True classes	1	c_{11}	c_{12}	. . .	c_{1G}	n_1
	2	c_{21}	c_{22}	. . .	c_{2G}	n_2
	. . .	. . .	. . .	. . .	. . .	. . .
	G	c_{G1}	c_{G2}	. . .	c_{GG}	n_G
		n'_1	n'_2		n'_G	n

Classification indices are metrics that synthesize the information of the confusion matrix. Three well-known class-based measures (sensitivity, precision and specificity) can be used to estimate the classification performance achieved for each class (Ballabio et al., 2018; Brown, 2018). They are calculated on each class separately and encode different aspects of the classification. These parameters are called *primary measures of single classes*.

The *sensitivity* of the gth class (Sn_g) represents the ability of a given classifier to correctly identify the samples of the gth class:

$$Sn_g = \frac{c_{gg}}{n_g} \tag{2.12}$$

where n_g and c_{gg} are the number of samples of the gth class and the number of samples correctly classified in the gth class, respectively.

The *precision* of the gth class (Pr_g) defines the purity of a class, that is, the classifier ability to avoid wrong assignments to that class:

$$Pr_g = \frac{c_{gg}}{n'_g} \tag{2.13}$$

where n'_g is the number of samples that are assigned to the gth class.

The *specificity* of the gth class (Sp_g) represents the ability of a classifier to reject samples of other classes; it is calculated as the ratio of samples not belonging to the gth class, which were not classified in the gth class, over the total number of samples not belonging to the gth class ($n - n_g$):

$$Sp_g = \frac{\sum_{\substack{k=1 \\ k \neq g}}^{G} \left(n_k - c_{kg}\right)}{n - n_g} \tag{2.14}$$

Sensitivity, precision and specificity range between 0 (no class discrimination) and 1 (perfect class discrimination). When dealing with binary classification (e.g., authentic/non-authentic food products), these measures are defined in terms of true (T)/false (F) positives (P)/negatives (N), and the sensitivity of one class always corresponds to the specificity of the other class.

The *non-error rate* (also called *balanced accuracy* or *recall*) is the simplest measure of the quality of a classification model and is calculated as the arithmetic mean of class sensitivities:

$$NER = \frac{\sum_{g=1}^{G} Sn_g}{G} \tag{2.15}$$

The *NER* complementary index is called *error rate* (*ER*). From a statistical point of view, the performance of a classification model is considered satisfactory if the model *ER* is significantly smaller than that obtained by random assignments to the classes. Thus, the classification efficiency can be evaluated by comparing *ER* with the *no-model error rate* (*NOMER*), which represents the error rate obtained by assigning all the samples to the largest class or by a random assignment.

Finally, another common classification measure is the *accuracy*, which corresponds to the ratio of correctly assigned samples:

$$ACC = \frac{\sum_{g=1}^{G} c_{gg}}{n} \tag{2.16}$$

However, consider that accuracy can be a biased estimate when unbalanced classes are modeled, that is, samples are not equally distributed in the considered classes, but some of these are more represented.

To calculate the classification metrics, errors can have a different weight according to how severe the classification error is perceived. For instance, for the authenticity test of a food product, positive (authentic) and negative (non-authentic) samples can be recognized by means of a two-class model. In this case, failing to classify positive samples can be perceived differently as failing to recognize negative samples (i.e., when non-authentic samples are classified as authentic). To address this issue, a penalty matrix, called *loss matrix*, with diagonal elements equal to zero and the off-diagonal elements representing the user-defined costs of classification errors, can be used to calculate the *misclassification risk* (*MR*), which is an estimate of the misclassification probability that takes into account the error costs defined by the user.

Some graphical tools allow to easily analyze the classification results and select the optimal complexity of classification models. For two-class problems, the most common graphical tool is the ROC (receiver operating characteristic) curve, which is obtained by plotting the false positive rate ($1 - Sp$) against the true positive rate (Sn) while the discrimination threshold is changed (Fawcett, 2006). The best classification method would yield a point in the upper left corner of the ROC space, which represents the best combination of high sensitivity and high specificity, while a random classification provides solutions along the diagonal line from the left bottom to the top right corner.

2.3 UNSUPERVISED ANALYSIS

Unsupervised data exploration is the key step in most of the applications of multivariate analysis, as data exploration is always useful to catch relevant information in the data and search for meaningful patterns before calculating any statistic model (Basilevsky, 1994; Jackson, 1991; Jolliffe, 1986; Krzanowski, 1988; Varmuza and Filzmoser, 2009). This preliminary analysis aims to obtain a first insight into the data structure, information about the data distribution, the presence of outliers and clusters, to disclose relationships and correlations between samples and/or variables.

Principal component analysis (PCA) is the most popular technique for data exploration thanks to its wide applicability, the simple calculation algorithm, the capability to extract sequential information and visualize the results by simple plots. Cluster analysis (CA), multidimensional scaling (MSD), minimum spanning tree (MST) and self-organizing map (SOM) are alternative/complementary methods to get information on the data structure in an unsupervised way, that is, without taking into account any qualitative or quantitative response to be modeled. Most of these techniques rely on the use of distance metrics to define the similarity relationships between samples, thus the results can vary depending on the type of metric used and the application of specific variable scaling functions that avoid the bias due to the different measurement scales of variables.

2.3.1 Principal Component Analysis (PCA)

PCA (Basilevsky, 1994) projects the multivariate data in a hyperspace defined by orthogonal principal components (PCs). By retaining only the most relevant PCs, one can reduce the data dimension, eliminate spurious and redundant information and, therefore, clean the data to facilitate the interpretation of the subsequent modeling outcomes.

The PCs are orthogonal vectors calculated as linear combinations of the original variables and ordered by the amount of explained variance. The coefficients of the variables to determine the principal components are called *loadings*. Given the data matrix $\mathbf{X}$ ($n \times p$), the loadings are calculated by spectral decomposition of the correlation matrix $\mathbf{C}$ as shown in Figure 2.1.

The application of the spectral decomposition to the correlation matrix $\mathbf{C}$ (but also to the covariance matrix $\mathbf{S}$) provides two mathematical objects: the eigenvalue matrix Λ ($p \times p$) and the eigenvector matrix $\mathbf{L}$ ($p \times p$), usually called loading matrix. This means that the correlation matrix can be viewed as follows:

$$\mathbf{C} = \mathbf{L} \cdot \Lambda \cdot \mathbf{L}^{\mathrm{T}} \tag{2.17}$$

The loading matrix $\mathbf{L}$ is a matrix whose rows represent the p original variables ($\mathbf{x}_1, \mathbf{x}_2, \mathbf{x}_3, \ldots, \mathbf{x}_p$), and the p columns are the eigenvectors (i.e., the principal components $\mathbf{PC}_1, \mathbf{PC}_2, \ldots, \mathbf{PC}_p$). The

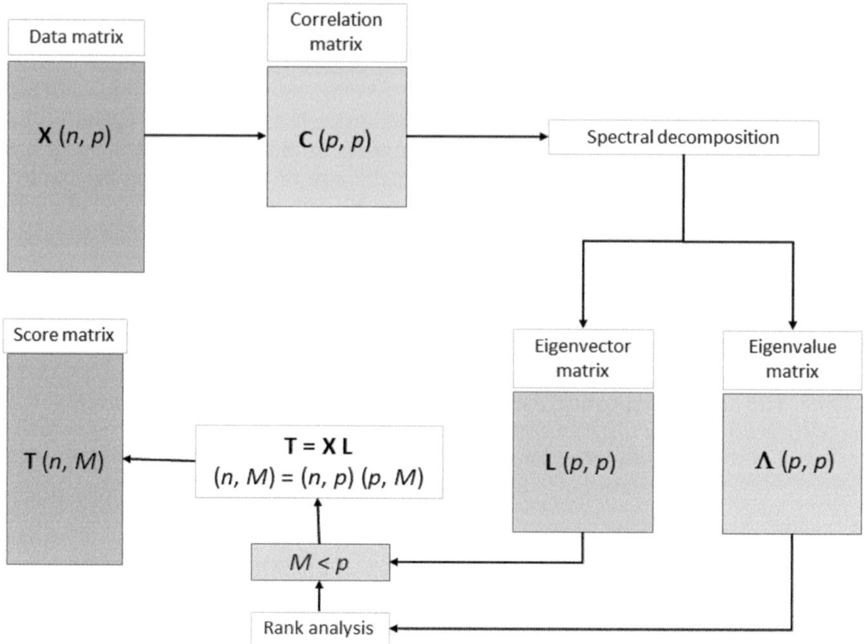

Figure 2.1 Outline of principal component analysis and the relationships among the involved matrices.

values of the matrix **L** are called *loadings* and represent the coefficients in the linear combinations of the original variables to obtain the new orthogonal principal components:

$$\mathbf{L} \equiv \begin{array}{c|cccc|} & PC_1 & PC_2 & \dots & PC_p \\ \hline x_1 & \ell_{11} & \ell_{12} & \dots & \ell_{1p} \\ x_2 & \ell_{21} & \ell_{22} & \dots & \ell_{2p} \\ \dots & \dots & \dots & \dots & \dots \\ x_p & \ell_{p1} & \ell_{p2} & \dots & \ell_{pp} \\ \hline \end{array} \tag{2.18}$$

The loadings are linear standardized coefficients with the following properties:

$$-1 \le \ell_{jm} \le +1 \quad \sum_{j=1}^{p} \ell_{jm}^2 = 1 \quad j = 1, p \quad m = 1, p \tag{2.19}$$

The eigenvalue matrix Λ is a diagonal matrix (i.e., a matrix whose values different from zero are only the diagonal elements), which contains the *eigenvalues*:

$$\Lambda = \begin{vmatrix} \lambda_1 & 0 & \dots & 0 \\ 0 & \lambda_2 & \dots & 0 \\ \dots & \dots & \dots & \dots \\ 0 & 0 & \dots & \lambda_p \end{vmatrix} \tag{2.20}$$

The main properties of the eigenvalues are:

$$\lambda_1 \ge \lambda_2 \ge \lambda_3 \ge \dots \ge \lambda_p \tag{2.21}$$

$$\sum_{j=1}^{p} \lambda_j = p = \sum_{j=1}^{p} c_{ii} \qquad c_{ii} \in \mathbf{C} \qquad \prod_{j=1}^{p} \lambda_j = det(\mathbf{C}) \tag{2.22}$$

$$EV_m\% = \frac{\lambda_m}{\sum_{m=1}^{p} \lambda_m} \cdot 100 \qquad CEV\ \% = \frac{\sum_{m=1}^{M} \lambda_m}{\sum_{m=1}^{p} \lambda_m} \cdot 100 \tag{2.23}$$

where **C** is the correlation matrix and *EV*% denotes the percentage of explained variance.

The *p* eigenvalues, ranked in decreasing order, measure the variances of the principal components. Thus, the sum of the eigenvalues is the total data variance, which is preserved in the new space; in the particular case of diagonalization of the correlation matrix **C**, whose all diagonal elements equal 1, the eigenvalue sum is equal to the number of variables (e.g., *p*). The explained variance (*EV*) of each PC is its normalized eigenvalue, while the cumulative explained variance (*CEV*) is the total data variance explained by the first *M* significant components, which can be selected by the so-called rank analysis.

Considering only the *M* significant components that are associated with the largest eigenvalues, the dimension of the loading matrix **L** decreases from $(p \times p)$ to $(p \times M)$ and the samples can thus be projected into a lower-dimensional space defined by the significant principal components in the following way:

$$\mathbf{T} = \mathbf{XL} \tag{2.24}$$

where **T** $(n \times M)$ is the score matrix, which collects the new coordinates of the *n* samples in the principal component space.

2.3.2 Cluster Analysis (CA)

CA is an unsupervised analysis technique that searches for similarity relationships among samples and defines the clusters where samples can be grouped: similar samples are grouped in the same cluster while less similar samples are in different clusters (Massart and Kaufman, 1983; Willett, 1987). CA is based on the evaluation of the similarity/diversity of all the pairs of samples of a data set. Similarities among samples are estimated by means of distance metrics: similar samples are characterized by small distances while dissimilar samples are associated with large distances. The clustering methods are divided into two main categories: partitioning and hierarchical methods. The crisp clustering methods require that the clusters are non-overlapping, while the fuzzy methods allow every sample to be assigned to more than one cluster since they assign cluster membership probabilities; some other clustering methods are based on the graphical analysis of the sample relationships (e.g., the minimum spanning tree).

Hierarchical clustering methods use linkage approaches (i.e., average linkage, complete linkage, single linkage, weighted average linkage) to quantify the distances between groups of samples: distance measures are used to pair samples into binary clusters and the new clusters are then grouped into larger clusters until a hierarchical dendrogram is obtained. The dendrogram encodes the structure of the data and the final partition of samples is obtained by cutting the dendrogram at the desired level of similarity.

The most popular partitioning method is the k-means method (MacQueen, 1967), also known as the relocation method since the samples are reallocated in different clusters during the iterative algorithm.

The number of clusters *G* has to be chosen a priori by the user, together with the type of variable scaling and the distance metric. Mathematically, the k-means method consists of the minimization of the following cost function *c*:

$$c = min\left[\sum_{g=1}^{G}\sum_{i=1}^{n_g}\left\|\mathbf{x}_i - \mathbf{c}_g\right\|\right] \tag{2.25}$$

where $\mathbf{c}_g$ is the cluster centroid of the *g*th cluster (i.e., the vector of the variable means of the cluster samples) and $\left\|\mathbf{x}_i - \mathbf{c}_g\right\|$ represents the distance between the *i*th sample and the *g*th cluster centroid. The advantages of the k-means algorithm are the computational efficiency, the simple mathematically background and the quick implementation; however, the major limitations are the low accuracy of the initial random centroids and the user-defined number of clusters.

To avoid an a priori selection of the number of clusters, several cluster validity indices were defined. Cluster validity indices (CVI) are used to validate the results of clustering methods and find the cluster set that best fits the natural partitions of the data (Arbelaitz et al., 2013; Desgraupes, 2013). In general, the optimal number G of clusters is estimated on the basis of the trend of the cluster validity index, which is calculated by increasing the number of clusters from 2 to a maximum K.

The Jarvis-Patrick method (Jarvis and Patrick, 1973) is another partitioning method based on the definition of the neighbor matrix and two user-defined parameters: the size L of the neighbor matrix (e.g., $L = n/3$) and the number of common neighbors k, commonly chosen as $k = n/4$ (Frank and Todeschini, 1994). Given a distance matrix $\mathbf{D}$ ($n \times n$), calculated for a selected metric, the first step of the algorithm is the generation of the neighbor matrix ($n \times L$), which collects for each sample the list of the first L neighbors (i.e., the L most similar samples), which are ordered from the most to the least similar. Then, two samples s and t are joined in the same cluster if the following three conditions hold: (1) sample t is among the neighbors of sample s; (2) sample s is among the neighbors of sample t; and (3) they have k common neighbors (other than s and t). The advantage of this method is that the optimal number of clusters results automatically from the algorithm and thus has not to be defined a priori or selected by means of cluster validity indices.

2.3.3 Multidimensional Scaling (MDS)

MDS, also known as non-linear mapping (NLP), takes into account the mutual relationships of samples by reproducing the data structure encoded in the distance (similarity) matrix into a low-dimensional space (Kruskal, 1964a; Kruskal, 1964b; Mardia, 1978). The advantage of MDS is to compress the entire information into a few dimensions (i.e., usually two), providing a visual representation of the original data structure.

The optimization algorithm minimizes one of the following functions, called *stress functions* (Seber, 1984):

$$S_1 = \sum_{s=1}^{n-1} \sum_{t=s+1}^{n} \left| d_{st} - \hat{d}_{st} \right|$$
$$S_2 = \sum_{s=1}^{n-1} \sum_{t=s+1}^{n} \left(d_{st} - \hat{d}_{st} \right)^2 \qquad (2.26)$$
$$S_3 = \frac{1}{C} \cdot \sum_{s=1}^{n-1} \sum_{t=s+1}^{n} \left(\frac{d_{st} - \hat{d}_{st}}{d_{st}} \right)^2$$

where d_{st} is the distance between the sth and tth samples in the p-dimensional space and $\hat{d}_{st}$ the corresponding calculated distance that minimizes the function in the low-dimensional space; the quantity C in the function S_3 is defined as:

$$C = \sum_{s=1}^{n-1} \sum_{t=s+1}^{n} \hat{d}_{st} \qquad (2.27)$$

The function S_1 and S_2 are the Manhattan and Euclidean metrics, respectively, while the *Sammon maps* minimize the function S_3, which privileges small errors for small distances. For this reason, the functions S_1 and S_2 are usually considered more interesting, since the results are more focused on the minimization of the errors on the larger distances.

The limitations of MDS are the lack of information on the role of each variable in determining the observed data patterns and the risk of distortion of the real data structure if the degree of approximation of the low-dimensional space is high.

2.3.4 Minimum Spanning Tree (MST)

The MST is a clustering algorithm that provides a graphical tree representation of the data by exploiting local information to link pairs of samples (Gower and Ross, 1969). The MST algorithm generates a graph where the vertices correspond to the samples and the edges represent the pairwise proximity relationships between samples. These similarity relationships are collected into the pairwise distance matrix $\mathbf{D}$ ($n \times n$). The graph is calculated from the distance matrix according to the following rules: (1) every vertex is connected to every other vertex by a set of edges (i.e., a path); (2) each vertex is connected to at least one edge, that is, the vertex degree is at least equal to 1; and (3) no closed circuits appear in the graph, that is, the graph is a tree.

Among the several algorithms proposed to build an MST, the most popular is an iterative algorithm that elaborates two different sets A and B of pairwise distances between samples: the set A contains the edges (i.e., sample pairs, the pairwise links) of the MST (A is initially an empty set); the set B contains the sample pairs that have not been assigned to the MST yet. The algorithm iteratively assigns to A the sample pair of B, which has the shortest distance and does not form a closed loop with any of the edges already present in A. The algorithm stops when the set A contains $n - 1$ edges.

The MST can be thought of as a distance selection approach, that exploits a subset of $n - 1$ distances out of $n \times (n - 1)/2$ distances (i.e., a percentage equal to $2/n$) to build the tree. As each edge joins a pair of similar samples, MST is used, de facto, as a clustering method. MST was also proposed as an unsupervised classification method (Todeschini and Valsecchi, 2022) by calculating the classification performance after the clustering procedure. For the same data set, several MSTs can be obtained by different distance measures.

2.3.5 Self-Organizing Map (SOM)

In the last few years, the application of artificial neural networks (ANNs) has rapidly grown in different fields requiring data mining and modeling due to its inherent ability to handle non-linear relationships in the data and multitask issues, when multiple experimental responses are considered at the same time (Zupan et al., 1997). ANNs can solve both supervised and unsupervised problems and is very efficient in the analysis of the so-called big data. These methods are supposed to mimic the action mechanism of biological networks of neurons, where each neuron accepts and simultaneously elaborates different signals from neighboring neurons. Through the network neurons, the information encoded in the input vector is fed to one or more hidden layers of neurons, processed and transferred to the output vector.

Among the different ANN algorithms, the Self-Organizing Map (SOM), also known as the Kohonen map, is one of the most popular methods for unsupervised data analysis (Kohonen, 1982; Kohonen, 2001). SOM is usually a squared toroidal topological space of $N \times N$ neurons, N being the number of neurons for each side of the space, where the samples are projected and their mutual proximity relationships can be easily visualized. If the multivariate data set contains n samples described by p variables (i.e., the network input), each neuron is associated with p weights, that is, it contains as many weights as the number of variables. The neuron weights are initialized between 0 and 1 and updated on the basis of the n samples that are seen by the network for a defined number of iterations (i.e., the training epochs). The map learns from the input by means of sequential or batch training algorithms.

In the case of sequential training, for each epoch, the samples are randomly introduced to the network, one at a time. The most similar neuron (i.e., the winning neuron) that minimizes the Euclidean distance from the sample $\mathbf{x}_i$ is selected. Then, the weights of all the neurons are updated as a function of the difference between their values and the variable values of the input sample and the topological distance from the winning neuron:

$$\Delta \mathbf{w}_r^t = \eta^t \cdot \left(1 - \frac{d_{ri}}{d_{max} + 1}\right) \cdot \left(\mathbf{x}_i - \mathbf{w}_r^{t-1}\right) \tag{2.28}$$

where $\mathbf{w}_r^{t-1}$ is the weight vector of the rth neuron at the previous epoch $t - 1$, η a tunable parameter called learning rate and d_{max} the size of the considered neighborhood, which decreases during the training phase. The topological distance d_{ri} is defined as the number of neurons between the rth neuron and the ith winning neuron. The learning rate η changes during the training phase, as follows:

$$\eta^t = \left(\eta^0 - \eta^{tot}\right) \cdot \left(1 - \frac{t}{t_{tot}}\right) + \eta^{tot} \tag{2.29}$$

where t is the current epoch, t_{tot} is the total number of epochs, and η^0 and η^{tot} are the learning rate at the beginning and at the end of the training, respectively.

In the batch training algorithm, all of the samples are presented to the network before the neuron weight update; once the winner neuron of each sample $\mathbf{x}_i$ has been found, the weights $\mathbf{w}_r$ of each neuron r are updated on the basis of the joint effect of all the samples seen by the network as follows:

$$\mathbf{w}_r^t = \frac{\sum_{i=1}^{n} u_{ir}^t \cdot \mathbf{x}_i}{\sum_{i=1}^{n} u_{ir}^t} \quad (2.30)$$

where u_{ir}^t is a smoothing parameter accounting for the learning rate at epoch t and the topological distance d_{ri} of the rth neuron from the winning neuron of the ith sample:

$$u_{ir}^t = \eta^t \cdot \left(1 - \frac{d_{ri}}{d_{max} + 1}\right) \quad (2.31)$$

After the network training, the projection of the input samples into the SOM space is performed by assigning each sample to the most similar neuron of the map; in this way, the data structure is easily visualized in a two-dimensional space and the role of experimental variables in defining the data structure can be elucidated on the basis of the neuron weights.

2.4 CLASSIFICATION APPROACHES

Classification approaches can be distinguished into discriminant and class modeling methods depending on whether they aim at the discrimination of all the considered classes or the recognition of the samples of a target class, respectively. The discriminant strategy classifies samples to one among a number of predefined classes, at least two in the binary framework or more than two in the multiclass framework. Instead, methods based on class modeling, also known as one-class classifiers (Brereton, 2011; Varmuza and Filzmoser, 2009), verify whether a sample is compliant or not with a single class of interest, called target class.

2.4.1 Class Modeling Methods

Class modeling methods exploit the analogies among the samples of the target class to define a decision boundary that distinguishes a specific class from the rest of the chemical hyperspace.

In the food authenticity field, most of the classification problems can be properly addressed by class modeling approaches, however, these methods are not often applied (Brereton, 2011; Oliveri, 2017). Usually, the verification of a given specification, such as the protected designations of geographical origin for food samples, is frequently modeled as a two-class problem, with samples labeled as compliant and noncompliant and the target class being composed of the compliant samples. The target class is usually appropriately sampled, while the non-target class is often unsuitably defined and not well sampled (Khan and Madden, 2014). In this framework, discriminant methods are not recommended as they can lead to biased results, while class modeling approaches can better verify the compliant samples, because they only need a representative sample set for the target class.

2.4.1.1 *SIMCA and UNEQ*

Soft independent modeling of class analogy (SIMCA) is one of the most popular class modeling approaches (Wold, 1976). Given a target class, SIMCA performs the PCA on the samples of the target class. When a new (unknown) sample is projected into the PCA class model, its distance from the class space is calculated on the basis of normalized Q residuals and normalized Hotelling's T^2 values.

Hotelling's T^2 measures the variation of each sample within the model (Bro and Smilde, 2014) and is calculated as the sum of the normalized squared scores as follows:

$$T_i^2 = \sum_{m=1}^{M} \frac{t_{im}^2}{\lambda_m} \quad (2.32)$$

where t_{im} and λ_m are the score of the ith sample and the eigenvalue on the mth component, respectively, and M is the number of significant principal components.

The Q statistic indicates how well each sample conforms to the model, it being a function of the sample residuals:

$$Q_i = \mathbf{e}_i \cdot \mathbf{e}_i^{\mathrm{T}} \tag{2.33}$$

$\mathbf{e}_i$ being the ith row of the residual matrix $\mathbf{E}$:

$$\mathbf{E} = \mathbf{X} - \mathbf{T}\mathbf{L}^{\mathrm{T}} \tag{2.34}$$

where $\mathbf{T}$ is the score matrix and $\mathbf{L}$ the loading matrix. Both Hotelling's T^2 and Q residuals can be associated with upper confidence limits (Bro and Smilde, 2014; Conlin et al., 2000). In particular, for SIMCA, Q residuals and Hotelling's T^2 are normalized over their 95% confidence limits (Oliveri, 2017).

The UNEqual class model (UNEQ) is a class modeling method (Derde and Massart, 1986) that, like SIMCA, exploits PCA on the samples of the target class to define the class space. Once the class model has been calculated, a new sample is projected into the PCA model and its distance from the class space is evaluated to decide whether classify the sample. Unlike SIMCA, UNEQ calculates the sample distance only on the basis of the normalized Hotelling's T^2.

Both for SIMCA and UNEQ, a classification threshold can be calculated by maximization of the class specificity and sensitivity (Vitale et al., 2018) and, finally, samples are classified in the target class space if their normalized distance (SIMCA) or Hotelling's T^2 (UNEQ) is lower than the defined threshold.

2.4.1.2 Potential Functions

Potential functions consider each sample of the target class as a point in the variable space. Around this point, a potential field is calculated, which decreases with the distance from the sample (Coomans et al., 1981). The assignment of a new sample to the target class is carried out by means of the cumulative potential of the class, which is obtained by summing up the individual potentials of the target class samples in the point of the variable space where the new sample is projected. A potential threshold can be defined on the basis of the percentiles of the distribution of class potentials (Forina et al., 1991). A sample is classified in the target class if its potential is higher than the potential class threshold. The shape of the potential field depends on the choice of the type of potential function (kernel) and smoothing parameter (Brereton, 2011).

2.4.1.3 One-Class SVM

Support vector machines (SVMs) are discriminant classifiers that find linear class boundaries either in the original space or in a projected kernel space (Vapnik, 1998). However, SVMs can be adapted to work as a class modeling approach: an alternative method, called support vector domain description (SVDD), has been in fact proposed to work as one-class classifier (Brereton, 2011) by using hyperspheres in the kernel space to define the target class. To calculate the SVDD model, a threshold to identify the class space, along with the kernel type, the cost and the smoothing parameters for the kernel calculation must be defined.

2.4.2 Discriminant Classification Methods

2.4.2.1 Discriminant Analysis (DA)

The Bayesian algorithms are classifiers based on Bayes's theorem, which states that, given a set of events that partition an event space, any event dependent on the event space enriches the knowledge of initial events. Among the Bayesian algorithms, discriminant analysis (DA) is probably the most known method (Fisher, 1936; Mardia et al., 1988; McLachlan, 1992) and can be considered the first multivariate classification technique.

The aim of DA is to find the directions in the data space, which maximize the ratio of the between-class to within-class variance. These directions (linear combinations of the original variables) are called discriminant functions or canonical variates and are in number equal to the number of classes minus one. A sample (x) is assigned to the class with the minimum discriminant score $d_g(\mathbf{x})$:

$$d_g(\mathbf{x}) = \left(\mathbf{x} - \bar{\mathbf{x}}_g\right)^{\mathrm{T}} \cdot \mathbf{S}_g^{-1} \cdot \left(\mathbf{x} - \bar{\mathbf{x}}_g\right) + \ln\left|\mathbf{S}_g\right| - 2\ln\left(P_g\right) \tag{2.35}$$

where P_g is the prior class probability, and $\bar{\mathbf{x}}_g$ and $\mathbf{S}_g$ are the centroid and the covariance matrix of the gth class, respectively. The quantity $d_g(\mathbf{x}) + 2\ln(P_g)$ is referred to as discriminant function, while $\left(\mathbf{x}-\bar{\mathbf{x}}_g\right)^{\mathrm{T}} \mathbf{S}_g^{-1}\left(\mathbf{x}-\bar{\mathbf{x}}_g\right)$ is the Mahalanobis distance between $\mathbf{x}$ and $\bar{\mathbf{x}}_g$.

Depending on the choice of the class-covariance matrix, two different discriminant analysis methods can be defined: quadratic discriminant analysis (QDA) separates the class spaces by quadratic boundaries, while linear discriminant analysis (LDA) defines linear boundaries between classes. QDA assumes that all the classes have a multivariate normal distribution, but the dispersion (represented by the class-covariance matrix $\mathbf{S}_g$) is different for each class, whereas LDA assumes that all the classes have an identical covariance matrix, which is a pooled estimate $\mathbf{S}_\mathrm{p}$ of the dispersion of all the classes.

When the number of samples is small compared to the dimension of the measurement space (i.e., the number of variables), the inversion of covariance matrices is a critical task. This is particularly true for QDA, since a large number of samples is required for each class to calculate the class-covariance matrix. When this requirement is not fulfilled, the methods based on the discriminant analysis can be combined with methods for the dimensionality reduction, such as variable selection approaches or PCA.

2.4.2.2 *Partial Least Squares–Discriminant Analysis (PLS-DA)*

PLS-DA is a classification method that benefits from the properties of PLS regression with the discrimination power of a classification technique (Wold, 1976). Like PLS, which nowadays is one of the most popular regression methods in chemistry (Wold, 1966), PLS-DA is widely applied in several fields for classification purposes. Barker and Rayens (2003) showed that PLS-DA matches the inverse least squares approach to LDA and produces fundamentally the same results but with the noise reduction advantages of PLS.

PLS-DA mainly exploits the PLS2 algorithm to search for the latent variables that have maximum covariance with the responses y. When dealing with binary classifications (i.e., samples are labeled in two classes), the response y encodes the class membership with binary codes equal to 1 for samples in the class of interest and 0 otherwise. The response vector calculated by the PLS-DA model does not have either 1 or 0 values, but a membership score in the range 0–1; therefore, a threshold for the membership score can be defined to decide whether a sample can or cannot be assigned to the class (Pérez et al., 2013). When dealing with multi-class problems (i.e., samples are labeled in G different classes), the PLS2 algorithm is applied to the unfolded class vector. For each sample, PLS-DA calculates a vector of length G with the membership score for each class in the range 0–1: a gth value close to zero indicates that the sample does not belong to the gth class, while a value close to 1 is the opposite. The classification rule can be twofold: (1) the sample is assigned to the class with the maximum membership score or (2) the sample is assigned to the class/classes with a score larger than a user-defined threshold. To select the optimal threshold value, ROC curves can be used to optimize the class specificity and sensitivity with different thresholds.

2.4.2.3 *Support Vector Machine (SVM)*

A support vector machine (SVM) is a binary classifier that identifies the hyperplane separating two classes by maximizing the distance between the closest training samples (Cortes and Vapnik, 1995). Although this method was originally proposed for binary classification, it was later extended to handle multi-class problems.

In this framework, the decision boundary is expressed in terms of a linear combination of functions that are parameterized by support vectors. Support vectors, which are a subset of training samples, are located in the proximity of the boundary between the classes. Consequently, the definition of the classification boundary depends only on a subset of samples. SVM algorithm searches for the boundary with maximal margin among all possible hyperplanes, the margin being the distance between the hyperplane and the support vectors. In the case of nonlinear classification problems, SVM can use nonlinear kernel functions for defining the nonlinear decision boundaries. To calculate the SVM model some parameters have to be optimized, such as the cost value for the margin definition and the smoothing parameter for the kernel calculation.

2.4.2.4 *Local Similarity-Based Algorithms*

Several classification approaches based on the analogies between samples have been proposed so far. These methods rely on the calculation of distance measures to provide a classification in terms of similarity among samples. As the similarity/diversity analysis is strongly influenced by the numerical scale of each variable, scaling procedures must be previously adopted, such as the *range scaling* (equation 35) and *autoscaling* (equation 36):

$$x'_{ij} = \frac{x_{ij} - x_j^{min}}{x_j^{max} - x_j^{min}} \qquad 0 \le x'_{ij} \le 1 \tag{2.36}$$

$$x'_{ij} = \frac{x_{ij} - \overline{x}_j}{s_j} \qquad \overline{x}' = 0 \quad s' = 1 \tag{2.37}$$

where x'_{ij} is the ith value of the jth variable after scaling; and $\overline{x}_j$, s_j, x_j^{min} and x_j^{max} are the mean, the standard deviation, the minimum and the maximum value of the jth variable, respectively.

The *k*-nearest neighbor (*k*NN) classifier is the simplest classification algorithm that is based on the similarity between samples (Cover and Hart, 1967). A sample is classified according to the most frequent class of its *k* most similar training samples. *k*NN is a nonlinear classification method, does not assume any underlying probability density functions and can easily handle multi-class problems.

The N3 (N-nearest neighbors) method is an evolution of the classical *k*NN approach (Todeschini et al., 2015). It is based on the same principle as the *k*NN method, that is, locally weighted information is considered to classify new samples. However, N3 considers all the training samples for the classification (and not only the *k* nearest ones), which are weighted by their similarity ranks to calculate the class membership function.

The binned nearest neighbors (BNN) method is similar to *k*NN with the difference that the prediction is based on a flexible number of neighbors (Todeschini et al., 2015). Similarity intervals (i.e., bins) are defined on the basis of a tuning parameter alpha and the training samples are then distributed into these intervals according to their similarity to the target. All the neighbors falling into the bin with the largest similarity are then considered for the prediction according to the criterion of majority vote.

2.4.2.5 *Tree-Based Algorithms*

Tree algorithms, or decision-making trees, rely on building a tree from the variables (nodes) and the possible values that they can take (strings) until one arrives at the leaves representing the class assigned to the sample. The path from the root node to a leaf node through the arch value determines the path that a particular sample must follow to reach the membership class. The data set is recursively divided into smaller subsets containing samples belonging to as few classes as possible. In each split, the partitioning is performed to reduce the entropy (maximize purity) of the new subsets and the final classification model consists of a collection of nodes (tree) that define the classification rule.

The most common tree-based approach is the classification and regression tree (CART), which is a form of binary recursive partitioning based on univariate rule induction (Breiman et al., 1984). This algorithm searches, at each node, for the single variable that provides the purest subsets of samples; the partitioning is formulated as a binary rule based on the selected variable and a defined threshold. All the samples that satisfy the rule are grouped in one subset, otherwise into the other subset. Alternatively, multivariate rule induction finds a partition of the data based on a linear combination of all the variables.

As the number of nodes in the tree can be very high, a simplification of the tree (pruning) is usually carried out by reducing the number of nodes through cross-validation procedures.

2.4.2.6 *Artificial Neural Networks (ANNs)*

ANNs are currently considered among the most important emerging tools in chemometrics. They are very versatile methods that can handle both unsupervised and supervised problems; moreover, they can be used to model both qualitative (classification) and quantitative (regression) responses.

Counter-propagation artificial neural networks (CP-ANNs) have been widely applied in chemometrics in the last few years (Hecht-Nielsen, 1987; Kohonen, 1982; Kohonen, 2001; Zupan et al., 1997). The CP-ANN algorithm extends the unsupervised Kohonen maps (the Self-Organizing Maps previously described) to handle classification problems by combining both supervised and unsupervised learning. In CP-ANN, the neuron weights of the Kohonen layer are modified in order to adapt to the input samples and an output layer is added, whose neurons have as many weights as the number of classes, which are adapted to the class labels of the input samples.

ANNs are generally useful and efficient for modeling classes separated by nonlinear boundaries. However, since they have several tunable parameters (e.g., the number of neurons and layers, the learning rate, the number of epochs), one of the major drawbacks and limitations of ANNs is the computational time required for the optimization of all the network parameters, which must be performed by reliable validation protocols to avoid overfitting.

2.5 EXAMPLE OF CHEMOMETRIC DATA ANALYSIS

The application of different methods for exploratory data analysis and classification in the framework of food authenticity is illustrated on the benchmark data set Itaoils, which was taken from the literature (Forina et al., 1983). It consists of 572 samples of olive oils from nine different Italian areas of production (North Apulia, Calabria, South Apulia, Sicily, Inland Sardinia, Coast Sardinia, Umbria, East Liguria and West Liguria). Each sample is described by the percentage composition of eight fatty acids (palmitic, palmitoleic, stearic, oleic, linoleic, linolenic, arachidic and eicosenoic) contained in the lipid fraction of the olive oils. The aim of the analysis is the evaluation of whether fatty acid composition can discriminate the geographical origin of Italian olive oil samples.

Data were first projected into the principal component space to allow an easy visualization of the main patterns, draw preliminary considerations on the information sources and evaluate whether the selected chemical variables are sufficiently representative and informative about the geographical provenance of the olive oils. Before the PCA calculation, the data were autoscaled (equation 36) to avoid bias due to the different measurement scales of the chemical variables. The first two principal components PC1 and PC2 explain together 69% of the total data variance (i.e., $EV\%$ = 47% for PC1 and $EV\%$ = 22% for PC2). As shown in the score plot of the first two PCs (Figure 2.2a), the groups of olive oils of different geographical origin are overall well separated from each other with some exceptions (e.g., the olive oils of Sicily and North Apulia are more spread and overlapped to the other groups). Furthermore, the percentage composition of the selected fatty acids generally seems to characterize well the different groups of olive oils and in particular some groups of homogeneous samples appear very compact (e.g., the two groups, InlandSard and CoastSard, of olive oils from Sardinia island). The role of the chemical variables in determining the PCs can be observed in the loading plot (Figure 2.2b).

Figure 2.3 shows specific representations of the data set Itaoils according to four different exploratory methods: multidimensional scaling (MDS), self-organizing map (SOM), method k-means of cluster analysis and minimum spanning tree (MST). Data patterns and similarity relationships of olive oils are visualized in a low-dimensional space.

Multidimensional scaling was calculated on autoscaled data by using the Manhattan distance as the distance metric. The data were projected in the two-dimensional space that provides the best approximation of the sample distance structure in the original space. The MDS plot (Figure 2.3a) is a visual representation of the data similar to that provided by PCA with the advantage of visualizing the entire data variance into a unique plot but with the limitation that it is not possible to deduce any information on the role played by the variables.

The Kohonen map (SOM) with an architecture of 15 × 15 neurons was trained for 1000 epochs (Figure 2.3b). Data were range scaled (equation 35) before the network training. The SOM plot highlights well-defined groups of homogeneous olive oils as in the score plot of PCA (Figure 2.2a). The olive oils of Sicily and North Apulia are not yet well grouped, while the olive oils of East Liguria are divided into two different sub-categories and the large group of South Apulia olive oils includes some anomalous samples. The influence of each variable in determining the different oil groups can be analyzed through the weight maps of the hidden layer of the network (Figure 2.4). Neurons are colored according to the variable weights (i.e., the darker the neuron, the higher the weight). In this way, one can deduce, for instance, that the olive oils of South Apulia (green class) are mainly characterized by high percentage of palmitic, palmitoleic and linoleic acids.

The k-means algorithm was carried out on the first two principal components provided by PCA by choosing the Euclidean distance as the metric to calculate the similarities between samples. The optimal number G of clusters (i.e., five) was selected by maximization of the validity index RS defined as (Sharma, 1996):

$$RS = \frac{BSS}{TSS} = \frac{\sum_{k=1}^{G} n_k \cdot \left\| \mathbf{c}_k - \mathbf{b} \right\|^2}{\sum_{i=1}^{n} \left\| \mathbf{x}_i - \mathbf{b} \right\|^2} \tag{2.38}$$

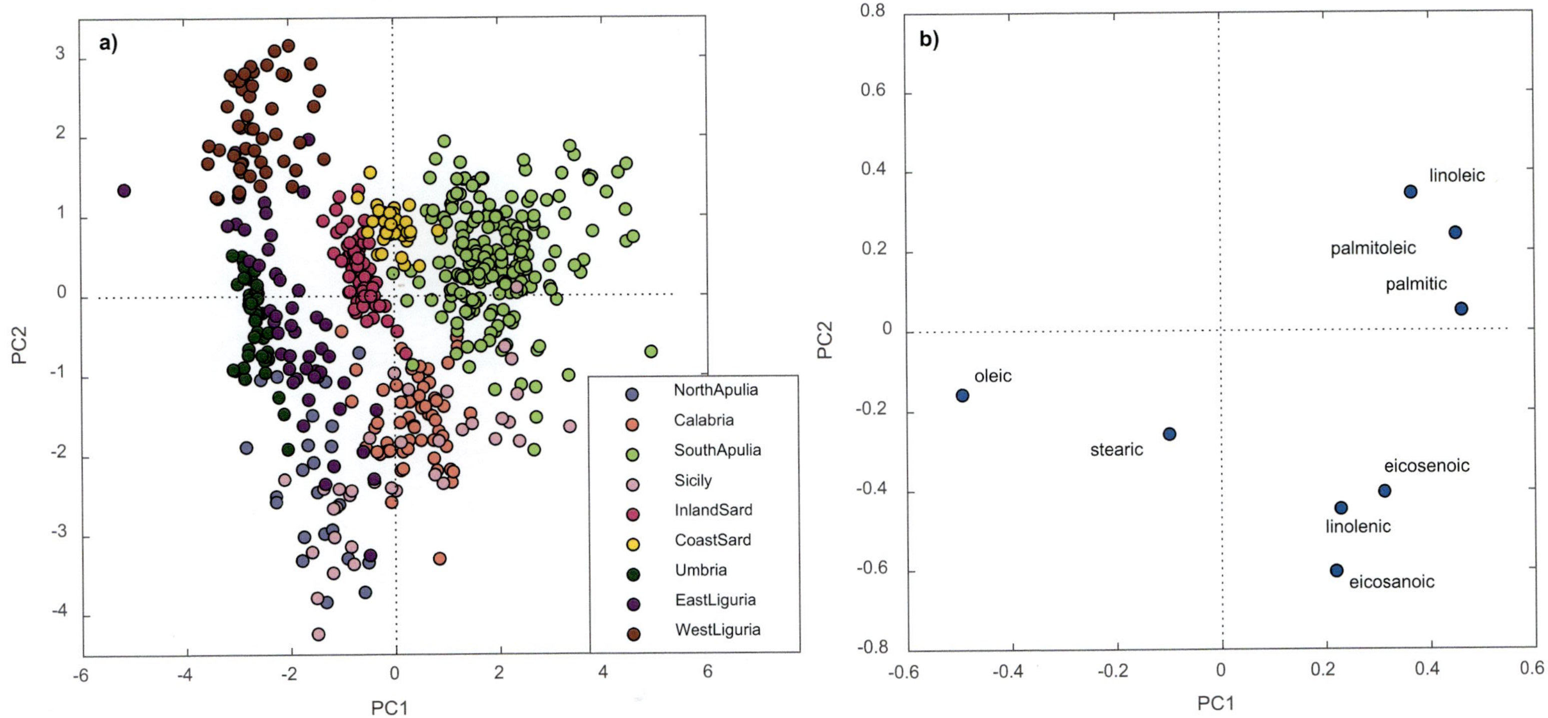

Figure 2.2 Representation of the Itaoils data in the space of the first two principal components (total explained variance of 69%: (a) Score plot; (b) loading plot). Dots represent the olive oils in the score plot (a) and the variables in the loading plot (b); olive oils are colored according to their known geographical origin.

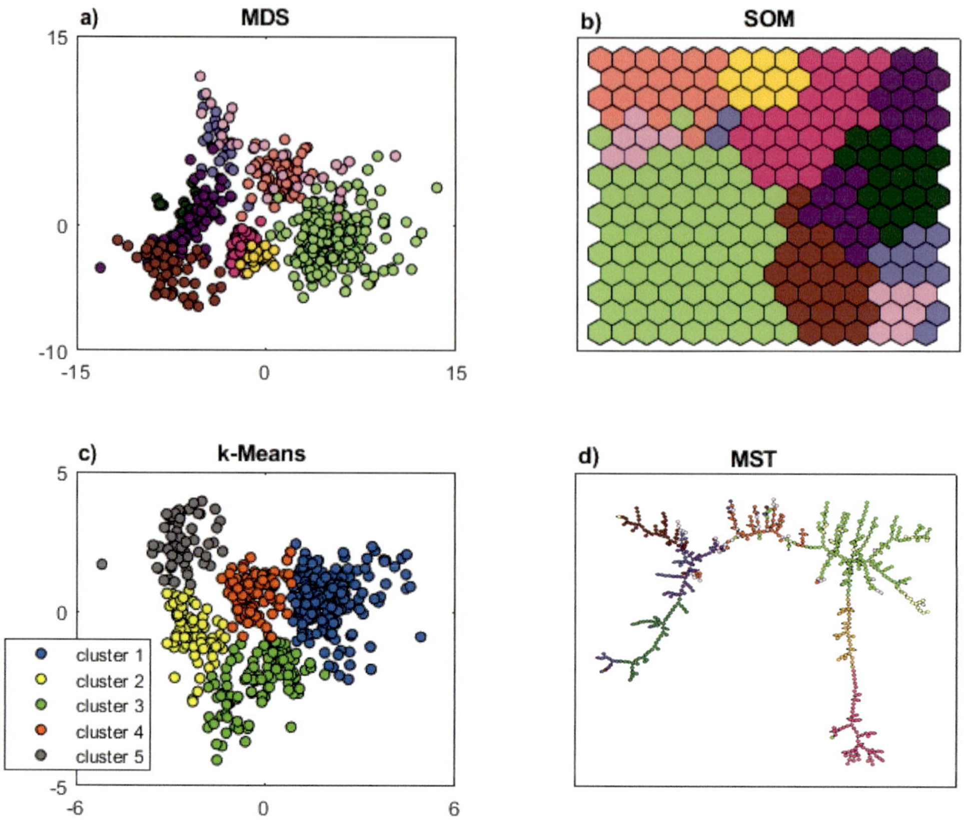

Figure 2.3 Representation of the data patterns and the similarity relationships of olive oils according to the chosen exploratory method on the data set Itaoils: data graphical visualization in a low-dimensional space by (a) Multidimensional scaling (MDS), (b) self-organizing map (SOM), (c) outcomes of cluster analysis by the method k-means and (d) minimum spanning tree (MST). Colors in MDS, SOM and MST represent the known geographical provenance of olive oils, while in k-means (c) colors represent the five clusters of samples, which were identified by the clustering algorithm.

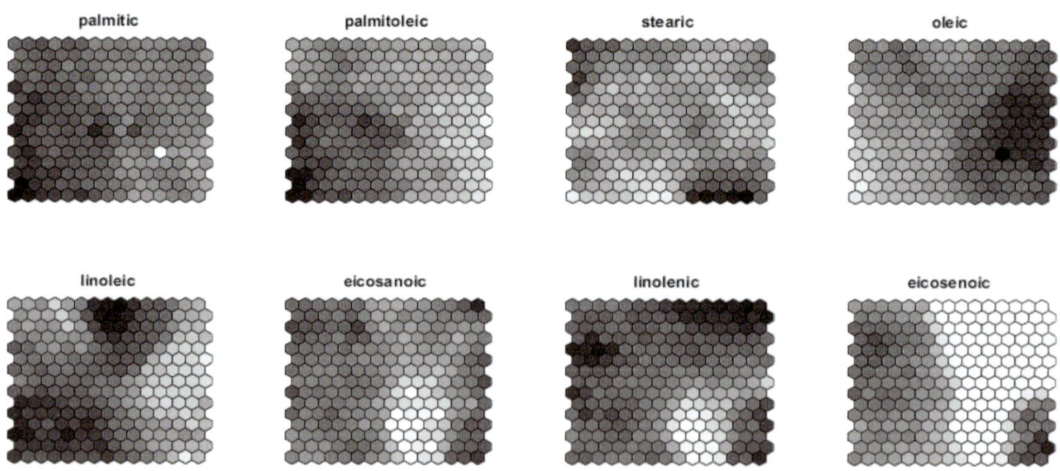

Figure 2.4 Representation of the hidden layer of the self-organizing map (SOM) after 1000 training epochs on the Itaoils data set. The hidden layer has been unfolded to show the distribution of the neuron weights of each variable (fatty acid) in the topological space of the model. Network neurons are colored according to the weights of the variables (i.e., the darker the neuron, the higher the weight).

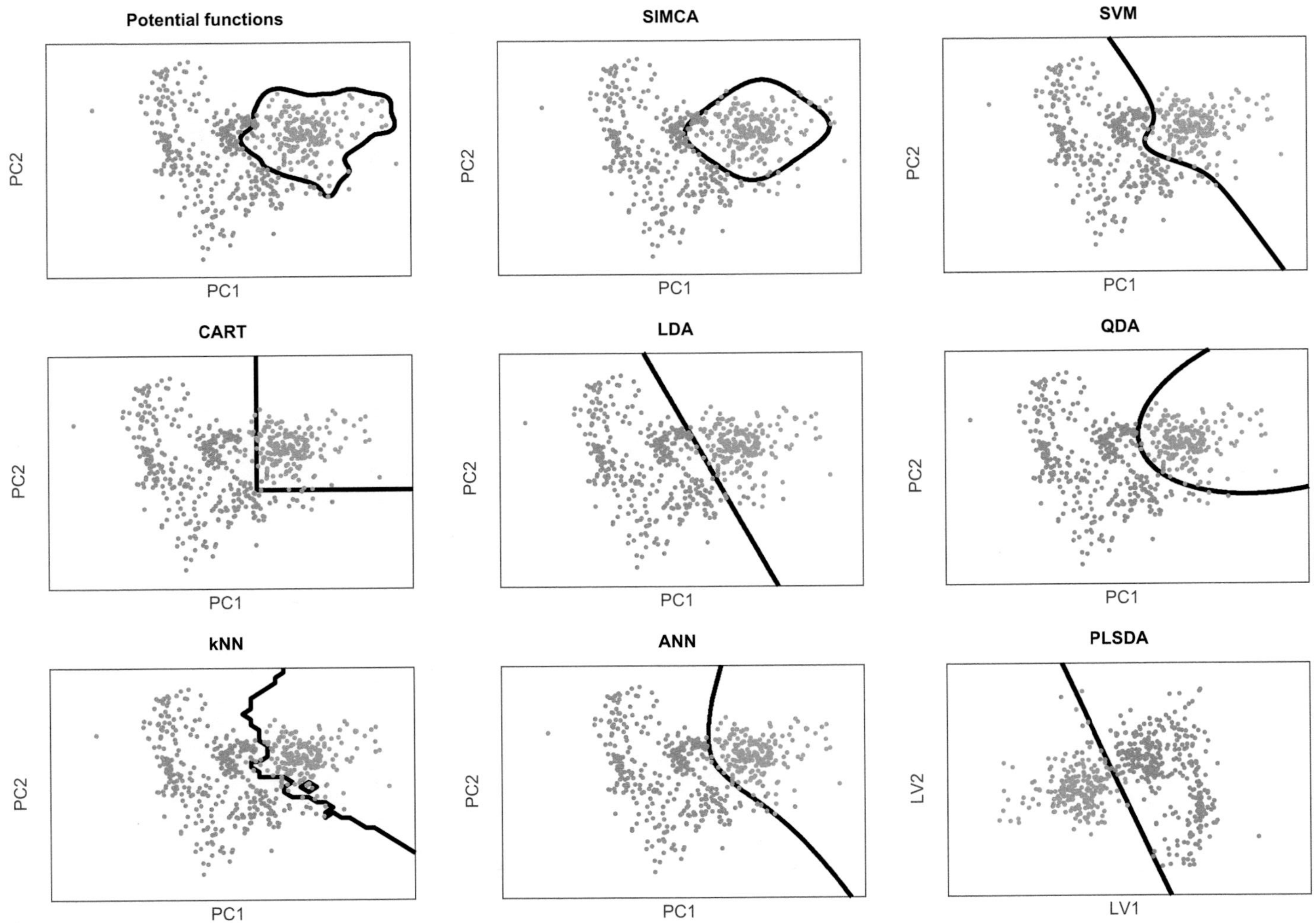

Figure 2.5 Representation of classification boundaries (black line) for the Itaoils data set in the space of the first two principal components (latent variables for PLSDA). South Apulia samples are colored in red, samples of other origins in blue. The results are presented for different classifiers.

where *BSS* is the between-cluster sum of squares and *TSS* is the total sum of squares of the whole data set; n_k is the number of samples in the *k*th cluster, $\mathbf{c}_k$ the cluster centroid and **b** the data set centroid. Figure 2.3c shows the partition of olive oils into the five clusters in the space defined by the first two PCs. Some of the clusters found by the k-means approach correspond to the known geographical origins of olive oils, such as those of South Apulia.

The minimum spanning tree (MST) allows to represent the similarities between the olive oils in the form of an acyclic graph, where the vertices are the olive oils and the edges are their similarity relationships (Figure 2.3d). MST was calculated utilizing the Euclidean distance on autoscaled data. The tree branches correspond to different clusters that contain distinct categories of oils.

After unsupervised data exploration, classification models were calibrated to distinguish one of the olive oil group from the others. In particular, binary classification tasks were carried out to discriminate South Apulia olive oils (red class) from those produced in the other eight geographical areas (blue class). All classification models were calculated on the first two principal components (PLS-DA on the first two latent variables). The aim of this analysis was to show in a 2D space the different class boundaries (Figure 2.5), which can be obtained by the different classification methods described in this chapter. Note that class modeling approaches (SIMCA and potential functions) define a boundary around the target class (South Apulia, red samples), while discriminant methods divide the entire chemical space into two sub-spaces, each associated to one of the two modeled classes, with linear or not-linear boundaries depending on the adopted classification algorithm.

2.6 CONCLUSION

Chemometrics has a fundamental role in the field of food authenticity, being able to find relationships between several chemical analytical features and food origin. Classification methods are the most used chemometric approach in this framework, since they deal with the supervised modelling of the labelling of food origins to distinguish authentic (positive) and non-authentic (negative) samples.

When dealing with supervised classification modelling, it is always fundamental to keep in mind the importance of validation strategies. Validation in fact enables the identification of potential overfitting and the actual capability of the model to classify new samples.

Beside classification, data exploration is another important modelling task, because it is also important to evaluate similarities and presence of clusters among samples, as well as correlations among variables. PCA is probably the most common and important method for multivariate data exploration to find patterns in the data.

REFERENCES

Andersson, M. A comparison of nine PLS1 algorithms. Journal of Chemometrics 23, 518–529. 2009.

Arbelaitz, O., Gurrutxaga, I., Muguerza, J., Pérez, J. M. & Perone, I. An extensive comparative study of cluster validity indices. Pattern Recognition 46, 243–256. 2013.

Ballabio, D., Grisoni, F. & Todeschini, R. Multivariate comparison of classification performance measures. Chemometrics and Intelligent Laboratory Systems 174, 33–44. 2018.

Barker, M. & Rayens, W. S. Partial least squares for discrimination. Journal of Chemometrics 17, 166–173. 2003.

Basilevsky, A. *Statistical Factor Analysis and Related Methods.* New York, Wiley. 1994.

Breiman, L. J., Friedman, J. H., Olsen, R. & Stone, C. *Classification and Regression Trees.* Belmont, Wadsworth. 1984.

Brereton, R. G. One-class classifiers. Journal of Chemometrics 25, 225–246. 2011.

Bro, R. & Smilde, A. Principal component analysis. Analytical Methods 6, 2812–2831. 2014.

Brown, J. B. Classifiers and their metrics quantified. Molecular Informatics 37, 1700127. 2018.

Cassotti, M., Grisoni, F. & Todeschini, R. Reshaped sequential replacement algorithm: An efficient approach to variable selection. Chemometrics and Intelligent Laboratory Systems 133, 136–148. 2014.

Clark, R. D. Boosted leave-many-out cross-validation: The effect of training and test set diversity on PLS statistics. Journal of Computer-Aided Molecular Design 17, 265–275. 2003.

Conlin, A. K., Martin, E. B. & Morris, A. J. Confidence limits for contribution plots. Journal of Chemometrics 14, 725–736. 2000.

Consonni, V., Ballabio, D. & Todeschini, R. Comments on the definition of the Q2 parameter for QSAR validation. Journal of Chemical Information and Modeling 49, 1669–1678. 2009.

Coomans, D., Massart, D. L., Broeckaert, I. & Tassin, A. Potential methods in pattern recognition: Part 1. Classification aspects of the supervised method ALLOC. Analytica Chimica Acta 133, 215–224. 1981.

Cortes, C. & Vapnik, V. Support-vector networks. Machine Learning 20, 273–297. 1995.
Cover, T. M. & Hart, P. E. Nearest neighbor pattern classification. IEEE Transactions on Information Theory 13, 21–27. 1967.
Derde, M. P. & Massart, D. L. UNEQ: A disjoint modelling technique for pattern recognition based on normal distribution. Analytica Chimica Acta 184, 33–51. 1986.
Desgraupes, B. Clustering Indices. 34. 2013. Package ClusterCrit for R. Online Source. https://cran.r-project.org/web/packages/clusterCrit/index.html
Draper, N. R. & Smith, H. *Applied Regression Analysis*, 3rd ed. ed. New York, Wiley & Sons. 1998.
Drgan, V., Župerl, Š., Vracko, M., Como, F. & Novic, M. Robust modelling of acute toxicity towards fathead minnow (Pimephales promelas) using counter-propagation artificial neural networks and genetic algorithm. SAR and QSAR in Environmental Research 27, 01–519. 2016.
Efron, B. *The Jackknife, the Bootstrap and Other Resampling Methods*, Philadelphia, Society for Industrial and Applied mathematics. 1982.
Efron, B. Better bootstrap confidence intervals. Journal of the American Statistical Association 82, 171–200. 1987.
Efroymson, M. A. "Multiple regression analysis", in *Mathematical Methods for Digital Computers*. New York, Wiley. 1960.
Fawcett, T. An introduction to ROC analysis. Pattern Recognition Letters 27, 861–874. 2006. 2006.
Fisher, R. A. The use of multiple measurements in taxonomic problems. Annals of Eugenics 7, 179–188. 1936.
Forina, M., Armanino, C., Lanteri, S. & Tiscornia, E. "Classification of olive oils from their fatty acid composition", in *Food Research and Data Analysis*, H. Martens & H. Russwurm Jr, eds., London, Applied Science Publishers, pp. 189–214. 1983.
Forina, M., Armanino, C., Leardi, R. & Drava, G. A class-modelling technique based on potential function. Journal of Chemometrics 5, 435–453. 1991.
Frank, I. E. & Todeschini, R. *The Data Analysis Handbook*. Amsterdam, The Netherlands, Elsevier. 1994.
Gower, J. C. & Ross, G. J. S. Minimum spanning trees and single linkage cluster analysis. Applied Statistics 18, 54–64. 1969.
Grisoni, F., Consonni, V. & Ballabio, D. Machine learning consensus to predict the binding to the androgen receptor within the CoMPARA project. Journal of Chemical Information and Modeling 59, 1839–1848. 2019.
Hastie, T., Tibshirani, R. & Friedman, J. H. *The Elements of Statistical Learning: Data Mining, Inference, and Prediction*, second edition ed. New York, Springer. 2009.
Hecht-Nielsen, R. Counter-propagation networks. Applied Optics 26, 4979–4984. 1987.
Hochreiter, S. & Schmidhuber, J. Long short-term memory. Neural Comput 9, 1735–1780. 1997.
Jackson, J. E. *A User's Guide to Principal Components*. New York, Wiley. 1991.
Jarvis, R. A. & Patrick, E. A. Clustering using a similarity measure based on shared nearest neighbors. IEEE Trans. Comput, C-22–1025. 1973.
Jolliffe, I. T. *Principal Component Analysis*. New York, Springer-Verlag. 1986.
Khan, S. S. & Madden, M. G. One-class classification: Taxonomy of study and review of techniques. The Knowledge Engineering Review 29, 345–374. 2014.
Kohonen, T. Self-organized formation of topologically correct feature maps. Biological Cybernetics 43, 59–69. 1982.
Kohonen, T. *Self-Organizing Maps*, third edition ed. Berlin, Springer. 2001.
Kruskal, J. B. Multidimensional scaling by optimizing goodness of t to a nonmetric hypothesis. Psychometrika 29, 1–27. 1964a.
Kruskal, J. B. Nonmetric multidimensional scaling: A numerical method. Psychometrika 29, 115–129. 1964b.
Krzanowski, W. J. *Principles of Multivariate Analysis. A User's Perspective*. Oxford, Oxford Univ. Press. 1988.
Lanteri, S. 1992. Full validation procedures for feature selection in classification and regression problems. Chemometrics & Intelligent Laboratory Systems 15, 159–169
Leardi, R. "Genetic algorithms in feature selection", in *Genetic Algorithms in Molecular Modeling*, J. Devillers, ed., Academic Press, p. 67. 1996. https://www.sciencedirect.com/book/9780122138102/genetic-algorithms-in-molecular-modeling#book-info
MacQueen, J. B. "Some methods for classification and analysis of multivariate observations", in *Proceedings of the Fifth Berkeley Symposium on Mathematical Statistics and Probability*, L. M. Le Cam & J. Neyman, eds., Vol. 1, University of California Press, pp. 281–297, 1967.
Mardia, K. V. Some properties of classical multidimensional scaling. Communications in Statistics—Theory and Methods A7, 1233–1241. 1978.
Mardia, K. V., Kent, J. T. & Bibby, J. M. *Multivariate Analysis*. London, Academic Press. 1988.

Marengo, E. & Todeschini, R. 1991. A fast method for the calculation of PLS coefficients. Chemometrics & Intelligent Laboratory Systems 12, 117–120

Massart, D. L. & Kaufman, L. *The Interpretation of Analytical Chemical Data by the Use of Cluster Analysis*. New York, Wiley. 1983.

Massy, W. F. Principal Component Regression in exploratory statistical research. Journal of American Statistical Association 60, 234–256. 1965.

McLachlan, G. *Discriminant Analysis and Statistical Pattern Recognition*. New York, Wiley. 1992.

Miller, A. J. *Subset Selection in Regression*. Lonson, Chapman and Hall. 1990.

Olivecrona, M., Blaschke, T., Engkvist, O. & Chen, H. Molecular de-novo design through deep reinforcement learning. Journal of Cheminformatics 9, 48. 2017.

Oliveri, P. Class-modelling in food analytical chemistry: Development, sampling, optimisation and validation issues—A tutorial. Analytica Chimica Acta 982, 9–19. 2017.

Pérez, N. F., Ferré, J. & Boqué, R. Calculation of the reliability of classification in discriminant partial least-squares binary classification. Chemometrics and Intelligent Laboratory Systems 95, 122–128. 2013.

Polishchuk, P. G., Muratov, E. N., Artemenko, A. G., Kolumbin, A. G., Muratov, N. N. & Kuz'min, V. Application of random forest approach to QSAR prediction of aquatic toxicity. Journal of Chemical Information and Modeling 49, 2181–2488. 2009.

Rakers, C., Reker, D. & Brown, J. B. Small random forest models for effective chemogenomic active learning. Journal of Computer Aided Chemistry 18, 124–142. 2017.

Seber, G. A. F. *Multivariate Observations*. Hoboken, Wiley & Sons. 1984.

Segler, M. H., Kogej, T., Tyrchan, C. & Waller, M. P. Generating focused molecule libraries for drug discovery with recurrent neural networks. ACS Central Science 4, 120–131. 2017.

Sharma, S. C. *Applied Multivariate Techniques*. New York, Wiley & Sons. 1996.

Svetnik, V., Liaw, A., Tong, C., Culberson, C., Sheridan, R. P. & Feuston, B. P. 2003. Random forest: A classification and regression tool for compound classification and QSAR modeling. *Journal of Chemical Information and Computer Sciences* 43, 1947–1958

Todeschini, R., Ballabio, D., Cassotti, M. & Consonni, V. N3 and BNN: Two new similarity based classification methods in comparison with other classifiers. Journal of Chemical Information and Modeling 55, 2365–2374. 2015.

Todeschini, R., Ballabio, D. & Grisoni, F. Beware of unreliable Q2! A comparative study of regression metrics for predictivity assessment of QSAR models. Journal of Chemical Information and Modeling 50, 1905–1913. 2016.

Todeschini, R. & Valsecchi, C. Evaluation of classification performances of minimum spanning trees by 13 different metrics. MATCH 87, 273–298. 2022.

Vapnik, V. *The Nature of Statistical Learning Theory*. New York, Springer. 1995.

Vapnik, V. "The support vector method of function estimation", in *Nonlinear Modeling: Advanced Black-Box Techniques*, J. A. K. Suykens & J. Vandewalle, eds., Boston (MA), Kluwer Academic Publishers. 1998.

Varmuza, K. & Filzmoser, P. *Introduction to Multivariate Statistical Analysis in Chemometrics*. Boca Raton, CRC Press. 2009.

Vitale, R., Marini, F. & Ruckebusch, C. SIMCA modeling for overlapping classes: Fixed or optimized decision thresholds? Analytical Chemistry 90, 10738–10747. 2018.

Willett, P. *Similarity and Clustering in Chemical Information Systems*. Letchworth (UK), Research Studies Press. 1987.

Wold, H. "Estimation of principal components and related models by iterative least squares", in *Multivariate Analysis*, P. R. Krishnaiah, ed., New York, Academic Press, pp. 391–420. 1966.

Wold, S. Pattern recognition by means of disjoint principal components models. Pattern Recognition 8, 127–139. 1976.

Wold, S. & Sjostrom, M. "SIMCA: A method for analyzing chemical data in terms of similarity and analogy", in *Chemometrics Theory and Application*, B. R. Kowalski, ed., Washington D.C., American Chemical Society, pp. 243–282. 1977.

Wu, W., Walczak, B., Massart, D. L., Heuerding, S., Erni, F., Last, I. R. & Prebble, K. A. Artificial neural networks in classification of NIR spectral data: Design of the training set. Chemometrics and Intelligent Laboratory Systems 33, 35–46. 1996.

Wythoff, B. J. Backpropagation neural networks: A tutorial. Chemometrics & Intelligent Laboratory Systems 18, 115–155. 1993.

Zupan, J., Novic, M. & Ruisánchez, I. Kohonen and counterpropagation artificial neural networks in analytical chemistry. Chemometrics and Intelligent Laboratory Systems 38, 1–23. 1997.

Žuvela, P., David, J. & Wong, M. W. Interpretation of ANN-based QSAR models for prediction of antioxidant activity of flavonoids. Journal of Computational Chemistry 39, 953–963. 2018.

3 Authenticity and Chemometrics of Olive Oil

Ioannis C. Martakos, Marios G. Kostakis, Marilena E. Dasenaki, and Nikolaos S. Thomaidis

CONTENTS

3.1 INTRODUCTION

Olive oil is the oil produced by the fruits of olive tree (*Olea europaea* L.) solely by mechanical physical extraction processes under controlled temperature conditions, which result in the preservation of its quality characteristics and bioactive compounds (Martakos *et al.*, 2020). It holds a special place on the Mediterranean diet, as it is the dominant oil consumed by Mediterranean people since ancient times. Olive oil's rich biochemical profile, containing numerous bioactive compounds (biophenols, antioxidant pigments, tocopherols, polyunsaturated fatty acids and many more), enlists olive oil as one of the most nutritious foods in nature (Guo *et al.*, 2017). Because of the widespread interest in the characterization of its content and quality, olive oil analysis is at the center of scientific research over the last decade.

Olive oil consists of a saponifiable fraction that includes various lipid and/or lipid soluble molecules such as triglycerides (TAGs), diglycerides (DAGs), monoglycerides (MAGs), free and esterified fatty acids and phospholipids, representing about 98% of its composition. The unsaponifiable fraction of olive oil consists of hydrocarbons such as squalene, tocopherols, free and esterified sterols, carotenoid and chlorophyllic pigments and poly- and simple phenolic compounds, and usually represents about 2% of its total weight (Sanchez de Medina *et al.*, 2015). Besides the nutritious and biological significance, these compounds contribute to the stability of olive oil, as the majority of them possess antioxidant properties (Borello and Domenici, 2019; Martakos *et al.*, 2020).

The EFSA (European Food Safety Authority) has established several health claims regarding olive oil and its constituents (e.g., phenols, fatty acids, tocopherols). In Commission Regulation (EU) 432/2012, three health claims were established. The first one refers to olive oil polyphenols regarding their contribution to the protection of blood lipids from oxidative stress (European Commission, 2012); the second to the contribution of alpha-linolenic and linoleic acids to the maintenance of normal blood cholesterol levels; and the third to the contribution of eicosapentaenoic acid (EPA) and docosahexaenoic acid (DHA) to the normal function of the heart. Furthermore, another health claim was established some years earlier regarding the contribution of vitamin E (tocopherols and tocotrienols) to the protection of cells from oxidative stress (European Commission, 2006a).

Food authenticity refers to the processes required to verify that a food's declared specifications are correct. The evaluation of food authenticity can facilitate the prevention of misdescription, adulteration and fraudulent origin labeling (Kalogiouri *et al.*, 2016). Special emphasis has been placed on the assurance of variety and geographical origin of olive oil, particularly for protected designation of origin (PDO) products, while efforts have also been made to differentiate olive oils in terms of agricultural practices (e.g., harvest time, irrigation system) and production procedures used (e.g., olive oil extraction, olive oil centrifugation system, addition of water during malaxation, malaxation time and temperature).

Geographical origin discrimination is of great importance for the agronomic sector. According to the EU, PDO products have been produced, processed and developed in a specific geographical area using the recognized know-how of local producers and ingredients from the region concerned (European Commission, 2006b). As a result, PDO products tend to have higher quality and higher

DOI: 10.1201/9781003266167-3

economical value, especially for the local agricultural authorities and associations. Protected geographical indication (PGI) is another geographical distinction established by EU in order to protect and promote products originating from a specific geographical region. PGI focuses on the correlation "between the specific geographic region and the name of the product, where a particular quality, reputation or other characteristic is essentially attributable to its geographical origin" (European Commission, 2006b). A product can be characterized as PGI, when at least one of the stages of production, processing or preparation takes place in the mentioned region.

Olive oils that have been characterized as PDO or PGI have a higher market value, mostly because the quality of the olive oil is ensured and validated through specific farming and processing procedures. In addition, they have been produced in specific regions, which has been linked with higher quality. PGI and PDO olive oils, often are subjected to adulteration incidents, mostly by using olive oils that do not fulfill the PDO/PGI criteria (e.g., origin of the olive fruits, variety, farming system, oil production, geographical region of cultivation) or oils of different plant origin. These kinds of incidents, referred as economically motivated adulteration (EMA), lead to customers' mistrust, while the whole agronomical sector is affected negatively. Moreover, the addition of low-quality plant and vegetable oils can often lead to negative health effects, such allergic incidents (Jurica *et al.*, 2021).

Food analysis can provide sufficient and accurate information regarding olive oil's authenticity. The simultaneous use of high-throughput analytical techniques and advanced chemometrics has been reported widely in the literature in the past ten years. Researchers of the field discriminated olive oils in terms of geographical and/or varietal origin and to investigate the addition of low-quality plant oils based on the chemical profile acquired by these techniques.

Various analytical techniques have been used to assess olive oil's authenticity. These techniques aim (1) to determine a specific compound or group of compounds (e.g., tocopherols, phenols) or (2) to acquire a specific chemical fraction (or "fingerprint") such as the polar fraction or non-polar fraction of olive oil. The most important techniques used in the olive oil authenticity assessment are chromatographic techniques, such as LC (Table 3.1) and GC (Table 3.2), and spectroscopic techniques such as fluorescence, IR and NMR (Table 3.3).

Chromatographic techniques are the dominant techniques used in food analysis, mostly due to the great variety of stationary phases, mobile phases and detectors that can separate, detect and quantify analytes with different physicochemical properties, providing robust, accurate and reproducible results. Liquid chromatography (LC) systems are commercially available with a variety of different detectors, with the most important being UV-Vis, diode array detector (DAD) and mass spectrometry analyzers (both low- and high-resolution MS). LC is a very reliable tool in olive oil authenticity assessment, and it is the most preferable technique used in the literature (Table 3.1). Gas chromatography has been also widely used for authentication of olive oil (Table 3.2). The use of GC is limited mostly because the analytes of interest must be volatile and thermally stable (Hajslova and Cajka, 2016). The main disadvantages of chromatographic techniques are the high cost and the need for time-consuming sample preparation.

Spectroscopic techniques (e.g., FTIR, NMR, NIR, IR and UV-Vis) can also provide reliable results regarding olive oil authenticity investigation (Table 3.3). Their huge advantages are the nondestructive nature of the analysis, the short analysis time, and the minimum- to non-sample preparation in comparison with chromatographic techniques. However, they are in part less sensitive in case of multiple analytes existing in the samples and are mainly used for specific analytes of interest. Among them, NMR has proved to be the most powerful spectroscopic technique, providing selective structural isomers and stereochemistry chemical screening. NMR has been widely used for the investigation of olive oil's authenticity presenting reliable, robust and accurate results. FTIR can also provide sensitive and reliable results regarding olive oil authenticity studies. The wavelength regions included in the infrared (IR) spectrum contain a high number of different stretching vibrations, thus major food components can be observed in those bands (Nawrocka and Lamorsk, 2013).

The development of analytical techniques follows closely on that of advanced chemometric approaches. Even though different chemometric techniques are used in various studies, there is a common workflow applied also to olive oil authenticity assessment. First, an unsupervised technique is used as a tool for a preliminary control of the data set, and then a supervised technique is applied to data (raw or scaled/normalized) in order to discriminate and classify their label or class, finding markers causing the discrimination and finally predicting the label or class of an unknown sample.

Principal component analysis (PCA) and partial least squares–discriminant analysis (PLS-DA) are the most commonly used unsupervised and supervised techniques, respectively. The overview

Table 3.1 Olive Oil Authenticity Studies by Liquid Chromatography

Aim of Study	Compounds	Detection	Chemometric Technique	Separation Power/Accuracy/ Variation Explained	Reference
Monovarietal olive oil's adulteration with vegetable oils	TAGs	HPLC-CAD	PCA	>93% of variation explained	(Green *et al.*, 2020)
Olive oil and other edible vegetable oils discrimination by multivariate classification method	Methyltransesterified fraction	HPLC-CAD	SIMCA and OPLS-DA (k-NN, PLS-DA, SVM)	95% of variation explained	(Jiménez-Carvelo *et al.*, 2017)
Vegetable oils discrimination (25 olive, 11 sunflower, 4 pomace, 3 rapeseed, 2 canola, 3 corn and 3 soybean samples)	Sterols fraction fingerprint	HPLC-UV	PCA, HCA, PLS-DA	Prediction ability 95%–100%	(Bagur-González *et al.*, 2015)
Geographical discrimination of olive oils of the Tunisian coast	10 phenolic compounds	HPLC-UV	ANOVA, PCA	–	(Hlima *et al.*, 2017)
Varietal and geographical origin investigation of 452 olive oils samples	Pigments and tocopherols	HPLC-DAD	One-way, parametric ANOVA	–	(Martakos *et al.*, 2020)
Classification of olive oil samples according to cultivar (Arbequina and non-Arbequina)	Epoxidized fraction	LC-DAD	PCA, PLS-DA, N-way PLS-DA, SIMCA	Sensibility: 1.00, specificity: 0.92, accuracy: 0.95	(Jiménez-Carvelo *et al.*, 2019)
Cultivar classification (5 cultivars from 4 different Albanian regions)	18 phenolic compounds	DAD-MS/MS	PCA	84.3% of variation explained	(Topi *et al.*, 2020)
Discrimination of 4 Moroccan and 9 Mediterranean olive oil varieties	32 phenolic compounds and phenolic profile	LC-MS/MS	PCA, LDA	>86% classification ability	(Bajoub, Medina-Rodríguez, Olmo-García *et al.*, 2017)
Cultivar classification of 25 Brazilian VOO samples from 8 different varieties	19 phenolic compounds	HPLC-ESI/ TOF-MS	PCA	64.2%, 77.1% and 79.9% of variation explained, respectively	(Ballus *et al.*, 2015)
Characterization of VOOs samples from 4 Italian cultivars and four different harvesting periods	18 phenolic compounds, tocopherols	DAD, Orbitrap, FLD	OPLS-DA	>85% reliability, >75% predictive ability	(Deiana *et al.*, 2019)
54 samples from 8 different countries for bitterness and pungency evaluation	25 phenolic compounds	LC-QTOF-MS, FTICR-MS	PLS-DA	Bitterness: 2PCs, $R^2 = 0.8$, SECV = 0.3, Pungency: 1PC, $R^2 = 0.7$, SECV = 0.4	(Dierkes *et al.*, 2012)

Discrimination between extra virgin olive oils and defective in Kolovi variety (19 samples from Lesvos, Greece)	14 target phenolic compounds and 60 suspect compounds (negative), 51 markers identified	QTOF-MS	PLS-DA, CP-ANNs	–	(Kalogiouri *et al.*, 2016)
Conventional and organic EVOOs discrimination (52 samples from Lesvos, Greece)	13 target and 24 suspect phenolic compounds (negative)	LC-QTOF-MS	ACO-LDA, ACO-RF/ RF	–	(Kalogiouri *et al.*, 2017)
Cultivar classification (51 EVOOs samples from 6 different varieties)	Non-target workflow (negative), suspect database of 1608 compounds	LC-QTOF-MS	ACO-RF, PCA	80.8% of variation explained	(Kalogiouri *et al.*, 2018)
EVOOs discrimination of 21 Tunisian from 5 Italian according to their phenolic and sterolic fingerprints	817 sterols and 140 polyphenols	ESI/QTOF	Cluster analysis, OPLS-DA	$R^2 = 0.98$, $Q^2 = 0.94$	(Mohamed *et al.*, 2018)
Varietal and geographical discrimination of EVOOs	Untargeted metabolomics	UPLC-MS	OPLS-DA and HCA	–	(Ghisoni *et al.*, 2019)
Geographical origin discrimination	Phenolic fraction	HPLC-ESI-IT-MS	PCA and LDA	Prediction ability > 94%	(Bajoub, Medina-Rodríguez, Olmo-García *et al.*, 2017)
Geographical origin discrimination	Polar a non-polar fractions	UPLC-QTOF	OPLS-DA and PLS-DA	87% classification ability	(Gil-Solsona *et al.*, 2016)
Geographical origin discrimination	Chlorophylls and carotenoids	HPLC-APCI-HRMS	PCA, PLS-DA, SIMCA	2 PCs explained the 77.9% of variance	(Arrizabalaga-Larrañaga *et al.*, 2020)
Geographical origin investigation of 452 EVOO samples	19 biophenols	LC-QTOF-MS	One-way parametric, ANOVA	–	(Kritikou *et al.*, 2021)

Table 3.2 Olive Oil Authenticity Studies by Gas Chromatography

Aim of Study	Compounds	Detection	Chemometric Technique	Separation Power/Accuracy	Reference
Varietal discrimination (Mastoeides and Koroneiki)	FAMES, sterols, and triterpene alcohols	GC-FID	ANOVA and PCA	81% of data variation explained	(Skiada *et al.*, 2020)
Geographical origin discrimination (Italian and non-Italian)	Sterol analysis	GC-FID	PCA and SIMCA	99.07% classification ability and 98.13% prediction ability	(Giacalone *et al.*, 2015)
Varietal and geographical origin discrimination of Turkish olive oils	Sterol and fatty acid analysis	GC-FID	PCA, SIMCA and PLS-DA	PLS-DA provided more accurate results	(Özdikicierler, 2021)
Geographical origin discrimination	Volatile compounds analysis	FGC-FID	PLS-DA and ANN	81%–89% and 88%–92% classification ability	(Palagano *et al.*, 2020)
Geographical origin discrimination	Sesquiterpene hydrocarbons profiling and fingerprinting	HS-SPME-GC-MS	PLS-DA	Classification ability 100% fingerprinting, 45.5%–100% profiling	(Quintanilla-Casas *et al.*, 2020)
Varietal authentication of Arbequina olive oils (Arbequina and non-Arbequina from 6 countries)	Sesquiterpene hydrocarbons profiling and fingerprinting	HS-SPME-GC-MS	PLS-DA	95.1% of overall correct classification	(Torres-Cobos *et al.*, 2021)
Geographical origin discrimination (4 Ionian islands)	Volatile compounds analysis	HS-SPME-GC-MS	LDA and MANOVA	Discrimination rate 100% and 85.7%	(Eriotou *et al.*, 2021)
Geographical origin discrimination (20 samples from Italy and 20 from Spain)	Volatile organic compound fractions	HS-GC-IMS	PCA and LDA	92%–98% correct classification	(Gerhardt *et al.*, 2017)
Geographical origin discrimination (92 Moroccan olive oils)	40 volatile compounds	HS-SPME/ GC-FID-MS	PCA, s-LDA	Correct classification 62.5%, 90.48%, 91.67% and 100%, respectively)	(Bajoub *et al.*, 2015)
Varietal discrimination (Four Italian varieties)	TAGs and volatile compounds	GC-FID/ SPME-GC-MS	LDA	97.8 and 54.8 of variance explained	(Blasi *et al.* 2019)
Adulteration investigation (refined pomace olive oil)	Volatile compounds analysis, TAGs and Fatty acids	GC-O-MS	PCA and LDA	EVOO correctly discriminated from adulterated	(Drira *et al.*, 2021)
Varietal discrimination of PDO olive oil	Volatile profile	SPME-GC-MS	PCA and kNN	Classification rates 94.3 and 94.7%	(Abbatangelo *et al.*, 2019)

Varietal differentiation of Greek olive oils	Volatile compound and fatty acid composition	HS-SPME-GC/MS and GC-FID	MANOVA, LDA	Classification rate 83.0% for VC, 92.1 for FA and 93% for a combination	(Kosma *et al.*, 2016)
Assessment of olive oil adulteration with soybean oil	Volatile oxidation compounds produced under 300°C	TGA-GC/MS	PCA	Three PCs explained the 56.4%, 16.8% and 3.5% of variance	(Zhou *et al.*, 2021)
Classification of olive oils based on their quality (EVOOs, VOOs, Lampante)	Non-target screening	GC–APCI-QTOF-MS	PCA, PLS DA, LPS-DA	70% correct classification	(Sales *et al.*, 2017)
Varietal and geographical origin discrimination	Volatile aroma compounds profile	GC-MS and GC × GC TOF-MS	ANOVA, PCA, SLDA	96.67%–100% correct classification	(Lukić *et al.*, 2019)
Geographical origin discrimination	Volatile fraction targeted and untargeted fingerprinting	MHS-SPME-GC × GC-MS/FID	PLS-DA	Average sensitivity 92.9% and specificity of 97.4%	(Stilo, Jiménez-Carvelo *et al.*, 2021)
Geographical origin discrimination (6 regions of Italy)	Volatile fraction targeted and untargeted fingerprinting	GC × GC TOF-MS	PCA and 2-level PLS	12 PCs explained total variance of 79.73%	(Stilo, Segura Borrego *et al.*, 2021)

Table 3.3 Olive Oil Authenticity Studies by Spectrometric Techniques

Aim of Study	Detection	Chemometric Technique	Separation Power/Accuracy	Reference
Olive oil adulteration investigation	FTIR	PLS	–	(Filoda *et al.*, 2019)
Geographical and varietal origin discrimination	FT-MIR	PCA, PLS-DA	Classification rates of 91.87% and 91.78%	(Zaroual, El Hadrami and Karoui, 2021b)
Geographical origin discrimination	ATR-FTMIR and UV-Vis	PCA, LDA, SVM	85.67%–100%	(El Orche *et al.*, 2020)
Olive oil adulteration investigation	FT-NIR	PLS-DA-OPS	Specificity and accuracy 99.7% and 99.9%, 96.0% and 95.5% for varieties, 98.2% and 98.8%	(Vieira *et al.*, 2021)
Discrimination of 4 Portuguese olive oil cultivars	FTIR	PCA, LDA	Correct classification 99% ± 7%	(Lamas *et al.*, 2021)
Geographical origin discrimination	IRMS and RAMAN	PCA, LDA	82% correct classification	(Portarena *et al.*, 2017)
Geographical origin discrimination	LIBS, absorption	PCA, LDA, SVC	Up to 100% correct classification	(Gyftokostas *et al.*, 2021)
Geographical origin discrimination	Absorption, fluorescence and Raman	PLS-DA	77.46%–97.18% successful classification	(Kontzedaki *et al.*, 2020)
Geographical and varietal origin discrimination	Front face fluorescence spectroscopy	FDA	96.72% and 95.12% correct classification	(Zaroual *et al.*, 2021a)
Geographical origin authentication	^{1}H-NMR	PCA, ANOVA	–	(Ün and Ok, 2018)
Authentication of 65 EVOOs (30 Maltese and 35 non-Maltese)	^{1}H-NMR and ^{13}C-NMR spectroscopy	PCA, PLS-DA, ANN	^{13}C-NMR and PLS-DA gave better classification	(Lia *et al.*, 2020)
Discrimination of botanical origin (38 samples)	^{1}H-NMR and ^{13}C-NMR	PCA	–	(Popescu *et al.*, 2015)
Geographical and varietal authentication of Italian EVOOs	^{1}H-NMR	LDA	77%–100% correct classification	(Ingallina *et al.*, 2019)
Varietal origin discrimination	^{1}H-NMR and ^{13}C-NMR	MANOVA, LSD, LDA	–	(Cabrita *et al.*, 2021)
Varietal origin discrimination	^{1}H-NMR	PCA, PLS-DA, OPLS-DA	$R^2 = 0.96$ and $Q^2 = 0.97$	(Girelli *et al.*, 2016)
Italian EVOOs blend characterization	^{1}H-NMR	OPLS-DA	Classification score over 0.35	(Girelli *et al.*, 2020)
Determine the content of vegetable oils in olive oil mixtures	^{1}H-NMR	PLS-DA	Recognition and prediction ability 84%–100%	(Alonso-Salces *et al.*, 2022)

of olive oil authenticity studies using these techniques is presented in Tables 3.1–3.4. The main scope of PCA is to reveal outliers to the data set and to make a preliminary visualization assessment of data (Lia *et al.*, 2020). Moreover, PCA is employed to access the quality control (QC) of the analysis (using several pool or QC samples) and also to provide justification on the subsequent use of methods that are based on the singular vector decomposition method (SVD) including PLS-DA. On the other hand, PCA cannot be used as stand-alone chemometric technique especially in case of multi-classification problems because the information of classes or labels of data are not used in relation to their PCs and it can be misinterpreted in cases that the variance is low (Kalogiouri *et al.*, 2020). Another problem with PCA is the dependency on the normalization methods and selection of optimal PCs. Owing to the complex olive oil matrix and large concentration range of analytes in relation to their classes, origin and so forth, the use of classes information to segment these ranges of response is inevitable.

PLS-DA is the most often applied supervised technique in olive oil authenticity studies. PLS-DA combines the properties of least squares regression with the discrimination power of classification method (Kalogiouri *et al.*, 2020). For this reason, PLS-DA has been applied to studies that distinguish samples for different varieties and/or various geographical origins. Alternatively, orthogonal projection to latent structure–discriminant analysis (OPLS-DA; Bylesjö *et al.*, 2006) is utilized as a modification of PLS-DA. In case of OPLS-DA, one class is faced against the other ones in turn, so that the pairwise variable importance for the projection (VIP) values could be derived. This way, class regulated markers can be derived more accurately than mean VIP values from PLS-DA model.

In addition to the previous chemometrics methodologies, hierarchical clustering analysis (HCA), k-nearest neighborhood (k-NN), linear discriminant analysis (LDA), soft independent modeling of class analogies (SIMCA), random forest (RF), artificial neural networks (ANN) and support vector machines (SVM) are applied in different cases depending on the analytical problem. Additionally, analysis of variance (ANOVA) is employed in some cases with more simple, targeted and univariate approaches (Hlima *et al.*, 2017; Lukić *et al.*, 2019; Martakos *et al.*, 2020; Skiada *et al.*, 2020; Eriotou *et al.*, 2021; Kritikou *et al.*, 2021).

Other than using different chemometric methods, sometimes the data from different analytical techniques are combined to capture complete chemical content of a sample (enriching X matrix) before developing the prediction/discrimination model. This way, as shown in Table 3.4, it is expected to have improved interpretation of results with better prediction accuracy (Biancolillo *et al.*, 2014). This approach is called *data fusion* (Silvestri *et al.*, 2013). Data fusion is not a new approach; however, this methodology has been applied in food authenticity studies only recently. In olive oils, data fusion has been employed between complementary techniques (LC-MS and

Table 3.4 Olive Oil Authenticity Studies by Multiple Techniques and Data Fusion

Aim of Study	Compounds	Detection	Chemometric Technique	Separation Power/ Accuracy	Reference
Botanical classification (5 different varieties)	Phenolic fraction	LC-ESI-QTOF MS and GC-APCI-QTOF-MS	PCA, PLS-DA	Correct classification	(Bajoub *et al.*, 2016)
Geographical origin investigation	FA, TAGs and fingerprint	MALDI-TOF-MS, GC-MS and NIR	PCA, PLS-DA	–	(Peršurić *et al.*, 2018)
Varietal discrimination (Morocco)	TAGs, FA, phytosterols, tocopherols and phenols	HPLC-DAD, HPLC-RI, GC-FID	ANOVA, PCA	6 PCs explained 100% of variation	(Mansouri *et al.*, 2019)
Geographical varietal and processing system origin of Chemlali and Arbequina olive oils	Quality indices, tocopherols, pigments, polar phenols	HPLC DAD, UV-Vis, GC-FID	ANOVA, PCA	–	(Hassine *et al.*, 2014)
Authentication of olive oil and other vegetable oils	FAMEs, tocopherols, sterols and squalene	GC-FID, GC-MS, HPLC-FLD	PCA, LDA, PLS-DA	Correct classification	(Shen *et al.*, 2021)
Geographical and varietal discrimination of Greek EVOOs	Squalene, tocopherol, and fatty acid composition	HPLC-FLD, GC-FID	PCA, OPLS-DA	–	(Mikrou *et al.*, 2020)

(Continued)

Table 3.4 (Continued)

Aim of Study	Compounds	Detection	Chemometric Technique	Separation Power/ Accuracy	Reference
Varietal and geographical traceability (49 samples of 4 municipalities of Huerva)	Tocopherols, fatty acids, NMR fingerprinting	HPLC-DAD	LDA, SIMCA, PLS-DA	>80% of variation explained	(Sayago *et al.*, 2019)
Varietal discrimination of three Tunisian varieties	FAMEs (for GC)	GC-FID and MIR	MB-PLS1-DA, data fusion	84%–99% correct classification	(Maléchaux *et al.*, 2020)
Geographical origin authentication (Greece, Italy and Spain)	Volatile compounds	ATR/FT-IR and HS-GC-IMS	PCA-LDA, PLS-DA, data fusion	Classification error 14%	(Schwolow *et al.*, 2019)
Geographical origin authentication (Arbequina variety)	Fingerprinting	HPLC-CAD and HT-GC-FID	PCA, PLS-DA, SIMCA, data fusion	Sensitivity and specificity were 100%	(Vera *et al.*, 2019)
Varietal authentication of 140 EVOOs	Fingerprint of phenolic fraction	HPLC-FLD and HPLC-DAD	PCA-PLS-DA, k-NN and SIMCA, data fusion	Sensitive and selective model, good classification	(Bajoub, Medina-Rodríguez, Gómez-Romero *et al.*, 2017)

GC-MS; Bajoub *et al.*, 2016; Olmo-García *et al.*, 2019), to boost the models by using the capabilities of high resolution techniques (Bajoub, Medina-Rodríguez, Gómez-Romero *et al.*, 2017; Kalogiouri *et al.*, 2020) or to combine data between two cheap and affordable techniques (Bajoub, Medina-Rodríguez, Gómez-Romero *et al.*, 2017; Maléchaux *et al.*, 2020).

A brief representation of the contents of this chapter is presented in Figure 3.1.

3.2 AUTHENTICATION OF OLIVE OIL BY SEPARATION TECHNIQUES

Separation techniques are the most dominant techniques used by researchers in the field of food analysis and authentication. Their widespread adoption by analysts is based on several factors such as ease of use, enhanced accuracy and precision, stability and universality (Tsagkaris *et al.*, 2021). Olive oil is a complex natural matrix containing hundreds of chemical substances with different chemical properties. Thus, separation techniques are used in order to separate the analytes of interest from other constituents of the matrix. Liquid chromatography (LC) and gas chromatography (GC) used for olive oil authentication will be presented next.

3.2.1 Liquid Chromatography

Liquid chromatography is the most frequent technique used in olive oil authenticity studies. The great number of different stationary and mobile phases, as well as the different detectors available in the market, provide researchers with different solutions for different needs (Cortés-Herrera *et al.*, 2019). In Table 3.1, studies regarding the use of LC in the field of olive oil authentication, are reviewed. The scope of the study, the targeted compound(s) and the chemometric approach followed are also presented.

Olive oil consists mainly of triglycerides (about 98%–99% of its total weight). In a study by Green *et al.* (2020), the TAG fraction of monovarietal olive oils was targeted in order to investigate the adulteration with other vegetable oils by HPLC-CAD. The results obtained were statistically processed using PCA, with a total of 93% of variance explained. Working in a similar direction, Jimenez Carvelo *et al.* (2017) targeted the methyl-transesterified fraction of olive oils; the authors explained the 95% of the variation by using SIMCA and constructing different chemometrics models such as OPLS-DA, k-NN, PLS-DA and SVM.

UV and DAD detectors can provide reliable results in terms of investigating the authenticity of olive oil samples due to the detector's stability. In a study published by Bagur-Gonzalez

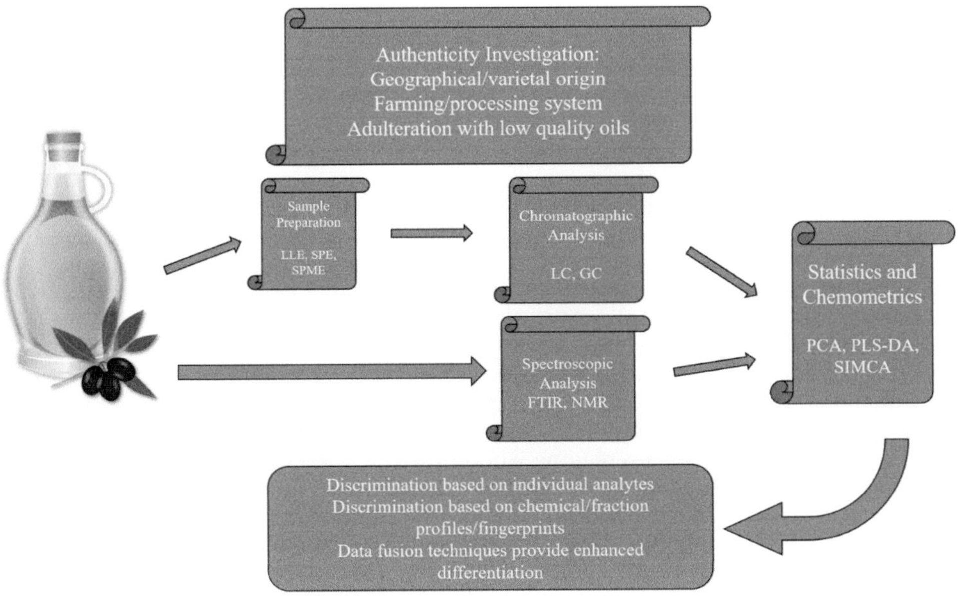

Figure 3.1 Olive oil authenticity investigation strategies by using instrumental analysis and chemometrics.

et al. (2015), the sterolic fingerprint of seven plant oils was acquired by HPLC-UV. By applying several different unsupervised and supervised pattern recognition procedures such as PCA and HCA, the authors discriminated the oils of different plant origin by applying a PLS-DA using the data obtained by HCA and PCA, with variant explanation of 76.88% and 95.47%, respectively. In another study, Martakos *et al.* (2020) analyzed 452 samples of extra-virgin olive oils (EVOOs) and virgin olive oils (VOOs) originating from five northeastern Aegean islands and belonging to six different varieties, in terms of pigments (chlorophyll and carotenoids), tocopherols and squalene. The samples were also of different agronomical background (e.g., different maturation stage at harvesting, addition of water during malaxation). The authors used one-way ANOVA to determine if there were any significant statistical differences in the pigment, tocopherol and squalene content between samples of different geographical origin, varietal origin and oil-producing parameters, comparing the different groups independently. The results obtained indicate that the geographical origin of the samples plays a crucial role in the carotenoid, tocopherol and squalene content, with samples originating from Lesvos Island presenting the higher content. Furthermore, the variety of the olive fruit seems to be a significant parameter, with Kolovi samples presenting the highest amounts of these compounds.

As previous stated, olive oil's phenolic compounds have drawn the attention of the scientific community the past years, mostly because of their bioactivity and the health benefits offered by their consumption. Several studies in the literature have been targeting the phenolic fraction of olive oils as a tool for authenticity assessment. In a study published by Hlima *et al.* (2017), the authors performed targeted analysis of ten phenolic compounds in order to discriminate Tunisian VOOs originating from different territories of Tunisia and different varieties. To achieve this, the authors used HPLC-UV and the results were statistically processed by ANOVA and PCA. The results suggested that the phenolic content of olive oil is strongly dependent on both the geographical origin and the variety of the olive fruit.

The use of mass spectrometry for the determination of phenolic compounds in olive oil has risen the past years. Several studies are published in the literature regarding the use of low- and high-resolution mass spectrometry (L/HRMS) for the characterization of different olive oil samples, based either on individual phenols or on the entire phenolic profile. Bajoub, Medina-Rodríguez, Olmo-García *et al.* (2017) used LC-MS/MS (microTOF-Q and Ion Trap MS) to determine 32 phenolic compounds in olive oils belonging in four Moroccan and nine Mediterranean olive oil varieties cultivated in Morocco in two consecutive crop seasons. The authors discriminated the samples based on their varietal origin by using both PCA and LDA. The classification ability of the model was greater than 86%, while the results suggest that the phenolic profiles of the studied oils are strongly dependent on the crop season. In a study published by Topi *et al.* (2020), the authors analyzed VOO samples originating from five different cultivars obtained from four regions of Albania. The samples were analyzed by HPLC-DAD-MS/MS by targeting 18 phenolic compounds. The results were subjected to chemometric analysis using PCA with two PCs explaining the 84.3% of variation. Furthermore, the authors suggest that the phenolic profile of olive oils is strongly affected by different varieties of olive fruits.

HRMS has been steadily evolving over the past decade, providing increased sensitivity and selectivity, higher speed and better mass resolution (Wang *et al.*, 2013; Ibáñez *et al.*, 2015; Cao *et al.*, 2020). HRMS can provide reliable results and unknown compound identification in complex food matrices like olive oil. HRMS combined with UPLC has been widely used in the literature for the study of authenticity of olive oil. In these studies, target, non-target or both screening strategies have been used. Kalogiouri *et al.* (2016) discriminated VOOs and defective olive oils by targeting 14 phenolic compounds, while suspect and non-target screening methodologies were also used. For the discrimination, the authors used PLS-DA and CP-ANNs. In another publication by the same team (Kalogiouri *et al.*, 2018), the authors classified 51 olive oil samples from six different olive varieties by applying a non-target screening workflow using a database of 1608 natural occurring compounds. The authors developed a new method called ACO-RF (ant colony optimization–random forest) not only to classify the samples but also to use nature-inspired methods to find markers behind the classification. Gil-Solsona *et al.* (2016) applied a metabolomic approach to the authentication of olive oil samples. The polar and non-polar fraction of olive oil samples originating from five territories of Spain were analyzed by UPLC-QTOF-MS, and the results were subjected to OPLS-DA and PLS-DA. The proposed model had a classification ability of 87%. An untargeted metabolomic approach was also used by Ghisoni *et al.* (2019) in order to discriminate EVOO samples originating from different territories of Italy and different varieties as well as blend olive oil samples. The authors used OPLS-DA and HCA in order to determine the analytes with the highest influence on the discrimination of the different samples

3.2.2 Gas Chromatography

Gas chromatography is the second most used chromatographic technique in food analysis. GC provides reliable results for a variety of volatile and semi-volatile compounds, usually responsible for the aroma of food. Gas chromatography offers great separation power, and owing to different detectors for various needs, it is an important and useful tool for food analysis (Hajslova and Cajka, 2016).

Olive oil has a unique flavor due to the existence of several volatile compounds. These compounds are strongly affected by the variety, geographical origin, crop year and many more agronomical and oil production parameters. As a result, the volatile fraction of olive oils can provide researchers with useful conclusions about the authenticity investigation of olive oil samples (Genovese *et al.*, 2021). The majority of the articles in the literature make use of GC-MS rather than other detectors such as flame ionization detector (FID), electron capture detector (ECD) and others. FID is used mostly for the determination of TAGs, FAMEs and sterols, while MS detectors have been used for the determination of several groups of volatile compounds (aldehydes, alcohols, esters, aromatic and aliphatic hydrocarbons, ketones). In Table 3.2, various studies regarding the use of GC in food authenticity investigation studies are presented.

In a study published by Skiada *et al.* (2020), two olive oil varieties (Mastoeides and Koroneiki) originating from Peloponnese, Greece, were examined in terms of FAMEs, sterols and triterpene alcohols, by GC-FID. The data obtained were processed by ANOVA and PCA, and 81% of the variation was explained. The sterol profile of olive oils was also determined in order to differentiate 107 samples of EVOOs into Italian and non-Italian origin (Giacalone *et al.*, 2015). The authors used SIMCA, resulting in 99.07% classification and 98.13% prediction ability for this model, respectively. In a publication by Özdikicierler (2021), the sterol and fatty acid contents were studied in order to discriminate Turkish olive oils in terms of variety and geographical origin. SIMCA, PCA and PLS-DA were used, with the author suggesting that PLS-DA provided better results than SIMCA, while PCA was used for the assessment of outliers in the data set.

GC coupled with MS analyzers has been widely used for the geographical and/or varietal discrimination of olive oil samples. Both low- and high-resolution mass spectrometers offer excellent results regarding olive oil authenticity studies. Casas *et al.* (2020) investigated the differences between sesquiterpene hydrocarbon (SH) profiling and fingerprinting and how the geographical origin investigation is affected. Eighty-two EVOO samples originating from seven countries, were analyzed by headspace solid phase microextraction–gas chromatography–mass spectrometry

(HS-SPME-GC-MS). The results obtained were subjected to PLS-DA to build a classification model. The classification ability based on the results received by the fingerprinting was higher (100%) than those received by the profiling (45.5%–100%). The authors concluded that the fingerprinting of sesquiterpene hydrocarbons and their use as geographical indicators is of great importance and can be useful for researchers of the field. The SH fingerprint of Arbequina and non-Arbequina EVOOs originating from six countries (404 samples: 178 Arbequina and 226 non-Arbequina), was also acquired by Torres-Cobos *et al.* (2021) in order to differentiate the samples in Arbequina and non-Arbequina samples. The authors used HS-SPME-GC-MS and the data were used to build a binary PLS-DA model. The model was externally validated and the external validation showed overall 95.1% correct classification, thus highlighting the suitability of SHs and GC-MS for the authentication of olive oil.

The same technique, but with different targeting compounds, was used by Eriotou *et al.* (2021). The authors targeted the volatile fraction of 137 olive oil samples of the Ntopia variety originating from four islands in the Ionian Sea, Greece. The samples were differentiated according to the origin for two of the islands (Zakynthos and Kerkyra) by using LDA and MANOVA (multivariate ANOVA), while several compounds were found to play a crucial role to the discrimination of EVOO samples and could possibly be used as geographical origin indicators/biomarkers. Volatile compounds were also targeted by Bajoub *et al.* (2015) for the discrimination of the Picholine Marocaine cultivar EVOOs originating from seven regions of North Morocco. By using HS-SPME/GC-FID-MS (headspace–solid phase microextraction), the authors managed to identify and quantify a total of 40 volatile compounds. Furthermore, the data were subjected to PCA and s-LDA resulting in a classification rate of 62.5%–100%. The authors suggest that the geographical origin of olive oil has a clear influence in 37 compounds among those identified. A similar methodology and technique was used by Blasi *et al.* (2019) in order to discriminate four Italian varieties (Dolce Agogia, Frantoio, Leccino and Moraiolo) based on TAG and FAME analysis and on the determination of 19 volatile compounds. The results were subjected to a LDA model, resulting in the discrimination of the studied varieties. Kosma *et al.* (2016) used a similar approach to differentiate six native Greek varieties based on fatty acid and volatile compound analysis. For the determination of fatty acids, GC-FID was used, while for the determination of volatiles the authors used HS-SPME-GC-MS. Furthermore, 56 volatile compounds have been identified, while the application of MANOVA and LDA resulted in 34 volatiles and nine fatty acids to be significant for the discrimination of the varieties. The authors report that the classification rate for fatty acids was 92.1%, while that of both volatile compounds and fatty acids was 93%.

HRMS can also provide very interesting results when hyphenated with GC. Sales *et al.* (2017) used a metabolomic-based approach in order to acquire the full mass spectrum of volatile fraction of olive oil samples of different quality (120 EVOO, 120 VOO, 60 Lampante and 125 of unknown quality) from different territories of Spain. The samples were subjected into to the GC-APCI-QTOF-MS (gas chromatography–atmospheric pressure–quadruple time-of-flight mass spectrometry) system after sample pretreatment. The acquired results were subjected in a PLS-DA and OPLS-DA model. The model was validated externally using the 125 "blind" samples, resulting in a classification rate of 70%.

HRMS can be a very helpful and powerful tool, especially when combined with two-dimensional gas chromatography (GC×GC), which provides solutions in very important analytical problems that researchers often come up against. Food samples (and olive oil in particular) are complicated natural products containing hundreds of chemical compounds. GC×GC offers better separation, greater resolution, resulting in cleaner chromatograms and identification of more analytes of interest (Dimandja, 2020). Lukic *et al.* (2019) used comprehensive two-dimensional GC hyphenated with a TOF-MS detector (HS-SPME-GC×GC-TOF-MS), combined with conventional GC-MS in order to discriminate VOOs originating from Croatia belonging to five different varieties. The authors present that more than 1000 compounds were detected, 256 of them were identified and 131 volatile compounds presented significant differences between samples of monovarietal EVOOs. The authors used ANOVA, PCA and s-LDA and managed to differentiate the five varieties by the first three discriminant functions, by using the data acquired by both techniques. In a similar study, Stilo *et al.* (Stilo, Jiménez-Carvelo *et al.*, 2021) used the same technique in order to differentiate EVOOs originating from different regions of Italy, different varieties and different crop years, based on their volatile fraction fingerprint. The authors applied a targeted and an untargeted workflow, resulting in the identification of 159 compounds. The results were subjected in a PCA model, and then six two-level PLS models were built. Overall, 86.8%–95.78% of variance between different varieties was explained. In another published paper by the same scientific group (Stilo, Segura Borrego *et al.*, 2021), the authors used GC×GC-MS and FID detection in order to comprehensively map targeted and untargeted constituents of PDO, PGI and organic EVOOs originating from three territories of Italy (Gard, Sicilia and Toscana). A PLS-DA model was constructed and the samples were differentiated according to their geographical origin, with an average model's sensitivity of 92.9% and specificity of 97.4%.

3.3 SPECTROSCOPIC TECHNIQUES

As stated above, the use of spectroscopic techniques is long established in food analysis. The ease of use (no specially trained personnel are needed), the short time needed for the analysis, the non-destructive nature of the analysis and the low-cost equipment (except NMR), as well as the robust, repeatable and accurate results obtained have established spectrometric techniques as powerful tools used for the authenticity investigation of food samples (Nawrocka and Lamorsk, 2013). Very interesting studies are available in the literature regarding the geographical and varietal discrimination of EVOOs, as well as studies investigating the olive oil adulteration, using spectrometric techniques. The most dominant techniques, according to the literature, are IR spectroscopy (FTIR, FT-NIR, FT-MIR), UV-Vis spectrometry and nuclear magnetic resonance (NMR). Raman spectroscopy and laser-induced breakdown spectroscopy (LIBS) have also been used in some applications. In Table 3.3, studies regarding the use of spectroscopic techniques in olive oil authentication are presented. The column referring to the targeted compounds has been erased, as spectroscopic techniques take advantage of the samples' full spectra rather than certain fractions/profiles/targeted compounds.

FTIR applications have been widely reported in food authenticity studies. Filoda *et al.* (2019) investigated the adulteration of olive oil samples by different vegetable oils (soybean, sunflower, corn and canola), by FTIR and PLS-DA. The authors managed to build a statistical model independent of the adulterating oil with low adulteration identification limits. Similarly, Vieira *et al.* (2021) used FT-NIR for the investigation of EVOOs adulteration using vegetable or low-quality olive oils (soybean oil, corn oil, canola oil, sunflower oil and virgin olive oil, lampante olive oil and olive oil). The authors prepared adulterated samples by adding the appropriate quantity of adulterant (1%, 2%, 5%, 10%, 15% and 20%) to each one of the samples studied. The acquired spectra were used to build a PLS-DA-OPS model to differentiate the adulterated samples. The model built achieved robust discrimination between the different EVOO varieties and their blend, as well as the differentiation of seven adulterants at low concentrations. The performance of classification models was assessed using confusion matrix, which involved computing percentages such as true positive, false positive, true negative and false negative. Zaroual *et al.* (2021b) used FT-MIR to acquire the spectrum of 41 VOO samples originating from five territories of Morocco and of four different varieties. The obtained spectra were subjected to a PCA model resulting in five PCs explaining more than 99% of the total variation. Furthermore, LDA was applied, achieving 91.87% and 91.78% of correct classification for geographic and varietal discrimination, respectively.

NMR spectroscopy has been extensively used in the field of food science and analysis, mostly for the chemical characterization of food samples and food extracts. The non-destructive nature of the technique, its high robustness and accuracy, and the capability of analyzing complex mixtures have established NMR as a very powerful and useful technique (Hatzakis, 2019).

In the field of olive oil authenticity studies, NMR has been used for the discrimination of EVOOs based on their geographical and/or varietal origin Ün and Ok (2018) used ^{1}H-NMR to differentiate 33 EVOO samples originating from five regions of Turkey. The NMR data were subjected to a PCA and ANOVA analysis. Samples from three regions were discriminated. Furthermore, samples originating from Turkey and samples originating from other European countries were distinguished based on the NMR data. ^{1}H-NMR was also used by Girelli *et al.* (2016) with the purpose of discriminating EVOO samples based on their varietal origin. More than 100 samples originating from two South Italian regions (Apulia and Calabria) and from four varieties were analyzed by ^{1}H-NMR. The obtained data were used to build a PCA, PLS-DA and OPLS-DA models. The authors suggest that this approach could be a useful tool for the discrimination of olive oils produced from specific olive fruit varieties and geographical regions but also by standard production parameters. Ingallina *et al.* (2019) applied ^{1}H-NMR for the analysis of 383 EVOO samples from nine Italian regions. The collected NMR data were subjected to a LDA model resulting in the discrimination of three groups (north, islands and center-south). Samples belonging to the three groups were more differentiated in sub-regions with very good classification rates. The authors present specific compounds responsible for the chemical characterization of said EVOOs samples.

The use of ^{1}H-NMR and ^{13}C-NMR has also provided reliable and robust results regarding olive oil geographical and varietal origin. Lia *et al.* (2020) applied both NMR methodologies for the authentication and discrimination of Maltese EVOO samples. Sixty-five EVOO samples were analyzed (30 samples of Maltese origin and 35 of other). As a chemometric approach, the authors applied PCA, PLS-DA and ANN on the data acquired by both methodologies. As the authors suggest, the application of ^{13}C-NMR was found to be more effective than the ^{1}H-NMR regarding the differentiation of Maltese and non-Maltese EVOOs. A similar approach was applied by Cabrita *et al.* (2021) to assess the varietal origin of olive oil samples of nine varieties. The authors applied MANOVA, least significant differences (LSD) and LDA to the acquired data, resulting in the differentiation of the samples. The suggested results agree with those reported by Lia *et al.* (2020), as the authors present that the ^{13}C-NMR model has a stronger performance than that based on the data acquired by ^{1}H-NMR.

3.4 MULTI-TECHNIQUE AND DATA FUSION APPROACHES

3.4.1 Multiple Techniques

As stated before, olive oil consists of many chemical compounds belonging in various classes with different physicochemical properties. In that direction, different analytical approaches are used by researchers for comparative and/or synergistic reasons. In Table 3.4, studies regarding the use of multiple analytical techniques and chemometrics are presented. For instance, Bajoub *et al.* (2016) compared two metabolomic approaches (LC-ESI-TOF-MS and GC-APCI-TOF-MS) for the botanical classification of 25 EVOO samples of five varieties. Both analyses targeted the phenolic fraction of the samples. The resulting data were subjected to a PCA model, with the results suggesting that two varieties (Arbequina and Frantoio) were clearly distinguished by the other varieties. Several two-class PLS-DA models were constructed for each data set (LC-MS and GC-MS), with the final results suggesting that the models built using the LC-MS data were superior to those built by the GC-MS data set.

Peršurić *et al.* (2018) applied three different techniques, targeting different compound classes for the geographical origin discrimination of EVOOs. Fatty acids were determined using GC-MS, TAGs using MALDI-TOF-MS, and a NIR fingerprint was acquired. The results of each methodology were subjected to PCA, PLS and PLS-DA models and were compared in terms of discrimination and classification. NIR and TAG profiling showed the greatest potential, according to the authors, while the highest classification power was presented by TAG profiling. In another study published by Shen *et al.* (2021), GC-FID, HPLC-FLD and GC-MS were used for the determination of FAMEs, tocopherols, sterols and squalene, respectively. The aim of this study was to assess the authentication of olive, camellia and other vegetable oils. Chemometric models of PCA, LDA and PLS-DA were constructed using the obtained data. The differentiation power of the different oil components was compared, with fatty acids presenting the better results for the discrimination of vegetable oils, followed by tocopherols and sterols, while squalene contributes best to the differentiation of olive and camelia oils. Sayago *et al.* (2019) used a similar approach by targeting tocopherols (by HPLC-DAD), aliphatic hydrocarbon fraction (GC-FID) and NMR fingerprinting for the varietal and geographical differentiation of 49 olive oil samples. Several chemometric techniques were applied (LDA, SIMCA, PLS-DA, RF, ANNs, SVM) with the goal of identifying differentiative patterns between the samples. The authors reported that tocopherols and squalene, alongside some aliphatic alcohols and sterols, were the most discriminant variants. Furthermore, some NMR resonances mostly related to fatty acid chains presented significant differences.

3.4.2 Data Fusion Approaches

In recent years there has been an increasing trend regarding the use of multiple analytical techniques, followed by data fusion and multivariate chemometric analyses. As a very complex matrix, olive oil often needs to be subjected to a holistic approach requiring the use of more than one analytical technique. As the need for more reliable, accurate and robust results for the authenticity of food products increases, data fusion chemometric approaches emerge to better exploit the different analytical signals acquired.

Vera *et al.* (2019) investigated the authenticity of 65 EVOOs of the Arbequina variety originating from three territories of Spain. The authors acquired the chromatographic fingerprints of the samples using two different chromatographic systems, a HPLC-CAD (charged aerosol detector) and a HT-GC-FID. The data of each chromatographic technique were subjected in a PCA model as an explanatory technique followed by a SIMCA and a PLS-DA model. The authors used two different data fusion strategies, high- and low-level, with the PLS-DA model of the low-level data fusion providing 100% sensitivity and specificity in almost all classes. Maléchaux *et al.* (2020) used data fusion on GC-FID chromatographic and MIR spectral data in order to differentiate EVOOs of three Tunisian varieties. PLS-DA models were constructed using the data obtained by each analysis individually. Furthermore, multiblock PLS-DA model was built using the data acquired from both analytical techniques. The results suggest that the GC data alone appear to achieve better classification than the MIR data alone, while the non-weighted MB-PLS1-DA models resulted in better quality parameters and higher sensitivity, specificity and total classification rates. In another study, Bajoub, Medina-Rodríguez, Gómez-Romero *et al.* (2017) applied two data fusion strategies (low- and mid-level) to discriminate 140 EVOO samples of seven olive oil varieties. The author acquired a chromatographic fingerprint of the phenolic fraction using two different HPLC detection systems (DAD and FLD) and the obtained data were separately, and also fused, subjected to a PCA, PLS-DA, SIMCA and kNN models. The results highlight that the combination of the different data obtained presented the most accurate classification between the monovarietal samples.

3.5 CONCLUSIONS AND FUTURE TRENDS

Olive oil is a food product of great importance, not only for its variety of bioactive chemicals but also for its economic value in the food market. Different analytical techniques and methodologies have been developed and applied over the years for the assessment of olive oil authenticity regarding the geographical origin (PDO, PGI), varietal origin, farming and/or processing parameters. The vast majority of researchers use LC and GC with different detectors for the investigation of olive oil's authenticity, mostly due to their ease of use, and potential of determining different compounds with different properties. Spectroscopic techniques also provide reliable and robust results, with NMR appearing to be the technique that has the highest potential, as it can detect a wide number of compounds. The statistical and chemometric approaches used gave the researchers the ability to distinguish olive oil samples based on their chemical profile or specific biomarkers. The dominating chemometric techniques for this aim are PCA and PLS-DA, with several other techniques individually or combined, to provide more reliable and sensitive results for the authenticity investigation.

The rising interest in food authenticity studies, especially for olive oil, establishes new needs for the researchers of the field, as they require new analytical and chemometric tools. Data mining, data fusion and feature selection are fundamental for the interpretation of the huge data set generated through various analytical methods. In this context, advances in data fusion could stimulate great advances in authentication and quality evaluation of food products, especially when the prediction of complex properties cannot be performed by a single technique, such as sensory properties, adulteration, fraud, origin, harvest and production type. Additionally, researchers should reinforce the chemometrics part with the measurement of uncertainty accompanied by multivariate methods that tend to be rare in the authenticity studies.

3.6 ACKNOWLEDGMENT

The authors would like to thank postdoctoral researcher Reza Aalizadeh for his kind contribution to this chapter.

REFERENCES

Abbatangelo, M. *et al.* (2019) 'Multidisciplinary approach to characterizing the fingerprint of Italian EVoO', *Molecules*, 24(8), pp. 1–12. doi: 10.3390/molecules24081457.

Alonso-Salces, R. M. *et al.* (2022) 'Stepwise strategy based on ^{1}H-NMR fingerprinting in combination with chemometrics to determine the content of vegetable oils in olive oil mixtures', *Food Chemistry*, 366(July 2021). doi: 10.1016/j.foodchem.2021.130588.

Arrizabalaga-Larrañaga, A. *et al.* (2020) 'Pigment profiles of Spanish extra virgin olive oils by ultra-high-performance liquid chromatography coupled to high-resolution mass spectrometry', *Food Additives and Contaminants—Part A Chemistry, Analysis, Control, Exposure and Risk Assessment*, 37(7), pp. 1075–1086. doi: 10.1080/19440049.2020.1753891.

Bagur-González, M. G. *et al.* (2015) 'Using the liquid-chromatographic-fingerprint of sterols fraction to discriminate virgin olive from other edible oils', *Journal of Chromatography A*, 1380, pp. 64–70. doi: 10.1016/j.chroma.2014.12.052.

Bajoub, A. *et al.* (2015) 'First comprehensive characterization of volatile profile of north Moroccan olive oils: A geographic discriminant approach', *Food Research International*, 76, pp. 410–417. doi: 10.1016/j.foodres.2015.05.043.

Bajoub, A. *et al.* (2016) 'Comparing two metabolic profiling approaches (liquid chromatography and gas chromatography coupled to mass spectrometry) for extra-virgin olive oil phenolic compounds analysis: A botanical classification perspective', *Journal of Chromatography A*, 1428, pp. 267–279. doi: 10.1016/j.chroma.2015.10.059.
Bajoub, A., Medina-Rodríguez, S., Gómez-Romero, M., *et al.* (2017) 'Assessing the varietal origin of extra-virgin olive oil using liquid chromatography fingerprints of phenolic compound, data fusion and chemometrics', *Food Chemistry*, 215, pp. 245–255. doi: 10.1016/j.foodchem.2016.07.140.
Bajoub, A., Medina-Rodríguez, S., Olmo-García, L., *et al.* (2017) 'In-depth two-year study of phenolic profile variability among olive oils from autochthonous and Mediterranean varieties in Morocco, as revealed by a LC-MS chemometric profiling approach', *International Journal of Molecular Sciences*, 18(1). doi: 10.3390/ijms18010052.
Ballus, C. A. *et al.* (2015) 'Profile of phenolic compounds of Brazilian virgin olive oils by rapid resolution liquid chromatography coupled to electrospray ionisation time-of-flight mass spectrometry (RRLC-ESI-TOF-MS)', *Food Chemistry*, 170, pp. 366–377. doi: 10.1016/j.foodchem.2014.08.054.
Biancolillo, A. *et al.* (2014) 'Data-fusion for multiplatform characterization of an Italian craft beer aimed at its authentication', *Analytica Chimica Acta*, 820, pp. 23–31. doi: 10.1016/J.ACA.2014.02.024.
Blasi, F., Pollini, L. and Cossignani, L. (2019) 'Varietal authentication of extra virgin olive oils by triacylglycerols and volatiles analysis', *Foods*, 8(2). doi: 10.3390/foods8020058.
Borello, E. and Domenici, V. (2019) 'Determination of pigments in virgin and extra-virgin olive oils: A comparison between two near UV-vis spectroscopic techniques', *Foods*, 8(1). doi: 10.3390/foods8010018.
Bylesjö, M. *et al.* (2006) 'OPLS discriminant analysis: Combining the strengths of PLS-DA and SIMCA classification', *Journal of Chemometrics*, 20(8–10), pp. 341–351. doi: 10.1002/cem.1006.
Cabrita, M. J. *et al.* (2021) 'Seeking a fast screening method of the varietal origin of olive oil: The usefulness of an NMR-based approach', *Foods*, 10(2), pp. 1–9. doi: 10.3390/foods10020399.
Cao, G. *et al.* (2020) 'Mass spectrometry for analysis of changes during food storage and processing', *Journal of Agricultural and Food Chemistry*, 68(26), pp. 6956–6966. doi: 10.1021/acs.jafc.0c02587.
Cortés-Herrera, C. *et al.* (2019) 'Liquid chromatography analysis of common nutritional components, in feed and food', *Foods*, 8(1), doi: 10.3390/foods8010001.
Deiana, P. *et al.* (2019) 'Multivariate approach to assess the chemical composition of Italian virgin olive oils as a function of variety and harvest period', *Food Chemistry*, 300, p. 125243. doi: 10.1016/j.foodchem.2019.125243.
Dierkes, G. *et al.* (2012) 'High-performance liquid chromatography-mass spectrometry profiling of phenolic compounds for evaluation of olive oil bitterness and pungency', *Journal of Agricultural and Food Chemistry*, 60(31), pp. 7597–7606. doi: 10.1021/jf3020574.
Dimandja, J. M. D. (2020) 'Introduction and historical background: The "inside" story of comprehensive two-dimensional gas chromatography', *Separation Science and Technology (New York)*. Elsevier, 12, pp. 1–40, doi: 10.1016/B978-0-12-813745-1.00001-5.
Drira, M. *et al.* (2021) 'Safe and fast fingerprint aroma detection in adulterated extra virgin olive oil using gas chromatography—olfactometry-mass spectrometry combined with chemometrics', *Food Analytical Methods*, 14(10), pp. 2121–2135. doi: 10.1007/s12161–021-02034-z.
El Orche, A. *et al.* (2020) 'Evaluation of the capability of horizontal ATR-FTMIR and UV-visible spectroscopy in the discrimination of virgin olive oils from the Moroccan region of Beni Mellal-Khenifra', *Journal of Spectroscopy*, 2020. doi: 10.1155/2020/9317350.
Eriotou, E. *et al.* (2021) 'Geographical origin discrimination of "Ntopia" olive oil cultivar from Ionian islands using volatile compounds analysis and computational statistics', *European Food Research and Technology*, 247(12), pp. 3083–3098. doi: 10.1007/s00217-021-03863-2.
European Commission (2006a) 'COMMISSION REGULATION (EC) No 1924/2006 on nutrition and health claims made on foods', *Official Journal of the European Union*, 404, p. 9.
European Commission (2006b) 'COMMISSION REGULATION on the protection of geographical indications and designations of origin for agricultural products and foodstuffs', *Official Journal of the European Union*, 93, pp. 12–25.
European Commission (2012) 'COMMISSION REGULATION (EU) No 432/2012. Establishing a list of permitted health claims made on foods, other than those referring to the reduction of disease risk and to children's development and health', *Official Journal of the European Union*, 136, pp. 4–40.

Filoda, P. F. *et al.* (2019) 'Fast Methodology for Identification of Olive Oil Adulterated with a Mix of Different Vegetable Oils', *Food Analytical Methods*, 12(1), pp. 293–304. doi: 10.1007/s12161-018-1360-5.

Genovese, A., Caporaso, N. and Sacchi, R. (2021) 'Flavor chemistry of virgin olive oil: An overview', *Applied Sciences (Switzerland)*, 11(4), pp. 1–21. doi: 10.3390/app11041639.

Gerhardt, N. *et al.* (2017) 'Resolution-optimized headspace gas chromatography-ion mobility spectrometry (HS-GC-IMS) for non-targeted olive oil profiling', *Analytical and Bioanalytical Chemistry*, 409(16), pp. 3933–3942. doi: 10.1007/s00216-017-0338-2.

Ghisoni, S. *et al.* (2019) 'Discrimination of extra-virgin-olive oils from different cultivars and geographical origins by untargeted metabolomics', *Food Research International*, 121(September 2018), pp. 746–753. doi: 10.1016/j.foodres.2018.12.052.

Giacalone, R. *et al.* (2015) 'Origin assessment of EV olive oils by esterified sterols analysis', *Food Chemistry*, 188, pp. 279–285. doi: 10.1016/j.foodchem.2015.04.139.

Gil-Solsona, R. *et al.* (2016) 'Metabolomic approach for extra virgin olive oil origin discrimination making use of ultra-high performance liquid chromatography—Quadrupole time-of-flight mass spectrometry', *Food Control*, 70, pp. 350–359. doi: 10.1016/j.foodcont.2016.06.008.

Girelli, C. R. *et al.* (2020) '^{1}H NMR spectroscopy to characterize Italian extra virgin olive oil blends, using statistical models and databases based on monocultivar reference oils', *Foods*, 9(12). doi: 10.3390/foods9121797.

Girelli, C. R., Del Coco, L. and Fanizzi, F. P. (2016) '^{1}H NMR spectroscopy and multivariate analysis as possible tool to assess cultivars, from specific geographical areas, in EVOOs', *European Journal of Lipid Science and Technology*, 118(9), pp. 1380–1388. doi: 10.1002/ejlt.201500401.

Green, H. S. *et al.* (2020) 'A rapid method for the detection of extra virgin olive oil adulteration using UHPLC-CAD profiling of triacylglycerols and PCA', *Food Control*, 107(April 2019), p. 106773. doi: 10.1016/j.foodcont.2019.106773.

Guo, Z. *et al.* (2017) 'Chemical composition and nutritional function of olive (Olea europaea L.): a review', *Phytochemistry Reviews*, 17(5), pp. 1091–1110. doi: 10.1007/s11101-017-9526-0.

Gyftokostas, N. *et al.* (2021) 'Classification of Greek olive oils from different regions by machine learning-aided laser-induced breakdown spectroscopy and absorption spectroscopy', *Molecules*, 26(5). doi: 10.3390/molecules26051241.

Hajslova, J. and Cajka, T. (2016) 'Gas chromatography in food analysis', *Handbook of Food Analysis Instruments*, (January 2008), pp. 119–144. doi: 10.1093/jaoac/62.1.208a.

Hassine, K. Ben *et al.* (2014) 'Consumer discrimination of chemlali and arbequina olive oil cultivars according to their cultivar, geographical origin, and processing system', *European Journal of Lipid Science and Technology*, 116(7), pp. 812–824. doi: 10.1002/ejlt.201300254.

Hatzakis, E. (2019) 'Nuclear magnetic resonance (NMR) spectroscopy in food science: A comprehensive review', *Comprehensive Reviews in Food Science and Food Safety*, 18(1), pp. 189–220. doi: 10.1111/1541–4337.12408.

Hlima, H. Ben *et al.* (2017) 'Geographical discrimination of virgin olive oils from the Tunisian coasts by combining fatty acids and phenolic acids profiles within a multivariate analysis', *Journal of Oleo Science*, 66(9), pp. 963–971. doi: 10.5650/jos.ess17080.

Ibáñez, C. *et al.* (2015) 'The role of direct high-resolution mass spectrometry in foodomics', *Analytical and Bioanalytical Chemistry*, 407(21), pp. 6275–6287. doi: 10.1007/s00216-015-8812-1.

Ingallina, C. *et al.* (2019) 'Extra-virgin olive oils from nine Italian regions: An ^{1}H NMR-chemometric characterization', *Metabolites*, 9(4). doi: 10.3390/metabo9040065.

Jiménez-Carvelo, A. M. *et al.* (2017) 'Fast-HPLC fingerprinting to discriminate olive oil from other edible vegetable oils by multivariate classification methods', *Journal of AOAC International*, 100(2), pp. 345–350. doi: 10.5740/jaoacint.16-0411.

Jiménez-Carvelo, A. M. *et al.* (2019) 'Classification of olive oils according to their cultivars based on second-order data using LC-DAD', *Talanta*, 195(November 2018), pp. 69–76. doi: 10.1016/j.talanta.2018.11.033.

Jurica, K. *et al.* (2021) 'Unauthorized food manipulation as a criminal offense: Food authenticity, legal frameworks, analytical tools and cases', *Foods*, 10(11). doi: 10.3390/foods10112570.

Kalogiouri, N. P., Aalizadeh, R. and Thomaidis, N. S. (2018) 'Application of an advanced and wide scope non-target screening workflow with LC-ESI-QTOF-MS and chemometrics for the classification of the Greek olive oil varieties', *Food Chemistry*, 256, pp. 53–61. doi: 10.1016/j.foodchem.2018.02.101.

Kalogiouri, N. P. *et al.* (2016) 'Olive oil authenticity studies by target and nontarget LC—QTOF-MS combined with advanced chemometric techniques', *Analytical and Bioanalytical Chemistry*, 408(28), pp. 7955–7970. doi: 10.1007/s00216-016-9891-3.

Kalogiouri, N. P., Aalizadeh, R. and Thomaidis, N. S. (2017) 'Investigating the organic and conventional production type of olive oil with target and suspect screening by LC-QTOF-MS, a novel semi-quantification method using chemical similarity and advanced chemometrics', *Analytical and Bioanalytical Chemistry*, 409(23), pp. 5413–5426. doi: 10.1007/s00216-017-0395-6.

Kalogiouri, N. P. *et al.* (2020) 'Application of high resolution mass spectrometric methods coupled with chemometric techniques in olive oil authenticity studies—A review', *Analytica Chimica Acta*, 1134, pp. 150–173. doi: 10.1016/j.aca.2020.07.029.

Kontzedaki, R. *et al.* (2020) 'Verifying the geographical origin and authenticity of Greek olive oils by means of optical spectroscopy and multivariate analysis', *Molecules*, 25(18). doi: 10.3390/molecules25184180.

Kosma, I. *et al.* (2016) 'Differentiation of Greek extra virgin olive oils according to cultivar based on volatile compound analysis and fatty acid composition', *European Journal of Lipid Science and Technology*, 118(6), pp. 849–861. doi: 10.1002/ejlt.201500293.

Kritikou, E. *et al.* (2021) 'Geographical characterization of olive oils from the north Aegean region based on the analysis of biophenols with UHPLC-QTOF-MS', *Foods*, 10(9). doi: 10.3390/foods10092102.

Lamas, S. *et al.* (2021) 'Fourier transform infrared spectroscopy-chemometric approach as a non-destructive olive cultivar tool for discriminating Portuguese monovarietal olive oils', *European Food Research and Technology*, 247(10), pp. 2473–2484. doi: 10.1007/s00217-021-03809-8.

Lia, F. *et al.* (2020) 'Application of ^{1}H and 13C NMR fingerprinting as a tool for the authentication of Maltese extra virgin olive oil', *Foods*, 9(6), pp. 1–14. doi: 10.3390/foods9060689.

Lukić, I. *et al.* (2019) 'Combined targeted and untargeted profiling of volatile aroma compounds with comprehensive two-dimensional gas chromatography for differentiation of virgin olive oils according to variety and geographical origin', *Food Chemistry*, 270(July 2018), pp. 403–414. doi: 10.1016/j.foodchem.2018.07.133.

Maléchaux, A. *et al.* (2020) 'Multiblock chemometrics for the discrimination of three extra virgin olive oil varieties', *Food Chemistry*, 309(October 2019), p. 125588. doi: 10.1016/j.foodchem.2019.125588.

Mansouri, F. *et al.* (2019) 'Quality and chemical profiles of virgin olive oils of three European cultivars suitable for super-high-density planting conditions in eastern Morocco', *Materials Today: Proceedings*, 13, pp. 998–1007. doi: 10.1016/j.matpr.2019.04.065.

Martakos, I. *et al.* (2020) 'Simultaneous determination of pigments, tocopherols, and squalene in Greek olive oils: A study of the influence of cultivation and oil-production parameters', *Foods*, 9(1). doi: 10.3390/foods9010031.

Mikrou, T. *et al.* (2020) 'Varietal and geographical discrimination of Greek monovarietal extra virgin olive oils based on squalene, tocopherol, and fatty acid composition', *Molecules*, 25(17). doi: 10.3390/molecules25173818.

Mohamed, M. Ben *et al.* (2018) 'Discrimination of Tunisian and Italian extra-virgin olive oils according to their phenolic and sterolic fingerprints', *Food Research International*, 106(December 2017), pp. 920–927. doi: 10.1016/j.foodres.2018.02.010.

Nawrocka, A. and Lamorsk, J. (2013) 'Determination of food quality by using spectroscopic methods', in Grundas, S. and Stepniewski, A. (eds) *Advances in Agrophysical Research*. InTech, pp. 1854–1858. doi: 10.5772/52722.

Olmo-García, L. *et al.* (2019) 'Exploring the capability of LC-MS and GC-MS multi-class methods to discriminate virgin olive oils from different geographical indications and to identify potential origin markers', *European Journal of Lipid Science and Technology*, 121(3), pp. 1–13. doi: 10.1002/ejlt.201800336.

Özdikicierler, O. (2021) 'Chemometric discrimination of Turkish olive oils by variety and region using PCA and comparison of classification viability of SIMCA and PLS-DA', *European Food Research and Technology*, 247(1), pp. 157–168. doi: 10.1007/s00217-020-03614-9.

Palagano, R. *et al.* (2020) 'Compliance with EU vs. extra-EU labelled geographical provenance in virgin olive oils: A rapid untargeted chromatographic approach based on volatile compounds', *Lwt*, 130(May), p. 109566. doi: 10.1016/j.lwt.2020.109566.

Peršurić, Ž. *et al.* (2018) 'Comparison of triacylglycerol analysis by MALDI-TOF/MS, fatty acid analysis by GC-MS and non-selective analysis by NIRS in combination with chemometrics for determination of extra virgin olive oil geographical origin. A case study', *Lwt*, 95(December 2017), pp. 326–332. doi: 10.1016/j.lwt.2018.04.072.

Popescu, R. *et al.* (2015) 'Discrimination of vegetable oils using NMR spectroscopy and chemometrics', *Food Control*, 48, pp. 84–90. doi: 10.1016/j.foodcont.2014.04.046.

Portarena, S., Baldacchini, C. and Brugnoli, E. (2017) 'Geographical discrimination of extra-virgin olive oils from the Italian coasts by combining stable isotope data and carotenoid content within a multivariate analysis', *Food Chemistry*, 215, pp. 1–6. doi: 10.1016/j.foodchem.2016.07.135.

Quintanilla-Casas, B. *et al.* (2020) 'Profiling versus fingerprinting analysis of sesquiterpene hydrocarbons for the geographical authentication of extra virgin olive oils', *Food Chemistry*, 307(March 2019). doi: 10.1016/j.foodchem.2019.125556.

Sales, C. *et al.* (2017) 'Quality classification of Spanish olive oils by untargeted gas chromatography coupled to hybrid quadrupole-time of flight mass spectrometry with atmospheric pressure chemical ionization and metabolomics-based statistical approach', *Food Chemistry*, 216, pp. 365–373. doi: 10.1016/j.foodchem.2016.08.033.

Sanchez de Medina, V., Priego-Capote, F. and de Castro, M. D. (2015) 'Characterization of monovarietal virgin olive oils by phenols profiling', *Talanta*, 132, pp. 424–432. doi: 10.1016/j.talanta.2014.09.039.

Sayago, A. *et al.* (2019) 'Combination of vintage and new-fashioned analytical approaches for varietal and geographical traceability of olive oils', *Lwt*, 111, pp. 99–104. doi: 10.1016/j.lwt.2019.05.009.

Schwolow, S. *et al.* (2019) 'Data fusion of GC-IMS data and FT-MIR spectra for the authentication of olive oils and honeys—is it worth to go the extra mile?', *Analytical and Bioanalytical Chemistry*, 411(23), pp. 6005–6019. doi: 10.1007/s00216-019-01978-w.

Shen, M. *et al.* (2021) 'Characterization and authentication of olive, camellia and other vegetable oils by combination of chromatographic and chemometric techniques: role of fatty acids, tocopherols, sterols and squalene', *European Food Research and Technology*, 247(2), pp. 411–426. doi: 10.1007/s00217-020-03635-4.

Silvestri, M. *et al.* (2013) 'Application of data fusion techniques to direct geographical traceability indicators', *Analytica Chimica Acta*, 769, pp. 1–9. doi: 10.1016/J.ACA.2013.01.024.

Skiada, V., Tsarouhas, P. and Varzakas, T. (2020) 'Comparison and discrimination of two major monocultivar extra virgin olive oils in the southern region of Peloponnese, according to specific compositional/traceability markers', *Foods*, 9(2). doi: 10.3390/foods9020155.

Stilo, F., Jiménez-Carvelo, A. M., *et al.* (2021) 'Chromatographic fingerprinting enables effective discrimination and identitation of high-quality Italian extra-virgin olive oils', *Journal of Agricultural and Food Chemistry*, 69(31), pp. 8874–8889. doi: 10.1021/acs.jafc.1c02981.

Stilo, F., Segura Borrego, M. D. P., *et al.* (2021) 'Delineating the extra-virgin olive oil aroma blueprint by multiple headspace solid phase microextraction and differential-flow modulated comprehensive two-dimensional gas chromatography', *Journal of Chromatography A*, 1650, p. 462232. doi: 10.1016/j.chroma.2021.462232.

Topi, D. *et al.* (2020) 'Comparative elucidation of phenolic compounds in Albanian olive oils using LC-DAD-ESI-MS/MS', *Journal of Liquid Chromatography and Related Technologies*, 43(5–6), pp. 203–212. doi: 10.1080/10826076.2019.1711117.

Torres-Cobos, B. *et al.* (2021) 'Varietal authentication of virgin olive oil: Proving the efficiency of sesquiterpene fingerprinting for Mediterranean Arbequina oils', *Food Control*, 128, p. 108200. doi: 10.1016/j.foodcont.2021.108200.

Tsagkaris, A. S. *et al.* (2021) 'Honey authenticity: analytical techniques, state of the art and challenges', *RSC Advances*, pp. 11273–11294. doi: 10.1039/d1ra00069a.

Ün, İ. and Ok, S. (2018) 'Analysis of olive oil for authentication and shelf life determination', *Journal of Food Science and Technology*, 55(7), pp. 2476–2487. doi: 10.1007/s13197-018-3165-3.

Vera, D. N. *et al.* (2019) 'Authentication of the geographical origin of extra-virgin olive oil of the Arbequina cultivar by chromatographic fingerprinting and chemometrics', *Talanta*, 203(February), pp. 194–202. doi: 10.1016/j.talanta.2019.05.064.

Vieira, L. S. *et al.* (2021) 'Building robust models for identification of adulteration in olive oil using FT-NIR, PLS-DA and variable selection', *Food Chemistry*, 345(December 2020). doi: 10.1016/j.foodchem.2020.128866.

Wang, X., Wang, S. and Cai, Z. (2013) 'The latest developments and applications of mass spectrometry in food-safety and quality analysis', *TrAC—Trends in Analytical Chemistry*, 52, pp. 170–185. doi: 10.1016/j.trac.2013.08.005.

Zaroual, H., El Hadrami, E. M. and Karoui, R. (2021a) 'A preliminary study on the potential of front face fluorescence spectroscopy for the discrimination of Moroccan virgin olive oils and the prediction of their quality', *Analytical Methods*, 13(3), pp. 345–358. doi: 10.1039/d0ay01746a.

Zaroual, H., El Hadrami, E. M. and Karoui, R. (2021b) 'Preliminary study on the potential application of Fourier-transform mid-infrared for the evaluation of overall quality and authenticity of Moroccan virgin olive oil', *Journal of the Science of Food and Agriculture*, 101(7), pp. 2901–2911. doi: 10.1002/jsfa.10922.

Zhou, X. *et al.* (2021) 'In-situ assessment of olive oil adulteration with soybean oil based on thermogravimetric-gas chromatography/mass spectrometry combined with chemometrics', *Food Control*, 130(May), p. 108251. doi: 10.1016/j.foodcont.2021.108251.

4 Authenticity and Chemometrics of Olives

Sofia Agriopoulou, Maria Tarapoulouzi, Jose Manuel Lorenzo Rodriguez, Mirian Pateiro, Xenofontas Stampelos, Theodoros Varzakas, and Sevastianos Roussos

CONTENTS

4.1 INTRODUCTION

The cultivation of emblematic evergreen olive tree (*Olea europaea* L.) started 5000 years ago, having a great connection with Greece, since wreaths from the olive branches were the prize of the winners of the Olympic Games in ancient Greece (Papadaki and Tsimidou, 2022; Valvez et al., 2021; Russo et al., 2020; Bitsani et al., 2019). The olive tree is the most symbolic tree of civilization and part of the humanized Mediterranean landscape (Schicchi et al., 2021; Tarapoulouzi et al., 2022). In the Mediterranean basin, it is estimated that more than 2000 varieties are grown, either for the production of olive oil or for the production of table olives (Breton et al., 2009; Jurado-Campos et al., 2021).

According to the International Olive Oil Council (IOOC):

> Table olives are the product prepared from the sound fruits of varieties of the cultivated olive tree that are chosen for their production of olives whose volume, shape, flesh-to-stone ratio, fine flesh, taste, firmness and ease of detachment from the stone make them particularly suitable for processing; treated to remove its bitterness and preserved by natural fermentation, or by heat treatment with or without the addition of preservatives; packed with or without covering liquid.
>
> *(International Olive Oil Council, 2004)*

Spain, Greece and Italy produce the largest quantities of table olives in Europe. Outside Europe, Egypt, Algeria, Turkey and Morocco produce significant quantities of table olives, while Syria, Peru and the United States have begun to appear on the market. Global total production exceeded 2.5 million tonnes in the 2018–2019 season, and is expected to exceed 3.0 million tonnes in the 2020–2021 season (International Olive Oil Council, 2022).

Food authenticity is one of the biggest issues throughout Europe (Bryła, 2018). What characterizes the authenticity of a food is its production or cultivation in such a way as to give it special characteristics related to its nutritional value, sensory properties and chemical composition, as well as to be produced by traditional techniques or with advanced techniques (Jurica et al., 2021; Danezis et al., 2016).

Food authenticity and nutritional quality is a top priority for the food industry, producers, legislation authorities, scientific community, traders and of course consumers, focusing on the nutritional value, origin, cultivation system and production processes of food. Selling a food at a satisfactory price presupposes its compliance with what it claims on the label regarding its origin as well as its ingredients. Validated analytical techniques can demonstrate whether food is authentic or not, and these checks can be carried out either by independent inspection authorities or by consumer's organizations (Granato et al., 2018; Erban et al., 2019; Tahir et al., 2022).

DOI: 10.1201/9781003266167-4

The 2020 annual report from the Agri-food Fraud Network (FFN) from the European Union (EU) revealed that almost the 40% of the total non-compliances were due to mislabeling issues (European Union, 2021).

Since 1992, the EU has enacted specific regulations to provide legal protection to local foods through the protected designation of origin (PDO) and protected geographical indication (PGI) labels. PGI and traditional specialties guaranteed (TSG), with the number of these protected products increasing every year. These regulations cover products that are produced, processed and manufactured in a specific geographical area using specific materials and methodology. The main purpose of this legislation is to ensure the safety and quality of regional products that protect them from fraud and counterfeiting; therefore, it is absolutely necessary to develop tools suitable for the scientific certification of the authenticity of PDO, PGI and TSG products. Such an act also encourages the production of unique foods based on characteristics related to the soil, their composition and their sensory characteristics, as well as their preparation methods. In addition, these foods have a higher value, both in the domestic and international markets (EEC Regulations, 2081/92, 2082/92, 509/2006). Optional indications related to quality are the indications "mountain product" and "product of island farming" (EU Regulation, 1151/2012; Regulation, 1992a, 1992b, 2006). The benefit derived from the use of these measures is significant for many traditional Mediterranean products, with their intrinsic value exceeding their nutritional quality. Peculiar characteristics related to the microclimate, soil, production process, plant varieties and animal races determine the inherent values. These foods usually have a higher quality and nutritional value, and at the same they time contribute to the preservation of traditional production methods, stimulating the economy of local communities and protecting their cultural and natural heritage (Krajnc et al., 2020). Figure 4.1 presents the PDO, PGI and TSG symbols with definitions.

The explanation for the appearance of fraud in organic foods generally starts with the higher prices they achieve in the market (Hong et al., 2017). Organic foods have been constantly gaining the attention and interest of consumers in recent years. The turn of consumers toward this choice comes from the fact that these foods have a lower burden of plant protection, are considered safer and have a smaller environmental footprint (Popović-Djordjević et al., 2022). Organically produced foods have benefits for the health of consumers as they are produced without chemical fertilizers and pesticides, which could create various problems causing diseases related to carcinogenesis, teratogenesis and mutagenesis (dos Santos et al., 2019).

The table olives are among the most popular fermented food worldwide, with their taste to be associated with the qualitative and quantitative composition of volatile compounds and to be

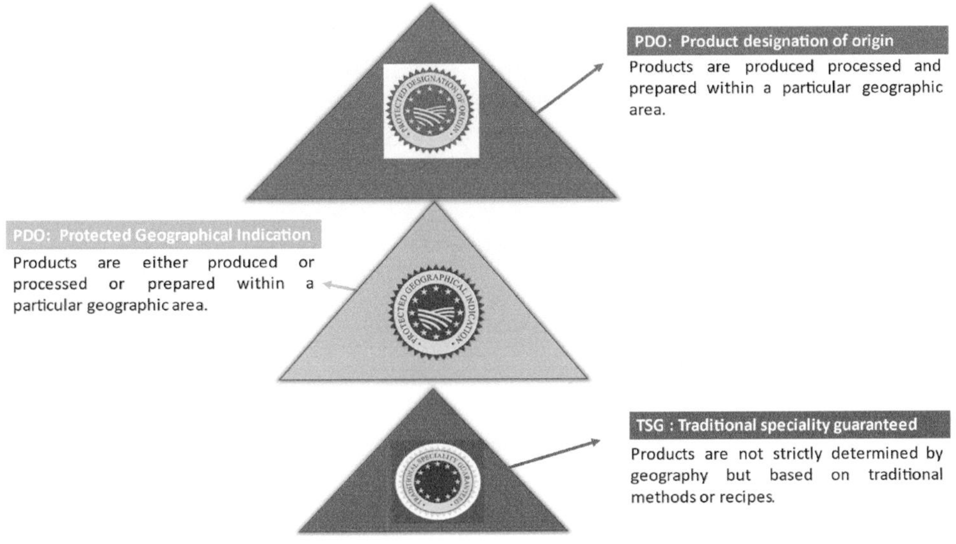

Figure 4.1 The PDO, PGI and TSG symbols with definitions.

considered an important indicator of their quality, playing an important role in consumer acceptance (Sabatini et al., 2009). Carotenoids, tocopherols, anthocyanins and oleosides, and a rich source of antioxidants and various other secondary metabolites consist of the chemical composition of olive fruits (Yadav et al., 2021). Among the chemical ingredients of olives and olive oil, oleocanthal (OLEO), oleacein (OLEA), oleuropein (OLEU) and hydroxytyrosol (HT) are the main polyphenols or biophenols in olives and olive oil with positive effects in health (Li et al., 2022). In addition, olive oil polyphenols, including the aforementioned, have been recognized by the European Food Safety Authority (EFSA) for the positive correlation of the protection of low-density lipoproteins (LDL; European Food Safety Authority, 2011).

The aim of this chapter is to present the application of different analytical techniques and chemometric approaches for the investigation of olive authenticity.

4.2 AUTHENTICITY AND FRAUD IN OLIVES

Greece, Italy, France, Spain, and Portugal have been included in the list of PDO table olives. **The cultivars** Nocellara del Belice, Bella della Daunia, Ascolana del Piceno and Oliva di Gaeta are the four Italian PDO table olives (Pasqualone et al., 2013). The cultivars Aceituna Aloreña de Málaga and Aceituna de Mallorca/Mallorquina are the two Spanish PDO table olives. The cultivars Olives Noires de Nyons, Olives Noires de la Vallée des Baux de Provence, Olives Cassées de la Vallée des Baux de Provence and Olives de Nice are the four French PDO table olives. The cultivars Azeitona de Conserva Negrinha de Freixo and Azeitona de Conserva de Elvas e Campo Maior are the two Portuguese PDO table olives (Concepción et al., 2019).

Today, ten Greek PDO cultivars of table olives exist: Elia Kalamatas, Konservolia Atalantis, Throumba Thassos, Konservolia Rovion, Konservolia Amfissis, Konservolia Stylidas, Throumba Ampadias Rethymnis Kritis, Throumba Chiou, Konservolia Piliou Volou and Prasines Elies Chalkidikis, according to the Greek Ministry of Agricultural Development and Food (Greek Ministry of Agricultural Development and Food, 2022). Table 4.1 presents the PDO and PGI of table olives in European countries.

Table 4.1 PDO and PGI of Table Olives in European Countries

Country	PGI/PDO	Registered Name	Date	Trade Preparation
Italy	PDO	Nocellara del Belice	1998	Treated and natural olives
		Bella della Daunia	2000	Treated and oxidized olives
		Ascolana del Piceno	2005	Treated and natural olives
		Oliva di Gaeta	2016	Natural olives
France	PDO	Olives Noires de Nyons	1996	Natural olives
		Olives Noires de la Vallée des Baux de Provence	1999	Natural olives
		Olives Cassées de la Vallée des Baux de Provence	1999	Treated olives
		Olive de Nice	2005	Natural olives
Spain	PDO	Aceituna Aloreña de Málaga	2012	Natural olives
		Aceituna de Mallorca/Mallorquina	2014	Natural olives
	PGI	Aceituna Manzanilla y Gordal de Sevilla	2019	Treated olives
Portugal	PDO	Azeitona de Conserva Negrinha de Freixo	1996	Treated, oxidized and natural
		Azeitona de Conserva de Elvas e Campo Maior	2007	Natural olives
Greece	PGI	Konservolia Artas	1996	Treated and natural olives
	PDO	Elia Kalamatas	1996	Natural olives
		Konservolia Atalantis	1996	Treated and natural olives
		Throumba Thassos	1996	Natural olives
		Konservolia Rovion	1996	Treated olives
		Konservolia Amfissis	1996	Natural olives
		Konservolia Stylidas	1996	Treated and natural olives
		Throumba Ampadias Rethymnis Kritis	1996	Natural olives
		Throumba Chiou	1996	Natural olives
		Konservolia Piliou Volou	1997	Treated and natural olives
		Prasines Elies Chalkidikis	2012	Treated olives

Source: Adapted from Concepción et al. (2019).

The geographical characterization of olives is an important component of checking their authenticity. The trace elements that are inherent in every soil environment, in various concentrations, characterize it, are unique and depict the geochemical imprint, which is transferred to the olives, helping in their geographical authentication (Cabrera-Vique et al., 2012; Nasr et al., 2022).

One type of authenticity in olives concerns their production, which can be conventional or organic. It is well established that the agricultural production system, whether conventional or organic, affects the composition of their nutrients, both in olives and in the olive oil produced. There is an increasing shift to organic farming and production, which is based on the avoidance of agrochemicals. Mineral fertilizers, hormones and pesticides are harmful to human health, and in the case of organic cultivation of olives are avoided (Jurado-Campos et al., 2021; Pleguezuelo et al., 2018). In addition, financial incentives have been provided in the countries of the EU in order to further develop organic production, and new organic legislation 2018/848 on organic production and labeling of organic products is applicable from January 1, 2022 (EC, 2022; EU Regulation, 2018/848).

Up to now, Portugal has registered six olive oils and two table olives as PDO. There is a lack of physicochemical data for Portuguese PDO olive products, especially for table olives. Nonetheless, differences linked to the origin of the olive oils are reported, including within the same region of origin. For instance, a range of 95.9 to 260 mg kg^{-1} for α-tocopherol was reported for PDO Azeite de Trás-os-Montes, while for PDO Azeites do Norte Alentejano it varied from 92.6 to 198 mg kg^{-1} (Albuquerque et al., 2019).

4.3 APPLICATION OF CHROMATOGRAPHIC AND HYPHENATED TECHNIQUES IN COMBINATION WITH CHEMOMETRICS FOR AUTHENTICITY IN OLIVES

The interpretation of the huge data load is achieved by the science of chemometrics, through which measurements are processed to produce the overall model that visualizes the samples (Tarapoulouzi et al., 2020; Tarapoulouzi et al., 2021; Vanloot et al., 2014; Beyaz et al., 2017; Martínez et al., 2018; Blazakis et al., 2017; Piras et al., 2016). To study authenticity of table olives, different chemical and biological methodologies have been applied so far. There are several chemical techniques that can indicate food authenticity (Pacholczyk-Sienicka et al., 2021; Hao et al., 2021; Górska-Horczyczak et al., 2022; Rodríguez-Bermúdez et al., 2018; Modupalli et al., 2021). The enormous and remarkable capabilities of chromatographic techniques, in addition to their high sensitivity, reproducibility and robustness, allow the receipt of an analytical signal with the highest information content. Chromatographic methods for the detection of specific metabolites in food matrices are based on liquid chromatography (LC) or gas chromatography (GC) in combination with different detectors, such as diode, fluorescence, flame ionization or mass spectrometry (Rasheed et al., 2021). They generally require laborious sample preparation and sophisticated and expensive equipment, which can limit the application of these techniques outside the laboratory environment. In addition, the time required to analyze each sample, as well as to obtain the results, can be very long, thus limiting the performance of the assay (Ross et al., 2021).

GC and high-performance liquid chromatography (HPLC) are the most common methods of chromatography. Over the past decade, gas chromatography-flame ionization detector (GC-FID) has been used successfully as an established, inexpensive and rapid process for fingerprint development. However, the basic disadvantage of GC is that it often requires derivatization. On the contrary, HPLC can be used to analyze some polar and non-polar food components without derivatization (Esteki et al., 2018). Most methods developed for carbohydrates, carotenoids, amino acids, phenolic, food residues and contaminants, such as mycotoxins or other organic compounds, use HPLC or GC (Herrero et al., 2017; Agriopoulou et al., 2016).

Currently, huge gains in efficiency, selectivity, efficiency, speed and resolution are possible with the transition from HPLC to UHPLC due to the reduced particle size of the static phase, while sample size and mobile phases are also significantly reduced. Information obtained by chromatographic methods includes peak intensity (area or height) and their positions (retention time) and, if coupled to MS, peak m/z. However, one can use the entire profile instead of discrete top-notch information (Viaene et al., 2018).

4.3.1 Liquid Chromatography

The geographical and botanical origin of olives plays a significant role in their authentication, traceability and the detection of their adulteration (Alves et al., 2016; Alves et al., 2018). According to mass spectrometry, liquid chromatography coupled to mass spectrometry (LC-MS) is usually applied for olive authenticity (Lukić et al., 2020; Moreno-González et al., 2020; Antonelli et al., 2020).

Polar lipids are a group of minor components of olives (Boskou, 2015). Data from the isolation, identification and characterization of secondary components, such as polar lipids, contribute significantly to the molecular footprint for traceability and authenticity (Perri et al., 2012). The main analyses performed for the profile of the main chemical components of olives are for triacylglycerols and total fatty acids, but these determinations are insufficient to distinguish the olives per se, and the simultaneous analysis of minor components is necessary (Montealegre, 2010).

The use of HPLC-MS has several advantages, as a fast and reproducible analysis in a short time frame can take place (Boukhchina et al., 2004; Montealegre et al., 2013) with high sensitivity (Verardo et al., 2013). In addition, detailed structural information on the polar head and fatty acyl composition of each class of glycerophospholipids is provided through the analysis and interpretation of MS/MS data (Boukhchina et al., 2004). However, some problematic analytical points of this methodology are the chemical noise in the MS/MS spectra and the impossible MS/MS determination of the fatty acyl chains for all glycerophospholipids due to their low quantity in the samples (Boukhchina et al., 2004) and the absence of fragmentation data, and possibly due to their low concentration and matrix complexity (Montealegre et al., 2013), as well as the need for further validation of quantification values by absolute quantitative methods and the need for further validation of quantification values with absolute quantitative methods (Montealegre et al., 2013).

4.3.2 Gas Chromatography

According to mass spectrometry, gas chromatography coupled to mass spectrometry (GC-MS) is usually applied for the volatile profile, key odorants and phenolic compounds of table olives (Cortés-Delgado et al., 2016; Selli et al., 2018; Sánchez et al., 2020). The unique and pleasant taste of table olives is perhaps the most appreciated feature for consumers. Their taste is closely related to both the qualitative and quantitative composition of the volatile compounds. Various factors affect the presence of this particular taste such as the variety of olive, the stage of ripening of the fruit and the method of processing (Cortés-Delgado et al., 2016). Analysis of volatile compounds by chromatography may also help to identify compounds responsible for abnormal fermentations and to predict olive fruit quality after harvest (Martínez Gila et al., 2020).

4.3.3 Spectroscopic Techniques

Regarding spectroscopy techniques in the field of food authentication, Fourier transform infrared spectroscopy (FTIR; Terouzi et al., 2011; Kavdır et al., 2018) and nuclear magnetic resonance spectroscopy (NMR; Beteinakis et al., 2020) are very often used in olive studies.

FTIR has been successfully used to characterize a number of olive oil parameters (Tapp et al., 2003). FTIR data have been often handled with chemometric techniques to develop methods of classification and characterization through the building of multivariate models. This approach has demonstrated to be very useful in many applications because of the ability of these methods in achieving the spectral resolution of the FTIR signals (Zagonel et al., 2004).

De Luca et al. (2012) investigated a new approach to classify different varieties of olives according to the analytical information carried out from their endocarp by means of attenuated total reflection coupled with FTIR (ATR-FTIR).

The potential of FTIR combined with chemometrics was studied to classify five Moroccan varieties of olives by analysis on the endocarps. Attenuated total reflectance (ATR) enabled the samples to be examined directly in the solid state. The spectral data were subjected to a preliminary derivative elaboration based on the Norris gap algorithm to reduce the noise and extract larger analytical information. Linear discriminant analysis (LDA) was adopted as classification method, and principle component analysis (PCA) was employed to compress the original data set into a reduced new set of variables before LDA. The calibration set was built by using the IR data from 75 samples scanned in reflectance mode, and the ranges 3000–2400 and 2300–600 cm^{-1} were selected because furnishing the most useful analytical information. PCA allowed clustering the samples in five classes by using the first two principal components with an explained variance of 98.16%.

In addition to MS-based approaches, NMR spectroscopy is a potentially valuable technique for analyzing phospholipids in olives (Dais and Hatzakis, 2013). Compared to HPLC-MS and NMR (specifically ^{31}P-NMR), higher selectivity, ease of performance and faster analysis are achieved. Obtaining a single signal in the glycerophospholipid spectrum is an advantage of using ^{31}P-NMR, while different molecules of these lipids are characterized by specific resonance frequencies derived from their distinct chemical structures. Therefore the separation of the components in the sample resonance is not necessary, while in the case of HPLC-MS it is necessary (Hatzakis et al., 2008; Vigli et al., 2003).

Kavdır et al. (2018) classified green olives (*Olea europaea* L. 'Ayvalik') based on their surface features such as existence of bruise and fly defect using two NIR spectrometer readings of reflectance and transmittance, and classifiers such as artificial neural networks (ANN) and statistical (Ident and Cluster). Spectral readings were performed in the ranges of 780–2500 and 800–1725 nm for reflectance and transmittance modes, respectively. The classification performance in detecting bruised olives using diameter-corrected reflectance features and ANN classifier was 99%, while it was 98% for Ident and Cluster classification approaches using regular reflectance features (Kavdır et al., 2018).

NMR spectroscopy in combination with chemometrics has been successfully used in the authentication of a wide range of food commodities, such as olive oil, meat, honey and saffron (Jakes et al., 2015; Spiteri et al., 2015; Böhme et al., 2019; Rotondo et al., 2019). Beteinakis et al. (2020) used the NMR-based metabolic profiling tool for the determination of quality parameters of edible olives from the Konservolia, Kalamon and Chalkidikis cultivars from different areas of Greece (Beteinakis et al., 2020).

The application of mass spectrometry (MS) coupled to an array of separation techniques, including gas chromatography (GC) or liquid chromatography (LC), enables simultaneous analysis of a relatively large number of metabolites in a high throughput manner (Ibáñez et al., 2013).

The role of variety and effects of pickling on Egyptian olive fruit metabolome was determined using mass spectrometry-based metabolomics targeting nutrients and bioactive metabolites by Fayek et al. (2021). The analyzed fresh olive fruit varieties included Manzanilo, Picual, Koroneiki, and Coratina, while the pickled samples included the Manzanilo and Picual varieties. Profiling of primary and secondary metabolites resulted in the detection of 201 metabolites. Variation between varieties was mostly observed among sugars, sugar alcohols, secoiridoids, and flavonoids. An abundance of carbohydrates and O-glycosides in Picual and Manzanilo versus enrichment of secoiridoids in Picual and Coratina olives (namely, dehyro-oleuropein) could account for the difference in palatability and health benefits among varieties.

4.3.4 PCR-Based DNA Markers

PCR-based DNA markers such as RAPD (Sesli and Yegenoglu, 2010) SSR (Ercisli et al., 2011), AFLPs (Pafundo et al., 2005) and SNP (Reale et al., 2006), and marker technology today offers powerful tools to analyze the plant genome.

Beiki et al. (2012) used different bioinformatics algorithms to propose new tools to classify ten cultivars of olive based on RAPD and ISSR genetic markers data sets generated from PCR reactions. Five RAPD markers and five ISSR markers selected as the most important markers by all attribute weighting models. K-Medoids unsupervised clustering run on SVM data set was fully able to cluster each olive cultivar to the right classes. All trees (176) induced by decision tree models generated meaningful trees, and the UBC841a4 attribute clearly distinguished between foreign and domestic olive cultivars with 100% accuracy.

Flavor is closely connected with the quali-quantitative composition of volatile organic compounds (VOCs). It is considered a quality index of olive products (Sabatini and Marsilio, 2008). Headspace–solid phase microextraction (HS-SPME) combined with gas chromatography–mass spectrometry (GC-MS) is currently one of the most popular techniques for the analysis of VOCs in foods. In the case of Spanish-style table olives, this technique was used to study the effect of post-fermentation and packaging stages on the VOC composition (Cortés-Delgado et al., 2016).

A multi-statistical approach (standard and compositional data analysis) was used by Garrido-Fernandez et al. (2021) for the study of the VOCs associated with altered (butyric, sulfidic and putrid) and nonaltered (normal) Manzanilla Spanish-style table olive fermentations. Samples were collected from two industrial fermentation yards in Seville (Spain) in the 2019–2020 season. The VOC profiles of altered (n = 4) and non-altered (n = 6) samples were obtained by headspace–solid phase microextraction combined with gas chromatography–mass spectrometry (HS-SPME-GC-MS). Ninety-one VOCs were identified and grouped into alcohols (30), esters (21), carbonyl compounds (12), acids (10), terpenes (6), phenols (6), sulfur compounds (2), and others (4). The association of the VOCs with spoilage samples depended on the standard or compositional statistical methodology used. However, butyric spoilage was strongly linked by several techniques to methyl butanoate, ethyl butanoate and butanoic acid; sulfidic spoilage with 2-propyl-1-pentanol, dimethyl sulfide, methanol, 2-methylbutanal, 2-methyl-2-butenal, ethanol, 2-methyl-3-buten-2-ol and isopentanol; and putrid with D-limonene and 2-pentanol.

4.4 DIFFERENT APPROACHES FOR AUTHENTICITY IN OLIVES

4.4.1 Morphological Characteristics

Agriopoulou et al. (2021) investigated the authentication of six Greek PDO olive varieties according to the morphological characteristics of their stones and produced a reliable chemometric model for their discrimination (Agriopoulou et al., 2021). The cultivars Prasines Chalkidikis (PX), Kalamata Olive (KO), Konservolia Stylidas (KS), Konservolia Amfissis (KA), Throumba Thassos (TT) and Throumba Chios (TC) were the studied olive varieties. The supervised orthogonal partial least squares–discriminant analysis (OPLS-DA) procedure was followed to discriminate and classify the samples. The OPLS-DA model proved to be an effective tool for the certification of Greek table olive varieties by their morphological characteristics, achieving 98.33%. The OPLS-DA model proved to be an effective tool for the certification of Greek table olive varieties by their morphological characteristics, achieving 98.33% variety distinction. In Figure 4.2, an overall OPLS-DA model is presented, which was constructed after chemometric interpretation of the data.

New effective techniques for automatic detection and classification of external olive fruits defects based on image processing techniques were presented by Hassan and Nashat (2019). The proposed techniques can separate between the defected and the healthy olive fruits, and then detect and classify the actual defected area. The proposed techniques are based on texture analysis and the homogeneity texture measure. The results reveal that proposed techniques have the highest accuracy rate among other techniques (Hassan and Nashat, 2019).

Vanloot et al. (2014) achieved the discrimination of five French varieties (Aglandau, Bouteillan, Lucques, Picholine and Tanche) through artificial vision and chemometric analysis of olive stones with 100% correct classification, working with the data obtained from front and profile pictures (Vanloot et al., 2014).

A summarized overview of the analytical techniques and chemometric methods recently studied for certification of authenticity of olives is presented in Table 4.2.

An article by Aguilera Puerto et al. (2019) presents an online computer vision system that automatically classifies olive lots according to their quality level in different categories. The classification is based on the differentiation between olives that have been picked up from the ground or

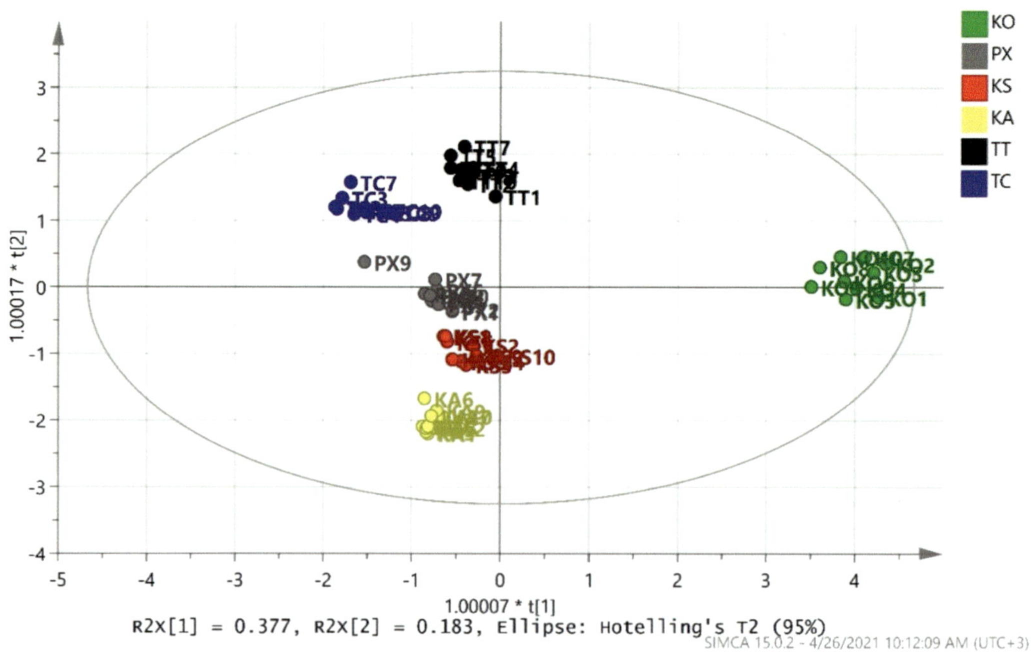

Figure 4.2 An OPLS-DA model for authentication and chemometric discrimination of six Greek PDO table olive varieties through morphological characteristics of their stones.

Table 4.2 Analytical Techniques and Chemometric Methods for Olive Authenticity

Aim of Study	Analytes/ Compounds	Analytical Technique	Chemometric Technique	Separation Power/Accuracy/ Variation Explained	Reference
Discrimination organic from conventional olives	Na, K, Ca, Fe, Mg, Cu, Zn, Se, S and P	HPLC-CAD	PCA	>93% of variation explained	(Hidalgo et al., 2018)
Profiling of pentacyclic triterpenes and polyphenols in Arbequina and Empeltre table olives	Pentacyclic triterpenes	LC-APCI-MS	–	3.28 g kg^{-1} in Arbequina and 2.77 g kg^{-1} in Empeltre	(Moreno-González et al., 2020)
	Polyphenols	LC-ESI-MS/ MS	–	1.04 g kg^{-1} in Arbequina and 0.83 g kg^{-1} in Empeltre	
Triacylglycerols and polar lipids in Portuguese olives	Lipid and polar lipid fractions	HPLC-MS/ MS	–	350 triacylglycerols 107 phospholipids, glyceroglycolipids, glycosphingolipids, betaine lipids	(Alves et al., 2019)
Phenolic compounds in fresh and processed Kalamata table olives	Phenolic compounds	HPLC-DAD LC-(ESI)-MS/ MS	–	Hydroxytyrosol, tyrosol, verbascocide, rutin, oleuropein and luteolin	(Salis et al., 2021)
Olive fruit and olive oil from varieties Carolea and Ottobratica, both from the Calabria region (Italy)	Polar lipid fractions	HPLC-UV	–	Digalactosyldiacyl-glycerol	(Bianco et al., 1998)
The volatile profiles of Spanish-style green table olives	*p*-creosol, phenylethyl alcohol, acetic acid, ethanol, benzyl alcohol, ethyl acetate, (Z)-3-hexen-1-ol	GC-MS	PCA		(Cortés-Delgado et al., 2016)
Key odorants and phenolic compounds in black dry-salted olives	Alcohol, carboxylic acids, lactones, ketones, esters, terpenes, aldehydes	GC-MS-O	–	39 volatile compounds (in a total concentration of 29 459 μg kg^{-1})	(Selli et al., 2018)
Aroma profile and volatile composition of black ripe olives (Manzanilla and Hojiblanca cultivars)	Acids, alcohols, aldehydes, heterocyclic compounds, esters, hydrocarbons, phenols, terpenes	GC-MS	PLS	92% of variance explained	(Sánchez et al., 2020)
Ayvalik variety olives	Surface feature properties such as defect free, with bruise or fly defects	FT-NIR	ANN	99% classification	(Kavdır et al., 2018)

from the tree (The tree fruit category includes olives harvested directly from the tree, containing some waste like leaves and branches.) Two full setups have been installed before and after the olive washing process. The suggested approach uses a feature vector that concatenates the olive image histograms from different color spaces and the texture values of three algorithms (image entropy, gray level co-occurrence matrix and statistics like contrast, correlation, energy and homogeneity). As classifiers, an artificial neural network (ANN) and a support vector machine (SVM) were used. For the experimental validation, 6325 images from 100 batches were analyzed, showing good classification results (success ratios of 98.4% before the washing stage working with an SVM and 98.8% after cleaning using ANN algorithm).

The computer vision systems are widely utilized in food product inspection and classification, improving the general quality of the goods, being more reliable and less intensive in labor forces (Eissa et al., 2012; Brosnan and Sun, 2004, 2002). The introduction of learning techniques such as neural networks, statistical learning, fuzzy logic, decision tree, genetic algorithms or support vector machines (Du and Sun, 2006) combined with computer vision have also improved the performance of these systems. In particular, ANN and SVM have been widely used as machine learning algorithms in food (Du and Sun, 2005; Unay and Gosselin, 2005; Avila et al., 2015; Zhang et al., 2014; Funes et al., 2015). Other statistical models combined with advanced sensor have been employed (Di Rosa et al., 2017, 2018; Tretola et al., 2017a, 2017b; Diaz et al., 2004).

The application of machine vision in the olive sector is not widespread; most of the studies focus on the detection of defective olives. In Diaz et al. (2004), an automatic inspection system based on computer vision that takes into account aspects such as the color of the skin or the presence of defects on the surface of the olive was developed for table olives.

Convolutional neural networks (CNN) have made a breakthrough in image processing and analysis, and nowadays they make up the main framework for image classifier development and pattern recognition.

An artificial-vision algorithm was created by Aquino et al. (2020) able to detect visible olives in digital images of olive trees captured directly in the field, at nighttime and with artificial illumination. These images were taken in an intensive olive orchard of *Olea europaea* L. 'Picual' in September 2018 (two months prior to harvesting). Regarding the methodology, first, the images are pre-processed to generate a set of sub-images with high probability of containing an olive, thus reducing the search space by a magnitude of 10^3. Next, these sub-images are classified by a CNN as *olive*, if they are centered in an olive fruit, or as *other* in any other case (even if they contain peripheral fruits). To train and validate the CNN, a special database called OLIVEnet was compiled, with two disjoint sets integrating these sub-images.

This is the first image analysis approach to automatically identify olive fruits in an image of the entire tree directly taken in the field. The obtained results constitute a first and solid step toward the implementation of an automatic system for yield estimation of olive orchards.

Ben Ayed et al. (2017) examined 18 worldwide table olive cultivars (*O. europaea* L.) based on morphological, biological and physicochemical markers analyzed by bioinformatic and biostatistic tools. Accordingly, they assessed the relationships between the studied varieties and the potential productivity–quantitative parameter links. The bioinformatic analysis based on the graphical representation of the matrix of Euclidean distances, the principal components analysis, unweighted pair group method with arithmetic mean, and principal coordinate analysis (PCoA) revealed three major clusters that were not correlated with the geographic origin. The statistical analysis based on Kendall's and Spearman's correlation coefficients suggests two highly significant associations with both fruit color and pollinization and the productivity character. These results were confirmed by the multiple linear regression prediction models. Only two direct influences are detected: effect of tolerance on fruit and stone symmetry on side and effect of tolerance on stone form and oil content on the other side.

Ponce et al. (2019) developed a methodology for olive-fruit variety classification, approaching it as an image classification problem. To that purpose, 2800 fruits belonging to seven different olive varieties were photographed. After processing these initial captures by means of image processing techniques, the resulting set of images of individual fruits was used to train, and continuedly to externally validate, the implementations of six different CNN architectures in order to compute the classifiers with which perform the variety categorization of the fruits. Remarkable hit rates were obtained after testing the classifiers on the corresponding external validation sets: the Inception-ResnetV2 architecture yielded a top accuracy of 95.91%. The results suggest that the proposed methodology, once integrated into industrial conveyor belts, promises to be an advanced solution to postharvest olive-fruit processing and classification.

4.4.2 E-Nose Discrimination

Sánchez et al. (2021) used an electronic nose (e-nose) to assess the abnormal fermentation defects of Spanish-style green olives that were previously classified by a tasting panel according to the International Oil Council protocol (zapateria, butyric, putrid, and musty or humidity). The mixing of olives with different defects highlighted the putrid defect as the defect with the greatest sensory impact and the butyric defect as the defect with the least sensory effect. Forty-nine volatile compounds were identified by gas chromatography and each defect was characterized by a specific profile. The e-nose data were analyzed using principal component analysis (PCA) and partial least squares–discriminant analysis (PLS-DA). Moreover, the e-nose differentiated control olives from table olives with combined sensory defects despite the dilution effect resulting from the combination. These results demonstrate that e-nose can be used as an olfactory sensor for the organoleptic classification of table olives and can successfully support the tasting panel (Sánchez et al., 2021).

E-noses are electromechanical systems formed by a chamber or array of gas sensors and a pneumatic circuit that conducts the air including volatiles from the product to be analyzed to the sensor chamber (Zhang et al., 2018). The sensors generate an electrical signal in the presence of certain molecular chains. In this work the e-nose PEN3 (Airsense) was used to detect defects in oils and olives. In the scientific literature we can find the use of this same nose to classify kiwi fruit (Ma et al., 2014) and plants (Wang et al., 2019a, 2019b) according to their geographical origin; for detecting the freshness of the strawberry during storage (Xing et al., 2018); classifying apples based on their cultivars (Wu et al., 2017); detection of adulterants (Wang et al., 2019a, 2019b); and contaminations (Concina et al., 2009). E-noses also have been applied to the detection of fungal contaminations and infestations (Liu et al., 2019; Wen et al., 2019) and to assess quality (Jiang and Wang, 2016; Raigar et al., 2017) and fruit maturity (Chen et al., 2019; Huang et al., 2017).

Martinez-Gila et al. (2020) developed a methodology based on an e-nose sensor and pattern recognition algorithms to predict the quality of the oil to be processed from measurements on freshly harvested olive fruit. The pattern recognition algorithms applied were the naïve Bayes (NB) classifier, the PLSDA and a multilayer perceptron (MLP) artificial neural network. Using the measurements performed on 82 samples of olives, the best result was obtained with the MLP network, with 90.2% success obtained in the classification of the virgin and extra virgin olive oil quality by applying tenfold cross-validation. Integration of this methodology to virgin olive oil production allows prediction of the quality of the final oil from the olive fruit received from the farmer.

4.4.3 Other Innovative Approaches

Recently, the use of compositional data analysis (CoDa) integrated a battery of exploratory tools in the study of table olive fatty acids (Garrido-Fernández et al., 2018a, 2018b). A similar approach could also apply to olive fat minor components. Besides, multiple factor analysis (MFA), which allows for testing of the variables both independently and by group, could be a potential tool that is so far scarcely used in chemometrics (López-López et al., 2011).

López-López et al. (2020) aimed to study the effect of green Spanish-style table olive processing and extraction method of fat on its minor components. For this purpose, they used standard multivariate analysis (developed for Euclidean space), compositional data (CoDa) analysis (for data in the simplex) and MFA. Overall, processing had a scarce effect on most of the minor components except ethyl and methyl esters and diacylglycerols, which markedly increased during fermentation; however, these compounds in table olives do not have the negative connotations that those in olive oil do because they are normal metabolites from the yeast microflora habitually present during the process. The work also showed that the quantification of minor components in table olive fat was an extraction-dependent method because Soxhlet increased the concentrations of fatty alcohols, triterpene dialcohols, sterols, waxes and polar compounds. Regarding statistical methods, CoDa analysis strategies were successfully applied to produce more appropriate clustering and PCA segregation than standard tools.

4.5 CONCLUSIONS AND FUTURE TRENDS

Different analytical techniques and chemometric methods for olive authenticity have been addressed and need to be explored more adequately in order to investigate more deeply the characteristics and morphology of olives. More powerful tools will be employed in the future in conjunction with the classical tools to avoid fraud and implement authenticity in order to satisfy consumers but also safeguard quality and safety.

REFERENCES

Agriopoulou, S., Koliadima, A., Karaiskakis, G. and Kapolos, J., 2016. Kinetic study of aflatoxins' degradation in the presence of ozone. *Food Control, 61*, 221–226. https://doi.org/10.1016/j.foodcont.2015.09.013.

Agriopoulou, S., Tarapoulouzi, M., Bedine Boat, M.A., Rébufa, C., Dupuy, N., Theocharis, C.R., Varzakas, T., Roussos, S. and Artaud, J., 2021. Authentication and chemometric discrimination of six Greek PDO table olive varieties through morphological characteristics of their stones. *Foods, 10*, 1829. https://doi.org/10.3390/foods10081829

Aguilera Puerto, D., Cáceres Moreno, Ó., Martínez Gila, D.M., Gómez Ortega, J. and Gámez García, J., 2019. Online system for the identification and classification of olive fruits for the olive oil production process. *Journal of Food Measurement and Characterization, 13*(1), 716–727.

Albuquerque, T.G., Costa, H.S. and Oliveira, M.B.P., 2019. An overview of Portuguese olive oils and table olives with protected designation of origin. *European Journal of Lipid Science and Technology, 121*(5), 1800129.

Alves, E., Domingues M.R.M. and Domingues P., 2018. Polar lipids from olives and olive oil: A review on their identification, significance and potential biotechnological applications. *Foods, 7*, 109, doi:10.3390/foods7070109.

Alves, E., Melo T., Barros M.P., Domingues M.R.M. and Domingues P., 2019. Lipidomic Profiling of the Olive (*Olea europaea* L.) Fruit towards its valorisation as a functional food: In-depth identification of triacylglycerols and polar lipids in Portuguese olives. *Molecules, 24*, 2555; doi:10.3390/molecules24142555.

Alves, E., Melo, T., Rey, F., Moreira, A.S., Domingues, P. and Domingues, M.R.M., 2016. Polar lipid profiling of olive oils as a useful tool in helping to decipher their unique fingerprint. *LWT Food Science and Technology, 74*, 371–377.

Antonelli, M., Benedetti, B., Cannazza, G., Cerrato, A., Citti, C., Montone, C.M., Piovesana, S. and Laganà, A., 2020. New insights in hemp chemical composition: A comprehensive polar lipidome characterization by combining solid phase enrichment, high-resolution mass spectrometry, and cheminformatics. *Analytical and Bioanalytical Chemistry, 412*(2), 413–423.

Aquino, A., Ponce, J.M. and Andújar, J.M., 2020. Identification of olive fruit, in intensive olive orchards, by means of its morphological structure using convolutional neural networks. *Computers and Electronics in Agriculture, 176*, 105616.

Avila, F., Mora, M., Oyarce, M., Zuñiga, A. and Fredes, C., 2015. A method to construct fruit maturity color scales based on support machines for regression: Application to olives and grape seeds. *Journal of Food Engineering, 162*, 9–17.

Beiki, A.H., Saboor, S. and Ebrahimi, M., 2012. A new avenue for classification and prediction of olive cultivars using supervised and unsupervised algorithms. *PLOS One 7*(9), e44164.

Ben Ayed, R., Ennouri, K., Ben Amar, F., Moreau, F., Triki, M.A. and Rebai, A., 2017. Bayesian and phylogenic approaches for studying relationships among table olive cultivars. *Biochemical genetics, 55*(4), 300–313.

Beteinakis, S., Papachristodoulou, A., Gogou, G., Katsikis, S., Mikros, E. and Halabalaki, M., 2020. NMR-based metabolic profiling of edible olives—Determination of quality parameters. *Molecules, 25*(15), 3339.

Beyaz, A., Özkaya, M.T., Duygu, I. 2017. Identification of some Spanish olive cultivars using image processing techniques. *Scientia Horticulturae, 225*, 286–292.

Bianco, A., Mazzei, R.A., Melchioni, C., Scarpati, M.L., Romeo, G., Uccella, N., 1998. Microcomponents of olive oil. Part II: Digalactosyldiacylglycerols from Olea europaea. *Food Chemistry, 62*, 343–346.

Bitsani, E., Agriopoulou, S. and Athanasopoulou, C., 2019. The cultural nutritional and socio-economic value of Greek Messinian olive oil. In *Strategic Innovative Marketing and Tourism. Springer Proceedings in Business and Economics*; Kavoura, A., Kefallonitis, E. and Giovaris, A. Eds. Springer, Cham.

Blazakis, K.N., Kosma, M., Kostelenos, G., Baldoni, L., Bufacchi, M., Kalaitzis, P. 2017. Description of olive morphological parameters by using open access software. *Plant Methods, 13*, 1–15.

Böhme, K., Calo-Mata, P., Barros-Velázquez J, Ortea I. 2019. Recent applications of omics-based technologies to main topics in food authentication. *TrAC—Trends in Analytical Chemistry, 110*, 221–232. https://doi.org/ 10.1016/j.trac.2018.11.005.

Boskou, D., 2015. 1-Olive oil: Properties and processing for use in food. In *Specialty Oils and Fats in Food and Nutrition*, 1st ed.; Talbot, G., Ed.; Woodhead, pp. 3–38, ISBN 9781782423768.

Boukhchina, S., Sebai, K., Cherif, A. and Kallel, H., 2004. Mayer, P.M. Identification of glycerophospholipids in rapeseed, olive, almond, and sunflower oils by LC-MS and LC-MS-MS. *Canadian Journal of Chemistry, 82*, 1210–1215.

Breton, C., Terral, J.F., Pinatel, C., Médail, F., Bonhomme, F. and Bervillé, A., 2009. The origins of the domestication of the olive tree. *Comptes Rendus Biologies, 332*(12), 1059–1064.

Brosnan, T. and Sun, D.W., 2002. Inspection and grading of agricultural and food products by computer vision systems—A review. *Computers and Electronics in Agriculture, 36*(2–3), 193–213.

Brosnan, T. and Sun, D.W., 2004. Improving quality inspection of food products by computer vision—A review. *Journal of Food Engineering, 61*(1), 3–16.

Bryła, P., 2018. The impact of obtaining a European quality sign on origin food producers. *Quality Assurance and Safety of Crops & Foods, 10*(2), 155–164.

Cabrera-Vique, C., Bouzas, P.R. and Oliveras-Lopez M.J., 2012. Determination of trace elements in extra virgin olive oils: A pilot study on the geographical characterization. *Food Chemistry, 134*, 434–439.

Chen, H.Z., Zhang, M. and Guo, Z., 2019. Discrimination of fresh-cut broccoli freshness by volatiles using electronic nose and gas chromatography-mass spectrometry. *Postharvest Biology and Technology, 148*, 168–175.

Concepción, R., García, P., Medina, E. and Brenes, M., 2019. The PDO and PGI table olives of Spain. *European Journal of Lipid Science and Technology, 121*(3), 1800136.

Concina, I., Falasconi, M., Gobbi, E., Bianchi, F., Musci, M., Mattarozzi, M., Pardo, M., Mangia, A., Careri, M. and Sberveglieri, G., 2009. Early detection of microbial contamination in processed tomatoes by electronic nose. *Food Control, 20*, 873–880.

Cortés-Delgado, A., Sánchez, A.H., de Castro, A., López-López, A., Beato, V.M. and Montaño, A., 2016. Volatile profile of Spanish-style green table olives prepared from different cultivars grown at different locations. *Food Research International, 83*, 131–142.

Dais, P. and Hatzakis, E., 2013. Quality assessment and authentication of virgin olive oil by NMR spectroscopy: A critical review. *Analytica Chimica Acta, 765*, 1–27.

Danezis, G.P., Tsagkaris, A.S., Camin, F., Brusic, V. and Georgiou, C.A., 2016. Food authentication: Techniques, trends & emerging approaches. *TrAC Trends in Analytical Chemistry, 85*, 123–132.

De Luca, M., Terouzi, W., Kzaiber, F., Ioele, G., Oussama, A. and Ragno, G., 2012. Classification of Moroccan olive cultivars by linear discriminant analysis applied to ATR—FTIR spectra of endocarps. *International Journal of Food Science & Technology, 47*(6), 1286–1292.

Di Rosa, A.R., Leone, F., Cheli, F. and Chiofalo, V., 2017. Fusion of electronic nose, electronic tongue and computer vision for animal source food authentication and quality assessment—A review. *Journal of Food Engineering, 210*, 62–75.

Di Rosa, A.R., Leone, F., Scattareggia, C. and Chiofalo, V., 2018. Botanical origin identification of Sicilian honeys based on artificial senses and multi-sensor data fusion. *European Food Research and Technology, 244*(1), 117–125.

Diaz, R., Gil, L., Serrano, C., Blasco, M., Moltó, E. and Blasco, J., 2004. Comparison of three algorithms in the classification of table olives by means of computer vision. *Journal of Food Engineering, 61*(1), 101–107.

Dos Santos, A.M., Lima, J.S., Dos Santos, I.F., Silva, E.F., de Santana, F.A., de Araujo, D.G. and Dos Santos, L.O., 2019. Mineral and centesimal composition evaluation of conventional and organic cultivars sweet potato (*Ipomoea batatas* (L.) Lam) using chemometric tools. *Food Chemistry, 273*, 166–171. https://doi.org/10.1016/j.foodchem.2017.12.063.

Du, C.J. and Sun, D.W., 2005. Pizza sauce spread classification using colour vision and support vector machines. *Journal of Food Engineering, 66*(2), 137–145.

Du, C.J. and Sun, D.W., 2006. Learning techniques used in computer vision for food quality evaluation: A review. *Journal of Food Engineering, 72*(1), 39–55.

EC. 2022. *EU funding under the new CAP. Agriculture and rural development.* Organic Farming. https://ec.europa.eu/agriculture/organic/eu-funding/eu-funding-and-the-new-cap_en. (Accessed on February 8, 2022).

Eissa, A.A., Khalik, A.A. and Abdel, A.A., 2012. Understanding color image processing by machine vision for biological materials. *Structure and Function of Food Engineering*, 227–274.

Erban, A., Fehrle, I., Martinez-Seidel, F., Brigante, F., Más, A.L., Baroni, V., Wunderlin, D. and Kopka, J., 2019. Discovery of food identity markers by metabolomics and machine learning technology. *Scientific Reports, 9*(1), 1–19. https://doi.org/10.1038/s41598-019-46113-y.

Ercisli, S., Ipek, A. and Barut, E., 2011. SSR marker-based DNA fingerprinting and cultivar identification of olives (Olea europaea). *Biochemical Genetics, 49*(9), 555–561.

Esteki, M., Simal-Gandara, J., Shahsavari, Z., Zandbaaf, S., Dashtaki, E. and Vander Heyden, Y., 2018. A review on the application of chromatographic methods, coupled to chemometrics, for food authentication. *Food Control, 93*, 165–182.

European Commission, 2018. Regulation (EU) 2018/848 of the European parliament and of the council of 30 may 2018 on organic production and labelling of organic products and repealing council regulation (EC) No 834/2007. *Official Journal of the European Union, 150*(61), 1–92.

European Food Safety Authority (EFSA), 2011. Scientific Opinion on the substantiation of health claims related to polyphenols in olive and protection of LDL particles from oxidative damage (ID 1333, 1638, 1639, 1696, 2865), maintenance of normal blood HDL-cholesterol concentrations (ID 1639), maintenance of normal blood pressure (ID 3781), "anti-inflammatory properties" (ID 1882), "contributes to the upper respiratory tract health" (ID 3468), "can help to maintain a normal function of gastrointestinal tract" (3779), and "contributes to body defences against external agents" (ID 3467) pursuant to Article 13(1) of Regulation (EC) No 1924/2006. *ESFA Journal, 9*, 2033–2058.

European Union., 2021. *2020 Annual Report. The EU Agri-food fraud network and the administrative assistance and cooperation system.* https://doi.org/10.2875/20163. https://food.ec.europa.eu/system/files/2021-09/ff_ffn_annual-report_2020_1.pdf

Fayek, N.M., Farag, M.A. and Saber, F.R., 2021. Metabolome classification via GC/MS and UHPLC/MS of olive fruit varieties grown in Egypt reveal pickling process impact on their composition. *Food chemistry, 339*, 127861.

Funes, E., Allouche, Y., Beltrán, G. and Jiménez, A., 2015. A review: Artificial neural networks as tool for control food industry process. *Journal of Sensor Technology, 5*(01), 28.

Garrido-Fernández, A., Cortés-Delgado, A. and López-López, A., 2018a. Tentative application of compositional data analysis to the fatty acid profiles of green Spanish-style Gordal table olives. *Food Chemistry, 241*, 14–22.

Garrido-Fernández, A., Cortés-Delgado, A. and López-López, A., 2018b. Data on fatty acid profiles of green Spanish-style Gordal table olives studied by compositional analysis. *Data in Brief, 16*, 231–238.

Garrido-Fernández, A., Montaño, A., Cortés-Delgado, A., Rodríguez-Gómez, F. Arroyo-López, F.N., 2021. Multi-statistical approach for the study of volatile compounds of industrial spoiled manzanilla Spanish-style table olive fermentations. *Foods, 10*(6), 1182.

Górska-Horczyczak, E., Zalewska, M. and Wierzbicka, A., 2022. Chromatographic fingerprint application possibilities in food authentication. *European Food Research and Technology*, 1–15.

Granato, D., Putnik, P., Kovačević, D.B., Santos, J.S., Calado, V., Rocha, R.S., Cruz, A.G.D., Jarvis, B., Rodionova, O.Y. and Pomerantsev, A., 2018. Trends in chemometrics: Food authentication, microbiology, and effects of processing. *Comprehensive Reviews in Food Science and Food Safety, 17*(3), 663–677.

Greek Ministry of Agricultural Development and Food, 2022. www.minagric.gr/index.php/el/for-farmer-2/2012-02-02-07–52–07/ellinikaproionta/1182-elaioladaelies-2 (Accessed on February 16, 2022).

Hao, X., Gao, F., Wu, H., Song, Y., Zhang, L., Li, H. and Wang, H., 2021. From soil to grape and wine: Geographical variations in elemental profiles in different Chinese regions. *Foods, 10*, 3108.

Hassan, N.M.H. and Nashat, A.A., 2019. New effective techniques for automatic detection and classification of external olive fruits defects based on image processing techniques. *Multidimensional Systems and Signal Processing, 30*, 571–589.

Hatzakis, E., Koidis, A., Boskou, D., Dais, P. 2008. Determination of phospholipids in olive oil by 31P-NMR spectroscopy. *Journal of Agricultural and Food Chemistry, 56*, 6232–6240.

Herrero, M., Castro-Puyana, M., Ibáñez, E. and Cifuentes, A., 2017. Compositional analysis of foods. In *Liquid chromatography*, pp. 359–380 (2nd ed.). Elsevier. https://doi.org/10.1016/B978-0-12-805392-8.00012-8.

Hidalgo, M.J., Pozzi, M.T., Furlong, O.J., Marchevsky, E.J. and Pellerano, R.G., 2018. Classification of organic olives based on chemometric analysis of elemental data. *Microchemical Journal, 142*, 30–35.

Hong, E., Lee, S.Y., Jeong, J.Y., Park, J.M., Kim, B.H., Kwon, K. and Chun, H.S., 2017. Modern analytical methods for the detection of food fraud and adulteration by food category. *Journal of the Science of Food and Agriculture, 97*(12), 3877–3896.

Huang, L., Meng, L., Zhu, N. and Wu, D., 2017. A primary study on forecasting the days before decay of peach fruit using near-infrared spectroscopy and electronic nose techniques. *Postharvest Biology and Technology, 133*, 104–112.

Ibáñez, C., García-Cañas, V., Valdés, A., Simó, C., 2013. Novel MS-based approaches and applications in food metabolomics. *TrAC Trends in Analytical Chemistry, 52*, 100–111.

International Olive Oil Council (IOC), 2004. *Trade Standard Applying to Table Olives*. International Olive Oil Council.

International Olive Oil Council (IOC). World Table Olive Figures, 2021. Available online: www.internationaloliveoil.org/what-we-do/economic-affairs-promotion%20unit/#figures (accessed on February 5, 2022).

Jakes, W., Gerdova, A., Defernez, M., Watson, A.D., McCallum, C., Limer, E., Colquhoun, I.J., Williamson, D.C. and Kemsley, E.K., 2015. Authentication of beef versus horse meat using 60 MHz ^{1}H NMR spectroscopy. *Food Chemistry, 175*, 1–9.

Jiang, S. and Wang, J., 2016. Internal quality detection of Chinese pecans (Carya cathayensis) during storage using electronic nose responses combined with physicochemical methods. *Postharvest Biology and Technology, 118*, 17–25.

Jurado-Campos, N., García-Nicolás, M., Pastor-Belda, M., Bußmann, T., Arroyo-Manzanares, N., Jiménez, B., Viñas, P. and Arce, L., 2021. Exploration of the potential of different analytical techniques to authenticate organic vs. conventional olives and olive oils from two varieties using untargeted fingerprinting approaches. *Food Control, 124*, 107828.

Jurica, K., Brčić Karačonji, I., Lasić, D., Bursać Kovačević, D. and Putnik, P., 2021. Unauthorized food manipulation as a criminal offense: Food authenticity, legal frameworks, analytical tools and cases. *Foods, 10*(11), 2570.

Kavdır, İ., Burak Büyükcan, M. and Kurtulmuş, F., 2018. Classification of olives using FT-NIR spectroscopy, neural networks and statistical classifiers. *Journal of Food Measurement and Characterization, 12*(4), 2493–2502.

Krajnc, B., Bontempo, L., Luis Araus, J., Giovanetti, M., Alegria, C., Lauteri, M., Augusti, A., Atti, N., Smeti, S., Taous, F. and Amenzou, N.E., 2020. Selective methods to investigate authenticity and geographical origin of Mediterranean food products. *Food Reviews International*, 1–27.

Liu, Q., Sun, K., Zhao, N., Yang, J., Zhang, Y., Ma, C., Pan, L. and Tu, K., 2019. Information fusion of hyperspectral imaging and electronic nose for evaluation of fungal contamination in strawberries during decay. *Postharvest Biology and Technology, 153*, 152–160.

Li, W., Chountoulesi, M., Antoniadi, L., Angelis, A., Lei, J., Halabalaki, M., Demetzos, C., Mitakou, S., Skaltsounis, L.A. and Wang, C., 2022. Development and physicochemical characterization of nanoliposomes with incorporated oleocanthal, oleacein, oleuropein and hydroxytyrosol. *Food Chemistry*, 132470.

López-López, A., Cortés-Delgado, A. and Garrido-Fernández, A., 2011. Chemometric characterisation of the fats released during the conditioning processes of table olives. *Food chemistry, 126*(4), 1620–1628.

López-López, A., Cortés-Delgado, A. and Garrido-Fernández, A., 2020. Impact of processing and extraction on the minor components of green Spanish-style Gordal table olive fat, as assessed by innovative approaches. *Foods, 9*(12), 1907.

Lukić, I., Da Ros, A., Guella, G., Camin, F., Masuero, D., Mulinacci, N., Vrhovšek, U., Mattivi, F., 2020. Lipid profiling and stable isotopic data analysis for differentiation of extra virgin olive oils based on their origin. *Molecules, 25*, 4.

Ma, Y., Guo, B., Wei, Y., Wei, S. and Zhao, H., 2014. The feasibility and stability of distinguishing the kiwi fruit geographical origin based on electronic nose analysis. *Food Science and Technology Research, 20*(6), 1173–1181.

Martínez Gila, D.M., Gámez García, J., Bellemont, A., Mencarelli, F., Gómez Ortega, J. 2020. Fast tool based on electronic nose to predict olive fruit quality after harvest. *Postharvest Biology and Technology, 160*, 111058.

Martínez, S.S., Gila, D.M., Beyaz, A., Ortega, J.G., García, J.G. 2018. A computer vision approach based on endocarp features for the identification of olive cultivars. *Computers and Electronics in Agriculture, 154*, 341–346.

Modupalli, N., Naik, M., Sunil, C.K. and Natarajan, V., 2021. Emerging non-destructive methods for quality and safety monitoring of spices. *Trends in Food Science and Technology, 108*, 133–147.

Montealegre, C., Marina Alegre, M.L., García-Ruiz, C., 2010. Traceability markers to the botanical origin in olive oils. *Journal of Agricultural and Food Chemistry, 58*, 28–38.

Montealegre, C., Sanchez-Hernandez, L., Crego, A., Marina, M., 2013. Determination and characterization of glycerophospholipids in olive fruit and oil by nonaqueous capillary electrophoresis with electrospray-mass spectrometric detection. *Journal of Agricultural and Food Chemistry, 61*, 1823–1832.

Moreno-González, R., Juan, M.E. and Planas, J.M., 2020. Profiling of pentacyclic triterpenes and polyphenols by LC-MS in Arbequina and Empeltre table olives. *LWT, 126*, 109310.

Nasr, E.G., Epova, E.N., de Diego, A., Souissi, R., Hammami, M., Abderrazak, H. and Donard O.F.X., 2022. Trace elements analysis of Tunisian and European extra virgin olive oils by ICP-MS and chemometrics for geographical discrimination. *Foods, 11*(1), 82.

Pacholczyk-Sienicka, B., Ciepielowski, G. and Albrecht, Ł., 2021. The application of NMR spectroscopy and chemometrics in authentication of spices. *Molecules, 26*, 382.

Pafundo, S., Agrimonti, C. and Marmiroli, N., 2005. Traceability of plant contribution in olive oil by amplified fragment length polymorphisms. *Journal of Agricultural and Food Chemistry, 53*(18), 6995–7002.

Papadaki, E. and Tsimidou, M.Z., 2022. Edible oils from olive drupes as a source of bioactive pentacyclic triterpenes. Is there a prospect for a health claim authorization? *Food Chemistry*, 132286.

Pasqualone, A., Di Rienzo, V., Nasti, R., Blanco, A., Gomes, T. and Montemurro, C., 2013. Traceability of Italian protected designation of origin (PDO) table olives by means of microsatellite molecular markers. *Journal of Agricultural and Food Chemistry, 61*(12), 3068–3073.

Perri, E., Benincasa, C. and Muzzalupo, I., 2012. Chapter 13. Olive oil traceability. In *Olive Germplasm-the Olive Cultivation, Table Olive and Olive Oil Industry in Italy*; Muzzalupo, I., Ed.; InTechOpen, pp. 265–286, ISBN 978-953-51-0883-2.

Piras, F., Grillo, O., Venora, G., Lovicu, G., Campus, M., Bacchetta, G. 2016. Effectiveness of a computer vision technique in the characterization of wild and farmed olives. *Computers and Electronics in Agriculture, 122*, 86–93.

Pleguezuelo, C.R.R., Zuazo, V.H.D., Martínez, J.R.F., Peinado, F.J.M., Martín, F.M. and Tejero, I.F.G., 2018. Organic olive farming in Andalusia, Spain. A review. *Agronomy for Sustainable Development, 38*(2), 20.

Ponce, J.M., Aquino, A. and Andújar, J.M., 2019. Olive-fruit variety classification by means of image processing and convolutional neural networks. *IEEE Access, 7*, 147629–147641.

Popović-Djordjević, J.B., Kostić, A.Ž., Rajković, M.B., Miljković, I., Krstić, Đ., Caruso, G., Siavash Moghaddam, S. and Brčeski, I., 2022. Organically vs. conventionally grown vegetables: Multi-elemental analysis and nutritional evaluation. *Biological Trace Element Research, 200*(1), 426–436.

Raigar, R.K., Upadhyay, R. and Mishra, H.N., 2017. Storage quality assessment of shelled peanuts using non-destructive electronic nose combined with fuzzy logic approach. *Postharvest Biology and Technology, 132*, 43–50.

Rasheed, D.M., Serag, A., Abdel Shakour, Z.T. and Farag, M., 2021. Novel trends and applications of multidimensional chromatography in the analysis of food, cosmetics and medicine bearing essential oils. *Talanta, 223*(P1), 121710.

Reale, S., Doveri, S., Díaz, A., Angiolillo, A., Lucentini, L., Pilla, F., Martín, A., Donini, P. and Lee, D., 2006. SNP-based markers for discriminating olive (*Olea europaea* L.) cultivars. *Genome, 49*(9), 1193–1205.

Regulation (EU) No 1151/2012 of the European Parliament and of the Council of 21 November 2012. *On Quality Schemes for Agricultural Products and Foodstuffs*. European Union.

Regulation, C., 1992a. Council Regulation (EEC) No. 2081/92 of 14 July 1992 on the protection of geographical indications and designations of origin for agricultural products and foodstuffs. *Official Journal of the European Union, 208*, 1–8.

Regulation, C., 1992b. No 2082/92 of 14 July 1992 on certificates of specific character for agricultural products and foodstuffs. *Official Journal of the European Union, 208*, 9–14.

Regulation, C., 2006. No 509/2006 of 20 March 2006 on agricultural products and foodstuffs as traditional specialities guaranteed. *OJ L, 93*(31.3).

Rodríguez-Bermúdez, R., López-Alonso, M., Miranda, M., Fouz, R., Orjales, I. and Herrero-Latorre, C., 2018. Chemometric authentication of the organic status of milk on the basis of trace element content. *Food Chemistry, 240*, 686–693.

Ross, A., Brunius, C., Chevallier, O., Dervilly, G., Elliott, C., Guitton, Y., Prenni, J. E., Savolainen, O., Hemeryck, L., Vidkjaer, N. H., Scollan, N., Stead, S. L., Zhang, R. and Vanhaecke, L., 2021. Making complex measurements of meat composition fast: Application of rapid evaporative ionisation mass spectrometry to measuring meat quality and fraud. *Meat Science, 181*, 108333. https://doi.org/10.1016/j.meatsci.2020.108333.

Rotondo, A., Mannina, L. and Salvo, A., 2019. Multiple assignment recovered analysis (MARA) NMR for a direct food labeling: The case study of olive oils. *Food Analytical Methods, 12*(5), 1238–1245. https://doi.org/ 10.1007/s12161-019-01460-4.

Russo, G., Beritognolo, I., Bufacchi, M., Stanzione, V., Pisanelli, A., Ciolfi, M., Lauteri, M. and Brush, S.B., 2020. Advances in biocultural geography of olive tree (*Olea europaea* L.) landscapes by merging biological and historical assays. *Scientific Reports, 10*(1), 1–12.

Sabatini N., Perri E. and Marsilio V., 2009. An investigation on molecular partition of aroma compounds in fruit matrix and brine medium of fermented table olives. *Innovative Food Science and Emerging Technologies, 10*, 621–626.

Sabatini, N. and Marsilio, V., 2008. Volatile compounds in table olives (Olea Europaea L., Nocellara del Belice cultivar). *Food Chemistry, 107*(4), 1522–1528.

Salis C., Papadakis I. and Hagidimitriou M., 2021. Identification and quantification of phenolic compounds in fresh and processed table olives of cv. 'Kalamata'. *Notulae Botanicae Horti Agrobotanici Cluj-Napoca 49*(2), 12394 DOI:10.15835/nbha49212394.

Sánchez, A.H., López-López, A., Cortés-Delgado, A., de Castro, A. and Montaño, A., 2020. Aroma profile and volatile composition of black ripe olives (*Manzanilla* and *Hojiblanca* cultivars). *Food Research International, 127*, 108733.

Sánchez, R., Martín-Tornero, E., Lozano, J., Boselli, E., Arroyo, P., Meléndez, F. and Martín-Vertedor, D., 2021. E-Nose discrimination of abnormal fermentations in Spanish-Style Green Olives. *Molecules, 26*(17), 5353.

Schicchi, R., Speciale, C., Amato, F., Bazan, G., Di Noto, G., Marino, P., Ricciardo, P. and Geraci, A., 2021. The monumental olive trees as biocultural heritage of Mediterranean landscapes: The case study of Sicily. *Sustainability, 13*(12), 6767.

Selli, S., Kelebek, H., Kesen, S. and Sonmezdag, A.S., 2018. GC-MS olfactometric and LC-DAD-ESI-MS/MS characterization of key odorants and phenolic compounds in black dry-salted olives. *Journal of the Science of Food and Agriculture, 98*(11), 4104–4111.

Sesli, M. and Yeğenoğlu, E.D., 2010. RAPD assay of wild-type olives in Turkey. *Genetics and Molecular Research, 9*(2), 966–972.

Spiteri, M., Jamin, E., Thomas, F., Rebours, A., Lees, M., Rogers, K.M. and Rutledge, D.N., 2015. Fast and global authenticity screening of honey using ^{1}H-NMR profiling. *Food Chemistry, 189*, 60–66.

Tahir, H.E., Arslan, M., Komla Mahunu, G., Adam Mariod, A., B.H. Hashim, S., Xiaobo, Z., Jiyong, S., El-Seedi, H.R. and Musa, T.H., 2022. The use of analytical techniques coupled with chemometrics for tracing the geographical origin of oils: A systematic review (2013–2020). *Food Chemistry, 366*, 130633. https://doi.org/10.1016/j.foodchem.2021.130633.

Tapp, H.S., Defernez, M. and Kemsley, E.K., 2003. FTIR Spectroscopy and multivariate analysis can distinguish the geographical origin of extra virgin olive oils. *Journal of Agricultural and Food Chemistry, 51*, 6110–6115.

Tarapoulouzi, M., Agriopoulou, S., Koidis, A., Proestos, C., Enshasy, H.A.E. and Varzakas, T., 2022. Recent advances in analytical methods for the detection of olive oil oxidation status during storage along with chemometrics, authenticity and fraud studies. *Biomolecules, 12*, 1180.

Tarapoulouzi, M., Kokkinofta, R. and Theocharis, C.R., 2020. Chemometric analysis combined with FTIR spectroscopy of milk and Halloumi cheese samples according to species' origin. *Food Science and Nutrition, 8*, 3262–3273.

Tarapoulouzi, M., Skiada, V., Agriopoulou, S., Psomiadis, D., Rébufa, C., Roussos, S., Theocharis, C.R., Katsaris, P. and Varzakas, T., 2021. Chemometric discrimination of the geographical origin of three Greek cultivars of olive oils by stable isotope ratio analysis. *Foods, 10*, 336.

Terouzi, W., De Luca, M., Bolli, A., Oussama, A., Patumi, M., Ioele, G. and Ragno, G., 2011. A discriminant method for classification of Moroccan olive varieties by using direct FT-IR analysis of the mesocarp section. *Vibrational Spectroscopy, 56*(2), 123–128.

Tretola, M., Di Rosa, A.R., Tirloni, E., Ottoboni, M., Giromini, C., Leone, F., Bernardi, C.E.M., Dell'Orto, V., Chiofalo, V. and Pinotti, L., 2017b. Former food products safety: Microbiological quality and computer vision evaluation of packaging remnants contamination. *Food Additives & Contaminants: Part A, 34*(8), 1427–1435.

Tretola, M., Ottoboni, M., Di Rosa, A.R., Giromini, C., Fusi, E., Rebucci, R., Leone, F., Dell'Orto, V., Chiofalo, V. and Pinotti, L., 2017a. Former food products safety evaluation: Computer vision as an innovative approach for the packaging remnants detection. *Journal of Food Quality, 2017*, 1–6. https://doi.org/10.1155/2017/1064580.

Unay, D. and Gosselin, B., 2005. Artificial neural network-based segmentation and apple grading by machine vision. In *IEEE International Conference on Image Processing*, 2, II-630. IEEE.

Valvez, S., Maceiras, A., Santos, P. and Reis, P.N., 2021. Olive stones as filler for polymer-based composites: A review. *Materials*, *14*(4), 845.

Vanloot, P., Bertrand, D., Pinatel, C., Artaud, J. and Dupuy, N. 2014. Artificial vision and chemometrics analyses of olive stones for varietal identification of five French cultivars. *Computers and Electronics in Agriculture*, *102*, 98–105.

Verardo, V., Gómez-Caravaca, A., Montealegre, C., Segura-Carretero, A., Caboni, M., Fernández-Gutiérrez, A. and Bendini, A., 2013. Optimization of a solid phase extraction method and hydrophilic interaction liquid chromatography coupled to mass spectrometry for the determination of phospholipids in virgin olive oil. *Food Research International*, *54*, 2083–2090.

Viaene, J., Vander Heyden, Y., Komsta, Ł. and Sherma, J., 2018. Introduction to herbal fingerprinting by chromatography. In *Chemometrics in chromatography*; Sherma, J. Komsta, L. and Vander Heyden Y. Eds., pp. 351–369, CRC Press. http://dx.doi.org/10.1201/ 9781315154404.

Vigli, G., Philippidis, A., Spyros, A. and Dais, P., 2003. Classification of edible oils by employing ^{31}P and ^{1}H NMR spectroscopy in combination with multivariate statistical analysis. A proposal for the detection of seed oil adulteration in virgin olive oils. *Journal of Agricultural and Food Chemistry*, *51*, 5715–5722.

Wang, Q., Li, L., Ding, W., Zhang, D., Wang, J., Reed, K. and Zhang, B., 2019a. Adulterant identification in mutton by electronic nose and gas chromatography-mass spectrometer. *Food Control 98*, 431–438.

Wang, Z.-C., Yan, Y., Nisar, T., Sun, L., Zeng, Y., Guo, Y., Wang, H. and Fang, Z., 2019b. Multivariate statistical analysis combined with e-nose and e-tongue assays simplifies the tracing of geographical origins of Lycium ruthenicum Murray grown in China. *Food Control 98*, 457–464.

Wen, T., Zheng, L., Dong, S., Gong, Z., Sang, M., Long, X., Luo, M. and Peng, H., 2019. Rapid detection and classification of citrus fruits infestation by Bactrocera dorsalis (Hendel) based on electronic nose. *Postharvest Biology and Technology*, *147*, 156–165.

Wu, H., Yue, T., Xu, Z. and Zhang, C., 2017. Sensor array optimization and discrimination of apple juices according to variety by an electronic nose. *Analytical Methods*, *9*(6), 921–928.

Xing, M., Sun, K., Liu, Q., Pan, L. and Tu, K., 2018. Development of novel electronic nose applied for strawberry freshness detection during storage. *International Journal of Food Engineering*, *14*(7–8).

Yadav, S., Carvalho, J., Trujillo, I., Prado, M., 2021. Microsatellite markers in olives (*Olea europaea* L.): Utility in the cataloging of germplasm, food authenticity and traceability studies. *Foods*, *10*, 1907. https://doi.org/10.3390/ foods10081907

Zagonel, G.F., Peralta-Zamora, P. and Ramos, L.P., 2004. Multivariate monitoring of soybean oil ethanolysis by FTIR spectrometry. *Talanta*, *63*, 1021–1025.

Zhang, X., Cheng, J., Wu, L., Mei, Y., Jaffrezic-Renault, N. and Guo, Z., 2018. An overview of an artificial nose system. *Talanta*, *184*, 93–102.

Zhang, Y., Wang, S., Ji, G. and Phillips, P., 2014. Fruit classification using computer vision and feed-forward neural network. *Journal of Food Engineering*, *143*, 167–177.

5 Authenticity and Chemometrics of Coffee and Tea

Leandro S. Oliveira and Adriana S. França

CONTENTS

5.1 INTRODUCTION

Authenticity is certainly one of the most relevant food quality criteria for the food industry and trading sectors and also for consumers. Regarding authenticity, foods can be classified by their botanical and/or geographic origin (traceability), by the type of process they were subjected to and by their possible intentional adulteration with cheaper, lower-quality products and ingredients.

The traceability of botanical and genotypic origin is of relevance to products for which different species and genotypes present distinct quality traits and where one is considered of superior quality than the others (e.g., *Coffea arabica* is considered of superior organoleptic quality compared to *C. canephora*). Traceability of geographic origin is of relevance to products for which the edaphoclimatic and agricultural and processing practices of a certain region impart unique and distinct quality traits to a product that are not observed in the same type of product from other regions (e.g., organoleptic traits of wines produced in southern France are distinct from organoleptic traits of wines produced from the same type of grapes in regions other than southern France). Traceability of the type of process the food was subjected to is relevant to products for which different types of processing steps can be applied, thus generating products with distinct quality attributes (e.g., hot or cold pressing of olive oil that generates an oil with high acidity in the former and low acidity in the latter, with further undesirable alterations of chemical composition in the former due to the use of high temperatures in the processing). Adulteration is the intentional and fraudulent alteration of a food product with low-quality, cheap, nonedible or toxic products for profit gain (Kamruzzaman, 2016), which renders the food product unauthentic, thus requiring the development of authentication methods that allow for its proper detection, identification and quantification.

A plethora of analytical techniques are currently being used (or being proposed for use) for the authentication of food products. Analytical methodologies employed for food authentication, in its many facets, usually rely on the identification of a single or a few chemical markers that allow for discrimination of authentic and unauthentic products or on the determination of the chemical profile (fingerprint) that is unique to a certain product, thus allowing differentiation of distinct products via distinct chemical profiles. When chemical profiles are used for authentication, due to the large number of variables involved therein, there is a need for the employment of multivariate statistical methods, commonly termed "chemometrics", to accomplish the desired discrimination of food products (Kemsley, Defernez, & Marini, 2019).

In a recent study on the advances on food authenticity techniques and chemometric approaches, Medina et al. (2019) have concluded that gas chromatography coupled to mass spectrometry (GC-MS), liquid chromatography hyphenated to real-time quadrupole time-of-flight mass spectrometry (LC-q-TOF-MS) and nuclear magnetic resonance spectroscopy (NMR) combined with principal component analysis (PCA) and partial least squares–discriminant analysis (PLS-DA) are the most frequently employed analytical and chemometric techniques for authentication of foodstuffs using chemical fingerprints. However, these sophisticated analytical techniques are usually time-consuming, labor-intensive, and costly. Rapid, non-destructive, reliable, and less demanding analytical techniques (e.g., vibrational and fluorescence spectroscopies) combined with chemometrics are thus being intensively studied and used for food authentication (Valand, Tanna, Lawson, & Bengtström, 2019).

Although all types of food may require verification of authenticity, commodities such as coffee, tea, cocoa and vegetable oils are among the most studied food products for proper authentication

DOI: 10.1201/9781003266167-5

by distinct analytical techniques and multivariate statistical methods, be it for authentication of botanical species and varieties, geographic provenance, processing practices or adulteration.

Coffee is the most widely consumed biologically active hot beverage in the world, thus it is a food product of significant relevance from an economic standpoint (Franca & Oliveira, 2019). Coffee beverage results from the brewing of roasted and ground beans, which in turn are the product of roasting crude coffee beans (usually termed "green beans"). Hence, the green beans are the commodity that is largely produced and internationally commercialized, ranking second only to petroleum in terms of currency traded worldwide (Illy, 2002). The green beans are the seeds of the coffee plant, which belongs to the genus *Coffea* of the Rubiaceae family. Although more than 100 species of coffee have already been catalogued, only two species are of commercial relevance worldwide, *C. arabica* and *C. canephora* (Farah & dos Santos, 2015), with *C. arabica* (commonly known as Arabica) representing approximately two-thirds of the world coffee production and *C. canephora* (commonly known as Robusta) the remaining one third (Franca & Oliveira, 2019).

Brews of Arabica coffees usually present more fruity notes, buttery odor, and roasty aroma than brews of Robusta coffees, with the latter tending to be more bitter (Wang, Lim, & Fu, 2020). Hence, Arabica coffees are commercialized at higher market values than Robustas for they are considered of higher quality. However, when roasted and ground, Robustas tend to produce more soluble solids, imparting more body to the brews than the Arabica counterparts. Therefore, a fair portion of the roasted and ground coffee commercialized in the world is a blend of Arabica and Robusta coffees, and the resulting product is priced accordingly depending on the relative amounts of Arabica and Robusta coffees. In their crude state, Arabica and Robusta are easily discriminated by shape, color and density. Arabica beans are greyish or blueish green, shaped like a half triaxial ellipsoid with a sigmoidal longitudinal slit on its flat side, whereas Robustas are yellowish-brown and of a rounder shape with a straighter central slit (Franca & Oliveira, 2019). However, after roasting, due to the intense physicochemical changes the beans undergo (e.g., color, volume, density), it becomes impossible to visually discriminate the two species. Therefore, industry, traders and consumers have a keen interest in the development of reliable analytical methods that can identify and quantify the presence of a specific species in roasted and ground coffee to attest for its authenticity regarding the species present therein.

Commercial production of coffee beans occurs exclusively within the regions of the world delimited by the Tropic of Cancer (23.4° N) and the Tropic of Capricorn (23.4° S), commonly termed the "Coffee Belt" (Franca & Oliveira, 2019). Although global commercial cultivation is confined within this delimited region, the quality of coffee beans of the same species and variety differ from microregion to microregion within the larger region because it is heavily influenced by the local edaphoclimatic conditions of each microregion and by the agricultural practices applied therein. Aside from the specific quality of a coffee from a certain region, other aspects that have become relevant to the consumers is the sustainability and ecofriendliness of the agricultural practices applied therein and whether coffee is organically cultivated. Thus, certification of authenticity of geographic provenance has become increasingly important, and the development of analytical methodologies for such a feat is of significant relevance (Kamiloglu, 2019).

Coffee is one of the most adulterated food products for the purpose of fraudulent financial gain due to its high market value (Ferreira, Galluzzi, de Paulis, & Farah, 2021). Ground roasted coffee is rather vulnerable to adulteration for it presents physical characteristics, such as particle size, texture and color, and some organoleptic properties that can be easily reproduced by roasting and grinding a variety of biological materials, such as cereals, seeds, roots, parchments and others (Reis, Franca, & Oliveira, 2013a). Coffee has been the target of fraudulent admixtures with a diversity of cheaper extraneous products (Wang, Lim, & Fu, 2020). Agricultural residues (e.g., twigs, coffee husks and parchments, and spent coffee grounds) and other roasted biological products (e.g., maize, barley, chicory, rye, triticale, and soybean) are the most common extraneous materials employed in coffee fraudulent strategies to reduce costs and maximize financial gains (Milani, et al., 2020). Hence, the detection and identification of adulterants for the verification of coffee authenticity is of utmost necessity for value assessment and consumer protection and has prompted the development of analytical techniques for such endeavor (Perez, Domínguez-López, López-Yerena, & Queralt, 2021).

Tea is also one of the most consumed beverages in the world, both in traditional form (hot or cold infusions) as well as constituents of ready-to-drink beverages. The production and consumption of this beverage, a hot water infusion of the leaves of *Camellia sinensis* L., has been continuously increasing over the years, and it is estimated that over two-thirds of the world's population drinks or has drunk tea (Ding, et al., 2019). Several plant infusions that are also termed "teas" are also very popular, including rooibos (*Aspalathus linearis*), rose hip (*Rosa* spp.), chamomile (*Matricaria recutita* L.), mate (*Ilex paraguariensis*) and many others (da Silva Pinto, 2013; Boaventura, et al., 2013). However, it is important to highlight that the term "herbal tea" is best applied to such beverages, and they will not be covered in the present chapter.

Tea, in the traditional sense, is classified into six major categories according to processing techniques: green tea (non-fermented); yellow and white teas (lightly fermented, 10%–20%); oolong tea (semi-fermented, 30%–60%); black tea (fully fermented, 80%–100%); and dark tea (post-fermented, ≥100%). The popularity of tea is not only due to its attractive taste and aroma but also because of its many pharmacological benefits, including the suppression of tumor cell formation, anti-obesity, prevention of cardiovascular diseases and potential for lowering the risk of atherosclerosis (Bag, Mondal, Majumder, & Banik, 2022). It is noteworthy to point out that the number of studies dealing with potential health benefits of tea or its constituents is continuously increasing (Engelhardt, 2020). Such beneficial health attributes are associated to different phytochemicals, including amino acids, vitamins, flavonoids, caffeine and polysaccharides, among others (Khan & Mukhtar, 2007). Given that variations in tea geographic origin and processing conditions have a significant impact not only on its flavor but also its composition, the establishment of authenticity parameters is of paramount importance.

A recent report indicates that coffee and tea are shown as the seventh in the top ten products that are most at risk of food fraud (Burns, Tweed, & Walker, 2017), and thus an overview of recent studies related to the establishment of authenticity parameters for such products is presented in the following sections. As previously stated, the analytical methodologies employed for the authentication of both coffee and tea mostly rely on the identification of chemical markers or a specific chemical profile that will allow for the discrimination of authentic and unauthentic products. Given the large number of variables involved in these methods, the analyses rely on the employment of multivariate statistical methods (chemometrics), so a brief description of the most used terms and abbreviations is presented in the following section.

5.2 BRIEF OVERVIEW OF CHEMOMETRIC METHODS

Chemometrics can be defined as the application of mathematical and statistical methods for the design and analysis of chemical data, experiments and simulation to improve its interpretation. Chemometric methods can be divided into two major groups: (1) supervised and (2) unsupervised methods. In supervised methods, each analyzed sample comes from an already established class, and this information is used during data analysis. Unsupervised methods, on the other hand, do not make use of such information, so they do not require any prior knowledge with respect to sample classification (Fulgencio, Resende, & Botelho, 2022).

The two most employed unsupervised methods are principal component analysis (PCA) and hierarchical cluster analysis (HCA). PCA is a pattern recognition method that is widely used for visualization of sample class-based differences in a data set. PCA projects a given data set onto a new set of axes that provide a more accurate description of the occurring variations. It requires the input of a 2D data matrix in which rows represent different samples and columns represent different variables. PCA then finds linear combinations of latent variables referred to as principal components (PCs) in the data matrix. If sample classes are very different from each other, they should form distinct clusters on a scores plot, usually a plot of PC2 scores on the y-axis versus PC1 scores on the x-axis (Sudol, et al., 2020). HCA is a classification method based on data grouping by clusters of similar samples and differentiation by their distances. The general idea in this type of analysis is the clustering of samples by the smaller distances, or in other words, by the closest similarities of the samples. The major limitation of HCA is that it does not show the overall relationship between all the samples but only between the ones that are close together (Callao & Ruisánchez, 2018).

Supervised methods provide the type of binary response that is required by qualitative analysis (e.g., yes/no, belongs/does not belong) by applying a classification technique that requires classes (or categories) to be defined. Each class consists of a set of samples with a common property (i.e., authentic samples) and different from the other classes (i.e., non-authentic samples). These classification techniques can be divided into (1) discriminant and (2) class modeling. Discriminant techniques include linear discriminant analysis (LDA), quadratic discriminant analysis (QDA), k-nearest neighbors (kNN), partial least squares–discriminant analysis (PLS-DA) and support vector machines (SVM). The main modeling techniques are soft independent modeling of class analogy (SIMCA) and unequal dispersed classes (UNEQ; Fulgencio, Resende, & Botelho, 2022; Callao & Ruisánchez, 2018).

5.3 COFFEE

The recent applications of analytical methodologies coupled with chemometrics to verify and certify the authenticity of coffees will be presented in this section. It is not the aim of this section to be exhaustive of the literature on the topic but rather to present the more recently published works. Scientific literature on coffee authenticity is abundant, and the reader is referred to excellent recent reviews on the topic (Burns, Tweed, & Walker, 2017; Kamiloglu, 2019; Wang, Lim, & Fu, 2020; Ferreira, Galluzzi, de Paulis, & Farah, 2021; Perez, Domínguez-López, López-Yerena, & Queralt, 2021).

Analytical techniques coupled to chemometric methods were firstly applied to coffee in the early 1980s (Liardon, Ott, & Daget, 1984) for the analysis of coffee headspace profiles. These types of techniques were timidly employed during the 1990s for coffee analysis (Kemsley, Defernez, & Marini, 2019; Downey & Boussion, 1996; Martín, Pablos, & González, 1998; Martín, Pablos, & González, 1998) and, from the early 21st century, applications have been steadily spreading to all areas of coffee analysis. Techniques have been successfully applied to the identification and discrimination of geographic provenance of coffees for both macro (countries and continents) and micro (states, provinces and municipalities) regions of the planet (Alves, Casal, Oliveira, & Ferreira, 2003; Bertrand, et al., 2008; Wang, et al., 2009; Wei, et al., 2012; Valentin & Watling, 2013; Link, et al., 2014; da Silva Taveira, et al., 2014; Babova, Occhipinti, & Maffei, 2016; Marquetti, et al., 2016; Mehari, et al., 2015; de Toledo, et al., 2017; Botelho, Oliveira, & Franca, 2017; Bona, et al., 2017; Firmansyah, Winingsih, & Soebara, 2017; Monteiro, et al., 2018; Peng, et al., 2019; Giraudo, et al., 2019; Worku, et al., 2019; Núñez, Martínez, Saurina, & Núñez, 2020; Badmos, Fu, Granato, & Kuhnert, 2020; Marek, et al., 2020; Ongo, et al., 2020; Abdelwareth, Zayed, & Farag, 2021; Núñez, Martínez, Saurina, & Núñez, 2020; Robert, de Gois, Rocha, & Luna, 2022). Geographic provenance, due to the associated specific edaphoclimatic conditions and farming practices, plays a major role in defining the organoleptic properties of coffees (Bertrand, et al., 2012).

Coffea arabica and *C. canephora* are the major coffee species cultivated and commercialized in the world. Although each species has its own specific desirable traits, globally *C. arabica* is considered of higher quality and thus is usually commercialized at higher prices than *C. canephora*. Hence, there is a need for the development of analytical methods for the certification of authenticity of coffees regarding the species. Discrimination of coffee species has been a fertile research area in which analytical and chemometric techniques have been and still are frequently applied (Kemsley, Defernez, & Marini, 2019; Downey & Boussion, 1996; Martín, Pablos, & González, 1998; Alves, Casal, Oliveira, & Ferreira, 2003; Alves, Casal, Alves, & Oliveira, 2009; Keidel, et al., 2010; El-Abassy, Donfack, & Materny, 2011; Wermelinger, D'Ambrosio, Klopprogge, & Yeretzian, 2011; Pacetti, Boselli, Balzano, & Frega, 2012; Wei, et al., 2012; Cagliani, Pellegrino, Giugno, & Consonni, 2013; Monakhova, et al., 2015; Babova, Occhipinti, & Maffei, 2016; Cossignani, Montesano, Simonetti, & Blasi, 2016; Servillo, et al., 2016; Colzi, et al., 2017; Bressanello, et al., 2017; Dankowska, Domagała, & Kowalewski, 2017; Craig, Botelho, Oliveira, & Franca, 2018; Assis, et al., 2019; Assis, et al., 2020; Núñez, Martínez, Saurina, & Núñez, 2020; Badmos, Fu, Granato, & Kuhnert, 2020; Ongo, et al., 2020; Abdelwareth, Zayed, & Farag, 2021; Núñez, et al., 2020; Baqueta, et al., 2021). Chemometric techniques were also applied to the discrimination of coffee varieties (genotypes), considering specifically Arabica genotypes (Bertrand, et al., 2008; Link, et al., 2014; da Silva Taveira, et al., 2014; Marquetti, et al., 2016; Mehari, et al., 2015; Bona, et al., 2017; Figueiredo, et al., 2019; Malta, et al., 2020; Malta, et al., 2021) and Robusta genotypes (Dong, et al., 2017).

Coffee has been a subject of adulteration since its early times of production and commercialization, with reports on identified cases of coffee adulteration dating back to the late 1800s (Ferreira, Galluzzi, de Paulis, & Farah, 2021). Although research has been conducted on methods to detect, identify and quantify adulterations in coffee since the beginning of the 20th century, it was only in the 1990s that such type of research has gained momentum because of an increased awareness of researchers, merchants and consumers on the subject of widespread coffee adulteration (Prodolliet, et al., 1995) and because of the significant advancements in analytical techniques and associated apparatuses at the time. Initially, the majority of research on the subject was directed to identifying specific chemical markers that would allow for the detection and identification of adulterants in coffee but, given the complexity of coffee chemical composition and of its common adulterants, it rapidly became clear it is a rather difficult task. However, with the advancements in chemometrics, research on coffee adulteration shifted to attempts to detect, identify and quantify adulterants by their discriminating chemical profiles in relation to those of coffees. Research on detection, identification and quantification of adulterants in coffee using chemometric approaches intensified by the end of the first decade of the 21st century (Oliveira, Oliveira, Franca, & Augusti, 2009; Garcia, et al., 2009; Ebrahimi-Najafabadi, et al., 2012; Reis, Franca, & Oliveira, 2013a, 2013b; Cebi, Yilmaz, & Sagdic, 2017; de Moura Ribeiro, et al., 2017; Brondi, Torres, Garcia, & Trevisan, 2016; Yulia & Suhandy, 2017; Milani, et al., 2020; Cheah & Fang, 2020; Flores-Valdez, et al., 2020; Núñez, Pons, Saurina, & Núñez, 2021).

Aside from the use for discrimination of coffees by geographic provenance, botanical species and genotypes, and adulteration, chemometrics were also employed for discrimination of defective and non-defective beans in both crude and roasted and ground coffee (Agresti, Franca, Oliveira, & Augusti, 2008; Mendonça, Franca, Oliveira, & Nunes, 2008; Craig, Franca, & Oliveira, 2011, 2012a, 2012b; Santos, Sarraguça, Rangel, & Lopes, 2012; Craig, et al., 2014; Dias, et al., 2018); and discrimination of coffees by their sensory attributes (Bertrand, et al., 2012; Bressanello, et al., 2017; Craig, Botelho, Oliveira, & Franca, 2018; Belchior, Botelho, Oliveira, & Franca, 2019; Baqueta, et al., 2021; Barrios-Rodríguez, et al., 2021; Malta, et al., 2021; Silva, et al., 2021).

An overview of the analytical techniques and chemometric methods recently studied for certification of the authenticity of coffees is presented in Table 5.1.

Table 5.1 Overview of Analytical Techniques and Chemometric Methods Studied for Certification of Authenticity of Coffees

Coffee	Purpose of Analysis/Sample Differentiation	Analytical Technique	Number of Samples	Chemometric Technique	Reference
Arabica	Adulteration with corn, coffee husks, barley and soybean	^{1}H-NMR	31	PCA	(de Moura Ribeiro, et al., 2017)
Arabica and Robusta	Species and geographic origin	GC-MS	8	PCA	(Bressanello, et al., 2017)
Arabica	Adulteration with corn	DSC and FTIR	33 for DSC and 99 for FTIR	PCA, PLS	(Brondi, Torres, Garcia, & Trevisan, 2016)
Robusta	Botanical varieties and roasting degree	Electronic nose and tongue sensors	126	PCA, PLS-DA, ANN, PLSR	(Dong, et al., 2017)
Arabica	Geographic origin	Fluorescence spectroscopy	110	PARAFAC, NPLS-DA, UPLS-DA	(Botelho, Oliveira, & Franca, 2017)
Robusta civet	Robusta civet and Robusta non-civet	UV-Vis spectroscopy	100	SIMCA, PLS-DA	(Yulia & Suhandy, 2017)
Arabica and Robusta	Geographic origin and farming system	PTR-MS, NIRS	45	PLS-DA, LDA-kNN	(Monteiro, et al., 2018)
Arabica and Robusta	Cup quality	FTIR	30	PCA, PLS-DA	(Craig, Botelho, Oliveira, & Franca, 2018)
Arabica and Robusta	Defective and non-defective	FTIR-PAS	154	PLS-DA	(Dias, et al., 2018)
Arabica and Robusta	Sensory attributes	FTIR	24	PLS-DA	(Belchior, Botelho, Oliveira, & Franca, 2019)
Arabica and Robusta	Geographic origin and farming system	Stable isotope	67	LDA, kNN SVM	(Peng, et al., 2019)
Arabica and Robusta	Composition of species blends	FTIR/PS-MS	120	PLS, GA, OPS	(Assis, et al., 2019)
Arabica	Genotypes	Raman spectroscopy	63	PCA, PLS-DA	(Figueiredo, et al., 2019)

(*Continued*)

Table 5.1 (Continued)

Coffee	Purpose of Analysis/Sample Differentiation	Analytical Technique	Number of Samples	Chemometric Technique	Reference
Arabica and Robusta	Composition of species blends	NIRS/TXRF	40	PLS, GA, OPS	(Assis, et al., 2020)
Arabica and Robusta	Geographic origin, variety and roasting degree	HPLC-UV	306	PLS-DA, PLSR	(Núñez, Martínez, Saurina, & Núñez, 2020)
Arabica and Robusta	Geographic origin and sensory attributes	GC-MS	10	PCA, HCA, OPLS-DA	(Abdelwareth, Zayed, & Farag, 2021)
Arabica	Sensory attributes	ATR-FTIR	27	PCA, LDA	(Barrios-Rodríguez, et al., 2021)
Arabica and Robusta	Adulteration with chicory	HPLC-UV-FLD	88	PLS-DA	(Núñez, Pons, Saurina, & Núñez, 2021)
Arabica and Robusta	Sensory attributes	FTIR	90	PCA, LDA	(Silva, Pérez-Rodríguez, et al., 2021)

Abbreviations: ANN, artificial neural networks; FTIR, Fourier transform infrared spectroscopy; FTIR-PAS, infrared–photoacoustic spectroscopy; GA, genetic algorithm; GC-MS, gas chromatography–mass spectrometry; HCA, hierarchical cluster analysis; HPLC-UV-FLD, high-performance liquid chromatography with ultraviolet and fluorescence detection; LDA, linear discriminant analysis; LDA-kNN, linear discriminant analysis–k-nearest neighbor; NIRS, near-infrared spectroscopy; NPLS, superior order or N partial least squares; OPLS-DA, orthogonal projection to latent structures discriminant analysis; OPS, ordered predictors selection; PARAFAC, parallel factor analysis; PCA, principal component analysis; PLS, partial least squares; PLS-DA, partial least squares–discriminant analysis; PLSR, partial least squares regression; SIMCA, soft independent modeling of class analogy; SVM, support vector machine; TXRF, total reflection X-ray fluorescence; U-PLSDA, unfolded partial least squares discriminant analysis; UV-Vis, ultraviolet-visible.

5.4 TEA

This section discusses recent applications of chemometrics for the authenticity analysis of tea, focusing on studies published in the last five years (2017–2021). Reference to earlier studies can be found on previously published reviews on the authenticity of tea and other beverages (Kamiloglu, 2019; Engelhardt, 2020).

From a commercial point of view, teas can be classified into the following categories according to the degree of fermentation: non-fermented (green, white, and green pu-erh teas), partially fermented (oolong and paochong teas), and fully fermented (black and pu-erh teas). Green and oolong teas are mostly consumed in Asia and northern Africa, white tea in Asia and Europe, pu-erh tea in Asia, and black tea worldwide (Zhao, Zhang, & Zhang, 2017). The major concerns in terms of tea authenticity or adulteration are similar to those of coffee and include the use of tea from other geographical origins, lower-quality grades, reduced or extended aging, and substitution with similar but less valued varieties. An overview of recent studies that employ chemometrics for the analysis of tea is presented in Table 5.2.

Ongoing consumer demands about labeling specifications regarding not only tea but also the geographic origin of foods in general have driven research efforts toward this goal, given that recent regulations require the provision of origin information on the label (Zhao, Zhang, & Zhang, 2017). Tea produced in specific locations has found favorable acceptance among customers, who are willing to pay more for a genuine and typical product. However, the differences between similar products are quite subtle, so mislabeling of teas by unscrupulous producers is of concern. Although the geographic origin of some tea varieties could be assessed by an expert sensory panel, such analysis can be subjective and sometimes inconsistent and unpredictable due to individual variability and the influence of the panelist's physiological or psychological condition. Therefore, many studies have focused on the development of accurate and robust analytical methods for such endeavors. The major discrimination techniques used for authentication of teas in terms of geographic provenance include the employment of analytical tools such as analysis of mineral composition, spectroscopy and chromatographic methods.

Mineral elements are usually stable and less influenced by processing methods and storage time compared with other chemical components; they have been extensively employed to establish the

Table 5.2 Recent Applications of Chemometrics for the Analysis of Tea

Tea	Purpose of Analysis/Sample Differentiation	Analytical Technique	Number of Samples	Chemometric Technique	Reference
White	Geographical origin	ICP-MS, ICP-OES	64	LDA, SVM, kNN	(Ye, et al., 2016)
Fresh tea leaves	Geographical origin	ICP-MS	91	LDA	(Zhao, et al., 2017)
Black, oolong, green and white	Geographical origin and tea type	HS-SPME/ Es-GC–MS	23	PCA, PLS-DA	(Mu, et al., 2018)
Black	Geographical origin, processing and botanical varieties	HPLC-MS	60	HCA, PLS-DA	(Shevchuk, Jayasinghe, & Kuhnert, 2018)
Black	Geographical origin	NIRS	250	PLS-DA, SIMCA	(Firmani, et al., 2019)
Black	Geographical origin	HS-GC/MS	306	OPLS-DA *k*-NN, RF	(Yun, et al., 2021)
Longjing	Drying process, maturity and harvest period	Stable isotope	48	OPLS-DA	(Xia, et al., 2021)
Green	Cultivar, elevation and processing variations	UPLC–Q-TOF/ MS	66	PCA, PLS-DA, OPLS-DA	(Wang, Cao, Yuan, & Guo, 2021)
Black	Geographical origin	ICP-MS	104	LDA, OPLS-DA	(Ren, et al., 2021)
Green	Geographical origin	Stable isotopic and elemental profiles	143	PCA, LDA, OPLS-DA	(Liu, et al., 2021)
Black, green, white and herbal	Geographical origin	XRF spectroscopy	75	PCA, SIMCA, PLS-DA	(Lim, Carey, Williams, & Koidis, 2021)
White	Storage time	UHPLC–QTOF/ MS	26	PCA, PLS-DA, PLS	(Dai, et al., 2018)
Dark (An)	Storage time	UHPLC–QTOF/ MS	21	PCA, HCA, PLS-DA	(Shen, et al., 2022)
Green	Storage time	VNIR hyperspectral imaging	150	PLS-DA, SVM	(Li, et al., 2020)
Black, oolong, green, dark, white, yellow and scented	Adulteration (addition of caffeine)	LC-IRMS	116	PLS-DA	(Ding, et al., 2019)
Green	Adulteration (sugar and rice flour)	Micro NIRS	475	SVR	(Li, et al., 2021)

Abbreviations: Es-GC–MS, enantioselective gas chromatography–mass spectrometry; HCA, hierarchical cluster analysis; HPLC-MS, high-performance liquid chromatography–mass spectrometry; HS-GC/MS, headspace gas chromatography–mass spectrometry; HS-SPME, headspace–solid phase microextraction; ICP-MS, inductively coupled plasma–mass spectrometry; ICP-OES, inductively coupled plasma optical emission spectrometry; LC-IRMS, liquid chromatography–isotope ratio mass spectrometry; LDA, linear discriminant analysis; NIRS, near-infrared spectroscopy; OPLS-DA, orthogonal projection to latent structures discriminant analysis; PCA, principal component analysis; SIMCA, soft independent modeling of class analogy; SVR, support vector machine regression; UPLC–Q-TOF/MS, ultra-high-performance liquid chromatography quadrupole time-of-flight mass spectrometry; VNIR, visible near-infrared; XRF, X-ray fluorescence.

geographic origin of tea (Ye, et al., 2016; Zhao, Zhang, & Zhang, 2017; Liu, et al., 2021; Ren, et al., 2021; Moreda-Piñeiro, Fisher, & Hill, 2003; Pilgrim, Watling, & Grice, 2010; Ma, et al., 2016). The theoretical basis for mineral fingerprinting is that the elemental signature is transferred from the environment during plant growth, thus giving rise to differences in the elemental signatures of plants depending on variables including underlying geology and soil composition, climate and environmental conditions.

Zhao, Zhang, & Zhang (2017) proposed the use of ICP-MS for evaluation of the mineral profile of fresh tea leaves collected from three different tea producing regions in China. The concentrations of 20 isotopes were determined for tea samples of different varieties: Anjibai, Wuniuzao and Longjing. Results obtained from one-way ANOVA showed that there were statistically significant differences between elemental levels of the tea samples from different regions, except for K, Mn, Fe, Se and Mo concentrations. Soil samples were also analyzed, and the results showed a positive correlation between elemental composition of tea leaves and the respective provenance soils for the following elements: Na, Mg, Ca, Cr, Fe, Ni, Rb, Sr, Cd, Tl and Pb. Stepwise linear discriminant analysis (SLDA) was then carried out to establish the ability of the previously cited elements to provide discrimination according to geographical origin. A robust discriminant model (100% correct classification) was built with Na, Mg, Ca, Ni, Rb, Sr and Pb levels as variables.

Ren and collaborators (2021) also used elemental fingerprinting for evaluation of the geographic origin of black tea coming from three different locations. The concentrations of 27 isotopes of elements were determined using ICP-MS. PCA and HCA did not provide a good separation of the clusters. The most important mineral elements for discrimination of samples determined according to OPLS-DA were Sr, Rb, Ni, Cd, Ba, Zn, P, Cu and Mo. While OPLS-DA had accuracy rates over 82%, LDA and SVM models exhibited better discrimination performances, with 100% predictive ability. These results indicate that elemental profiling seems to be able to provide satisfactory discrimination in terms of the geographic provenance of teas.

In the study developed by Mu, et al. (2018), the goal was to verify the relationships between tea cultivars and the enantiomeric distributions of their volatile constituents according to their chromatographic profiles. Headspace–solid phase microextraction (HS-SPME) coupled to enantioselective gas chromatography–mass spectrometry (Es-GC–MS) was able to quantify 15 volatile components in 22 tea cultivars from different locations within China. Fresh tea leaves from these cultivars were processed into suitable tea types (green, black, oolong and white) according to their suitability for manufacture. PLS-DA of the fresh leaves provided separation of the samples into three classes according to their geographical origins: Guangdong, Fujian and Zhejiang provinces. The key components identified as most important for sample differentiation were *S*-*E*-nerolidol, (2*R*, 5*R*)-linalool oxide A, (2*S*, 5*S*)-linalool oxide A, *S*-citronellol, *R*-linalool, (1*R*, 2*R*)-methyl jasmonate and *S*-α-ionone. In addition, a comparison of processed teas and fresh leaves showed that the processing methods had significant influence on the formation of volatile enantiomers, and that the trends in their concentrations differed significantly depending on the type of tea.

In a recent study by Yang and collaborators (2021), black tea samples collected from ten different geographic origins in China, India and Sri Lanka (Kandy) were analyzed by headspace volatilization followed by GC-MS (HS-GC-MS). Forty-eight volatile compounds were identified by comparison of mass spectra (MS) and retention index (RI) with reference analytical standards. The data were analyzed by the *k*-nearest neighbor (*k*NN) and random forest (RF) classification methods. The discrimination rates based on the full spectra were 100% and 95% for *k*NN and RF, respectively. Key aroma analysis identified linalool and geraniol as potential markers for geographic discrimination, so analysis was also performed using only these two substances. The *k*NN discrimination rates were 100% for most origins, except for 89% discrimination for one of the areas (a misclassification of three samples). Such results indicate that the volatile profile based on GC methods coupled to chemometrics provides robust models for tea provenance discrimination.

As previously discussed, both mineral content and chromatographic methods have been successfully used for tea authentication with respect to provenance. Despite their efficiency, these methods present some drawbacks, including time-consuming sample preparation and analysis, high costs and environmental impact. Spectroscopic methods, mostly NIR and FTIR, have been extensively employed for food authentication in general, and their use has been increasing over the last decades because of their many advantages compared with the use of other methods, including no destruction of the sample, minimal or no sample preparation, and the fact that they do not involve the use of any hazardous chemicals during the analysis (Franca & Nollet, 2017).

Darjeeling tea, valued for its quality in the world beverage market, is produced in the eponymous district in the West Bengal state in India. This specific type of black tea is part of the register of protected designations of origin (PDO) and protected geographical indications (PGI; Das, 2006). Because of its high added value, it can be particularly subject to adulterations and frauds, mostly by the addition of black teas with lower commercial value (Firmani, Luca, Bucci, Marini & Biancolillo, 2019). A recent study developed by Firmani and collaborators (2019) has established the potential of NIR for the authentication of black Darjeeling tea. In the first part of the work, black tea samples of different provenances were analyzed: Ceylon, English breakfast and Darjeeling. PLS-DA was able to provide 100% and 92.3% correct classifications for Darjeeling and other teas, respectively. SIMCA had a similar but slightly inferior performance, with correct classification values of 97.4% and 84.61% for Darjeeling and others, respectively. Mixed samples of Darjeeling and Ceylon teas were also tested, with adulteration levels (amount of Ceylon) in the range of 10% to 50%. Both PLS-DA and SIMCA were used to build classification models. As in the case of pure teas, the highest classification rates were provided by PLS-DA, with correct classification of 138 over 140 test samples. The results indicated PLS-DA applied to NIR data to be the most suitable methodology for the authentication of Darjeeling teas.

Storage time is a crucial parameter that affects the quality of teas because it can change its aroma, taste, bioactivity and chemical profile (Ning, et al., 2016). White and dark fermented teas stored for long periods are considered to have higher quality and commercial value. On the other hand, green teas stored for long periods are regarded as not fresh and umami (Dai, et al., 2018). Given that storage time will affect several tea quality parameters, a wide variety of techniques have been employed for tea discrimination, such as electronic tongue and eye, and chromatographic (Dai, et al., 2018; Zhou, et al., 2020; Meng, et al., 2021; Shen, et al., 2022) and spectroscopic (Li, et al., 2020; Hong, et al., 2021) methods.

Dai and collaborators (2018) employed an ultra-high-performance liquid chromatography–quadrupole time-of-flight/mass spectrometry (UHPLC–QTOF/MS)-based metabolomics approach to verify the effect of storage time in the non-volatile compound profiles of white tea. The levels of flavan-3-ols, procyanidins, theasinensins, theaflavins, flavonol-*O*-glycosides, flavone-*C*-glycosides, and most of the amino acids were shown to decrease after long-term storage (>4 years). Also, the contents of 8-C *N*-ethyl-2-pyrrolidinone-substituted flavan-3-ols (EPSFs), formed from theanine and flavan-3-ols during storage, were positively correlated with storage time. Both PCA and PLS-DA were able to provide satisfactory discrimination, with the samples being separated into three groups based on storage time: (1) 1 year, (2) 2–4 years and (3) over 4 years. The results demonstrated that the formation of EPSFs from theanine and flavan-3-ols was linearly correlated to the storage time, identifying EPSFs as potential markers for the discrimination of the storage duration of white teas. Zheng et al. (2021) also employed metabolomics and were able to identify a total of 47 compounds as critical metabolites responsible for quality changes during storage of Qingzhuan (post-fermented) tea. These were associated to methylation of catechins, glycosylation of flavonoids, degradation of flavoalkaloids, biosynthesis of triterpenoids, and formation of theabrownins, resulting in the reduction of bitterness and astringency and a general increase in tea quality with time.

As previously mentioned, in the case of green tea, increasing the storage time will have a negative effect on quality. As storage time increases chlorophyll is oxidized and degraded into pheophytin. With this, the green color gradually changes to olive brown, thus damaging the tea appearance. Li, et al. (2020) proposed the use of visible–near-infrared (VIS-NIR) hyperspectral imaging for identification of green tea storage period under two different temperatures, 4°C and 25°C. The optimal wavelengths were selected by different methods: regression coefficients (RCs) of partial least squares (PLS)-based models, second derivative (2-Der), and successive projections algorithm (SPA). For samples stored at 4°C, 11, 11, and 15 wavelengths were selected by RCs, 2-Der and SPA, respectively. For samples stored at 25°C, RCs, 2-Der and SPA selected 10, 12, and 11 wavelengths, respectively. A comparison of the discriminant rates of PLS- DA and support vector machine (SVM) models for evaluation of green tea storage period was also performed. For the samples stored at 4°C, the SVM model based on optimal wavelengths selected by SPA provided the best results, with 100% and 98% recognition accuracies for the calibration and prediction sets, respectively. For samples stored at 25°C, the SVM model based on optimal wavelengths selected by 2-Der had the best discrimination results, with recognition accuracies of 98% for both correction and prediction sets. It was observed that, regardless of the storage temperature, the SVM models provided better results than PLS-DA models.

There are a few recent studies that deal with tea authenticity regarding adulteration. Ding and collaborators (2019) employed liquid chromatography coupled to isotope ratio mass spectrometry

(LC-IRMS) for evaluation of commercial teas available in the Chinese market. The main components of tea include caffeine, theophylline, polyphenols, organic acids and amino acids, among others. Caffeine levels are of particular interest and may influence tea consumption (McKay & Blumberg, 2002), so the goal of the study was to differentiate samples with naturally occurring or synthetic (added) caffeine. Samples comprised different categories of tea, including black, oolong, dark, green, white, yellow and scented. Although PLS-DA models could not discriminate among such categories, the method was also applied to verify the stated natural caffeine sources in 19 tea drinks, out of which seven samples were found to be mislabeled.

One of the problems in terms of adulteration of green tea is the addition of sugar and glutinous rice flour to improve flavor by reducing the inherent astringency and bitterness that is associated to some tea components, including tannins, caffeine and catechins. With that in mind, Li, et al. (2021) employed micro near-infrared spectroscopy (NIRS) to scan a total of 475 samples of pure teas and teas adulterated with sugar and with glutinous rice flour. In the qualitative stage, PVS was used to verify the classification trend. PLS-DA and SVM were used to differentiate pure and adulterated tea. Furthermore, in a quantitative stage, PLSR and support vector machine regression (SVR) methods were used to predict the concentrations of adulterants. Models were built using both the full spectra and wavelength selection methods that provided a significant reduction in the number of model variables. A simplified SN–IRIV–SVR model (comprising 43 wavelengths for tea adulterated with sugar and 18 wavelengths for tea adulterated with glutinous rice flour) had a classification accuracy similar to that of the full-wavelength SNV-SVR models.

5.5 CONCLUSIONS

Coffee and tea are natural products, present rather complex physical structures and chemical compositions, and the hot beverages prepared from coffee and tea are also complex. The quality of these natural products and their associated hot beverages is defined by a diversity of factors such as botanical species and genotypes of the product, the specific edaphoclimatic conditions at the cultivation regions, farming and processing practices, and type of brewing method employed for the beverage preparation. Research on analytical methods for attesting the authenticity of these products has been timidly carried out throughout the 20th century and more intensively in the first decades of the 21st century due to the significant advancements in the analytical techniques and in the chemometric methods. The employment of chemometric methods to the analysis of coffee and tea has allowed the use of their complex chemical profiles as a basis for discrimination of species and genotypes, of geographic provenance, of organoleptic attributes and of the product and its eventual adulterants. Robust and sophisticated analytical techniques, such as gas and liquid chromatography coupled to mass spectrometry and nuclear magnetic resonance, as well as simple and reliable ones, such as Fourier transform infrared spectroscopy and fluorescence spectroscopy, have been extensively studied as means for the certification of coffee and tea authenticity. Although these analytical techniques have been successfully employed for such tasks with the support of chemometric methods, there is still a need for further research before the developed methodologies can be implemented as routine methods for authentication of coffee and tea by regulatory and monitoring agencies throughout the world. One major aspect is that some of these techniques are labor-intensive, require intricate sample preparations, are expensive and require qualified personnel to operate them. However, the major precluding aspect is the fact that, regardless of the recent advancements in chemometric methods, there is still need for extensive research on calibration transfer before these methodologies can be safely and reliably implemented as routine methods for authentication of coffee and tea products. Calibration transfer requires numerous and distinct strategies and mathematical techniques for implementing a single calibration database comprising samples, reference data and calibration equations to multiple instruments.

5.6 ACKNOWLEDGMENT

The authors acknowledge financial support from Conselho Nacional de Desenvolvimento Científico e Tecnológico (CNPq), Brazil.

REFERENCES

Abdelwareth, A., Zayed, A., & Farag, M. A. (2021). Chemometrics-based aroma profiling for revealing origin, roasting indices, and brewing method in coffee seeds and its commercial blends in the Middle East. *Food Chemistry, 349*, 129162. doi:10.1016/j.foodchem.2021.129162

Agresti, P. D., Franca, A. S., Oliveira, L. S., & Augusti, R. (2008). Discrimination between defective and non-defective Brazilian coffee beans by their volatile profile. *Food Chemistry, 106*, 787–796. doi:10.1016/j.foodchem.2007.06.019

Alves, M. R., Casal, S., Oliveira, M. B., & Ferreira, M. A. (2003). Contribution of FA profile obtained by high-resolution GC/chemometric techniques to the authenticity of green and roasted coffee varieties. *Journal of the American Oil Chemists Society, 80*, 511–517. doi:10.1007/s11746-003-0730-0

Alves, R. C., Casal, S., Alves, M. R., & Oliveira, M. B. (2009). Discrimination between Arabica and Robusta coffee species on the basis of their tocopherol profiles. *Food Chemistry, 114*, 295–299. doi:10.1016/j.foodchem.2008.08.093

Assis, C., Gama, E. M., Nascentes, C. C., de Oliveira, L. S., Anzanello, M. J., & Sena, M. M. (2020). A data fusion model merging information from near infrared spectroscopy and X-ray fluorescence. Searching for atomic-molecular correlations to predict and characterize the composition of coffee blends. *Food Chemistry, 325*, 126953. doi:10.1016/j.foodchem.2020.126953

Assis, C., Pereira, H. V., Amador, V. S., Augusti, R., de Oliveira, L. S., & Sena, M. M. (2019). Combining mid infrared spectroscopy and paper spray mass spectrometry in a data fusion model to predict the composition of coffee blends. *Food Chemistry, 281*, 71–77. doi:10.1016/j.foodchem.2018.12.044

Babova, O., Occhipinti, A., & Maffei, M. E. (2016). Chemical partitioning and antioxidant capacity of green coffee (Coffea arabica and Coffea canephora) of different geographical origin. *Phytochemistry, 123*, 33–39. doi:10.1016/j.phytochem.2016.01.016

Badmos, S., Fu, M., Granato, D., & Kuhnert, N. (2020). Classification of Brazilian roasted coffees from different geographical origins and farming practices based on chlorogenic acid profiles. *Food Research International, 134*, 109218. doi:10.1016/j.foodres.2020.109218

Bag, S., Mondal, A., Majumder, A., & Banik, A. (2022). Tea and its phytochemicals: Hidden health benefits & modulation of signaling cascade by phytochemicals. *Food Chemistry, 371*, 131098. doi:10.1016/j.foodchem.2021.131098

Baqueta, M. R., Coqueiro, A., Março, P. H., Mandrone, M., Poli, F., & Valderrama, P. (2021). Integrated ^{1}H NMR fingerprint with NIR spectroscopy, sensory properties, and quality parameters in a multi-block data analysis using ComDim to evaluate coffee blends. *Food Chemistry, 355*, 129618. doi:10.1016/j.foodchem.2021.129618

Barrios-Rodríguez, Y. F., Reyes, C. A., Campos, J. S., Girón-Hernández, J., & Rodríguez-Gamir, J. (2021). Infrared spectroscopy coupled with chemometrics in coffee post-harvest processes as complement to the sensory analysis. *LWT, 145*, 111304. doi:10.1016/j.lwt.2021.111304

Belchior, V., Botelho, B. G., Oliveira, L. S., & Franca, A. S. (2019). Attenuated Total Reflectance Fourier Transform Spectroscopy (ATR-FTIR) and chemometrics for discrimination of espresso coffees with different sensory characteristics. *Food Chemistry, 273*, 178–185. doi:10.1016/j.foodchem.2017.12.026

Bertrand, B., Boulanger, R., Dussert, S., Ribeyre, F., Berthiot, L., Descroix, F., & Joët, T. (2012). Climatic factors directly impact the volatile organic compound fingerprint in green Arabica coffee bean as well as coffee beverage quality. *Food Chemistry, 135*, 2575–2583. doi:10.1016/j.foodchem.2012.06.060

Bertrand, B., Villarreal, D., Laffargue, A., Posada, H., Lashermes, P., & Dussert, S. (2008). Comparison of the Effectiveness of Fatty Acids, Chlorogenic Acids, and Elements for the Chemometric Discrimination of Coffee (Coffea arabica L.) Varieties and Growing Origins. *Journal of Agricultural and Food Chemistry, 56*, 2273–2280. doi:10.1021/jf073314f

Boaventura, B. C., Murakami, A. N., Prudêncio, E. S., Maraschin, M., Murakami, F. S., Amante, E. R., & de Mello Castanho Amboni, R. D. (2013). Enhancement of bioactive compounds content and antioxidant activity of aqueous extract of mate (Ilex paraguariensis A. St. Hil.) through freeze concentration technology. *Food Research International, 53*, 686–692. doi:10.1016/j.foodres.2012.07.042

Bona, E., Marquetti, I., Link, J. V., Makimori, G. Y., da Costa Arca, V., Lemes, A. L., . . . Poppi, R. J. (2017). Support vector machines in tandem with infrared spectroscopy for geographical classification of green arabica coffee. *LWT, 76*, 330–336. doi:10.1016/j.lwt.2016.04.048

Botelho, B. G., Oliveira, L. S., & Franca, A. S. (2017). Fluorescence spectroscopy as tool for the geographical discrimination of coffees produced in different regions of Minas Gerais State in Brazil. *Food Control, 77*, 25–31. doi:10.1016/j.foodcont.2017.01.020

Bressanello, D., Liberto, E., Cordero, C., Rubiolo, P., Pellegrino, G., Ruosi, M. R., & Bicchi, C. (2017). Coffee aroma: Chemometric comparison of the chemical information provided by three different samplings combined with GC—MS to describe the sensory properties in cup. *Food Chemistry, 214*, 218–226. doi:10.1016/j.foodchem.2016.07.088

Brondi, A., Torres, C., Garcia, J., & Trevisan, M. (2016). Differential Scanning Calorimetry and Infrared Spectroscopy Combined with Chemometric Analysis to the Determination of Coffee Adulteration by Corn. *Journal of the Brazilian Chemical Society*. doi:10.21577/0103–5053.20160296

Burns, D. T., Tweed, L., & Walker, M. J. (2017). Ground Roast Coffee: Review of Analytical Strategies to Estimate Geographic Origin, Species Authenticity and Adulteration by Dilution. *Food Analytical Methods, 10*, 2302–2310. doi:10.1007/s12161-016-0756-3

Cagliani, L. R., Pellegrino, G., Giugno, G., & Consonni, R. (2013). Quantification of Coffea arabica and Coffea canephora var. robusta in roasted and ground coffee blends. *Talanta, 106*, 169–173. doi:10.1016/j.talanta.2012.12.003

Callao, M. P., & Ruisánchez, I. (2018). An overview of multivariate qualitative methods for food fraud detection. *Food Control, 86*, 283–293. doi:10.1016/j.foodcont.2017.11.034

Cebi, N., Yilmaz, M. T., & Sagdic, O. (2017). A rapid ATR-FTIR spectroscopic method for detection of sibutramine adulteration in tea and coffee based on hierarchical cluster and principal component analyses. *Food Chemistry, 229*, 517–526. doi:10.1016/j.foodchem.2017.02.072

Cheah, W. L., & Fang, M. (2020). HPLC-based chemometric analysis for coffee adulteration. *Foods, 9*, 880. doi:10.3390/foods9070880

Colzi, I., Taiti, C., Marone, E., Magnelli, S., Gonnelli, C., & Mancuso, S. (2017). Covering the different steps of the coffee processing: Can headspace VOC emissions be exploited to successfully distinguish between Arabica and Robusta? *Food Chemistry, 237*, 257–263. doi:10.1016/j.foodchem.2017.05.071

Cossignani, L., Montesano, D., Simonetti, M. S., & Blasi, F. (2016). Authentication of Coffea arabica according to triacylglycerol stereospecific composition. *Journal of Analytical Methods in Chemistry, 2016*, 1–7. doi:10.1155/2016/7482620

Craig, A. P., Botelho, B. G., Oliveira, L. S., & Franca, A. S. (2018). Mid infrared spectroscopy and chemometrics as tools for the classification of roasted coffees by cup quality. *Food Chemistry, 245*, 1052–1061. doi:10.1016/j.foodchem.2017.11.066

Craig, A. P., Franca, A. S., & Oliveira, L. S. (2011). Discrimination between immature and mature green coffees by attenuated total reflectance and diffuse reflectance Fourier transform infrared spectroscopy. *Journal of Food Science, 76*, C1162—C1168. doi:10.1111/j.1750–3841.2011.02359.x

Craig, A. P., Franca, A. S., & Oliveira, L. S. (2012a). Discrimination between defective and non-defective roasted coffees by diffuse reflectance infrared Fourier transform spectroscopy. *LWT, 47*, 505–511. doi:10.1016/j.lwt.2012.02.016

Craig, A. P., Franca, A. S., & Oliveira, L. S. (2012b). Evaluation of the potential of FTIR and chemometrics for separation between defective and non-defective coffees. *Food Chemistry, 132*, 1368–1374. doi:10.1016/j.foodchem.2011.11.121

Craig, A. P., Franca, A. S., Oliveira, L. S., Irudayaraj, J., & Ileleji, K. (2014). Application of elastic net and infrared spectroscopy in the discrimination between defective and non-defective roasted coffees. *Talanta, 128*, 393–400. doi:10.1016/j.talanta.2014.05.001

da Silva Pinto, M. (2013). Tea: A new perspective on health benefits. *Food Research International, 53*, 558–567. doi:10.1016/j.foodres.2013.01.038

da Silva Taveira, J. H., Borém, F. M., Figueiredo, L. P., Reis, N., Franca, A. S., Harding, S. A., & Tsai, C.-J. (2014). Potential markers of coffee genotypes grown in different Brazilian regions: A metabolomics approach. *Food Research International, 61*, 75–82. doi:10.1016/j.foodres.2014.02.048

Dai, W., Tan, J., Lu, M., Zhu, Y., Li, P., Peng, Q., . . . Lin, Z. (2018). Metabolomics investigation reveals that 8-C N-ethyl-2-pyrrolidinone-substituted flavan-3-ols are potential marker compounds of stored white teas. *Journal of Agricultural and Food Chemistry, 66*, 7209–7218. doi:10.1021/acs.jafc.8b02038

Dankowska, A., Domagała, A., & Kowalewski, W. (2017). Quantification of Coffea arabica and Coffea canephora var. robusta concentration in blends by means of synchronous fluorescence and UV-Vis spectroscopies. *Talanta, 172*, 215–220. doi:10.1016/j.talanta.2017.05.036

Das, K. (2006). International protection of India\textquotesingles geographical indications with special reference to "Darjeeling" tea. *The Journal of World Intellectual Property, 9*, 459–495. doi:10.1111/j.1422–2213.2006.00300.x

de Moura Ribeiro, M. V., Boralle, N., Pezza, H. R., Pezza, L., & Toci, A. T. (2017). Authenticity of roasted coffee using 1 H NMR spectroscopy. *Journal of Food Composition and Analysis, 57*, 24–30. doi:10.1016/j.jfca.2016.12.004

de Toledo, P. R., de Melo, M. M., Pezza, H. R., Toci, A. T., Pezza, L., & Silva, C. M. (2017). Discriminant analysis for unveiling the origin of roasted coffee samples: A tool for quality control of coffee related products. *Food Control, 73*, 164–174. doi:10.1016/j.foodcont.2016.08.001

Dias, R. C., Valderrama, P., Março, P. H., dos Santos Scholz, M. B., Edelmann, M., & Yeretzian, C. (2018). Quantitative assessment of specific defects in roasted ground coffee via infrared-photoacoustic spectroscopy. *Food Chemistry, 255*, 132–138. doi:10.1016/j.foodchem.2018.02.076

Ding, B., Zeng, G., Wang, Z., Xie, J., Wang, L., & Chen, W. (2019). Authenticity determination of tea drinks in the Chinese market by liquid chromatography coupled to isotope ratio mass spectrometry. *Microchemical Journal, 144*, 139–143. doi:10.1016/j.microc.2018.09.001

Dong, W., Zhao, J., Hu, R., Dong, Y., & Tan, L. (2017). Differentiation of Chinese Robusta coffees according to species, using a combined electronic nose and tongue, with the aid of chemometrics. *Food Chemistry, 229*, 743–751. doi:10.1016/j.foodchem.2017.02.149

Downey, G., & Boussion, J. (1996). Authentication of coffee bean variety by near-infrared reflectance spectroscopy of dried extract. *Journal of the Science of Food and Agriculture, 71*, 41–49. doi:10.1002/(sici)1097–0010(199605)71:1<41::aid-jsfa546>3.0.co;2-i

Ebrahimi-Najafabadi, H., Leardi, R., Oliveri, P., Casolino, M. C., Jalali-Heravi, M., & Lanteri, S. (2012). Detection of addition of barley to coffee using near infrared spectroscopy and chemometric techniques. *Talanta, 99*, 175–179. doi:10.1016/j.talanta.2012.05.036

El-Abassy, R. M., Donfack, P., & Materny, A. (2011). Discrimination between Arabica and Robusta green coffee using visible micro Raman spectroscopy and chemometric analysis. *Food Chemistry, 126*, 1443–1448. doi:10.1016/j.foodchem.2010.11.132

Engelhardt, U. H. (2020). Tea chemistry—What do and what don't we know?—A micro review. *Food Research International, 132*, 109120. doi:10.1016/j.foodres.2020.109120

Farah, A., & dos Santos, T. F. (2015). *The Coffee Plant and Beans*. Elsevier. doi:10.1016/b978-0-12-409517-5.00001–2

Ferreira, T., Galluzzi, L., de Paulis, T., & Farah, A. (2021). Three centuries on the science of coffee authenticity control. *Food Research International, 149*, 110690. doi:10.1016/j.foodres.2021.110690

Figueiredo, L. P., Borém, F. M., Almeida, M. R., de Oliveira, L. F., de Carvalho Alves, A. P., dos Santos, C. M., & Rios, P. A. (2019). Raman spectroscopy for the differentiation of Arabic coffee genotypes. *Food Chemistry, 288*, 262–267. doi:10.1016/j.foodchem.2019.02.093

Firmani, P., Luca, S. D., Bucci, R., Marini, F., & Biancolillo, A. (2019). Near infrared (NIR) spectroscopy-based classification for the authentication of Darjeeling black tea. *Food Control, 100*, 292–299. doi:10.1016/j.foodcont.2019.02.006

Firmansyah, A., Winingsih, W., & Soebara Y, S. (2017). The utilization of FTIR (Fourier transform infra-red) method combined with chemometrics for authentication of Indonesian coffee powder. *International Journal of Pharmaceutical and Clinical Research, 9*. doi:10.25258/ijpcr.v9i3.8320

Flores-Valdez, M., Meza-Márquez, O. G., Osorio-Revilla, G., & Gallardo-Velázquez, T. (2020). Identification and quantification of adulterants in coffee (coffea arabica l.) using FT-MIR spectroscopy coupled with chemometrics. *Foods, 9*, 851. doi:10.3390/foods9070851

Franca, A. S., & Nollet, L. M. (2017). *Spectroscopic Methods in Food Analysis*. CRC Press. doi:10.1201/9781315152769

Franca, A. S., & Oliveira, L. S. (2019). *Coffee*. Elsevier. doi:10.1016/b978-0-12-814138-0.00017–4

Fulgencio, A. C., Resende, G. A., & Botelho, B. G. (2022). Chemometrics fundamentals and its application to infrared spectroscopy. In A. S. Franca, & L. S. Oliveira (Eds.), *FTIR Spectroscopy: Advances in Research and Applications*. Nova Publishers.

Garcia, L. M., Pauli, E. D., Cristiano, V., da Camara, C. A., Scarminio, I. S., & Nixdorf, S. L. (2009). Chemometric evaluation of adulteration profile in coffee due to corn and husk by determining carbohydrates using HPAEC-PAD. *Journal of Chromatographic Science, 47*, 825–832. doi:10.1093/chromsci/47.9.825

Giraudo, A., Grassi, S., Savorani, F., Gavoci, G., Casiraghi, E., & Geobaldo, F. (2019). Determination of the geographical origin of green coffee beans using NIR spectroscopy and multivariate data analysis. *Food Control, 99*, 137–145. doi:10.1016/j.foodcont.2018.12.033

Hong, Z., Zhang, C., Kong, D., Qi, Z., & He, Y. (2021). Identification of storage years of black tea using near-infrared hyperspectral imaging with deep learning methods. *Infrared Physics & Technology, 114*, 103666. doi:10.1016/j.infrared.2021.103666

Illy, E. (2002). The Complexity of Coffee. *286*, 86–91. doi:10.1038/scientificamerican0602–86

Kamiloglu, S. (2019). Authenticity and traceability in beverages. *Food Chemistry, 277*, 12–24. doi:10.1016/j.foodchem.2018.10.091

Kamruzzaman, M. (2016). *Food Adulteration and Authenticity*. Springer International Publishing. doi:10.1007/978-3-319-39253-0_7
Keidel, A., von Stetten, D., Rodrigues, C., Máguas, C., & Hildebrandt, P. (2010). Discrimination of Green Arabica and Robusta Coffee Beans by Raman Spectroscopy. *Journal of Agricultural and Food Chemistry, 58*, 11187–11192. doi:10.1021/jf101999c
Kemsley, E. K., Defernez, M., & Marini, F. (2019). Multivariate statistics: Considerations and confidences in food authenticity problems. *Food Control, 105*, 102–112. doi:10.1016/j.foodcont.2019.05.021
Khan, N., & Mukhtar, H. (2007). Tea polyphenols for health promotion. *Life Sciences, 81*, 519–533. doi:10.1016/j.lfs.2007.06.011
Li, L., Huang, J., Wang, Y., Jin, S., Li, M., Sun, Y., . . . Zhang, Z. (2020). Intelligent evaluation of storage period of green tea based on VNIR hyperspectral imaging combined with chemometric analysis. *Infrared Physics & Technology, 110*, 103450. doi:10.1016/j.infrared.2020.103450
Li, L., Jin, S., Wang, Y., Liu, Y., Shen, S., Li, M., . . . Zhang, Z. (2021). Potential of smartphone-coupled micro NIR spectroscopy for quality control of green tea. *Spectrochimica Acta Part A: Molecular and Biomolecular Spectroscopy, 247*, 119096. doi:10.1016/j.saa.2020.119096
Liardon, R., Ott, U., & Daget, N. (1984). Analysis of coffee headspace profiles by multivariate statistics. In *Analysis of Volatiles*. Analysis of coffee headspace profiles by multivariate statistics, Ed. Peter, S., Analysis of Volatiles: Methods. Applications. Proceedings. International Workshop Würzburg, Federal Republic of Germany, September 28–30, 1983, Berlin, De Gruyter. https://doi.org/10.1515/9783110855944.447.
Lim, C. M., Carey, M., Williams, P. N., & Koidis, A. (2021). Rapid classification of commercial teas according to their origin and type using elemental content with X-ray fluorescence (XRF) spectroscopy. *Current Research in Food Science, 4*, 45–52. doi:10.1016/j.crfs.2021.02.002
Link, J. V., Lemes, A. L., Marquetti, I., dos Santos Scholz, M. B., & Bona, E. (2014). Geographical and genotypic classification of arabica coffee using Fourier transform infrared spectroscopy and radial-basis function networks. *Chemometrics and Intelligent Laboratory Systems, 135*, 150–156. doi:10.1016/j.chemolab.2014.04.008
Liu, W., Chen, Y., Liao, R., Zhao, J., Yang, H., & Wang, F. (2021). Authentication of the geographical origin of Guizhou green tea using stable isotope and mineral element signatures combined with chemometric analysis. *125, Food Control,* 107954. doi:10.1016/j.foodcont.2021.107954
Ma, G., Zhang, Y., Zhang, J., Wang, G., Chen, L., Zhang, M., . . . Lu, C. (2016). Determining the geographical origin of Chinese green tea by linear discriminant analysis of trace metals and rare earth elements: Taking dongting biluochun as an example. *Food Control, 59*, 714–720. doi:10.1016/j.foodcont.2015.06.037
Malta, M. R., de Oliveira Fassio, L., Liska, G. R., Carvalho, G. R., Pereira, A. A., Botelho, C. E., . . . Pereira, R. G. (2020). Discrimination of genotypes coffee by chemical composition of the beans: Potential markers in natural coffees. *Food Research International, 134*, 109219. doi:10.1016/j.foodres.2020.109219
Malta, M. R., Oliveira, A. C., Liska, G. R., Carvalho, G. R., Pereira, A. A., Silva, A. D., . . . Mota, D. M. (2021). Selection of elite genotypes of coffee arabica l. to produce specialty coffees. *Frontiers in Sustainable Food Systems, 5*. doi:10.3389/fsufs.2021.715385
Marek, G., Dobrzański, B., Oniszczuk, T., Combrzyński, M., Ćwikła, D., & Rusinek, R. (2020). Detection and differentiation of volatile compound profiles in roasted coffee Arabica beans from different countries using an electronic nose and GC-MS. *Sensors, 20*, 2124. doi:10.3390/s20072124
Marquetti, I., Link, J. V., Lemes, A. L., dos Santos Scholz, M. B., Valderrama, P., & Bona, E. (2016). Partial least square with discriminant analysis and near infrared spectroscopy for evaluation of geographic and genotypic origin of arabica coffee. *Computers and Electronics in Agriculture, 121*, 313–319. doi:10.1016/j.compag.2015.12.018
Martín, M. J., Pablos, F., & González, A. G. (1998). Characterization of green coffee varieties according to their metal content. *Analytica Chimica Acta, 358*, 177–183. doi:10.1016/s0003–2670(97)00610-7
McKay, D. L., & Blumberg, J. B. (2002). The role of tea in human health: An update. *Journal of the American College of Nutrition, 21*, 1–13. doi: 10.1080/07315724.2002.10719187
Medina, S., Perestrelo, R., Silva, P., Pereira, J. A., & Câmara, J. S. (2019). Current trends and recent advances on food authenticity technologies and chemometric approaches. *Trends in Food Science & Technology, 85*, 163–176. doi:10.1016/j.tifs.2019.01.017
Mehari, B., Redi-Abshiro, M., Chandravanshi, B. S., Atlabachew, M., Combrinck, S., & McCrindle, R. (2015). Simultaneous determination of alkaloids in green coffee beans from Ethiopia: Chemometric evaluation of geographical origin. *Food Analytical Methods, 9*, 1627–1637. doi:10.1007/s12161-015-0340-2

Mendonça, J. C., Franca, A. S., Oliveira, L. S., & Nunes, M. (2008). Chemical characterisation of non-defective and defective green Arabica and Robusta coffees by electrospray ionization-mass spectrometry (ESI-MS). *Food Chemistry, 111*, 490–497. doi:10.1016/j.foodchem.2008.03.080

Meng, T., Zhipeng, X., Ai, H., Jiayu, C., Tianji, Y., & Zhengquan, L. (2021). Effect of 1–20 years storage on volatiles and aroma of Keemun congou black tea by solvent extraction-solid phase extraction-gas chromatography-mass spectrometry. *LWT, 136*, 110278. doi:10.1016/j.lwt.2020.110278

Milani, M. I., Rossini, E. L., Catelani, T. A., Pezza, L., Toci, A. T., & Pezza, H. R. (2020). Authentication of roasted and ground coffee samples containing multiple adulterants using NMR and a chemometric approach. *Food Control, 112*, 107104. doi:10.1016/j.foodcont.2020.107104

Monakhova, Y. B., Ruge, W., Kuballa, T., Ilse, M., Winkelmann, O., Diehl, B., . . . Lachenmeier, D. W. (2015). Rapid approach to identify the presence of Arabica and Robusta species in coffee using ^{1}H NMR spectroscopy. *Food Chemistry, 182*, 178–184. doi:10.1016/j.foodchem.2015.02.132

Monteiro, P. I., Santos, J. S., Brizola, V. R., Deolindo, C. T., Koot, A., Boerrigter-Eenling, R., . . . Granato, D. (2018). Comparison between proton transfer reaction mass spectrometry and near infrared spectroscopy for the authentication of Brazilian coffee: A preliminary chemometric study. *Food Control, 91*, 276–283. doi:10.1016/j.foodcont.2018.04.009

Moreda-Piñeiro, A., Fisher, A., & Hill, S. J. (2003). The classification of tea according to region of origin using pattern recognition techniques and trace metal data. *Journal of Food Composition and Analysis, 16*, 195–211. doi:10.1016/s0889–1575(02)00163-1

Mu, B., Zhu, Y., Lv, H.-P., Yan, H., Peng, Q.-H., & Lin, Z. (2018). The enantiomeric distributions of volatile constituents in different tea cultivars. *Food Chemistry, 265*, 329–336. doi:10.1016/j.foodchem.2018.05.094

Ning, J.-M., Ding, D., Song, Y.-S., Zhang, Z.-Z., Luo, X., & Wan, X.-C. (2016). Chemical constituents analysis of white tea of different qualities and different storage times. *European Food Research and Technology, 242*, 2093–2104. doi:10.1007/s00217-016-2706-0

Núñez, N., Collado, X., Martínez, C., Saurina, J., & Núñez, O. (2020). Authentication of the origin, variety and roasting degree of coffee samples by non-targeted HPLC-UV fingerprinting and chemometrics. application to the detection and quantitation of adulterated coffee samples. *Foods, 9*, 378. doi:10.3390/foods9030378

Núñez, N., Martínez, C., Saurina, J., & Núñez, O. (2020). High-performance liquid chromatography with fluorescence detection fingerprints as chemical descriptors to authenticate the origin, variety and roasting degree of coffee by multivariate chemometric methods. *Journal of the Science of Food and Agriculture, 101*, 65–73. doi:10.1002/jsfa.10615

Núñez, N., Pons, J., Saurina, J., & Núñez, O. (2021). Non-targeted high-performance liquid chromatography with ultraviolet and fluorescence detection fingerprinting for the classification, authentication, and fraud quantitation of instant coffee and chicory by multivariate chemometric methods. *LWT, 147*, 111646. doi:10.1016/j.lwt.2021.111646

Oliveira, R. C., Oliveira, L. S., Franca, A. S., & Augusti, R. (2009). Evaluation of the potential of SPME-GC-MS and chemometrics to detect adulteration of ground roasted coffee with roasted barley. *Journal of Food Composition and Analysis, 22*, 257–261. doi:10.1016/j.jfca.2008.10.015

Ongo, E. A., Montevecchi, G., Antonelli, A., Sberveglieri, V., & Fortunato Sevilla, I. I. (2020). Metabolomics fingerprint of Philippine coffee by SPME-GC-MS for geographical and varietal classification. *Food Research International, 134*, 109227. doi:10.1016/j.foodres.2020.109227

Pacetti, D., Boselli, E., Balzano, M., & Frega, N. G. (2012). Authentication of Italian Espresso coffee blends through the GC peak ratio between kahweol and 16-O-methylcafestol. *Food Chemistry, 135*, 1569–1574. doi:10.1016/j.foodchem.2012.06.007

Peng, C.-y., Zhang, Y.-l., Song, W., Cai, H.-m., Wang, Y., & Granato, D. (2019). Characterization of Brazilian coffee based on isotope ratio mass spectrometry and supervised chemometrics. *Food Chemistry, 297*, 124963. doi:10.1016/j.foodchem.2019.124963

Perez, M., Domínguez-López, I., López-Yerena, A., & Queralt, A. V. (2021). Current strategies to guarantee the authenticity of coffee. *Critical Reviews in Food Science and Nutrition*, 1–16. doi:10.1080/10408398.2021.1951651

Pilgrim, T. S., Watling, R. J., & Grice, K. (2010). Application of trace element and stable isotope signatures to determine the provenance of tea (Camellia sinensis) samples. *Food Chemistry, 118*, 921–926. doi:10.1016/j.foodchem.2008.08.077

Prodolliet, J., Bruelhart, M., Blanc, M. B., Leloup, V., Cherk, G., Donnelly, C. M., & Viani, R. (1995). Adulteration of soluble coffee with coffee husks and parchments. *Journal of AOAC International, 78*, 761–767. doi:10.1093/jaoac/78.3.761

Reis, N., Franca, A. S., & Oliveira, L. S. (2013a). Discrimination between roasted coffee, roasted corn and coffee husks by diffuse reflectance infrared Fourier transform spectroscopy. *LWT 50*, 715–722. doi:10.1016/j.lwt.2012.07.016

Reis, N., Franca, A. S., & Oliveira, L. S. (2013b). Quantitative evaluation of multiple adulterants in roasted coffee by diffuse reflectance infrared Fourier transform spectroscopy (DRIFTS) and chemometrics. *Talanta, 115*, 563–568. doi:10.1016/j.talanta.2013.06.004

Ren, Y.-f., Feng, C., Ye, Z.-h., Zhu, H.-y., Hou, R.-y., Granato, D., . . . Peng, C.-y. (2021). Keemun black tea: Tracing its narrow-geographic origins using comprehensive elemental fingerprinting and chemometrics. *Food Control*, 108614. doi:10.1016/j.foodcont.2021.108614

Robert, J. V., de Gois, J. S., Rocha, R. B., & Luna, A. S. (2022). Direct solid sample analysis using synchronous fluorescence spectroscopy coupled with chemometric tools for the geographical discrimination of coffee samples. *Food Chemistry, 371*, 131063. doi:10.1016/j.foodchem.2021.131063

Santos, J. R., Sarraguça, M. C., Rangel, A. O., & Lopes, J. A. (2012). Evaluation of green coffee beans quality using near infrared spectroscopy: A quantitative approach. *Food Chemistry, 135*, 1828–1835. doi:10.1016/j.foodchem.2012.06.059

Servillo, L., Giovane, A., Casale, R., Cautela, D., D'Onofrio, N., Balestrieri, M. L., & Castaldo, D. (2016). Homostachydrine (pipecolic acid betaine) as authentication marker of roasted blends of coffea arabica and coffea canephora (Robusta) beans. *Food Chemistry, 205*, 52–57. doi:10.1016/j.foodchem.2016.02.154

Shen, S., Huang, J., Li, T., Wei, Y., Xu, S., Wang, Y., & Ning, J. (2022). Untargeted and targeted metabolomics reveals potential marker compounds of an tea during storage. *LWT, 154*, 112791. doi:10.1016/j.lwt.2021.112791

Shevchuk, A., Jayasinghe, L., & Kuhnert, N. (2018). Differentiation of black tea infusions according to origin, processing and botanical varieties using multivariate statistical analysis of LC-MS data. *Food Research International, 109*, 387–402. doi:10.1016/j.foodres.2018.03.059

Silva, T. V., Pérez-Rodríguez, M., de Oliveira, N. R., de Santana, H., & de Almeida, L. C. (2021). Tracing commercial coffee quality by infrared spectroscopy in tandem with pattern recognition approaches. *Vibrational Spectroscopy, 116*, 103295. doi:10.1016/j.vibspec.2021.103295

Sudol, P. E., Pierce, K. M., Prebihalo, S. E., Skogerboe, K. J., Wright, B. W., & Synovec, R. E. (2020). Development of gas chromatographic pattern recognition and classification tools for compliance and forensic analyses of fuels: A review. *Analytica Chimica Acta, 1132*, 157–186. doi:10.1016/j.aca.2020.07.027

Valand, R., Tanna, S., Lawson, G., & Bengtström, L. (2019). A review of Fourier Transform Infrared (FTIR) spectroscopy used in food adulteration and authenticity investigations. *Food Additives & Contaminants: Part A, 37*, 19–38. doi:10.1080/19440049.2019.1675909

Valentin, J. L., & Watling, R. J. (2013). Provenance establishment of coffee using solution ICP-MS and ICP-AES. *Food Chemistry, 141*, 98–104. doi:10.1016/j.foodchem.2013.02.101

Wang, H., Cao, X., Yuan, Z., & Guo, G. (2021). Untargeted metabolomics coupled with chemometrics approach for Xinyang Maojian green tea with cultivar, elevation and processing variations. *Food Chemistry, 352*, 129359. doi:10.1016/j.foodchem.2021.129359

Wang, J., Jun, S., Bittenbender, H. C., Gautz, L., & Li, Q. X. (2009). Fourier transform infrared spectroscopy for Kona coffee authentication. *Journal of Food Science, 74*, C385—C391. doi:10.1111/j.1750-3841.2009.01173.x

Wang, X., Lim, L.-T., & Fu, Y. (2020). Review of analytical methods to detect adulteration in coffee. *Journal of AOAC International, 103*, 295–305. doi:10.1093/jaocint/qsz019

Wei, F., Furihata, K., Koda, M., Hu, F., Kato, R., Miyakawa, T., & Tanokura, M. (2012). 13C NMR-based metabolomics for the classification of green coffee beans according to variety and origin. *Journal of Agricultural and Food Chemistry, 60*, 10118–10125. doi:10.1021/jf3033057

Wermelinger, T., D'Ambrosio, L., Klopprogge, B., & Yeretzian, C. (2011). Quantification of the Robusta fraction in a coffee blend via Raman spectroscopy: Proof of principle. *Journal of Agricultural and Food Chemistry, 59*, 9074–9079. doi:10.1021/jf201918a

Worku, M., Upadhayay, H. R., Latruwe, K., Taylor, A., Blake, W., Vanhaecke, F., . . . Boeckx, P. (2019). Differentiating the geographical origin of Ethiopian coffee using XRF- and ICP-based multi-element and stable isotope profiling. *Food Chemistry, 290*, 295–307. doi:10.1016/j.foodchem.2019.03.135

Xia, W., Li, Z., Yu, C., Liu, Z., Nie, J., Li, C., . . . Yuan, Y. (2021). Understanding processing, maturity and harvest period effects to authenticate early-spring Longjing tea using stable isotopes and chemometric analyses. *Food Control, 124*, 107907. doi:10.1016/j.foodcont.2021.107907

Yang, Z., Gao, J., Wang, S., Wang, Z., Li, C., Lan, Y., . . . Li, S. (2021). Synergetic application of E-tongue and E-eye based on deep learning to discrimination of Pu-erh tea storage time. *Computers and Electronics in Agriculture, 187,* 106297. doi:10.1016/j.compag.2021.106297

Ye, X., Jin, S., Wang, D., Zhao, F., Yu, Y., Zheng, D., & Ye, N. (2016). Identification of the origin of white tea based on mineral element content. *Food Analytical Methods, 10,* 191–199. doi:10.1007/s12161-016-0568-5

Yulia, M., & Suhandy, D. (2017). Indonesian palm civet coffee discrimination using UV-visible spectroscopy and several chemometrics methods. *Journal of Physics: Conference Series, 835,* 012010. doi:10.1088/1742–6596/835/1/012010

Yun, J., Cui, C., Zhang, S., Zhu, J., Peng, C., Cai, H., . . . Hou, R. (2021). Use of headspace GC/MS combined with chemometric analysis to identify the geographic origins of black tea. *Food Chemistry, 360,* 130033. doi:10.1016/j.foodchem.2021.130033

Zhao, H., Zhang, S., & Zhang, Z. (2017). Relationship between multi-element composition in tea leaves and in provenance soils for geographical traceability. *Food Control, 76,* 82–87. doi:10.1016/j.foodcont.2017.01.006

Zheng, P. C., Qin, C. Y., Liu, P. P., Feng, L., Ling, T. J., Ning, J. M., Zhang. L., & Wan, X. C. (2021). Untargeted metabolomics combined with bioassay reveals the change in critical bioactive compounds during the processing of Qingzhuan tea. *Molecules, 26,* 6718. doi: 10.3390/molecules26216718

Zhou, B., Ma, C., Wu, T., Xu, C., Wang, J., & Xia, T. (November 2020). Classification of raw Pu-erh teas with different storage time based on characteristic compounds and effect of storage environment. *LWT, 133,* 109914. doi:10.1016/j.lwt.2020.109914

6 Authenticity and Chemometrics of Fruits and Vegetables

Raúl González-Domínguez, María Santos-Martín, Ana Sayago, and Ángeles Fernández-Recamales

CONTENTS

6.1 INTRODUCTION

The authentication of food and beverages is a very current topic of great interest due to its relationship with consumer protection. Nowadays, there is increasing attention on the identity and authenticity of food products by all the actors involved in the food chain (Esslinger et al., 2014). On one hand, consumers demand safe, honest and properly labeled foods. On the other, producers must ensure that their raw materials are safe and that their finished products have an added value that distinguishes them from competitors' ones. Finally, food authorities must ensure the quality and safety of all foods and beverages in the market.

In particular, plant-based food authentication can cover many different aspects, including mislabeling, adulteration and misleading claims about origin (e.g., species, geographical, genetic), production method (e.g., organic/conventional production) or processing technologies (e.g., irradiation, freezing, microwave heating) (Esteki et al., 2018a). Fruits and vegetables (F&Vs) are important food components in the diet of developed countries as they are related to healthy lifestyles. Observational studies have suggested that high intake of F&Vs reduces the risk of coronary heart disease, type 2 diabetes and obesity (World Health Organization, 2005). Their beneficial health effects have been attributed to the presence of dietary fibers, minerals, vitamins (e.g., provitamin A, carotenoids, vitamins C and E) and phytochemicals, including phenolic acids, flavonoids and anthocyanins. In this context, growing high-quality F&Vs has emerged as a quite profitable market niche, which requires powerful tools for assuring food authenticity. However, this is extremely difficult due to multiple factors related to the plant itself (e.g., cultivar, degree of maturity); the environment (e.g., climate, sunlight, soil conditions); and storage conditions and processing technologies. Thus, the identification and quantification of characteristic components, and the verification of quality parameters related to geographical origin and botanical variety (Cuadros-Rodríguez et al., 2016), are the main aspects to consider in the authentication of F&Vs.

In Europe, frauds related to geographical origin are one of the most important food authenticity problems, so much so that the European quality policy recognizes and protects the names of products that are related to a certain territory or to a specific production method (Official Journal of the European Union, 2012). This recognition is translated into quality logos that make it possible to clearly identify these products and which, through specific food controls, also guarantee their authenticity and quality. Among the geographical indications, the protected designation of origin (PDO) links the food products to the geographical area where they were produced, whereas the protected geographical indication (PGI) links the products to a geographical area where at minimum one of the three manufacturing processes (production, transformation, or elaboration) occurred. The traditional specialties guaranteed (TSG) designation protects traditional production methods (Drivelos and Georgiou, 2012; Kamiloglu, 2019). In this context, the European Union (EU) has registered 3304 protected food products from European countries and 46 products from countries outside the European Union (European Commission, 2022). Of these, 441 products correspond to the category "fresh or processed fruits, vegetables and cereals", which accounts for a 26% of the total, thus being the most represented category.

 DOI: 10.1201/9781003266167-6

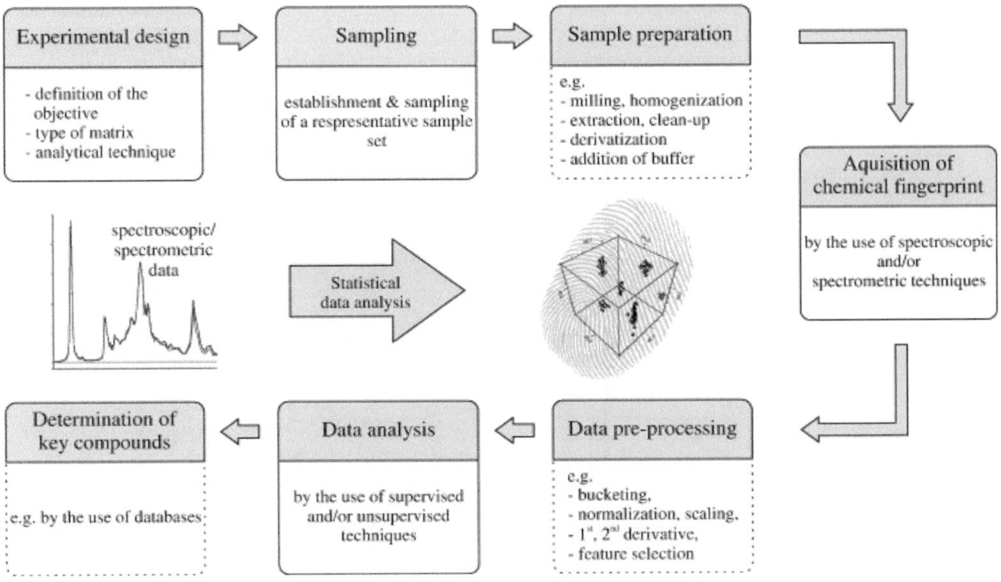

Figure 6.1 Typical workflow in food authentication research.

Source: Reproduced from Esslinger et al. (2014) with permission from Elsevier.

In this Chapter, we briefly describe the most common analytical and chemometric tools used for the authentication of F&Vs (Figure 6.1). Then, we provide a revision of recent literature on the application of both targeted and non-targeted analytical approaches, in combination with multivariate statistical analysis, for the detection of adulterations and the authentication of F&V food products in terms of variety, geographical origin and agricultural production method.

6.2 ANALYTICAL AND CHEMOMETRIC TOOLS IN FRUIT AND VEGETABLE AUTHENTICATION

Two complementary analytical strategies can be employed for the authentication of F&Vs: targeted and non-targeted approaches (Ballin and Laursen, 2019; Esslinger et al., 2014). The targeted methods rely on the analysis of the constituents that, based on prior knowledge, are expected to exist in the sample (e.g., minerals, antioxidant compounds, amino acids), whereas the non-targeted approach (i.e., omics technologies) is based on comprehensive chemical analyses, or fingerprinting, with the purpose of detecting as many substances as possible in the samples under study (Antignac et al., 2011; Esslinger et al., 2014). Among the omics-based fingerprinting strategies, metabolomics is nowadays one of the most suitable approaches for authentication purposes since it provides comprehensive information on the global metabolite composition (i.e., metabolome) derived from the plant metabolism (Creydt and Fischer, 2018). In this context, spectroscopic, spectrometric and chromatographic methods are among the most used techniques to assess the authenticity and adulteration of agri-food products. Chromatography coupled with spectroscopic or spectrometric techniques has demonstrated to be a powerful tool to evaluate and quantify individual components or a few compound classes in food samples (Danezis et al., 2016). Complementarily, non-destructive analytical techniques, such as near-infrared spectroscopy (NIR), mid-infrared spectroscopy (MIRS), Raman spectroscopy, ultraviolet-visible (UV-Vis) spectroscopy, and nuclear magnetic resonance (NMR), have also been described to generate fingerprints of a variety of food products (González-Domínguez et al., 2019, 2021; Sayago et al., 2019). New advanced analytical techniques, including thermal ionization mass spectrometry (TIMS), multi-collector–inductively coupled plasma mass spectrometry (MC-ICP-MS) and dynamic reaction cell–ICPMS (DRC-ICPMS), can generate multi-elemental and isotopic profiles to be used for food authentication purposes (Drivelos and Georgiou, 2012).

The large amount of data that is obtained by applying these analytical techniques requires the use of advanced multivariate chemometric tools for extracting as much valuable information as possible (Figure 6.2). Chemometrics is a discipline that uses statistical and mathematical methods

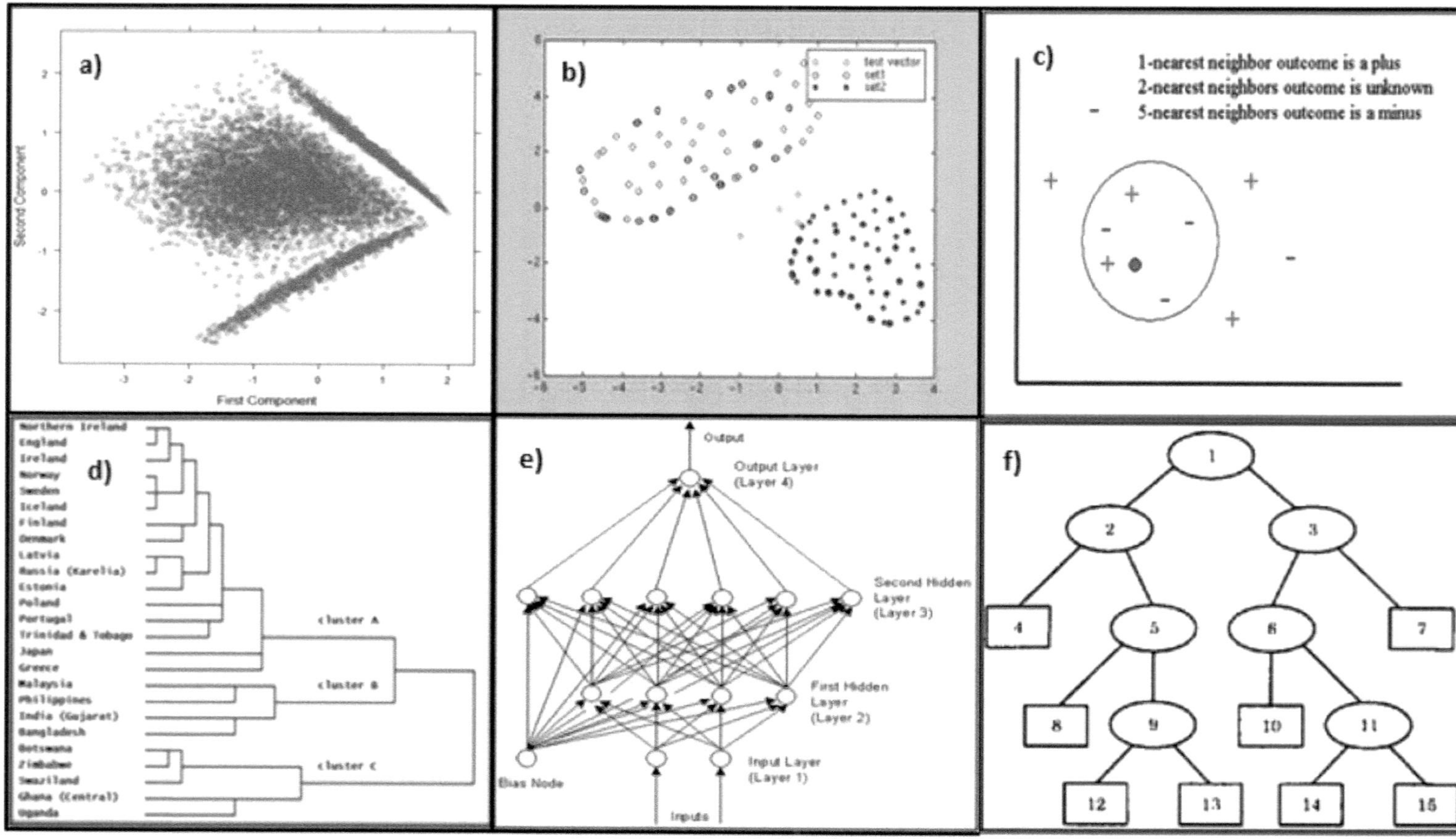

Figure 6.2 Representative graphs of the chemometric tools commonly used in food authentication research: (a) Principal component analysis (PCA); (b) linear discriminant analysis (LDA); (c) k-nearest neighbor (kNN); (d) cluster analysis (CA); (e) artificial neural networks (ANN); and (f) classification and regression trees (CARTs).

Source: Reproduced from Drivelos and Georgiou (2012) with permission from Elsevier.

to extract the maximum amount of chemical information from large and complex data sets with the aim of facilitating the detection of hidden relationships between variables. For this purpose, most multivariate data analysis methods are based on the construction of a model based on the available chemical data, which can interpret the characteristics of the system. The most common multivariate methods in food authentication research can be classified into three categories: (1) exploratory data analysis (EDA), (2) discrimination and classification and (3) regression and prediction. In turn, multivariate statistical analysis can be performed either by using unsupervised or supervised methods. Unsupervised approaches are based on the identification of clusters or relationships between samples, without any prior information about classes or groups (Berrueta et al., 2007). In contrast, supervised methods use a training stage, based on a priori assignment of samples to known groups, to build mathematical models that are subsequently employed to allocate new samples to the most likely group (Moncayo et al., 2015). In general, both approaches are normally combined in food science research, first by exploring the data with unsupervised pattern recognition methods, and then by the application of supervised learning approaches (Baker and Inventado, 2014). The choice of the most appropriate statistical tool to be applied depends on several factors, including the definition of the class criteria, the homogeneous distribution of the sample, the number of input variables, and the number of samples. Therefore, it is a common practice to apply more than one classification technique and evaluate their fitness for the problem under study (Callao and Ruisánchez, 2018; Sayago et al., 2018).

6.2.1 Exploratory Data Analysis

As a first step, EDA is normally employed to summarize the main characteristics of the data set and to facilitate the visualization of the data structure, often using two- and three-dimensional graphs. For this purpose, multiple unsupervised pattern recognition techniques are available, with principal component analysis (PCA) and cluster analysis (CA) being the most common ones. These techniques enable reducing the dimensionality of large data sets without losing information, so that samples can be projected into a low-dimensional space and graphically visualized (e.g., scatter plots, dendrograms) to inspect sample groupings/clusters, detect outliers and so on.

6.2.2 Discrimination and Classification

Discrimination and classification methods can be applied to predict the belonging of an object to a given category based on the characteristics of the object. To this end, the data matrix must contain at least one categorical variable, which indicates the class to which each object belongs and constitutes the response to be predicted, and one or more quantitative variables that describe the characteristics of the objects, which are used as predictor variables (Esteki et al., 2018a). A training set is first used to build the model, which should be composed of a set of objects whose category and values for the predictor variables are known a priori. Herein, the assignment of objects to categories must be exhaustive (i.e., all objects belong to some category) and mutually exclusive (i.e., no object belongs to more than one category). Once built, the model is used to predict the category of new objects from the measurement of the corresponding predictor variables.

A critical issue in any empirical modeling is to determine the correct complexity of the model by selecting an optimum number of variables, or components, to avoid overfitting (Roberts and Cozzolino, 2016). For this reason, validation procedures must be applied to assess the predictive ability of the model. The external validation method requires enough samples to create independent training, calibration and test sets, each one containing representative samples from each category. In this validation procedure, the test set must be completely independent of the model construction process (i.e., selection of variables and estimation of parameters) (Berrueta et al., 2007). In the absence of a real validation set, cross-validation is a practical and reliable way to test the goodness of the model to predict new data. For this, the data matrix is randomly divided into two sets, a training set that is used to construct the classification model and a test set to study the model performance, both containing the same percentage of samples within each class. Random division is applied to ensure that all samples have the same probability of belonging to the training set and, consequently, that the constructed model will consider the within-class variability, thereby learning the raw structure of the data (Palacios-Morillo et al., 2016). This procedure, called k-fold cross-validation or jackknife, is repeated several times to avoid conclusions by chance. Then, the model performance can be assessed by evaluating the recognition and prediction abilities of the model, which are equal to the percentage of the training and validation set members correctly classified, respectively (Berrueta et al., 2007). Another strategy is to compute the sensitivity (SENS) and specificity (SPEC) using the confusion matrix, where SENS refers to the percentage of cases belonging to a determinate class that are correctly classified, and SPEC refers to the percentage of cases not belonging to a class correctly that are not classified in this class (Ceballos-Magaña et al., 2012).

In this context, two different statistical approaches are available for classification and discrimination analysis (Kemsley et al., 2019). Discriminant analysis methods, also known as "hard modeling", are aimed at discriminating between classes, for which the data space would be divided into separate regions, each one corresponding to a class. In food science, the most common discriminant techniques include linear discriminant analysis (LDA), k-nearest neighbor (kNN), artificial neural network (ANN), wavelet neural network (WNN), support vector machine (SVM), partial least squares–discriminant analysis (PLS-DA) and orthogonal projections to latent structures–discriminant analysis (OPLS-DA) (Berrueta et al., 2007; Zielinski et al., 2014). On the other hand, class modeling analysis, or "soft modeling", is based on creating a separate model for each category and identifying a closed region of the multivariate space for each class, which represents the "class space". Within this category, soft independent modeling of class analogy (SIMCA) and unequal dispersed classes (UNEQ) are among the most used tools. In turn, these methods can also be classified as parametric versus non-parametric or deterministic versus probabilistic. Parametric or probabilistic methods (e.g., LDA, SIMCA, UNEQ, PLS-DA) assume that the data can be described with a probability density function, which determines its distribution. In contrast, non-parametric or deterministic methods (e.g., kNN, ANN) are used when the distribution of each class is not known, cannot be determined or is not normal (Ferreira, 2015).

6.2.3 Regression and Prediction

The aim of multivariate calibration methods, based on statistical regression and prediction, is to determine relationships between a response variable and several independent variables. For this, the first step is estimating efficient mathematical models able to interpret the performance of the system and to accurately predict the characteristics of future samples. Then, the quality and robustness of the models must be validated (e.g., external validation, cross-validation) as described above (Esteki et al., 2018b). Linear-based multivariate calibration methods include multiple linear regression (MLR), principal component regression (PCR), partial least squares regression (PLSR) and orthogonal projections to latent structures (O-PLS). Other techniques for regression and prediction purposes, such as artificial neural network (ANN), wavelet neural network (WNN) and support vector machine (SVM), are based on non-linear mathematical models (Yu et al., 2018).

6.3 AUTHENTICITY AND CHEMOMETRICS OF FRUITS AND VEGETABLES: A LITERATURE REVIEW

Plants contain numerous primary compounds, common to all or most plant species, which play essential roles in plant growth, development and reproduction. These compounds are intermediates and products of the central metabolism, involved in photosynthesis and other biosynthetic processes, which may belong to multiple chemical families including carbohydrates, organic and amino acids, nucleotides, lipids, mineral elements and others. In addition, plants also contain a wide variety of secondary metabolites that are not involved in primary functions but are essential for plant survival, such as phenolic compounds and carotenoids. For instance, polyphenols participate in defense mechanisms by acting as mediators of plant responses to biotic and abiotic stresses. Furthermore, these metabolites are crucial for the quality and organoleptic characteristics of F&Vs, and their presence and concentrations are influenced by many factors such as variety, environmental conditions, growing area and agronomical practices (Akhatou and Fernández-Recamales, 2014a, 2014b). Therefore, they can be used as chemical descriptors for authentication purposes.

Authenticity and traceability of F&V-based products is critical for the food industry and consumers, either considering frauds related to the origin (botanical or geographical) and to the food processing method (organic vs. conventional production, genetic modification). In particular, consumer awareness about the identity and authenticity of foods has considerably increased in the last years, mainly because of the increasing demand for organic and local products and the implementation of environmentally friendly and animal welfare production methods (Drivelos and Georgiou, 2012). However, this market niche of healthier, safer, autochthonous, environmentally friendly and consequently more expensive foodstuffs is more susceptible to adulterations and frauds.

Accordingly, F&V authentication requires powerful analytical techniques for the determination of specific compounds (e.g., polyphenols, carotenoids, organic acids, fatty acids, volatile compounds, trace minerals) or global fingerprinting (e.g., metabolomics, spectroscopic techniques) (Cifuentes, 2009), in combination with advanced chemometric tools.

6.3.1 Chromatography-Based Approaches for Fruit and Vegetable Authentication

Chromatography allows for the separation and quantification of macro- and micro-components in food products, with high-performance liquid chromatography (HPLC) and gas chromatography (GC) being the most widely used techniques for botanical and geographical authentication of F&Vs. In particular, liquid chromatography combined with ultraviolet-photodiode array detection (DAD) or mass spectrometry (MS) is commonly employed for the analysis of phenolic compounds and other metabolites because of its precision, low cost, and wide range of applicability (González-Domínguez et al., 2022). Furthermore, the vast development of ultra-high-performance liquid chromatography combined with MS (UHPLC-MS) has further improved the separation and detection performance, thus reducing the time required for qualitative and quantitative analysis and allowing the simultaneous determination of several compound classes in a single run (La Barbera et al., 2017). On the other hand, gas chromatography is frequently applied for profiling volatile substances (e.g., aroma-related compounds). Table 6.1 summarizes some recent studies on the application of chromatography-based approaches for F&V authentication.

Using high-performance liquid chromatography coupled to a diode array detector and atmospheric pressure chemical ionization mass spectrometry (HPLC-DAD-APCI-MS) and gas

Table 6.1 Selected Applications of Chromatographic-Based Approaches for Fruit and Vegetable Authentication

Fruit/ Vegetable	Geographical Origin	Parameters Measured	Analytical Technique	Chemometric Tool	Aim of the Study	Reference
Apricot	Romania	Carotenoids Fatty acids Volatile compounds	HPLC-DAD-APCI-MS GC-MS	PCA	Variety	(Pintea et al., 2020)
Goji berry and tomato	France	Carotenoids Phenolic compounds Sugars Amino acids Organic acids	LC-ESI-MS GC-MS	PCA PLS-DA HCA	Discrimination of Solanaceae species	(Akhatou et al., 2017)
Strawberry	Spain	Phenolic compounds	UHPLC-MS	PLS-DA RF	Variety Agronomic practices	(Akhatou et al., 2017)
Strawberry	Spain	Sugars Organic acids Polyphenols Mineral elements	HPLC-DAD ICP-OES	PCA LDA SIMCA PLS-DA	Variety Cultivation system (open/ closed soilless system)	(González-Domínguez et al., 2020a)
Date	Egypt	Volatile compounds	SPME-GC-MS	PCA HCA OPLS-DA	Variety	(Khalil et al., 2017)
Strawberry	Spain	Volatile compounds	SPME-GC-FID	PCA LDA	Variety Cultivation system (open/ closed soilless system)	(González-Domínguez et al., 2020b)
Tomato	United States	Volatile compounds	SPME-GC-MS	PLS-DA RF OPLS-DA	Cultivation system (net-house/ open field)	(Lee et al., 2019)

chromatography coupled to mass spectrometry (GC-MS), Pintea et al. (2020) analyzed the content of carotenoids, fatty acids, and volatiles compounds in 11 apricot cultivars (*Prunus armeniaca*). The authors applied PCA with cross-validation to detect and interpret the data structure and to reveal the similarities and differences between the 11 cultivars under study. The first two principal components explained 76% of the overall variance (58% and 18% for PC1 and PC2, respectively) and enabled to differentiate apricot cultivars according to four clusters. The variables with higher contribution to the first component were fatty acids (i.e., 18:1n-9, 15:0, 17:0) and carotenoids (i.e., β-carotene, 13-cis-β-carotene), whereas the most contributing variables to PC2 were volatile compounds (i.e., cis-2-heptenal, butyl acetate, butanoic acid 3-hexenyl ester, γ-decalactone and γ-dodecalactone).

Multi-targeted metabolic profiling was employed in another study to discover biomarkers in goji berry and tomato samples (Dumont et al., 2020). A set of 13 carotenoids, 46 phenolic compounds and 67 primary compounds (simple sugars, amino acids, organic acids) were profiled by LC-MS and GC-MS in 24 fruit samples of three Solanaceae species: *Lycium barbarum*, *Lycium chinense* and *Solanum lycopersicum*. PCA results indicated that the main source of variation (47%) was associated with the genus *Lycopersicum* versus *Lycium*, whereas the second source of variation (14.4%) was associated with the *Lycium* species. Then, PLS-DA enabled obtaining models with good discrimination performance (R^2_Y = 0.98 and 0.99 for components 1 and 2, respectively; Q^2 = 0.97 and 0.99 for components 1 and 2, respectively). The corresponding loadings plot revealed higher levels of lycopene in tomato, lycibarbarphenylpropanoids in *L. barbarum*, and zeaxanthin esters or phenolamides in *L. chinense*. These results were then corroborated by heatmap clustering, evidencing that lycopene, carotene, glutamate and γ-aminobutyric acid can serve as metabolic markers for discriminating *Lycium* and *Solanum* fruits; chlorogenic acids, asparagine and quinic acid were more abundant in *L. chinense*; and *L. barbarum* accumulated more lycibarbarphenylpropanoids A-B, coumaric acid, fructose and glucose.

A targeted metabolomics approach based on UHPLC-MS/MS was applied to identify and quantitate the main polyphenol compounds in strawberry fruits with the aim of discriminating strawberry varieties grown in soilless systems under different agronomic conditions, such as electrical conductivity, substrate type and coverage (Akhatou et al., 2017). Then, two data mining methods, partial least squares–discriminant analysis (PLS-DA) and random forest (RF), were applied for building predictive models to explore the differences in metabolite profiles between strawberry cultivars and to assess the influence of agronomic practices on polyphenol production. Three binary classification models were built by PLS-DA to compare strawberry varieties: model A, Camarosa versus Festival; model B, Camarosa versus Palomar; model C, Festival versus Palomar. The quality of these models was evaluated using sevenfold cross-validation, yielding in all cases good quality of fit (R_x^2 = 0.645, 0.702 and 0.659 for models A, B and C, respectively), validity (R_y^2 = 0.655, 0.751 and 0.733 for models A, B and C, respectively) and prediction (Q^2 = 0.527, 0.663 and 0.664 for models A, B and C, respectively). Discriminant metabolites were selected according to the variable importance for the projection parameter (VIP), and the results showed that Camarosa presented significantly higher levels of anthocyanins, ellagic acid rhamnoside, quercetin 3-O-glucuronide, and procyanidin B2; Festival was richer in catechin and chlorogenic acid; and Palomar in sanguiin H10. Complementarily, RF analysis was also applied to compare its statistical performance with that provided by PLS-DA models. Samples were perfectly classified using 1000 trees and two predictors for each node, with 97.4% accuracy rate for the training set, 93.3% accuracy rate for the testing set and a 2.56% error rate. Then, the mean decrease of the Gini index (MDGI) was used to select the most important metabolites for interclass discrimination, which indicated that the content of chlorogenic acid, ellagic acid rhamnoside, cyanidin 3-O-glucoside, quercetin 3-O-glucoside, epicatechin, sanguiin H10, pelargonidin 3-O-rutinoside and hyperoside were the most discriminant variables, in agreement with the results provided by PLS-DA. Finally, the statistical performance of the PLS-DA and RF models was assessed by computing the SENS and SPEC parameters. Interestingly, RF provide higher sensitivity compared with PLS-DA (24% increase for models A and B, 15.6% increase for model C), but slightly decreasing the specificity.

Also in strawberry, a multi-targeted profiling approach was employed to characterize the global chemical composition in terms of sugars, organic acids, polyphenols and mineral elements, which are related to sensory and health characteristics of this berry fruit (González-Domínguez et al., 2020a). This multi-chemical profile, combined with complementary pattern recognition procedures (PCA, LDA, SIMCA and PLS-DA), was investigated as a potential tool for authentication and traceability purposes, with the aim of discriminating strawberry varieties grown under different climatic and agronomic conditions (Figure 6.3). The results obtained showed that forward stepwise

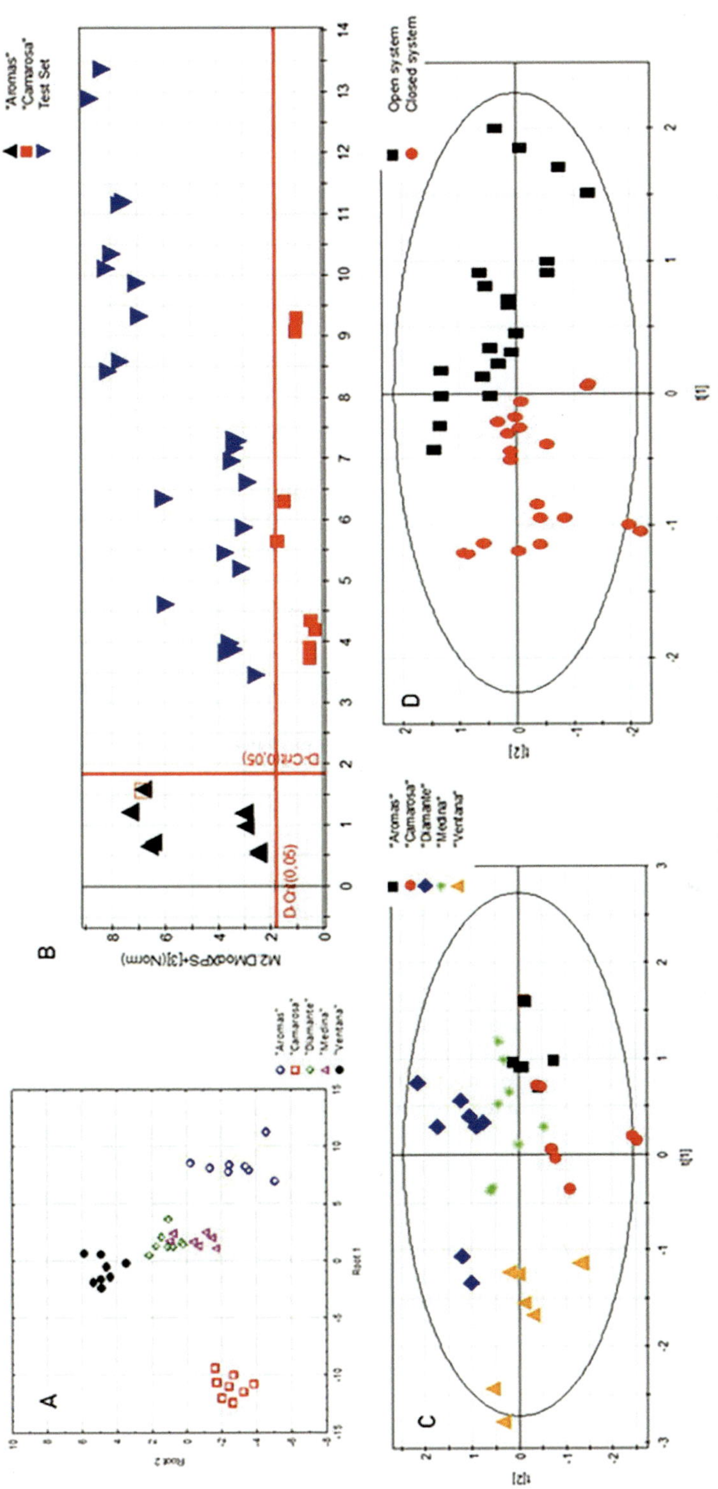

Figure 6.3 Comparison of chemometric tools for strawberry authentication according to the variety and cultivation system. (a) LDA using cultivar as the categorical variable; (b) SIMCA using cultivar as the categorical variable; (c) PLS-DA using cultivar as the categorical variable; and (d) PLS-DA using the cultivation system as the categorical variable.

Source: Reproduced from González-Domínguez et al. (2020a).

discriminant analysis (LDA) yielded a model capable of explaining 96% of the total variance with a 95% prediction ability. The Aromas and Camarosa cultivars were clearly differentiated from the rest of the samples along the first root, while the second one enabled the separation between the other three cultivars under study. The most discriminant variables among cultivars were cyanidin 3-O-glucoside, pelargonidin 3-O-rutinoside, p-coumaric acid, phosphorous, malic acid, caffeic acid and quercetin. SIMCA was subsequently applied to the same data matrix, and models based on three principal components were obtained with very good predictive ability (86.7%, 81.5%, 80.8%, 71.3% and 88.5% correct prediction for the five cultivars Aromas, Camarosa, Diamante, Medina and Ventana, respectively). However, neither LDA nor SIMCA provided good results for the classification of strawberry samples according to agronomic conditions. After applying PLS-DA, a five-component model was obtained with good quality of fit ($R^2_X = 0.744$) and predictive ability ($Q^2 = 0.413$) for the classification of strawberry samples according to the cultivar. The most important chemical descriptors driving this separation were anthocyanins and phenolic acids, in line with previous findings from ANOVA and LDA. Additionally, PLS-DA modeling also enabled the discrimination of samples grown in the two cultivation systems (i.e., open and closed soilless systems), being p-hydroxybenzoic acid, ferulic acid, unknown derivatives of pelargonidin, glucose, pelargonidin acetylglucoside, and cyanidin 3-O-glucoside the most discriminant variables in this model (explaining 70.2% of the variance).

Khalil et al. (2017) used solid phase microextraction (SPME) combined with GC-MS and subsequent multivariate data analysis to explore date (*Phoenix dactylifera* L.) fruit aroma composition from 13 varieties grown in Egypt. In this study, 89 volatile compounds were identified, which represented approximately 90% of the fruit aroma. The authors found varietal differences in terms of the amount rather than in the composition of volatiles, lipid-derived volatiles and phenylpropanoid derivatives being the major components of the characteristic date aroma. The PCA score plot evidenced that Siwi and Sheshi varieties, which presented negative values for PC1, exhibit the most distinct aroma. In turn, the volatile metabolites with major contribution to PC1 were cinnamaldehyde (negative contribution) and hexanal, hexanol and 2,3-butanediol (positive contribution). Similarly, hierarchical clustering analysis (HCA) also showed three clear clusters: group A1 (4 varieties), group A2 (7 varieties) and group B (Siwi and Sheshi). For this reason, OPLS-DA was used to test the validity of clustering and to sign markers characteristic for each class. The results showed that samples from the cluster A2 could be differentiated from the rest (clusters A1 and B), with a prediction goodness parameter of 68%. The corresponding S-plot identified cinnamaldehyde, 2,3-butanediol and hexanal as the main responsible variables for the differentiation, in agreement with PCA results.

González-Domínguez et al. (2020b) also investigated the potential of volatiles as chemical descriptors between strawberry cultivars (Festival, Candonga, Camarosa) grown in a soilless system. Methyl butyrate, hexyl hexanoate, linalool, geraniol and furaneol were the most abundant aromatic compounds detected in the three varieties of strawberries, Festival and, to a lesser extent, Candonga being the varieties characterized by richest aroma-related profile. As a first exploratory step, PCA was applied for a rapid exploration and visualization of the data trends. A PCA model with three principal components was able to explain up to 69.4% of the total variance. In the space defined by the two first principal components, the three varieties were distributed along the PC1, with Festival samples located in the right side of the PCA plot and Camarosa samples in the left side. The PC1 was positively associated with levels of ethyl butanoate, ethyl hexanoate, hexyl hexanoate, linalool, geraniol and furaneol, which could therefore serve as discriminators between varieties. Complementarily, LDA modeling also yielded satisfactory classification (92.18%) and prediction (75.68%) rates, thereby allowing a clear discrimination among the study groups. Two significant discriminant functions were obtained by stepwise LDA, which explained 68% and 32% of the total variability of the data, respectively. The first discriminant function enabled the separation of the three varieties, mainly between Festival and Camarosa cultivars with negative and positive loadings for root 1, respectively. The primary variables responsible for the discrimination of Festival strawberries with respect to the other two varieties were geraniol and hexyl hexanoate, with negative loadings in root 1.

Using the same analytical set-up (SPME-GC-MS), Lee et al. (2019) studied the influence of the production system (net-house and open field) on the volatile profile of three Texas A&M F1 hybrids (TAMU) and five commercial tomato varieties. PLS-DA evidenced considerable differences between TAMU and commercial tomato varieties in terms of hexanal, p-cymene and (E)-2-hexenal (characteristic for the three TAMU varieties); as well as (E,E)-2,4-nonadienal, benzothiazole, (E,E)-2,4-heptadienal, (E,Z)-2,4-decadienal, (E,E)-2,4-decadienal, (E)-2-octenal, and neral (characteristic for the five commercial varieties). Then, PLS-DA and OPLS-DA were also applied to investigate the influence of agronomical practices in the volatile composition of tomato samples. The S-plot showed that compounds such as D-limonene and geranylacetone could serve as potential markers of the production system, the first one being detected at higher concentrations in open-field grown tomatoes, whereas geranylacetone content was higher in net-house grown tomatoes.

6.3.2 Multi-Elemental Profiling and Isotopic Analysis for Fruit and Vegetable Authentication

Mineral elements are absorbed through the soil solution and then distributed across different plant organs. Thus, elemental profiles of plant-based foods are closely linked to the soil characteristics (i.e., geographical origin), environmental conditions of the growing area and crop management. Accordingly, the application of multi-elemental profiling approaches, such as atomic absorption spectroscopy (AAS), inductively coupled plasma–mass spectrometry (ICP-MS), and inductively coupled plasma–optical emission spectroscopy (ICP-OES), has been widely described for the authentication of F&Vs. Furthermore, it should be noted that many elements can be naturally present in soil and plants as different isotopic forms. As the ratio between isotopes of specific elements can change depending on several conditions (e.g., latitude, environmental conditions), isotopic ratios have also been proposed as candidate authenticity markers (Francois et al., 2020). In particular, stable isotope ratios of bio-elements ($^{2}H/^{1}H$; $^{18}O/^{16}O$; $^{13}C/^{12}C$; $^{15}N/^{14}N$) have successfully been used for authentication purposes in terms of geographical origin and agricultural practices. For instance, oxygen-18 ($\delta^{18}O$) and deuterium ($\delta^{2}H$) can serve to differentiate samples according to the origin, since these isotopes are highly affected by precipitation-related factors (hydrological cycle), such as distance to the sea and temperature. On the other hand, the carbon isotope ratio ($^{13}C/^{12}C$) is related to climatic conditions, plant type (i.e., C3 or C4 plants) and agricultural practice. Finally, the stable isotope ratio of nitrogen ($\delta^{15}N$) has been proposed as a marker to differentiate conventional from organic farming, because the nitrogen isotope content in synthetic nitrogen fertilizers is much lower than in fertilizers based on manure, compost and liquid/solid animal waste (Perini et al., 2018). To measure isotopic ratio profiles, various mass spectrometry–based techniques can be employed, such as isotopic ratio mass spectrometry (IRMS), thermal ionization mass spectrometry (TIMS), and multi-collector–inductively coupled plasma–mass spectrometry (MC-ICP-MS) (Laursen et al., 2016). As a proof of concept of the utility of multi-elemental profiling and isotopic measurements in F&V authentication, we provide below a revision of recent literature on this topic (Table 6.2).

Coelho et al. (2019) determined 24 elements, using ICP-MS and ICP-OES, to study the mineral content of a Portuguese variety of pear (Rocha) collected from two geographical regions. Relationships between elements were tested by correlation analysis and differentiation of samples was performed by applying pattern recognition techniques. A PCA model with six components, explaining 74.7% of total data variation, could identify relationships among samples collected from different trees in the same farm, but it was not able to identify relationships between all samples from the same geographical origin. In contrast, the application of HCA yielded two distinct clusters, fully separating the samples from each region under study. Finally, a predictive model was built by using LDA and cross-validation, which enabled to trace pear samples according to the origin based on 15 mineral elements, with 100% accuracy.

In another study, a multi-chemical approach was combined with chemometrics to classify Italian apples according to their provenance, both on a national and a regional scale (Aguzzoni et al., 2021). For this purpose, the authors applied multi-element profiling and isotopic analysis of hydrogen, oxygen and strontium ($\delta^{2}H$, $\delta^{18}O$, $^{87}Sr/^{86}Sr$) to apples of the variety Golden Delicious, which were collected from orchards within three protected areas (PDO Val di Non, PGI South Tyrol, PGI Valtellina) and one non-protected area, all in northern Italy. An LDA model was created to compare the four cultivation areas under study according to the multi-chemical composition of apples. In this model, Val di Non apples were clearly differentiated from the other groups by root 1, the ratios Ca/Ba, Sr/Ba and Ca/Sr being the variables that contributed most to the separation. Then, a second LDA model was built by limiting the analysis to the multi-chemical composition of South Tyrol samples, using the cultivation district (three groups) as the classification criteria. The variances explained by the two first linear discriminant (LD) functions were 63% and 37%, respectively, and the two-dimensional score plot showed a clear separation between the three study groups based on the $^{87}Sr/^{86}Sr$ ratio, Rb and Fe concentrations (LD1) and the Ca/Sr ratio, Sr and Cu concentrations (LD2).

To assess the geographical origin of blueberry samples from three regions of China, Kuang et al. (2021) determined the concentrations of K, Ca, Mg, Na, Fe, Cu, Mn, B, P and Zn by ICP-OES, and

Table 6.2 Selected Applications of Multi-Elemental Profiling and Isotopic Analysis for Fruit and Vegetable Authentication

Fruit/ Vegetable	Geographical Origin	Parameters Measured	Analytical Technique	Chemometric Tool	Aim of the Study	Reference
Pear	Portugal	Multi-elemental profile	ICP-MS ICP-OES	PCA HCA LDA	Geographical origin	(Coelho et al., 2019)
Apple	Italia	Multi-elemental profile Stable isotope profile: $^{87}Sr/^{86}Sr$ $\delta^2 H$, $\delta^{18}O$	ICP-MS IRMS	LDA	Geographical origin	(Aguzzoni et al., 2021)
Blueberry	China	Multi-elemental profile	ICP-OES	LDA DT MLP-NN SVM	Geographical origin	(Kuang et al., 2021)
Ginger	Malaysia	Multi-elemental profile	ICP-MS	PCA HCA ANN	Geographical origin	(Tabbassum et al., 2021)
Garlic	China	Stable isotope profile: ^{13}C, ^{15}N, ^{2}H, ^{18}O, ^{34}S	IRMS	PCA PLS-DA	Geographical origin	(Nie et al., 2021)
Carrot	Romania	Multi-elemental profile Stable isotope profile: $^{13}C/^{12}C$, $^{15}N/^{14}N$, $^{18}O/^{16}O$, $^{2}H/^{1}H$	ICP-MS IRMS	LDA	Geographical origin	(Magdas et al., 2018)
Potato	Romania	Multi-elemental profile Stable isotope profile: $^{2}H/^{1}H$, $^{18}O/^{16}O$, $^{13}C/^{12}C$, $^{15}N/^{14}N$	ICP-MS IRMS	LDA	Geographical origin Agronomic practices (organic/ conventional)	(Magdas et al., 2017)
Salad, tomato, potato, cabbage, garlic, leek	Romania	Multi-elemental profile Stable isotope profile: $^{2}H/^{1}H$, $^{18}O/^{16}O$, $^{13}C/^{12}C$, $^{15}N/^{14}N$	ICP-MS IRMS	LDA	Agronomic practices (greenhouse/ field, organic/ conventional)	(Feher et al., 2017)
Squash, radish	Romania	Multi-elemental profile Stable isotope profile: δ^2H, $\delta^{18}O$, $\delta^{13}C$	ICP-MS IRMS	LDA	Agronomic practices (greenhouse/ field) Geographical origin	(Cristea et al., 2017)

then applied several machine learning techniques, such as linear discriminant analysis (LDA), decision tree (DT), multilayer perceptron neural network (MLP-NN) and support vector machine (SVM). Among the statistical tools tested, MLP-ANN and SVM proved to be the more successful approaches for origin differentiation considering the average discrimination rates obtained for both the training and test sets: MLP-NN, 92.7% and 94.7% (training and test set, respectively); SVM, 91.8% and 94.7% (training and test set, respectively); DT, 91.8% and 86.8% (training and test set, respectively); and LDA, 85.8% and 84.5% (training and test set, respectively).

Multi-elemental fingerprinting by means of ICP-MS has also been proposed to combat adulteration and mislabeling of Bentong ginger grown in Malaysia (Tabbassum et al., 2021). To this end, the concentrations of Na, Mg, K, Ca, Cr, Mn, Fe, Ni, Cu, Zn, Rb, Ba and Pb were analyzed in 30 rhizomes of Bentong ginger and 18 non-Bentong ginger samples. The outcomes of both PCA and HCA revealed a distinct clustering pattern that separated Bentong from non-Bentong samples, where Rb and Pb levels were found mainly associated with Bentong gingers. Complementarily, an ANN model enabled distinguishing the Bentong gingers from other commercial samples with excellent predictive performance.

To differentiate PGI Jinxiang garlic to that produced in other exporting countries (Argentina, Southeast Asian countries) and other Chinese provinces (Guizhou, Hainan), Nie et al. (2021) tested different discrimination models based on stable isotope profiles (Figure 6.4). PCA modeling

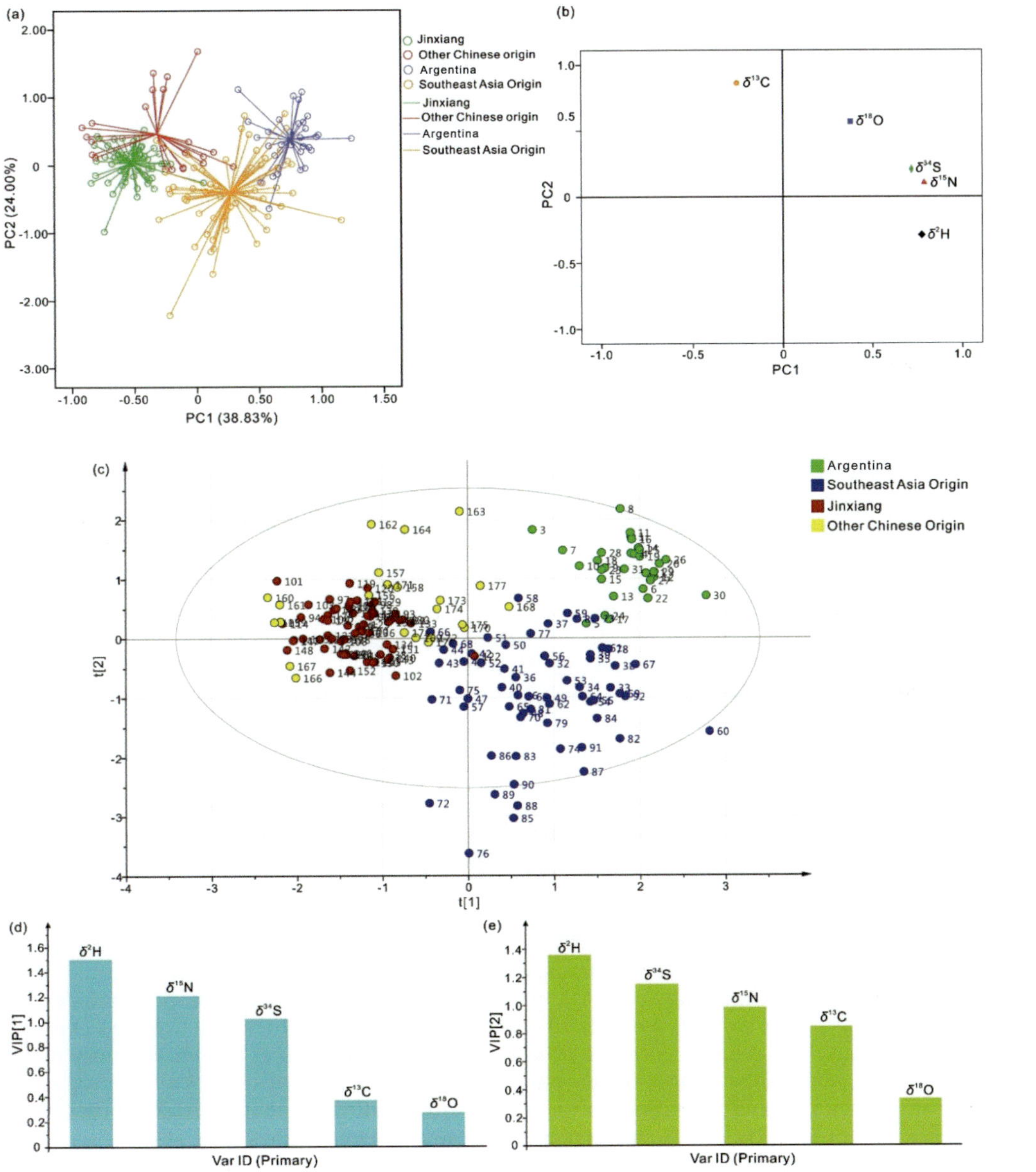

Figure 6.4 Chemometric discrimination of garlic samples according to the geographical origin based on stable isotope analysis. (a) PCA scores plot; (b) PCA loading plot; (c) PLS-DA scores plot; (d) VIP for the first component of PLS-DA model; and (e) VIP for the second component of PLS-DA model.

Source: Reproduced from Nie et al. (2021) with permission from Elsevier.

distinguished Jinxiang garlic from Southeast Asian and Argentinean samples along the PC1, whereas the PC2 enabled the separation between different Chinese provinces. Then, a four-component PLS-DA model was obtained with good quality of validity ($R^2_Y = 0.548$) and prediction ($Q^2 = 0.518$). As Jinxiang and other Chinese origin samples showed considerable overlap, pairwise PLS-DA classification models were then built to clearly differentiate PGI Jinxiang garlic from other Chinese and non-Chinese origins, with total accuracy rates of 89.5% and 100%, respectively.

The increasing consumer demand for organic food products, with higher market price, has led to a significant raise of frauds related to the mislabeling of conventionally grown F&Vs as organic products. In this context, Magdas et al. (2018) tested the potential of combining stable isotope ratios and elemental analysis for discriminating organic and conventionally grown carrot samples collected from Romania, other EU countries, and China. Stepwise LDA was applied to classify carrots according to the growing method and the origin, and the most significant variables involved in sample differentiation were selected by ANOVA. The best markers for the differentiation of organically versus conventionally grown carrots resulted to be the isotopic ratio of nitrogen ($^{15}N/^{14}N$) along with La and Ga contents, which yielded classification rates of 83.3% (initial classification) and 81% (cross-validation). For the differentiation of Romanian samples from others coming from other EU countries, the most discriminant variables were Mn and Tb concentrations (initial and cross-validated classification rate: 75.7%). On the other hand, the δ^2H isotopic ratio was the most powerful marker to discriminate between carrots grown in different Romanian areas, although the model yielded poor classification accuracy (66.1% for initial classification, 64.5% for cross-validation), probably due to their geographical proximity. Using a similar approach, the authors of this study also investigated the impact of the geographical origin (Romanian vs. non-Romanian) and/or growing conditions (open field vs. greenhouse; organic vs. conventional) on the isotope ratios and mineral content in other vegetables. They found that the $\delta^{15}N$ ratio and the concentrations of Cd, Ca, Cu and Zn allow to differentiate organically versus conventionally produced potatoes, with a 94.7% correct classification of original cases (91.2% in cross-validation) (Magdas et al., 2017). For geographical origin classification, LDA also provided good performance (96.5% for initial classification, 87% for cross-validation), with Ca, P, Co, Ni and $\delta^{13}C$ being the best predictor variables between Romanian and foreign potato samples. In another study, they investigated candidate mineral markers to differentiate organic versus conventionally cultivated vegetables, as well as vegetables grown in greenhouses versus those cultivated in the field (Feher et al., 2017). For plant foods with subterranean edible parts (potato, garlic), the best predictors for differentiating between organic and conventional products were Tl, Cu, Eu, δ^2H, Os, Sm, Pb, In and Ru; whereas for plants having edible parts above ground (tomato, salad, cabbage) the most discriminant variables were Zn and $\delta^{15}N$. On the other hand, Ni, Mo, Cd, Pb and $\delta^{13}C$ were the most important variables for differentiating vegetables grown in greenhouses from those produced in the field. With the aim of investigating the impact of the cultivation system (greenhouse vs. field) on the mineral and isotopic profile, Cristea et al. (2017) also analyzed squash and radish samples from different origins. In that case, the variables responsible for the discrimination in LDA models were Sn and $\delta^{13}C$ for radishes (94% correct classification) and Sn and Cu for squashes (98% correct classification). Regarding the origin, Romanian radishes and squashes presented higher contents of Na and Cu, respectively, whereas foreign samples were characterized by higher levels of Zn, Sr, Zr and Co.

6.3.3 Fingerprinting Approaches for Fruit and Vegetable Authentication

Metabolomics has proven to be a suitable strategy to characterize in a comprehensive manner the primary and secondary plant metabolome, which can reflect the impact of both endogenous and exogenous factors and thus serve as a powerful tool to monitor food authenticity (Medina et al., 2019). Different analytical techniques can be used to perform metabolomics analysis, mainly mass spectrometry (MS) and nuclear magnetic resonance (NMR). Each of these analytical techniques has their own advantages and limitations, so a single analytical technique is usually not capable of performing comprehensive untargeted screening (Gonzalez-Dominguez et al., 2017). Currently, MS-based platforms predominate due to their ability to identify and (semi-)quantitate a wide range of compounds with high sensitivity and high throughput capabilities. High-resolution mass spectrometry (HRMS) has become the gold standard technique for non-targeted metabolomics, but in recent years the use of large-scale targeted metabolomics platforms has emerged for comprehensive and quantitative metabolomics fingerprinting (González-Domínguez et al., 2020c). As an alternative, spectroscopic techniques have also demonstrated great potential for food fingerprinting due to their capacity to simultaneously detect many compounds in a fast, non-destructive,

Table 6.3 Selected Applications of Fingerprinting Approaches for Fruit and Vegetable Authentication

Fruit/ Vegetable	Geographical Origin	Analytical Technique	Chemometric Tool	Aim of the Study	Reference
Lingonberry Cranberry	Czech Republic	UHPLC-HRMS	PCA PLS-DA	Adulteration	(Hurkova et al., 2019)
Carrot	Belgium	LC-MS	OPLS-DA	Agronomic practices (organic/ conventional)	(Cubero-Leon et al., 2018)
Strawberry	Spain	GC-MS	PCA PLS-DA	Variety Agronomic practices	(Akhatou et al., 2016)
Persimmon	Spain	FT-NIR	SVM	Geographical origin	(Khanmohammadi et al., 2014)
Pineapple	Ghana	NIR	PCA kNN LDA	Agronomic practices (organic/ conventional)	(Amuah et al., 2019)
Apple	Italy	NIR	PCA QDA	Variety Geographical origin	(Eisenstecken et al., 2019)
Celery	Australia, Taiwan, China	^{1}H-NMR	PCA HCA PLS-DA	Geographical origin	(Lau et al., 2020)

non-invasive, solvent-free, and automatic manner, and at moderate instrumental costs (Danezis et al., 2016). In that case, the application of spectral pre-processing methods is usually required to extract relevant information from spectral data and to remove undesirable systematic variation due to light scattering effects, instrumental drift, baseline shift and slope variation (Arendse et al., 2021). In this sense, the most used preprocessing techniques include spectral derivatives and scatter correlation methods, such as multiplicative scatter correction (MSC), standard normal variate (SNV) and normalization. In this section, we provide an overview on the potential of metabolomics and spectroscopic-based fingerprinting in food authentication research based on recently published literature (Table 6.3).

Non-targeted metabolomic fingerprinting using UHPLC-HRMS has been proposed for the authentication of two *Vaccinium* species: lingonberries and cranberries (Hurkova et al., 2019). Although both species belong to the same botanical family and share many organoleptic characteristics, lingonberries have higher market value and, therefore, are susceptible to fraud. A PLS-DA model provided sufficient discrimination between cranberries and lingonberries, as indicated by the quality parameters: $R^2_X = 0.84$, $R^2_Y = 0.95$ and $Q^2 = 0.93$, for data acquired in the positive ionization mode; $R^2_X = 0.95$, $R^2_Y = 0.98$ and $Q^2 = 0.98$, for negative ion data. The most significant markers that were tentatively identified included various glycerophospholipids and polyphenols. In particular, peonidin 3-O-arabinoside and myricetin 3-O-glucoside could be regarded as specific markers to detect the addition of cranberries to lingonberries at concentrations below 1% (w/w), as these metabolites were not detected in lingonberries (Figure 6.5).

Using a similar analytical approach, Cubero-Leon et al. (2018) investigated the metabolomics differences in carrots (*Daucus carota* L.) cultivated under different agronomic environments. OPLS-DA models were built using a training set containing two-thirds on the samples and were then used to predict the origin of unknown samples by external validation. The models were able to correctly classify unknown samples (100% correct classification) and to provide a clear separation between conventionally grown and organic samples. Five metabolites showed significant differences between both cultivation systems using a univariate approach with Bonferroni correction, including a carbohydrate, fatty acid esters, a jasmonate derivative and glycosylated terpenes.

To focus on the primary metabolome, Akhatou et al. (2016) applied GC-MS–based metabolomics to investigate the effect of genotype and agronomic conditions on selected strawberry cultivars with different sensitivity to environmental stress. Using PCA and PLS-DA, the authors found that the cultivar (Palomar, Festival, Camarosa) and agricultural practices (electrical conductivity of irrigation, macrotunnel type, substrate) can affect the levels of numerous primary metabolites,

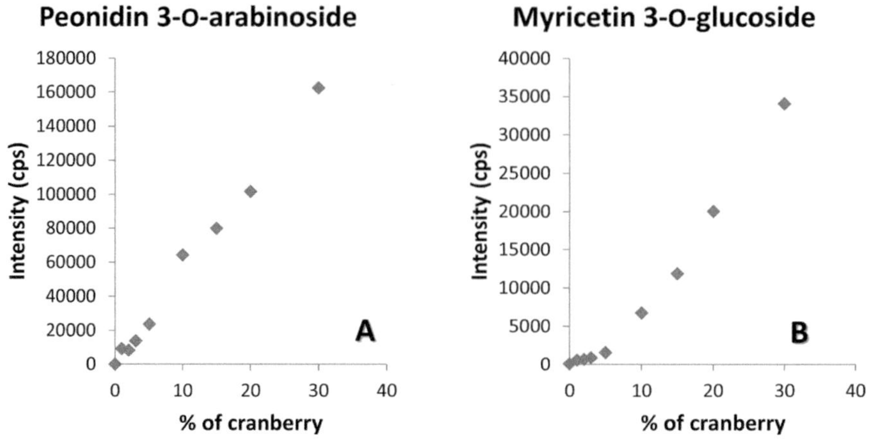

Figure 6.5 Potential of metabolomics to identify biomarkers (peonidin 3-O-arabinoside, panel A; myricetin 3-O-glucoside, panel B) for detecting the addition of cranberries to lingonberries.

Source: Reproduced from Hurkova et al. (2019) with permission from Elsevier.

including sugars (fructose, glucose), organic acids (malic acid, citric acid) and amino acids (alanine, threonine, aspartic acid), which could be related to differences in organoleptic properties and tolerance mechanisms against environmental stress.

Among spectroscopic techniques, infrared and NMR spectroscopy have been widely used for qualitative and quantitative analysis in a variety of agri-food products (Esteki et al., 2018a). Khanmohammadi et al. (2014) described the application of multi-class least squares SVM for processing NIR spectral data with the aim of discriminating Spanish persimmon samples according to the origin. They employed FT-NIR spectroscopy to analyze persimmon samples produced under the PDO Ribera del Xúquer and from other Spanish regions (Andalucía, Extremadura, Comunidad de Valencia). Chemometric modeling provided robust classification of samples with 0% training error and enabled the prediction of the origin of 50 samples from the independent test set with 2% prediction error.

In another study, NIR spectral fingerprints were used to differentiate organic from inorganic pineapple fruits, and to predict maturity by measuring total soluble solids (TSS) using a digital refractometer (Amuah et al., 2019). Two spectral preprocessing techniques, mean centering (MC) and multiplicative scatter correction (MSC), were compared before the application of multivariate calibration models based on PCA, kNN and LDA to identify the fruit class (i.e., organic and conventional pineapple fruits), while PLSR was used to predict the TSS content of the fruits. The MSC-PCA-LDA model provided the best classification performance to differentiate organic from conventionally produced fruits, with 100% correct classification rates. On the other hand, PLSR showed 85% accuracy to predict TSS values, both in the calibration and prediction set.

NIR spectroscopy has also successfully been used to classify Golden Delicious apples from three different orchard elevation levels (225, 650, and 1000 m above sea level) and nine cultivars (Braeburn, Coop39–Crimson Crisp, Fuji, Fujion, Gala, CIV323–Isaaq, Coop43–Juliet, SQ159–Natyra, UEB32642–Opal) (Eisenstecken et al., 2019). Before the construction of chemometric models, several mathematical preprocessing techniques were applied (MSC, SNV, second-order polynomial, and Savitzky-Golay derivatives). The best results in terms of overall correct classification were obtained when second-order polynomial followed by the second derivative was applied for discrimination of the orchard elevation, whereas second-order polynomial followed by the first derivative was best suited for cultivar discrimination. Then, PCA and quadratic discriminant analysis (QDA) based on PCA scores were applied to build prediction models. PCA-DA models correctly classified 93.6% and 77.9% of the apple samples from the mountain and valley origin, respectively, but only 57.1% correct classification rate was obtained for the intermediate orchard level. For cultivar classification, prediction accuracies ranged between 100% (SQ159–Natyra®,

CIV323–Isaaq®, Braeburn, Coop39–Crimson Crisp®) and 76.9% (Fujion, Coop43–Juliet®). When the models were validated using independent samples, correct classification rates of 87.5% and 86.3% were found for orchard elevation and cultivar, respectively. The most relevant spectral regions driving this discrimination were tentatively related to a broad spectrum of metabolites, including carbohydrates, organic acids, lipids, amino acids, and aromatic and aliphatic hydrocarbons.

Lau et al. (2020) demonstrated that proton NMR (^{1}H NMR) spectroscopy can effectively differentiate celery (*Apium graveolens* L. var. *dulce*) stem and leaf samples from different geographical origins (Australia, Taiwan, China). PCA and HCA were employed to visualize sample grouping and to identify the key metabolites responsible for the variability in the data, whereas PLS-DA was applied to build classification models. PCA and HCA plots showed a clear separation of samples in three distinct clusters depending on the geographical origin, with mannitol, sucrose, citric acid, glutamine, asparagine and 4-hydroxybenzoic acid being the major contributors to this differentiation. Then, PLS-DA models showed 100% prediction accuracy. In general, the authors found higher mannitol content in celery samples from Taiwan, while organic acids were higher in celery from Australia and China. On the other hand, amino acid metabolites were identified to be at lower concentrations in Australian celeries.

6.4 CONCLUDING REMARKS

The geographical origin, variety and agricultural practices can influence the chemical composition of plant-based foods, and therefore affect their organoleptic, nutritional and bioactive properties. For this reason, the food industry requires of powerful analytical methods with enough accuracy, robustness and high-throughput capacity to guarantee the authenticity and traceability of foods, and particularly fruits and vegetables (F&Vs). Chromatographic and spectroscopic approaches have traditionally been employed for the identification and quantification of specific compound classes (polyphenols, carotenoids, primary metabolites, mineral elements) as candidate markers for authentication purposes. In recent years, the development of global fingerprinting approaches (e.g., metabolomics) has enabled to get deeper understanding about the impact of endogenous and exogenous factors on food composition. The combination of these analytical tools with advanced chemometric techniques is mandatory to disentangle the high amount of data obtained, with the aim of building multivariate models for classification, discrimination and prediction, and for identifying candidate markers to differentiate F&V-based products based on their genotype (e.g., variety), geographical origin (e.g., PDO/PGI) or cultivation method (e.g., organic vs. conventionally grown conditions). For this purpose, common statistical tools in food science research include principal component analysis (PCA), hierarchical cluster analysis (HCA), linear discriminant analysis (LDA), partial least squares–discriminant analysis (PLS-DA), k-nearest neighbor (kNN), soft independent modeling of class analogy (SIMCA), support vector machine (SVM), random forest (RF), and artificial neural networks (ANN).

REFERENCES

Aguzzoni A, Bassi M, Pignotti E, Robatscher P, Scandellari F, Tirler W, Tagliavini M (2021) Multi-chemical analysis combined with chemometrics to characterize PDO and PGI Italian apples. J Sci Food Agric 101(12):5106–5115. https://doi.org/10.1002/jsfa.11156

Akhatou I, Fernández-Recamales Á (2014a) Influence of cultivar and culture system on nutritional and organoleptic quality of strawberry. J Sci Food Agric 94(5):866–875. https://doi.org/10.1002/jsfa.6313

Akhatou I, Fernández-Recamales Á (2014b) Nutritional and nutraceutical quality of strawberries in relation to harvest time and crop conditions. J Agric Food Chem 62(25):5749–5760. https://doi.org/10.1021/jf500769x

Akhatou I, González-Domínguez R, Fernández-Recamales Á (2016) Investigation of the effect of genotype and agronomic conditions on metabolomic profiles of selected strawberry cultivars with different sensitivity to environmental stress. Plant Physiol Biochem 101:14–22. https://doi.org/10.1016/j.plaphy.2016.01.016

Akhatou I, Sayago A, González-Domínguez R, Fernández-Recamales Á (2017) Application of targeted metabolomics to investigate optimum growing conditions to enhance bioactive content of strawberry. J Agric Food Chem 65(43):9559–9567. https://doi.org/10.1021/acs.jafc.7b03701

Amuah CLY, Teye E, Lamptey FP, Nyandey K, Opoku-Ansah J, Adueming POW (2019) Feasibility study of the use of handheld NIR spectrometer for simultaneous authentication and quantification of quality parameters in intact pineapple fruits. J Spectrosc 2019:5975461. https://doi.org/10.1155/2019/5975461

Antignac JP, Courant F, Pinel G, Bichon E, Monteau F, Elliott C, Le Bizec B (2011) Mass spectrometry-based metabolomics applied to the chemical safety of food. TrAC Trends Anal Chem 30(2):292–301. https://doi.org/10.1016/j.trac.2010.11.003

Arendse E, Nieuwoudt H, Magwaza LS, Nturambirwe JFI, Fawole OA, Opara UL (2021) Recent advancements on vibrational spectroscopic techniques for the detection of authenticity and adulteration in horticultural products with a specific focus on oils, juices and powders. Food Bioprocess Technol 14(1):1–22. https://doi.org/10.1007/s11947-020-02505-x

Baker RS, Inventado PS (2014) Educational data mining and learning analytics. In: Learning analytics: From research to practice. Springer, New York, NY: 61–75. https://doi.org/10.1007/978-1-4614-3305-7_4

Ballin NZ, Laursen KH (2019) To target or not to target? Definitions and nomenclature for targeted versus non-targeted analytical food authentication. Trends Food Sci Technol 86:537–543. https://doi.org/10.1016/j.tifs.2018.09.025

Berrueta LA, Alonso-Salces RM, Héberger K (2007) Supervised pattern recognition in food analysis. J Chromatogr A 1158(1–2):196–214. https://doi.org/10.1016/j.chroma.2007.05.024

Callao MP, Ruisánchez I (2018) An overview of multivariate qualitative methods for food fraud detection. Food Control 86:283–293. https://doi.org/10.1016/j.foodcont.2017.11.034

Ceballos-Magaña SG, Jurado JM, Muñiz-Valencia R, Alcázar A, de Pablos F, Martín MJ (2012) Geographical authentication of tequila according to its mineral content by means of support vector machines. Food Anal Methods 5(2):260–265. https://doi.org/10.1007/s12161-011-9233-1

Cifuentes A (2009) Food analysis and foodomics. J Chromatogr A 1216(43):7109–7358. https://doi.org/10.1016/j.chroma.2009.09.018

Coelho I, Matos AS, Teixeira R, Nascimento A, Bordado J, Donard O, Castanheira I (2019) Combining multielement analysis and chemometrics to trace the geographical origin of Rocha pear. J Food Compos Anal 77:1–8. https://doi.org/10.1016/j.jfca.2018.12.005

Creydt M, Fischer M (2018) Omics approaches for food authentication. Electrophoresis 39(13):1569–1581. https://doi.org/10.1002/ELPS.201800004

Cristea G, Feher I, Magdas DA, Voica C, Puscas R (2017) Characterization of vegetables by stable isotopic and elemental signatures. Anal Lett 50(17):2677–2690. https://doi.org/10.1080/00032719.2016.1263312

Cuadros-Rodríguez L, Ruiz-Samblás C, Valverde-Som L, Pérez-Castaño E, González-Casado A (2016) Chromatographic fingerprinting: An innovative approach for food "identitation" and food authentication—A tutorial. Anal Chim Acta 909:9–23. https://doi.org/10.1016/j.aca.2015.12.042

Cubero-Leon E, De Rudder O, Maquet A (2018) Metabolomics for organic food authentication: Results from a long-term field study in carrots. Food Chem 239:760–770. https://doi.org/10.1016/j.foodchem.2017.06.161

Danezis GP, Tsagkaris AS, Camin F, Brusic V, Georgiou CA (2016) Food authentication: Techniques, trends & emerging approaches. TrAC Trends Anal Chem 85:123–132. https://doi.org/10.1016/j.trac.2016.02.026

Drivelos SA, Georgiou CA (2012) Multi-element and multi-isotope-ratio analysis to determine the geographical origin of foods in the European Union. TrAC Trends Anal Chem 40:38–51. https://doi.org/10.1016/j.trac.2012.08.003

Dumont D, Danielato G, Chastellier A, Hibrand Saint Oyant L, Fanciullino AL, Lugan R (2020) Multi-targeted metabolic profiling of carotenoids, phenolic compounds and primary metabolites in goji (*Lycium* spp.) berry and tomato (*Solanum lycopersicum*) reveals inter and intra genus biomarkers. Metabolites 10(10):422. https://doi.org/10.3390/metabo10100422

Eisenstecken D, Stürz B, Robatscher P, Lozano L, Zanella A, Oberhuber M (2019) The potential of near infrared spectroscopy (NIRS) to trace apple origin: Study on different cultivars and orchard elevations. Postharvest Biol Technol 147:123–131. https://doi.org/10.1016/j.postharvbio.2018.08.019

Esslinger S, Riedl J, Fauhl-Hassek C (2014) Potential and limitations of non-targeted fingerprinting for authentication of food in official control. Food Res Int 60:189–204. https://doi.org/10.1016/j.foodres.2013.10.015

Esteki M, Shahsavari Z, Simal-Gandara J (2018a) Use of spectroscopic methods in combination with linear discriminant analysis for authentication of food products. Food Control 91:100–112. https://doi.org/10.1016/j.foodcont.2018.03.031

Esteki M, Simal-Gandara J, Shahsavari Z, Zandbaaf S, Dashtaki E, Vander Heyden Y (2018b) A review on the application of chromatographic methods, coupled to chemometrics, for food authentication. Food Control 93:165–182. https://doi.org/10.1016/j.foodcont.2018.06.015

European Commission (2022) eAmbrosia: the EU geographical indications register. Available at: https://ec.europa.eu/info/food-farming-fisheries/food-safety-and-quality/certification/quality-labels/geographical-indications-register/ (Accessed February 13, 2022).
Feher I, Magdas DA, Dehelean A, Cristea G, Voica C (2017) Application of light stable isotope ratios and elemental profile in the organic vegetables characterization. Rom J Phys 62:803
Ferreira M (2015) Quimiometria: conceitos, métodos e aplicações. Editora Unicamp, São Paulo.
Francois G, Fabrice V, Didier M (2020) Traceability of fruits and vegetables. Phytochemistry 173:112291. https://doi.org/10.1016/j.phytochem2020.112291
Gonzalez-Dominguez A, Duran-Guerrero E, Fernández-Recamales Á, Lechuga-Sancho AM, Sayago A, Schwarz M, Segundo C, Gonzalez-Dominguez R (2017) An overview on the importance of combining complementary analytical platforms in metabolomic research. Curr Top Med Chem 17(30):3289–3295. https://doi.org/10.2174/1568026618666171211144918
González-Domínguez R, Jáuregui O, Queipo-Ortuño MI, Andrés-Lacueva C (2020c) Characterization of the human exposome by a comprehensive and quantitative large-scale multianalyte metabolomics platform. Anal Chem 92(20):13767–13775. https://doi.org/10.1021/acs.analchem.0c02008
González-Domínguez R, Sayago A, Akhatou I, Fernández-Recamales Á (2020a) Multi-chemical profiling of strawberry as a traceability tool to investigate the effect of cultivar and cultivation conditions. Foods 9(1):96. https://doi.org/10.3390/foods9010096
González-Domínguez R, Sayago A, Akhatou I, Fernández-Recamales Á (2020b) Volatile profiling of strawberry fruits cultivated in a soilless system to investigate cultivar-dependent chemical descriptors. Foods 9(6):768. https://doi.org/10.3390/foods9060768
González-Domínguez R, Sayago A, Fernández-Recamales Á (2021) Potential of ultraviolet-visible spectroscopy for the differentiation of Spanish vinegars according to the geographical origin and the prediction of their functional properties. Foods 10(8): 1830. https://doi.org/10.3390/foods10081830
González-Domínguez R, Sayago A, Morales MT, Fernández-Recamales Á (2019) Assessment of virgin olive oil adulteration by a rapid luminescent method. Foods 8(8): 287. https://doi.org/10.3390/foods8080287
González-Domínguez R, Sayago A, Santos-Martín M, Fernández-Recamales Á (2022) High-Throughput Method for Wide-Coverage and Quantitative Phenolic Fingerprinting in Plant-Origin Foods and Urine Samples. J Agric Food Chem 70(25):7796-7804. https://doi.org/10.1021/acs.jafc.2c01453
Hurkova K, Uttl L, Rubert J, Navratilova K, Kocourek V, Stranska-Zachariasova M, Paprstein F, Hajslova J (2019) Cranberries versus lingonberries: A challenging authentication of similar Vaccinium fruit. Food Chem 284:162–170. https://doi.org/10.1016/j.foodchem.2019.01.014
Kamiloglu S. (2019) Authenticity and traceability in beverages. Food Chem 277:12–24. https://doi.org/10.1016/j.foodchem.2018.10.091
Kemsley EK, Defernez M, Marini F (2019) Multivariate statistics: Considerations and confidences in food authenticity problems. Food Control 105:102–112. https://doi.org/10.1016/j.foodcont.2019.05.021
Khalil MNA, Fekry MI, Farag MA (2017) Metabolome based volatiles profiling in 13 date palm fruit varieties from Egypt via SPME GC—MS and chemometrics. Food Chem 217:171–181. https://doi.org/10.1016/j.foodchem.2016.08.089
Khanmohammadi M, Karami F, Mir-Marqués A, Bagheri Garmarudi A, Garrigues S, de la Guardia M (2014) Classification of persimmon fruit origin by near infrared spectrometry and least squares-support vector machines. J Food Eng 142:17–22. https://doi.org/10.1016/j.jfoodeng.2014.06.003
Kuang L, Nie J, Zhang J, Xu G, Li J, Cheng Y, Shen Y, Saqui F (2021) Discrimination of geographical origin of blueberries from three major producing areas of China using mineral element analyses. At Spectrosc 42(2):91–98. https://doi.org/10.46770/AS.2021.004
La Barbera G, Capriotti AL, Cavaliere C, Montone CM, Piovesana S, Samperi R, Zenezini Chiozzi R, Laganà A (2017) Liquid chromatography-high resolution mass spectrometry for the analysis of phytochemicals in vegetal-derived food and beverages. Food Res Int 100:28–52. https://doi.org/10.1016/j.foodres.2017.07.080
Lau H, Laserna AKC, Li SFY (2020) ^{1}H NMR-based metabolomics for the discrimination of celery (*Apium graveolens* L. var. dulce) from different geographical origins. Food Chem 332:127424. https://doi.org/10.1016/j.foodchem.2020.127424
Laursen KH, Bontempo L, Camin F, Roßmann A (2016) Advances in isotopic analysis for food authenticity testing. In: Advances in food authenticity testing. Elsevier (Woodhead Publishing Series in Food Science, Technology and Nutrition), Amsterdam: 227–252. https://doi.org/10.1016/B978-0-08-100220-9.00009-6

Lee JHJ, Jayaprakasha GK, Avila CA, Crosby KM, Patil BS (2019) Metabolomic studies of volatiles from tomatoes grown in net-house and open-field conditions. Food Chem 275:282–291. https://doi.org/10.1016/j.foodchem.2018.09.091

Magdas DA, Dehelean A, Feher I, Radu S (2017) Isotopic and multielemental fingerprinting of organically and conventionally grown potatoes. Isotopes Environ Health Stud 53(6):610–619. https://doi.org/10.1080/10256016.2017.1335722

Magdas DA, Feher I, Dehelean A, Cristea G, Magdas TM, Puscas R, Marincaş O (2018) Isotopic and elemental markers for geographical origin and organically grown carrots discrimination. Food Chem 267:231–239. https://doi.org/10.1016/j.foodchem.2017.10.048

Medina S, Pereira JA, Silva P, Perestrelo R, Câmara JS (2019) Food fingerprints—A valuable tool to monitor food authenticity and safety. Food Chem 278:144–162. https://doi.org/10.1016/j.foodchem.2018.11.0461

Moncayo S, Manzoor S, Navarro-Villoslada F, Caceres JO (2015) Evaluation of supervised chemometric methods for sample classification by Laser Induced Breakdown Spectroscopy. Chemom Intell Lab Syst 146:354–364. https://doi.org/10.1016/j.chemolab.2015.06.004

Nie J, Shao S, Zhang Y, Li C, Liu Z, Rogers KM, Wu MC, Lee CP, Yuan Y (2021) Discriminating protected geographical indication Chinese Jinxiang garlic from other origins using stable isotopes and chemometrics. J Food Compos Anal 99:103856. https://doi.org/10.1016/j.jfca.2021.103856

Official Journal of the European Union (2012) Council Regulation (EC) No 1151/2012 of 21 November 2012 on quality schemes for agricultural products and foodstuff. J Eur Union:1–29.

Palacios-Morillo A, Jurado JM, Alcázar A, Pablos F (2016) Differentiation of Spanish paprika from Protected Designation of Origin based on color measurements and pattern recognition. Food Control 62:243–249. https://doi.org/10.1016/j.foodcont.2015.10.045

Perini M, Giongo L, Grisenti M, Bontempo L, Camin F (2018) Stable isotope ratio analysis of different European raspberries, blackberries, blueberries, currants and strawberries. Food Chem 239:48–55. https://doi.org/10.1016/j.foodchem.2017.06.023

Pintea A, Dulf FV, Bunea A, Socaci SA, Pop EA, Opriță VA, Giuffrida D, Cacciola F, Bartolomeo G, Mondello L (2020) Carotenoids, fatty acids, and volatile compounds in apricot cultivars from Romania—A chemometric approach. Antioxidants 9(7):562. https://doi.org/10.3390/antiox9070562

Roberts JJ, Cozzolino D (2016) An overview on the application of chemometrics in food science and technology—An approach to quantitative data analysis. Food Anal Methods 9(12):3258–3267. https://doi.org/10.1007/s12161-016-0574-7

Sayago A, González-Domínguez R, Beltrán R, Fernández-Recamales Á (2018) Combination of complementary data mining methods for geographical characterization of extra virgin olive oils based on mineral composition. Food Chem 261:42–50. https://doi.org/10.1016/j.foodchem.2018.04.019

Sayago A, González-Domínguez R, Urbano J, Fernández-Recamales Á (2019) Combination of vintage and new-fashioned analytical approaches for varietal and geographical traceability of olive oils. LWT 111:99–104. https://doi.org/10.1016/j.lwt.2019.05.009

Tabbassum M, Zeeshan F, Low KH (2021) Discrimination and recognition of Bentong ginger based on multi-elemental fingerprints and chemometrics. Food Anal Methods:1–10. https://doi.org/10.1007/s12161-021-02167-1

World Health Organization (2005) Fruit and vegetables for health: Report of a Joint FAO/WHO Workshop, September 1–3, 2004, Kobe, Japan. FAO/WHO

Yu P, Low MY, Zhou W. (2018) Design of experiments and regression modelling in food flavour and sensory analysis: A review. Trends Food Sci Technol 71:202–215. https://doi.org/10.1016/j.tifs.2017.11.013

Zielinski AAF, Haminiuk CWI, Nunes CA, Schnitzler E, van Ruth SM, Granato D (2014) Chemical composition, sensory properties, provenance, and bioactivity of fruit juices as assessed by chemometrics: A critical review and guideline. Compr Rev Food Sci Food Saf 13(3):300–316. https://doi.org/10.1111/1541-4337.12060

7 Authenticity and Chemometrics of Fruit Juices

David Psomiadis and Nikoleta Zisi

CONTENTS

7.1 FOOD FRAUD AND FOOD AUTHENTICITY

Food fraud poses a significant threat to the food industry. In today's complex global supply chains, the issue of brand reputation and protection has been in focus more than ever. Food fraud should not be viewed solely as a quality assurance problem for individual companies but as a global influencing factor for the entire food industry, shaping decisions, values, brand reputations and corporate strategies. Food fraud incidents have industry-wide impacts with potential to alter consumer behavior toward all brands (Meerza and Gustafson, 2019). In this environment, the food industry should look for innovative ways to communicate the principles of integrity to the consumer, and analytical authentication solutions have a definite role in this space. Verifying food authenticity offers an opportunity to prevent consumer distrust of food product claims.

In food authentication, we must first identify the product characteristics that are at food fraud risk. A critical and usually difficult point to address is defining an appropriate, available analytical method that satisfies the following questions:

- Is there really a suitable method, and if so, which is it?
- Is it official, accredited, globally recognized and properly applied?
- Is the detection limit adequate to identify adulteration?

When we talk about a testing program, it is of utmost importance to accurately define where, when, how often and for how long the analytical testing scheme should be carried out. Finally, the testing program can only be truly effective when it utilizes a wide spectrum of state-of-the-art analytical methods.

Accreditation is mandatory to ensure harmonization, credibility and integrity, as well as international recognition of the findings. Food authenticity testing does not always end with a specific qualitative or quantitative result; novel fingerprinting or screening methods are more like forensic investigations entailing studies that must point to a result—a conclusion based on evidence from the analysis that can support or reject a hypothesis. In recent years, new versions of food certification standards, following the guidelines of the Global Food Safety Initiative (GFSI), have included a comprehensive requirement for food fraud risk assessment in the food industry. Companies are required to deliver a risk mitigation plan, taking measures against the risk of food fraud in their supplies. In this approach, analytical testing is a key feature that leaves little space for assumptions and decreases the risk of food fraud incidents significantly. Food authenticity testing initially dealt with basic issues using the available technology. Today, however, we are equipped with a range of techniques, including both targeted and non-targeted analyses of food products. The need for low detection limits, as well as detection of sophisticated practices of adulteration, drives

DOI: 10.1201/9781003266167-7

us to develop compound- and subject-specific methods that can provide strong evidence for or against a claim in complex samples. This reflects the complexity of most supply chains, since most products comprise hundreds of major or minor ingredients, where only some or just one carries a premium claim and could be adulterated.

7.2 GENERAL MARKET FRAMEWORK OF FRUIT JUICES

With a global consumption volume of 44.1 billion liters in 2020 (Imarc, 2020) and an expected CAGR of 2.1% for the period 2022–2027 (EMR, 2020), fruit juices (100% fruit) and nectars (25%–99% fruit) are an important sector of the food industry. Asia Pacific, North America and Europe are the largest markets by volume consumption (AIJN, 2019). Regarding consumption per capita, North America has the highest volume (22.7 liters per person), followed by Western Europe (18.9 liters per person) and Eastern Europe (10.6 liters per person; AIJN, 2019).

In the EU, four countries account for over 60% of the total fruit juice and nectar consumption, with Germany coming out on top, followed by France, Poland and Spain. One of the key drivers of fruit juice consumption has always been its nutritional value, as a natural product high in nutrients such as vitamins, antioxidants and others. However, there are recent discussions over the consumption of juice products due to their relatively high content of sugar, which is one of the reasons responsible for the decline in juice, and particularly nectar, consumption.

The fruit juice market is extremely important to the world economy. Despite their narrow margins, the varied range of products spans a variety of categories that contribute both directly and indirectly to the value chain. Despite changing market dynamics and consumer mood, the market is expected to reach revenues of approximately US$695.6 billion by 2025 (Reportlinker, 2019), with the United States expected to retain a 4.9% growth rate. The market in Europe reached total revenues of US$160.7 billion in 2017, showing a 3.2% CAGR from 2013 to 2017 (Marketline, 2018), and the market in Japan is expected to reach US$37.6 billion in the near future. China has the potential to grow at an annual rate of 8.5% over the next few years, adding roughly US$121.9 billion in addressable opportunity (Reportlinker, 2019). The fruit juice industry, as one of the best-performing food and beverage businesses, plays an important role in the global economy across all markets (including developed, emerging and developing). It is seen as a promising business sector for the coming decades due to its long-term growth potential. According to the American Beverage Association (2019; www.ameribev.org), the beverage sector has a direct economic impact of US$182.6 billion in the United States.

Although orange and apple juices continue to be the most popular, the number and variety of fruit juices available on the market has changed dramatically in recent decades, with the constant arrival of new exotic fruit kinds, each claiming new health benefits. Other fruit juice kinds, such as pomegranate or berry-based drinks, have developed a high reputation and are being promoted as high-quality food items due to their outstanding health benefits, including protection of cancer and cardiovascular disease (Twohig et al., 2011). Almost half (49%) of 18- to 34-year-old consumers are motivated by curiosity to try new flavors and fragrances (Rinke and Jamin, 2018). This identifies young consumers as the most willing to experiment and try new products. Sweet complexions will likely catch the attention of young adults, as 42% of 18- to 34-year-olds are attracted by sweet flavored soft drinks—more than any other taste (AIJN, 2019). To combat the decline in juice consumption, efforts have centered on the NFC (not-from-concentrate) and smoothies sectors, with lower-sugar fruit and vegetable combinations and the addition of functional components like proteins. All of these variables, together with the industry's evolving landscape, make assuring the authenticity of what is accessible to consumers a continuing challenge for both the industry and authorities (Rinke and Jamin, 2018).

7.3 PRODUCT CHARACTERISTICS AND PRODUCTION PROCESSES

The processing method used to get the product from its growing location to the supermarket shelf, as well as the unique restrictions in force in the country where it is marketed, influence the type of fruit juice available on the market. Fruit juice is sold as either natural strength juice or purée, or as concentrated fruit juice or purée that has had the water removed. The latter significantly minimizes the amount of storage space required, as well as transportation costs. The concentrate is kept at extremely low temperatures or in aseptic drums and shipped in bulk from the manufacturing region to the main markets, where it is reconstituted into single-strength juice by adding water. After that, the juice is pasteurized, bottled and labeled as "100% fruit juice from concentrate" or "100% fruit juice from concentrated fruit juice" (Rinke and Jamin, 2018).

Natural fruit juice is extracted directly from the fruit, pasteurized and bottled for sale, or stored in sterile containers at low temperatures for later packing. It's marketed as "100% fruit juice" and "direct juice" on occasion.

In Europe, fruit nectars are also popular. Fruit juices (ranging from 25% to 90% juice content depending on the fruit type), water and sugar are combined to make these drinks.

Fruit purées and pulps are good raw materials for soft fruits that are susceptible to physical damage during transportation, such as strawberries and raspberries. Fruit juice mixes, beverages and nectars, as well as jam and marmalade, all contain such ingredients.

7.4 AUTHENTICITY AND FRAUD OF FRUIT JUICES

Fruit juices, and especially orange and apple juice, belong to the most targeted food commodities for adulteration and fraud (Moore et al., 2012). Given the huge quantities of fruit juice used worldwide, detecting juice adulteration and fraud is critical, and robust control measures are required to safeguard consumers against impure and fraudulently presented fruit liquids (Jandric et al., 2017). To highlight the vulnerability of fruit juices and their category to fraudulent activities, Syposs and Knight (2021) presented the top emerging hazards in nonalcoholic ready-to-drink products during 2017–2018 (Figure 7.1). The data indicate that fraud is topping the list of hazards in this category, while other common hazards such as allergens or foreign bodies fall behind.

The most common adulteration procedures in the fruit juice industry are dilution with water; the addition of sugars, pulp wash or other additives; and juice-to-juice adulteration, which is defined as the unreported addition of a juice of lower value to a product. All of these types of adulteration devalue the product for the user and may even endanger their health by producing allergic responses (Jandric et al., 2014). As a result, regulatory agencies and industry have developed quality criteria and identification standards to examine the quality and authenticity of fruit juices. The composition, reserved names, manufacture, and labeling aspects of fruit juices, concentrated fruit juices, dried fruit juices and fruit nectars are all subject to particular Community standards under Directive 2012/12/EU, which amends Council Directive 2001/112/EC. Furthermore, AIJN has produced the AIJN Code of Practices Reference Guidelines, which are based on pure, authentic juices and can be used to evaluate juices for quality, authenticity and identity (AIJN). The variety of adulteration tactics, the natural variation of fruits from various geographical and varietal origins, and the various storage and processing procedures make detecting and preventing juice adulteration a difficult operation. In this regard, effective, dependable and quick food authentication procedures are an invaluable instrument for authorities setting up control systems to ensure juice quality and safety while also promoting fraud mitigation.

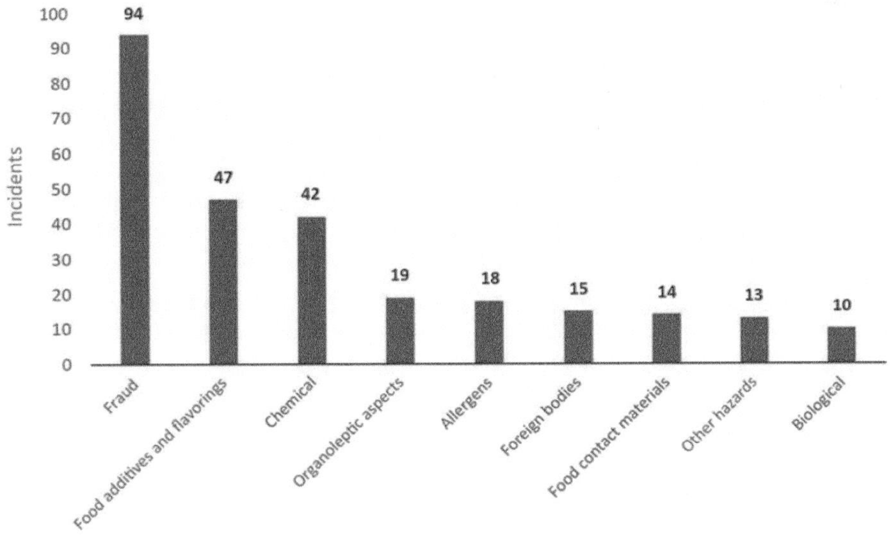

Figure 7.1 Top emerging hazards in nonalcoholic ready-to-drink beverages (2017–2018).
Sources: Foodakai (2019); Syposs and Knight (2021).

A comprehensive list of the main authenticity issues of fruit juices is given in the Food Integrity Handbook (Rinke and Jamin, 2018), as below:

- *Water addition*: Fruit juices have a natural fluctuation in the ratio of soluble dry matter to water. However, even if the defined minimum density or Brix is lower than any naturally derived product not from concentrate, water cannot be added. In addition, for reconstituted juice from concentrate, the amount of water supplied to reconstitute the juice must be the same as that removed during concentration, according to the regulation. As previously stated, the AIJN has established minimum density requirements and the accompanying Brix value, which is a measure of the soluble solids content. Dilution of a concentrate below the authorized minimum Brix is one of the most basic forms of adulteration.
- *Sugar addition*: Sugar is a far cheaper commodity than fruit juice, thus adding it to the latter to raise Brix values can be a cost-effective strategy. Sugar can be added in the form of beet, cane or corn sugars, as well as modified sugar syrups such HFCS (high-fructose corn syrup) or BMIS (beet medium invert syrup [beet, cane or corn syrup]).
- *Complete or partial replacement of juice by juices made from concentrate*: In several countries, customer preference for freshly squeezed or NFC juices has altered in recent years, resulting in a larger premium for these goods. Reconstituted juices may not be passed off as NFC, nor may a quantity of water or reconstituted juice be added to the direct product.
- *Added products from undeclared cheaper fruits*: Poor harvests, gluts and trade laws can cause fruit prices to change dramatically from one season to the next. Another prevalent kind of adulteration is the insertion of a cheaper fruit substitute to stretch one that is in low supply and/or high demand. Orange to passion fruit, apple to red fruit, and grape to pomegranate are just a few examples.
- *Addition of undeclared ascorbic acid/vitamin C*: Some fruit varieties are highly enriched in vitamin C, and this is marketed as a benefit. This results in interest to increase vitamin C content in poor juices by adding exogenous vitamin C, often from synthetic sources.
- *Addition of undeclared organic acids (e.g., citric acid, malic acid)*: In some cases, the acidity of a juice can be modified by increasing organic acids within the legal limitations, as long as the product is labeled appropriately. It is not permissible to use specific procedures such as adding malic acid to apple juice.
- *Addition of flavor compounds (natural or synthetic)*: When a product claiming to be "natural" has flavors that are not from the identified fruit or have been artificially synthesized, authenticity issues arise.
- *Colorings (e.g., anthocyanin extracts, cochineal red, beetroot)*: The color of fruit juices is a significant component of their attraction. Colorings may be added to meet color intensity criteria, to restore natural color lost during processing or storage, or to darken color when a less pigmented fruit type has been introduced to the legitimate product.
- *Addition or over-proportional use of fruit extracts that were produced by non-authorized technology (water extractable solids)*: The substance recovered after washing the leftover pulp and cell membrane material after extracting orange juice is known as water extractable solids. Its excessive use in juice is prohibited by the EU's legal definition of citrus fruits.
- *Texture influencing agents (e.g., pectins)*: Texturizers can improve the body and mouthfeel of liquids that aren't entirely made of fruit. This is only allowed for certain fruit kinds, and only if the compounds are listed on the product label.
- *Declaration of wrong origin*: If the country of origin of a fruit product is incorrectly claimed on the juice label or in the product's trade standards, it might constitute an authenticity issue. If the fruit origin in question is subject to favorable import taxes, this could be a source of worry for customs and excise authorities. On the market, certain geographical origins may also command a premium price.
- *Declaration of wrong fruit variety*: This could be a problem if a particular variety is favored for its flavor or processing characteristics.

The extent of vulnerability within this commodity group can be illustrated by a study presented by Syposs and Knight (2021) into the many varieties of food fraud. Adulteration of juice concentrates by adding HFCS and/or a variety of lower economic value juice blends, and the sale of product after the "best before" period with the dates manipulated are just a few examples of the product category's vulnerabilities. The research was based on publicly available data on food safety and fraud incidences from a variety of globally recognized food safety authorities and networks. Sources include the database of the European Commission's Rapid Alert System for Food and Feed (RASFF); regulatory and statutory authorities like the US Food and Drug Administration (FDA); the Japanese Ministry of Health and Food Standards Australia New Zealand (FSANZ); and professional portals like "Food Safety News" and "Food Safety Tech". Fraud was recognized as one of the top three food safety incident categories connected with fruit beverages between 2008 and 2018. During 2017–2018, the fruit juice business saw a dramatic spike in reported product fraud incidents. In comparison to 2008, there was a 200% rise in the number of reported events. The majority (almost 40%) of hazardous incidents reported in 2017–2018 involved fraud, and fraud was classified as the top incident category, with 94 cases worldwide.

Lord et al. (2017) highlighted the vulnerability of fruit beverages to fraud in their study of product tampering in Europe, where they attempted to better understand the scope of the crime, the perpetrators, and the skills, resources and knowledge involved. They concluded that market fraudsters were able to take advantage of an otherwise legitimate business transaction and an otherwise legitimate social/economic relationship in the food system (Lord et al., 2017) because they were aware of the ability to circumvent legally prescribed procedures in order to gain personal and/or organizational advantage. The lack of a regulatory structure and monitoring for fruit beverage fraud, as well as the lack of contractual duties and industry self-regulation in this market, gave enough possibilities for the fraud to take place with minimal danger of detection.

7.5 FRUIT JUICE AUTHENTICITY TESTING AND ANALYTICAL TECHNIQUES

Recently, Xu et al. (2021) analyzed 434 published papers related to fruit juice authenticity during 1996–2020 (Figure 7.2). In general, targeted analyses, which focus on the detection and identification of a specific compound or class of compounds, are the most commonly used analytical procedures (e.g., amino acids, organic acids, sugars). These components may act as biomarkers for genuine items or as indicators of adulteration (Navarro-Pascual-Ahuir, 2017). When paired with chemometrics, the targeted technique can be utilized to discover typical patterns of chemical parameters indicating sample adulteration or varietal and geographical origin (Borràs et al., 2015; Dasenaki and Thomaidis, 2019).

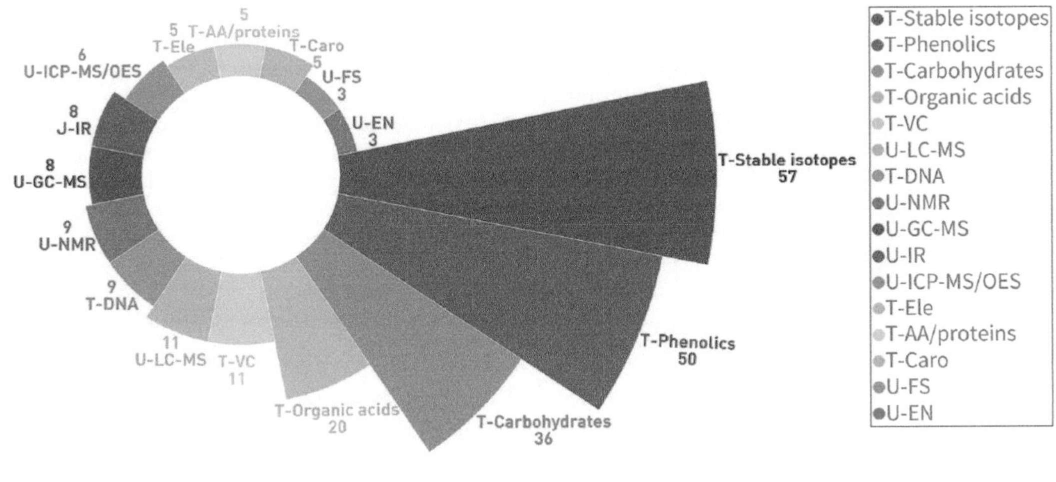

Figure 7.2 Statistics of the papers about fruit juice authenticity from 1996 to 2020. *Abbreviations*: AA, amino acids; Caro, carotenoids; Ele, elements; EN, electronic nose; FS, fluorescence spectra; T, targeted; U, untargeted; VC, volatile components.

Source: Xu et al. (2021).

Untargeted approaches are becoming increasingly common in this context, as researchers are required to examine more broadly to gain an integrated overview of the chemical composition of the material under investigation. Untargeted approaches are based on a holistic approach with the goal of identifying compounds from samples with unknown molecular content (Knolhoff and Croley, 2016). To offer the essential quality and authenticity information, advanced statistical algorithms and multivariate data analysis are applied. Metabolomics is a recent advanced analytical platform that determines and characterizes the whole composition of metabolites in a system or organism. Metabolites are thought to be the genome's ultimate products, which are modified by the genome's interaction with the environment, and hence the metabolic fingerprint is unique to each species (Alañón et al., 2015). The tremendous diversity of plant metabolite chemical structures, particularly in juices, makes metabolomics a difficult endeavor. Metabolomic studies have gotten a lot of attention in recent years because they provide useful information and unique analytical tools for ensuring juice authenticity and quality (Castro-Puyana et al., 2017; Cubero-Leon et al., 2014). Several review articles have been published on the evolution and application of novel analytical methodologies, as well as advanced statistical and mathematical tools, to assess food authenticity (Dasenaki and Thomaidis, 2019).

7.5.1 Physicochemical Parameters

Traditionally, physicochemical characteristics such as total soluble solids (TSS), titratable acidity (TA), pH, L-ascorbic acid content, formol number and ash have been used to verify the legitimacy of juice products. The International Federation of Fruit Juice Producers (IFU) recommends refractometry, titrimetry, gravimetry and potentiometry (Codex Alimentarius, 2005) as the most common methods of analysis. This method is often used by quality control labs since it is a quick and inexpensive way to determine the authenticity of a fruit juice. The results are contrasted to those given by the AIJN, which were calculated in 100% pure juices and are widely acknowledged by producers, users and regulatory authorities around the world.

Physicochemical characteristics have been determined to exploit probable adulteration or to classify juices according to their geographical and varietal origin (Maireva et al., 2013; Cautela et al., 2008; Voldrich et al., 2002; Lorente et al., 2014; Türkyılmaz, 2013; Dzugan et al., 2018). Brix value can reveal water and sugar addition to juice samples: a value below the necessary specification indicates over-dilution of the juice with water, whereas a higher Brix value shows the addition of sugar (Maireva et al., 2013). Different types of pomegranate have shown significant changes in pH, TA, and Brix values (Türkyılmaz, 2013), as well as between homemade and commercial samples (Dzugan et al., 2018). In addition, the percent maturity index (ratio of SSC to TA value) was determined, which is a very good indication of fruit maturity and, as a result, of juice flavor. Pomegranate varietals can be classed as sour, sour-sweet or sweet based on their maturity index (Türkyılmaz, 2013).

The assessment of these physicochemical properties can aid in the understanding of unique juice qualities and, ultimately, the establishment of reference recommendations. Lorente et al. (2014) presented a detailed analysis of commercial Spanish lemon juices, based on the determination of physicochemical properties in 92 samples. Mena et al. undertook phytochemical characterization of Spanish pomegranate juices from 15 different cultivars, using sophisticated chemometrics to identify the samples. The first three principal components may explain 74.3% of the total variation using principal component Analysis (PCA; Mena et al., 2011). However, the standard analytical procedures for determining these parameters are frequently inadequate to identify minor variations that could suggest low-level adulteration. Juice-to-juice adulteration may go undetected since the compositional variations between juices with similar features (e.g., color, aroma) aren't substantial enough to identify the fraudulent addition of one juice to another (Cautela et al., 2008).

7.5.2 Stable Isotope Ratios Analysis (SIRA)

Because their ratios alter with climatic circumstances, geographical origin, soil pedology and geology of the areas of food ingredient origin, and fruit evaporation intensity, stable isotopes are useful for food authenticity. It's tough to get around since altering the isotope ratios isn't cost-effective (Bong et al., 2008; Kelly et al., 2003). Isotopic methods, which are based on the determination of isotope ratios (δ^2H, $\delta^{13}C$, $\delta^{18}O$, $\delta^{15}N$, $\delta^{34}S$, and $\delta^{87}Sr$), are among the most effective techniques for determining the authenticity of fruit juice (Carter and Chesson, 2017) (Table 7.1). This explains why most AIJN reference guidelines for fruit juices suggest several isotopic parameters and thresholds in their authenticity criteria section. Isotope ratio mass spectrometry (IRMS) and site-specific

Table 7.1 List of Official Methods Based on Stable Isotope Ratio Analysis for Fruit Juice Authentication

Method	Technique	Isotope Ratios	Detection
AOAC Official Method 995.17	SNIF-NMR	(D/H)I, (D/H)II	C3-sugar addition
AOAC Official Method 2004.01	IRMS	$\delta^{13}C_{ethanol}$	C4-sugar addition
CEN (TC174 N108, ENV 12140)	IRMS	$\delta^{13}C_{sugars}$	C4(or C3)-sugar addition
CEN (TC174 N108, ENV 12141)	IRMS	$\delta^{18}O_{water}$	water addition
CEN (TC174 N108, ENV 13070)	IRMS	$\delta^{13}C_{pulp}$	C4-sugar addition

natural isotope fractionation–nuclear magnetic resonance (SNIF-NMR) are the most advanced and specific procedures for determining isotope ratios. Isotopic methods are based on the fact that natural products have measurable changes in specific proportions of various isotopes depending on their botanical origin, climatic and geographical conditions, and production processes (Jha et al., 2016; Rummel et al., 2010; Ogrinc et al., 2009; Bizjak Bat et al., 2016). Stable isotope analysis can also be used to detect additional sugar and water in juice, as well as to distinguish between directly pressed juice and juice from concentrates (Primrose et al., 2010; Schmutzer et al., 2016). $\delta^{13}C$ ratios, for example, can indicate the addition of sugars and sweeteners from C4 plants such as cane or corn in C3-plant juices, whereas site-specific (D/H) ratios in ethanol generated from sugar fermentation can be used to detect added beet sugar in C3-plant juices (Primrose et al., 2010). The $\delta^{18}O$ ratio (and occasionally $\delta^{15}N$), which can also disclose the illegal addition of pulp wash, can reveal the addition of tap water, as well as the geographical origin. $\delta^{15}N$ levels are impacted by soil nitrogen, which comes primarily from fertilizers and manure (Rummel et al., 2010).

Ogrinc et al. (2009) proposed a complementary application of IRMS and SNIF-NMR for the characterization of Slovenian and Cypriot fruit juices in 2009. IRMS was used to determine the values of $\delta^{13}C$ in sugars, pulp and ethanol (obtained from the fermentation of the juice), as well as $\delta^{18}O$ in water and (D/H)I and (D/H)II ratios by SNIF-NMR. In 40% of the juice samples from Cyprus and numerous Slovenian samples, undeclared cane sugar and beet sugar were found. Because of their varied precipitation regimes and climates, the stable isotope data were statistically assessed using PCA, and clear separation between juices from the two nations was accomplished. The identification of the botanical origin of fruit juices, on the other hand, proved to be more difficult. The isotopic and elemental characterization of 64 apple juice samples was used in a more recent investigation by the same group to distinguish between apple juices from different geographical regions of Slovenia and distinct apple cultivars (Idared, Golden Delicious, and Topaz; Bizjak Bat et al., 2016). When both isotope and elemental data were combined (12 variables: pulp $\delta^{15}N$ and $\delta^{13}C$, water $\delta^{2}H$ and $\delta^{18}O$, ethanol (D/H)I and (D/H)II, and S, Cl, Fe, Cu, Zn and Sr), an overall prediction ability of 83.9% for the geographical origin and 75.8% for the cultivars was attained. Rummel et al. (2010) evaluated the applicability of sulfur- and strontium-isotope analysis in determining the geographical origin of orange juice samples. Pulp $\delta^{2}H$, $\delta^{13}C$, $\delta^{15}N$, $\delta^{34}S$ and $^{87}Sr/^{86}Sr$ were measured in approximately 150 samples obtained from various orange cultivating regions around the world. Using all pulp isotope ratios as variables, discriminant analysis (DA) was used, and 88.5% of the samples were properly identified as samples from Mediterranean countries, South America, and Central America, with Florida having the highest discriminatory characteristics of $\delta^{2}H$ and $^{87}Sr/^{86}Sr$. The results of the $\delta^{34}S$ study can be utilized to distinguish between various orange-growing regions, such as Cuba and Mexico. Furthermore, Sr-isotope analysis has been shown to be a useful approach for detecting the unintentional addition of concentrate to single-strength juices (Rummel et al., 2010).

To distinguish between authentic and non-authentic orange and lemon juices, AIJN has established threshold limits for several isotopes ($\delta^{18}O_{water}$, (D/H)I, $\delta^{13}C_{sugars}$, $\delta^{13}C_{ethanol}$, $\delta^{13}C_{pulp}$, and $\delta^{13}C_{acids}$). The actual values measured in juices can vary significantly depending on geographical origin; accordingly, Bontempo et al. (2014) recommend revising AIJN reference values, pointing out that AIJN thresholds are not always fully applicable to samples of Italian origin, as evidenced by a thorough isotope characterization of more than 500 samples of Italian citrus juices.

The LC-IRMS approach has recently been applied in authenticity evaluation research, revealing its tremendous potential in this sector (Bononi et al., 2016; Guyon et al., 2014; Psomiadis et al., 2018). A liquid chromatograph is directly connected to a liquid oxidation interface (based on the wet oxidation of organic molecules in aqueous solution and the formation of CO_2), which is then connected to an isotope ratio mass spectrometer (Godin and McCullagh, 2011). With the invention of LC-IRMS, it was possible to determine the $\delta^{13}C$ of individual compounds directly from complex mixtures without having to go through any purification stages. Bononi et al. (2016) proposed the use of LC-IRMS to determine four parameters that relate the $\delta^{13}C$ values of citric acid, tartaric acid, malic acid, glucose and fructose to detect the undeclared addition of exogenous organic acids and sugars in lemon juice. Exogenous elements were found in significant quantities in Italian market juice samples. Psomiadis et al. (2018) used LC-IRMS to illustrate the significant improvement of the detection limit of exogenous sugars in coconut waters, allowing the method performance to match real levels of adulteration on the market.

7.5.3 Nuclear Magnetic Resonance (NMR)

For metabolomic research and food fingerprinting, NMR spectroscopy is one of the most commonly utilized analytical techniques (Esslinger et al., 2014). It entails examining the energy absorption of atomic nuclei with non-zero spins in the presence of a magnetic field. The energy absorptions of atomic nuclei are impacted by the nuclei of surrounding molecules, resulting in minor changes in the external magnetic field (Jha et al., 2016). NMR spectroscopy is a particularly powerful tool for structure identification since it can provide comprehensive information about a food sample's molecular structure, as well as high-throughput spectroscopic/structural information of the metabolites present in the sample (Sobolev et al., 2015). Its primary benefits include its non-invasive nature, great accuracy and precision, ease of use and quick data gathering. In comparison to other analytical techniques (FT-IR or MS), NMR spectroscopy has a poor sensitivity, and the instrumentation is relatively expensive with high operating expenses (Esslinger et al., 2014). Because of the increased sensitivity and shorter relaxation durations, most NMR studies use ^{1}H data rather than ^{13}C. Because it can provide more comprehensive information about the molecular structure of food metabolites, high-resolution NMR (frequencies more than 100 MHz) has been more widely used in food authentication than low-resolution NMR (frequencies of 10–40 MHz; Reid et al., 2006). Food NMR spectra are quite complex, having multiple signals linked to main and secondary metabolites. As a result, in most circumstances, multivariate analysis is required to extract the relevant data (Alañón et al., 2015). In 2015, Sobolev et al. evaluated the use of NMR methods for fruit and fruit juice characterization and reported on many metabolites discovered using untargeted NMR methodologies. Moreover, in recent years, several additional investigations utilizing NMR spectroscopy to determine the validity of fruit juice have been done (Spraul et al., 2009a; Vigneau and Thomas, 2012; Clausen et al., 2011; Koda et al., 2012; De Oliveira et al., 2014; Grandizoli et al., 2014; Longobardi et al., 2013).

Bruker BioSpin GmbH and SGF International e.V. developed a screening fruit juice profiling approach utilizing ^{1}H-NMR, which has been available since 2009 under the name Spin Generated Fingerprint (SGF) Profiling (Spraul et al., 2009a). Proton-NMR juice screening SGF Profiling (Rinke et al., 2007; Spraul et al., 2009a; Spraul et al., 2009b) is utilized for authenticity control as a powerful fingerprint method that allows for the screening of a wide spectrum of suspected adulterations. It's a non-targeted metabolomics application with a high throughput of quantification. The presence and quantity of natural fruit compounds with unknown chemical structures are employed in the same way as signals from characterized compounds are. Several factors are statistically evaluated and quantitatively determined using a single 400 MHz NMR experiment. The analysis time per sample is roughly 15 minutes, and the process can be automated to allow for a high sample throughput.

Centrifugation and diluting a juice or a concentrate with a buffer containing TSP (sodium salt of 3-(trimethylsilyl)-propionate acid-d4) as an internal standard and sodium azide for preservation are the steps involved in sample preparation. Depending on the variety of fruit, the pH of the buffer solution is 3.1 or 3.4. Due to the pH sensitivity of the chemical shifts of certain of the polar compounds, the use of these specialized buffer solutions is required to provide the required repeatability (e.g., organic acids). An automatic data treatment can involve baseline and phase corrections. The entire instrument arrangement, including the flow cell and individual tubes, can be built as a push-button system.

J-resolved NMR spectroscopy (JRES) is a typical procedure that allows for better signal assignment to molecular structures that show interferences with other compounds in the one-dimensional spectrum along the chemical shift axis (Ludwig and Viant, 2010; Rinke et al., 2007). Chemical shifts and J-coupling are separated in JRES spectra by suppressing J-coupling; both are projected on two different axes, resulting in a two-dimensional spectrum.

It is advised that additional analytical choices for juice authentication be based on the results of this screening for those products for which SGF Profiling models exist. It makes use of an accurate reference material spectral database. Different verification and classification models are used independently. In parallel, a number of analytical aspects such as product kind, origin and combination with other types are verified. It is also feasible to assess the fruit content in a sample for some fruits. However, not all items can be analyzed in the same way. The analytical possibilities and accessible models may differ depending on the type of fruit. With the growing number of reference samples collected, new and improved models are always being developed. Longobardi et al. (2013) used untargeted ^{1}H-NMR fingerprinting to characterize the geographical provenance of Italian sweet cherry juices. PCA was first used to distinguish samples from two different Italian areas (Emilia Romagna and Puglia). However, the first three principle components could only explain 21% of the data, hence PCA was deemed unable to identify samples from different origins. PLS-DA and LDA, on the other hand, were able to correctly identify cherry juices from the two origins using an external validation technique, with PLS-DA models demonstrating prediction abilities of 94.9% and 92.3%–94.9% for LDA models. Vigneau and Thomas (2012) published another investigation on the detection of juice-to-juice adulteration. Using ^{1}H-NMR spectroscopic profiling and PLSR, the classification of authentic orange juices and orange juices adulterated with clementine juice was examined. Various preprocessing tactics and variable selection procedures were examined, with logarithm and Pareto scaling, based on backward interval PLS or genetic algorithm, yielding the lowest error rate. Using ^{1}H-NMR spectra and a PLS-DA model with a prediction ability of more than 82%, several clones and cultivars of sour cherry juices were distinguished (De Oliveira et al., 2014; Grandizoli et al., 2014). Furthermore, the link between NMR data and sensory analysis revealed that the amount of malic acid in juices had a substantial impact on whether they were classified as sour or sweet. The presence of glucose had only a little impact on these characteristics, while fructose had no effect at all (Clausen et al., 2011). Nonetheless, this study has drawbacks, mostly due to the small number of samples (seven in total, one from each cultivar), which reduces the model's validity. De Oliveira et al. (2014) used specific metabolites and PCA to differentiate orange juice varieties from five different orange varieties: Pêra Rio, Bahia, Murcote, Lima and Persian Lima. The main goal of De Oliveira's research was to improve the quality of orange juices by determining and evaluating the principal chemical ingredients using NMR under various storage time and temperature settings. It was discovered that storing formic, fumaric, succinic, acetic and lactic acids for up to 24 hours resulted in the formation of formic, fumaric, succinic, acetic and lactic acids, as well as the appearance of ethanol, which was not identified at 0 hours.

Despite its growing popularity, proton-NMR screening is not yet recognized as an official procedure. In some laboratories, however, all measurements and statistical procedures are accredited according to ISO 17025.

7.5.4 Chromatographic and Separation Techniques

Liquid chromatography (LC), gas chromatography (GC), and thin layer chromatography (TLC) are widely used in food authenticity studies because they can provide rapid and reliable separation of molecules with very similar chemical characteristics, even in a complex matrix like food. They can be linked to a variety of detectors to detect and identify analytes, with mass spectrometric detection being the most common due to its superior compound identification capabilities (Dasenaki and Thomaidis, 2019).

7.5.4.1 Thin Layer Chromatography

Thin layer chromatography (TLC) is a simple, quick and cost-effective chromatographic technology that has been frequently employed in food analysis for decades. Its key advantages over column chromatography are its simplicity, low sample preparation and solvent consumption requirements, and the ability to analyze numerous samples simultaneously side by side. TLC has been used in many studies for fruit juice analysis (Hosu and Cimpoiu, 2017; Sherma and Rabel, 2018). The applications of TLC in fruit juice authentication are however less common. Based on anthocyanin and anthocyanidin fingerprints, Filip et al. (2012) used TLC densitometry to develop berry juice authentication. The procedure was successfully applied to juice samples collected from Romanian markets in order to ensure that they were in compliance with the declared label. The anthocyane patterns obtained were comparable for juices from the same plant source, but strikingly varied among plant sources (e.g., blackcurrant juice and elderberry juice were obviously distinct; Krüger et al., 2013).

7.5.4.2 *Gas and Liquid Chromatography–Mass Spectrometry*

In the field of food analysis, chromatography is the most extensively utilized separation technology. Its key advantage is that it can separate the parts of a complicated sample and identify them when used in conjunction with a powerful detection technique like mass spectrometry (MS; Dasenaki and Thomaidis, 2019). Gas chromatography (GC) is good for determining volatile and semi-volatile substances, whereas liquid chromatography (LC) is more adaptable and can be used to determine a wider range of compounds (Borràs et al., 2015).

Due to its remarkable sensitivity and specificity, MS has emerged as the leading technique in food authenticity research during the last decade. The extreme complexity and wide concentration range of food samples provide significant obstacles for any novel analytical technology, and the use of high-resolution and ultra-high-resolution mass spectrometers has greatly improved analytical performance (Ibáñez et al., 2013). Several fruit juice authenticity investigations have been published utilizing GC and LC, as well as MS detection and various detectors such as the flame ionization detector (FID), diode array detector (DAD), refractive index detector (RI) and fluorescence detector (FLD).

7.5.4.2.1 Gas Chromatography

The volatile component profile of fruit juices is determined using GC-MS approaches in the vast majority of fruit juice authenticity investigations, while GC procedures employing traditional (non-MS) detectors are rare. In order to distinguish the geographical origin, varietal origin, and probable adulteration of apple, orange, pomegranate, pear, lemon, and shiikuwasha juices, both targeted and untargeted GC-MS methods were used.

For the identification and quantification of specific groups of juice volatile compounds, targeted GC methodologies, primarily headspace–solid phase microextraction (HS-SPME) as extraction method, have been used (Schmutzer et al., 2016; Nuncio-Jáuregui et al., 2014; Yamamoto et al., 2012; Willems and Low, 2014). Various compounds in these classes have been identified as markers of juice-to-juice adulteration; for example, the presence of linalool and linalool oxide at concentrations greater than 3% and 0.10% in pomegranate juices indicates the addition of grape juice, whereas elevated concentrations of isoamyl butyrate, isobutyl butyrate, benzyl acetate and butyl acetate indicate the addition of peach juice (Nuncio-Jáuregui et al., 2014). Linalool and terpinene have also been found as shiikuwasha juice and calamondin juice detection indicators (Yamamoto et al., 2012; Yamamoto et al., 2013).

The oligosaccharide profile of pure and adulterated pear juices can be utilized to detect the inclusion of high-fructose corn syrup, total invert sugar, and hydrolyzed inulin syrup at low levels in legitimate pear juices, according to Willems and Low (Yamamoto et al., 2012). Using the juice-to-juice debasing of apple juice with pear juice, the capillary gas chromatography–flame ionization detection (CGC-FID) method for oligosaccharide identification can also be used for arbutin testing. In order to study the adulteration of organic with conventional lemon juice, GC-FID was also employed for an untargeted profile analysis of volatile organic compounds (VOCs) in lemon juice. A linear relationship was discovered between the percent of organic lemon juice and chromatogram features such as total peak area, peak height and peak number, indicating that this methodology might be used to detect organic lemon juice adulteration (Alizadeh et al., 2016).

Retention times variables were determined utilizing the chromatographic fingerprints of apple juice samples of six different types in an unique GC-MS untargeted technique (Guo et al., 2012). A stepwise linear discriminant analysis (SLDA) classification model was created and successfully used to distinguish apple juice kinds with 100% identification and prediction capabilities. SLDA chose ten retention periods that contributed significantly to variety differentiation, and these ten volatile chemicals were identified using GC-MS spectra. A similar approach was used to classify apple juices based on their geographical origin: SLDA was used to distinguish samples from four Chinese counties, with a recognition ability of 93.9% and a prediction ability of 89.8%. Butyl acetate, (E)-2-hexenal, hexyl 2-methylbutanoate and pentyl acetate have been found as informative markers for distinguishing apple juices by geographical origin (Guo et al., 2012). Wu et al. (2017) recently attempted to classify eight different apple juice varieties based on their volatile components (Fuji, Golden Delicious, Qin Guan Delicious, Pink Lady, Starkrimson, Jonagold, Gala, and Ralls Janet), and 44 volatile components were provisionally discovered. Schmutzer et al. (2016) developed a novel flavor index based on the total sesquiterpenes of a juice sample divided by total terpenes in this context (without limonene). A minimum of 40% is considered as a threshold value for classifying orange juices from concentrate as high or low value. The presence of benzoic acid and D-limonene in orange juice indicates the presence of food flavors (Bocharova et al., 2017), whereas total alcohols and total ketones can be used to distinguish freshly squeezed orange juices from commercial orange juices and orange nectars, as high concentrations of these compounds indicate sample processing and thermal treatment (Schmutzer et al., 2016). The volatile character of orange juices was also used in tandem with novel methodologies based on artificial neural networks (ANN) and aenetic algorithms to achieve varietal discrimination (GA). Individual markers (geranyl acetate for LDA and a-pinene and sabinene for the Kohonen network) that were able to discriminate the juices with a perfect match were identified when GA was paired with LDA and the Kohonen map (Rinaldi et al., 2009).

7.5.4.2.2 Liquid Chromatography

For fruit juice verification, MS hyphenated to liquid chromatography is emerging as a less expensive and more sensitive option. Chromatography's larger range of separation mechanisms makes it more ideal for developing profile approaches and expanding their applicability to a wider range of substances. LC-MS was used to identify the comprehensive metabolomic fingerprint of different fruit juices. High-pressure liquid chromatography (HPLC) and ultra-high-pressure liquid chromatography (UHPLC) have been widely employed in fruit juice authenticity tests since their application area is significantly broader than that of GC. For characterization, categorization, and authenticity evaluation, (U)HPLC has been used to separate and quantify fruit juice chemical constituents, primarily phenolic compounds, organic acids, sugars and amino acids (Dasenaki and Thomaidis, 2019). The growing employment of LC hyphenated with HRMS analyzers has resulted in a massive evolution of LC-MS–based authenticity techniques with suitable performance characteristics for high-throughput food authenticity applications (Ibáñez et al., 2013).

Obón et al. (2011) presented a polyphenol characterization study of red fruit and vegetable juices (strawberry, blueberry, red grape, red raspberry, blackcurrant, European cranberry, sour cherry, purple prickly pear, and purple carrot juice) using HPLC with simultaneous UV-Vis and fluorescence detection. The polyphenolic profiles of orange, mandarin, lemon, and grapefruit juices have also been intensively investigated with the goal of distinguishing them. Abad Garcia et al. published a study that used HPLC-DAD-ESI-MS/MS to identify 49 polyphenols in citrus juices from 18 different cultivars in Spain (Abad-García et al., 2014). Neo-hesperidin and naringin could be used as unique markers to trace adulteration of orange juice (Jaffa and Mosambi varieties) with Red Blush grapefruit juice at lowest adulteration level of 2% (Jandric et al., 2017) and naringin, neoeriocitrin, and neohesperidin can also be considered as indicative of the fraudulent addition of bergamot juice to authentic lemon juice (Cautela et al., 2008).

From one juice to another, the anthocyanin profile of pomegranate juice can also be used to detect adulteration: the anthocyanin profile of pomegranate juices, as assessed by HPLC-DAD, was very constant and was essentially unaffected by the fruit variety or geographic origin (Türkyilmaz, 2013). Borges and Crozier (2012) developed a simple methodology for pomegranate juice authenticity evaluation using HPLC-DAD-MS detection of red grape ingredients in a comparable investigation of pomegranate anthocyanins. The concentration levels of these components differed significantly between Chinese and non-Chinese juices, enabling for the geographical origin verification of pear juices (Willems and Low, 2018). The phenolic composition determination along with statistical data evaluation approaches has also been used to determine the geographical origin and varietal distinction of apple juices. Bizjak Bat et al. (2018) investigated the ability of regional and varietal differentiation of three apple cultivars from five different geographical locations in Slovenia by determining both primary and secondary metabolites of apple juices. Several flavanols and flavonols have been found as reliable markers for the geographical origin differentiation of juices, and an LDA model that included all of the determined components was able to distinguish the three apple cultivars studied (Bizjak Bat et al., 2018). Guo et al. reported similar results in 2013, when they used SLDA to achieve a suitable classification of apple juices based on cultivar and geographical origin, with a 98.3% and 91.2% prediction ability success rate, respectively (Guo et al., 2013). Finally, the neolignans isoamericanol A and isoamericanoic acid A, which can be used to distinguish fermented and freshly squeezed juices (Tolle et al., 2011), and 5,6- and 5,8-epoxycarotenoids, which can be used to estimate juice freshness (Dugo et al., 2009), have been confirmed to reveal the freshness and aging of fruit juices. Estimating the total amino acid value can identify different sorts of fruit juice adulteration; for example, a lower total amino acid concentration can indicate either dilution with water or the addition of sugar syrup (Asadpoor et al., 2014).

D'Orazio et al. (2008) consider the presence of D-amino acids in orange juices to be indicative of low-quality juices, because high-quality orange juices only contain L-amino acids, and

Nuncio-Jáuregui et al. (2014) have recently linked grape juice addition to pomegranate juice with a significant increase in proline concentration levels. Sugar profiles were employed in the same investigation to detect pomegranate juice adulteration with peach juice. Adulteration of pomegranate with peach juice, on the other hand, was detectable up to 10% (v/v), as sugar level increased dramatically. A successful classification of Merlin oranges from four different Greek regions was achieved using a combination of sugar profile, organic acids, and specific physicochemical parameters data (pH, titratable acidity, total dissolved solids [TDS] and electrical conductivity [EC]) with multivariate analysis of variance (MANOVA) and LDA (Nikolaou et al., 2017). Grape juice contains a large amount of tartaric acid, whereas apple juice contains a high amount of quinic acid. The addition of both juices in pomegranate juice can be recognized at levels as low as 5% based on their organic acid profile (Nuncio-Jáuregui et al., 2014; Zhang et al., 2009).

To produce massive data sets on thousands of analytes for which no prior knowledge exists, modern HRMS techniques (QTOF/MS, Orbitrap) are paired with UHPLC. Advanced data mining and data processing methods, as well as multivariate chemometric tools like PCA, LDA, OPLS-DA, and SVM, are required to process and evaluate these complicated data sets (Jandric et al., 2014). The combination of HRMS-untargeted analysis and chemometrics allows for the construction of discrimination models on an untargeted basis, followed by the identification of the specific marker molecules that have been found to contribute considerably to this discrimination (Jandric and Cannavan, 2017).

Borges et al. (2010) created a pomegranate juice polyphenolic fingerprint using HPLC-DAD-MS/MS and compared it to those obtained from blended drinks. Seventeen molecules were tentatively discovered in all 100% pomegranate juices using MS and MS/MS spectra, and they were semi-quantified as cyanidin-3-O-glucoside equivalents. Several untargeted authenticity studies have been reported with the goal of identifying new citrus juice authenticity markers and developing novel prediction models to evaluate authenticity and quality. The bulk of these approaches focus on the differentiation of citrus juices based on their HRMS metabolic profile and the detection of juice-to-juice adulteration (Vaclavik et al., 2011; Arbona et al., 2015), but there has also been work on the differentiation of varieties, origin and organic type of production (Jandric and Cannavan, 2017; Arbona et al., 2015).

Jandric et al. reported three significant studies demonstrating the successful discrimination of different fruit juices, as well as orange juice varietal and geographical origin categorization (Jandric et al., 2017; Jandric et al., 2014; Jandric and Cannavan, 2017). UHPLC-QTOF/MS was used to extract the metabolic profile of juices in both positive and negative ionization modes, and mass spectrometric data was analyzed using unsupervised and supervised pattern recognition approaches (PCA, OPLS-DA and PLS-DA). PCA was able to differentiate between pineapple, orange, grapefruit, apple, mandarin and pomelo juices (Jandric et al., 2017; Jandric et al., 2014), as well as oranges of various types (navel, lane late, Valencia and navelina) and geographical origins (Spain, Greece and Italy; Jandric and Cannavan, 2017). In negative ionization mode, discrimination was shown to be better explained, with higher model prediction values, and juice-to-juice adulteration could be identified at very low levels, as low as 1%. OPLS-DA and PLS-DA supervised methodologies were then utilized in all three investigations to assess the method's practicality and to identify the most influential indicators that contribute the most to juice classification. Vaclavik et al. (2011), Arbona et al. (2015), Cuevas et al. (2017), and Wang and Jablonski (2016) have all published untargeted metabolomic analyses evaluating fruit juice authenticity. To group orange, apple and grapefruit juice samples, Vaclavik et al. (2011) used low-resolution HPLC-QqQ/LIT-MS data paired with PCA. Overall, 24 principal components (PCs) were estimated, accounting for 90% of total variation. A supervised LDA model was constructed to investigate the possibility of detecting the addition of apple and grapefruit juice to orange juice. Arbona et al. (2015) used untargeted LC/ESI-QTOF-MS metabolite profiling to investigate the secondary metabolite composition of citrus fruits. Multivariate analysis (hierarchical cluster analysis [HCA] and PLS-DA) were used to separate different citrus juices, such as orange, lemon, mandarin and grapefruit juices, as well as varietal differentiation within a single fruit type (blonde and navel oranges). Cuevas et al. (2017) integrated metabolomic fingerprinting, volatile component profiles and advanced chemometrics to authenticate organic premium orange juices. Unsupervised PCA and HCA algorithms did not distinguish between organic and ordinary orange juices, but when alternative data fusion methodologies and PLS-DA were utilized, the model's discriminant power reached 100%.

7.5.4.2.3 Electronic Nose (EN)

EN employs a grouping of numerous sensitive elements to which volatile molecules bind, resulting in a change in the electrical signal of these elements. Chemometrics approaches might be used to process the outputs of these basic detectors (Roussel et al., 2003). The use of an EN allows for the quick examination of volatile chemicals in food samples. EN can be a strong method for authentication when combined with chemometrics since it gives the chemical profile of natural compounds (Rozanska et al., 2018). It was often employed to examine the entire volatile component combination, but it may also be used in conjunction with GC to increase resolving capabilities (Wu et al., 2018). EN conducted studies on the discriminating of fruit juices from various species and kinds. EN based on ultra-fast GC combined with supervised statistical approaches might allow juice samples having just 1% of adulteration to be distinguished from one other for the classification of 100% NFC orange and apple juices (Rozanska et al., 2018). There are numerous physiologically related species and kinds of fruits, including some hybrids. Due to allergen content and quality control requirements, rapid differentiation of fruit botanical origin is required (Lubinska-Szczygel et al., 2018; Wu et al., 2018). EN based on ultrafast GC was used to examine three citrus fruits. The findings suggested that it might be a unique analytical method for classifying hybrid fruits based on their botanical origin (Lubinska-Szczygel et al., 2018). EN was also used to categorize and describe apple juices by variety. It was able to distinguish between the juices made by eight distinct kinds. The results demonstrated a flawless differentiation of apple juices depending on varietals using LDA (Wu et al., 2017). However, according to one study, EN is not well adapted to the classification of white grape juice varieties, with a calibration error rate of 41.3% and a leave-one-out cross-validation prediction error rate of 47.1% in the created model (Roussel et al., 2003).

7.5.5 Spectroscopic Techniques

For food fingerprinting, spectroscopic techniques based on UV-Vis, fluorescence, infrared (IR), Raman, and NMR spectroscopy are widely utilized. The utility of spectroscopic approaches combined with multivariate data analysis for determining food authenticity and detecting adulteration has been widely shown in recent years (Abbas et al., 2018; Jha et al., 2016; Cozzolino, 2012; Zielinski et al., 2014). Non-targeted fingerprinting analysis obtains food fingerprints using quick spectroscopic techniques and is a viable alternative to chemical profiling approaches. These approaches have the advantages of requiring little or no sample preparation, being inexpensive, having good repeatability and being non-destructive (Valdés et al., 2018).

Two applications of UV spectroscopy in fruit juice authentication have been described in the last decade. In 2013, Boggia et al. (2013) suggested a simple and quick screening method based on UV-Vis spectroscopy and PCA to identify possibly contaminated pomegranate juice samples (diluted samples or samples containing undeclared cheaper juices like apple or grape). Pomegranate, apple, red and white grape juices were successfully distinguished, and pomegranate juice adulteration was detected down to 10%. Chang et al. (2016) recently used UV spectroscopy and chemometrics to successfully determine the variety, adulteration, quality and aging of apple juices. PCA proven to be a powerful tool for distinguishing between apple juice varieties and detecting adulteration. The effectiveness of principal component regression (PCR) and partial least squares regression (PLSR) models to predict the amount of adulterant, quality degradation owing to storage and storage period of apple juices was compared.

Fluorescence spectroscopy in food authenticity studies, along with multidimensional statistical approaches, has recently gotten a lot of attention (Karoui and Blecker, 2010). The use of fluorescence spectroscopy in fruit juice authentication has been described in two investigations. The first uses 3D-front-face fluorescence (3D-FFF) spectroscopy to determine orange juice adulteration with grapefruit juice (Ammari et al., 2015). The signal created by 3D-FFF spectroscopy comprises information about all of the fluorescent chemicals present in the sample. According to Ammari et al. (2015), when 3D-FFF spectroscopy is combined with independent components analysis (ICA), the fraudulent addition of grapefruit juice to orange juice can be detected down to 1%. The use of synchronous fluorescence spectra (SFS) and partial least squares–discriminant analysis (PLS-DA) in the discrimination of apple juice derived directly from fruit against apple juice reconstituted from concentrate is described in the second published application (Włodarska et al., 2018). When compared to standard emission spectra or excitation-emission matrices, the synchronous scanning fluorescence approach has a number of advantages in terms of producing food fluorescence fingerprints. Wodarska et al. (2018) used the PLS-DA method to create discriminating models for the two juice types; when pre-processed total SFS and specific

SFS were applied, good prediction ability was obtained. PCA was able to distinguish the juices to a degree (14.8% of total variation), but the majority of the spectral variability appeared to be independent of the distinct juice categories.

IR absorption and Raman scattering are two fundamentally separate phenomena that are used in vibrational spectroscopy techniques. The inelastic scattering of light by a sample is the subject of Raman spectroscopy (gas, liquid or solid). Irradiation causes molecules to change their vibrational state, which causes the energy of scattered photons to shift as well (Muro et al., 2015). IR spectroscopy, on the other hand, measures the absorbance of infrared radiation. In the IR region, each functional group of a molecule has its own vibrational frequency, which can be used to identify which functional groups are present in a sample. As a result, each sample has a unique molecular "fingerprint", which may be used to establish its authenticity (Cozzolino, 2012). Even though both IR absorption and Raman scattering report on molecular vibrational properties, the two phenomena are regulated by separate selection principles (Muro et al., 2015), so they provide complementary information.

The near-, mid-, and far-infrared (NIR, MIR and FIR, respectively) regions of the electromagnetic spectrum are separated into three zones based on their relationship to the visible spectrum. NIR spectroscopy (4000–10,000 nm) can be used to elucidate molecules by providing structural information on the vibrational behavior of combinations of bonds. MIR spectroscopy (2500–25,000 nm) yields less complex structural information than NIR, focusing on a molecule's chemical bonds and indicating the sorts of molecules present in a sample (Cozzolino, 2012). MIR spectroscopy, also known as FTIR (Fourier transform infrared) spectroscopy, is a type of spectroscopy that provides crucial chemical information about a food sample. The most common application of FTIR spectroscopy is to identify unknown materials by measuring significant fundamental vibrations (Abbas et al., 2018).

The applicability of NIR and, in particular, FTIR in conjunction with multivariate analysis for the assessment of fruit juice authenticity was established numerous times in the last decade. To distinguish pure bayberry juice from juice contaminated with water, NIR spectrometry was integrated with chemometric approaches (PCA, principal component–radial basis function neural networks [PC-RBFNN]). The results showed that PC-RBFNN can distinguish authentic bayberry juice samples from water-adulterated samples (recognition rate 97.6%) but cannot clearly estimate the levels of water in the adulterated bayberry juice after using multiplicative scatter correction (MSC) and standard normal variate (SNV) transformation to pre-process NIR spectra (Xie et al., 2008). Cozzolino et al. used NIR and MIR spectroscopy in combination with pattern recognition approaches to differentiate grape juice varietals. Using full cross-validation (leave-one-out) as a validation approach, PCA, LDA and PLS-DA were applied to grape juice samples categorized according to variety (Australian Chardonnay and Riesling) based on both NIR and MIR spectra. Using MIR and NIR, LDA models successfully identified 86% and 80% of the grape juices, respectively, whereas PLS-DA models obtained an overall rate of 86% correct classification (Cozzolino et al., 2011).

He et al. (2007) developed a straightforward technique for the classification of commercial apple, blueberry, cranberry, Concord grape and plum juices, as well as their blends, using FTIR spectroscopy. Following a SPE method, the sugar-rich and phenol-rich fractions of the juices were extracted, and their IR spectra were used to build the soft independent modeling by class analogy (SIMCA) model for pattern recognition. For cross-validation and external validation, the phenol-rich fraction model provided 100% proper categorization of the juices, as well as zero percent misclassification. Snyder et al. (2014) investigated the use of FTIR to verify Concord grape juice in grape juice blends, as well as other grape varietals (red, white and Niagara). SIMCA was able to distinguish between several grape juice varieties, owing to differences in IR bands between 1300 and 1800 cm^{-1}, which are related with particular vibrations of phenolic chemicals. FTIR spectroscopy combined with PCA can also be used to detect suspected adulteration of pomegranate juice concentrate with grape juice concentrate (Vardin et al., 2008). The largest difference between the pomegranate and grape concentrate spectra was seen in the 1700–1800 cm^{-1} range, which corresponds to the C=O stretching mode. PCA was able to distinguish between genuine pomegranate and grape juice concentrates and to detect pomegranate concentrate adulteration down to 2%.

FTIR spectrometry paired with chemometric analysis of spectrum data was able to predict the adulteration of mango juice by sugar addition in addition to juice-to-juice adulteration (Jha and Gunasekaran, 2010). The added sugar content detection limit ranged from 3% (for samples with low natural total soluble solids [TSS]) to 5% (for samples with natural TSS >10%). Except for TSS, which was best predicted using a multiple linear regression (MLR) model on three specific wavenumbers (1088, 1050, and 991 cm^{-1}; Jha and Gunasekaran, 2010), the most accurate prediction model was found when utilizing PLS regression in the range of 1476–912 cm^{-1} wavenumber. Shen et al. (2016) investigated the adulteration of freshly squeezed orange juices with 100% concentrated orange juices, and Ellis et al. (2016) investigated the applicability of FTIR in the detection of water and sugar addition in authentic orange juice. Adulterated orange juice samples were made with 0.5–20.0% water and with fructose, glucose or sucrose, and FTIR spectra combined with PLSR were able to accurately identify the levels of adulteration (mean error 1.7%). Freshly squeezed juice samples adulterated with various amounts of concentrated juice were also properly categorized using LDA (total prediction accuracy of 87.5%).

7.5.6 Molecular Techniques (DNA Analysis)

To verify the food source, qualitative and quantitative PCR techniques using species-specific genes can be used to confirm the relative content of certain species in a complex food matrix and to detect juice adulteration. DNA-based technologies, including PCR, real-time PCR, HRM analysis, microarrays and next-generation sequencing, have proven to be effective tools for a variety of food authentication methods (Madesis et al., 2014). Due to DNA's stability under various climatic circumstances, types of farming and production practices, they are deemed more dependable than chemical analysis (Madesis et al., 2014). Because DNA can preserve sequence-specific information during processing, it is an excellent authentication marker (Faria et al., 2013). DNA-based techniques may unequivocally identify a species or variety and determine if it is present in processed food.

Short sample preparation, high sample throughput, strong inter-laboratory repeatability and minimal operational costs are all characteristics of DNA-based techniques (Marieschi et al., 2016). Despite these benefits, the number of applications for determining the authenticity of fruit juice is quite restricted (Dasenaki and Thomaidis, 2019). Adulteration using mandarin juice has been exploited using dual-probe real-time PCR (qPCR) in the case of orange juice (Aldeguer et al., 2014; Pardo, 2015). To distinguish mandarin from orange DNA, Aldeguer et al. (2014) utilized a single-nucleotide polymorphism at the chloroplast chromosome trnL-trnF intergenic region, whereas Pardo (2015) used two TaqMan probes to investigate the detection of orange juice adulteration with mandarin. Adulteration was found at levels as low as 5% v/v and 1% v/v, respectively. The addition of mandarin and grapefruit juice to orange juice was also detected using PCR and laboratory-on-a-chip capillary electrophoresis (PCR-LOC) analysis: a PCR restriction fragment length polymorphism (RFLP) assay was used to detect grapefruit juice down to 10%, while a PCR heteroduplex assay was successfully applied to the LOC system for the detection of mandarin juice down to 2.5% (Scott and Knight, 2009).

Denaturing high-performance liquid chromatography (DHPLC) of previously amplified DNA fragments was utilized to distinguish seven fruit juices, demonstrating that it is a potent technology that might be employed for fruit juice labeling regulation (Han et al., 2012). Faria et al. (2013) evaluated the usefulness of HRM analysis to investigate the discrimination of orange, mango, peach, pear and pineapple liquids. The melt curve difference plots gave an easy way to see the various differences between these liquids. Marieschi et al. (2016) has developed a method based on sequence-characterized amplified region (SCAR) markers to detect pomegranate juice to pomegranate juice adulteration with ten different juices, with detection limits of up to 1% of each adulterant. SCAR markers were first created to detect and identify plants, but they have now been employed as authenticity markers for a variety of food goods (Madesis, 2014).

7.5.7 Elemental Analysis

Because it reflects the soil composition of the location where fruits and vegetables were produced, the multi-element composition of juices is a potent geographical indicator of a juice's origin (Bizjak Bat et al., 2016; Gaiad et al., 2016). However, extensive chemometric analysis is required to evaluate the collected information due to the huge number of variables involved when the elemental profile is used as a chemical descriptor. Gaiad et al. (2016) conducted a thorough investigation of the relationship between the geographical origin of Argentinean lemon juices and their elemental makeup; inductively coupled plasma–mass spectrometry (ICP-MS) was used to determine 25 trace elements, and five different multivariate analytic procedures were evaluated to classify the samples according to their geographical origin. The highest classification rate was found to be 76% for support vector machine (SVM), followed by 71% for random forest (RF). S, Cl, Fe, Cu, Zn and Sr have been found to be the most discriminatory trace elements in Slovenian apple juices; however, both isotopes and elemental data must be combined using linear discriminant analysis (LDA) to achieve efficient classification of apple juices according to geographical origin (Bizjak Bat et al., 2016). Borges et al. (2016) show that organic and conventional grape juices may be distinguished based on their elemental profiles. The organic-conventional grape juice discrimination was shown to be dominated by 11 factors using PCA and SIMCA models. Maione et al. (2016) took a step further by developing predictive models that could identify organic-conventional grape juice based on their multi-element composition using various data mining methods such as SVM, neural networks (NNs) and decision trees. The most significant variables were Na, K, P, Sn, Sm and Nd, and all of the models obtained had an accuracy of >85%.

Individual values of certain elements such as K, Na, Ca and Mg can be employed as markers for the identification of suspected adulteration of fruit juices in addition to trace element patterns. Orange juice with a low K content is most likely the result of considerable water dilution, given that oranges have high K concentrations (Maireva et al., 2013; Schmutzer et al., 2016). The same is true for pomegranate juice: drinks with a K content of less than 2000 mg L^{-1} are very likely to have been tampered with, either by adding water or substituting cheaper juices like grape or peach (Nuncio-Jáuregui et al., 2014). Low K/Mg ratios in orange juice can indicate the presence of sweeteners (Cozzolino, 2012), whereas a high Na concentration can disclose the presence of preservatives or adulterants in the form of salts (Schmutzer et al., 2016). Finally, high calcium readings may indicate that orange juice has been tampered with, as pulp wash contains far more calcium than juice. In every case, the presence of K, Na, Ca and Mg alone is insufficient to define a juice as non-authentic; it must be paired with additional parameters (Nuncio-Jáuregui et al., 2014).

7.6 CONCLUSIONS

Not-from-concentrate (NFC) fruit juice is becoming increasingly popular, and several functional fruit juices are growing. As a result, more complex counterfeiting methods are emerging during processing and shipping, such as adulteration of NFC juice with FC juice, non-thermally processed juice with thermally processed juice, and functional juice frauds. The distinction between these liquids might be considerably smaller. Above all, there are just a few ways or focused indicators to rely on. If only the authenticating methodology can keep up with the counterfeiting methods, the fruit juice sector might thrive.

Because of the complexities of fruit juice authenticity assessment, an analytical authenticity control is practically a mix of multiple procedures, techniques and methods. Fruit juice authenticity testing aims to detect and prevent fraudulent practices, especially frauds that are on the rise and also might be related to health impacts, such as undeclared sugar addition or use of synthetic ingredients. The current trend is the use of untargeted screening methods that utilize powerful fingerprinting techniques, while the recommended official methods combined with established acceptance criteria for authenticity and state-of-the-art techniques are the trusted means of quality and authenticity control.

The interpretation of analytical results will become more challenging in the future due to the expanding number of production zones for semi-finished goods, agricultural growth and changing environment, and particular reference databases will become more necessary.

REFERENCES

Abad-García, B.; Garmón-Lobato, S.; Sánchez-Ilárduya, M.B.; Berrueta, L.A.; Gallo, B.; Vicente, F.; Alonso-Salces, R.M., 2014 Polyphenolic contents in Citrus fruit juices: Authenticity assessment. Eur. Food Res. Technol., 238, 803–818.

Abbas, O.; Zadravec, M.; Baeten, V.; Mikus, T.; Lesic, T.; Vulic, A.; Prpic, J.; Jemersic, L.; Pleadin, J., 2018. Analytical methods used for the authentication of food of animal origin. Food Chem., 246, 6–17.

AIJN, 2019. AIJF European Fruit juice Association—Annual report 2019. Available at: www.aijn2019report.com/aijn2019report/the_fruit_juice_industry_overall_fruit_juice_consumption

AIJN. Available at: www.aijn.org/publications/code-of-practice/the-aijn-code-of-practice/ (accessed on March 7, 2022).

Alañón, M.E.; Pérez-Coello, M.S.; Marina, M.L., 2015. Wine science in the metabolomics era. Trac-Trend Anal. Chem., 74, 1–20.

Aldeguer, M.; Lopez-Andreo, M.; Gabaldon, J.A.; Puyet, A., 2014. Detection of mandarin in orange juice by single-nucleotide polymorphism qPCR assay. Food Chem., 145, 1086–1091.

Alizadeh, M.; Pirsa, S.; Faraji, N., 2016. Determination of lemon juice adulteration by analysis of gas chromatography profile of volatile organic compounds extracted with nano-sized polyester-polyaniline fiber. Food Anal. Method, 10, 2092–2101.

American Beverage Association, 2019 Annual review. BBC. 2019. Zambia bans 'Viagra' energy drink. www.ameribev.org/

Ammari, F.; Redjdal, L.; Rutledge, D.N., 2015. Detection of orange juice frauds using front-face fluorescence spectroscopy and Independent Components Analysis. Food Chem., 168, 211–217.

Arbona, V.; Iglesias, D.J.; Gomez-Cadenas, A., 2015. Non-targeted metabolite profiling of citrus juices as a tool for variety discrimination and metabolite flow analysis. BMC Plant Biol., 15, 38.

Asadpoor, M.; Ansarin, M.; Nemati, M., 2014. Amino Acid profile as a feasible tool for determination of the authenticity of fruit juices. Adv. Pharm. Bull., 4, 359–362.

Bizjak Bat, K.; Eler, K.; Mazej, D.; Mozetic Vodopivec, B.; Mulič, I.; Kump, P.; Ogrinc, N., 2016. Isotopic and elemental characterisation of Slovenian apple juice according to geographical origin: Preliminary results. Food Chem., 203, 86–94.

Bizjak Bat, K.; Vodopivec, B.M.; Eler, K.; Ogrinc, N.; Mulič, I.; Masuero, D.; Vrhovšek, U., 2018. Primary and secondary metabolites as a tool for differentiation of apple juice according to cultivar and geographical origin. Lwt-Food Sci. Technol., 90, 238–245.

Bocharova, O.; Reshta, S.; Bocharova, M.; Eshtokin, V., 2017. Evaluation of orange juice authenticity in respect of added food flavors using dilution index. J. Food Process. Preserv., 41, 13221–13226.

Boggia, R.; Casolino, M.C.; Hysenaj, V.; Oliveri, P.; Zunin, P., 2013. A screening method based on UV-Visible spectroscopy and multivariate analysis to assess addition of filler juices and water to pomegranate juices. Food Chem., 140, 735–741.

Bong, Y.-S.; Lee, K.-S.; Shin, W.-J.; Ryu, J.-S., 2008. Comparison of the oxygen and hydrogen isotopes in the juices of fast-growing vegetables and slow-growing fruits. Rapid Communications in Mass Spectrometry 22 (18):2809–2812. doi: 10.1002/rcm.3663.

Bononi, M.; Quaglia, G.; Tateo, F., 2016. Preliminary LC-IRMS characterization of Italian Pure Lemon juices and evaluation of commercial juices distributed in the Italian market. Food Anal. Method, 9, 2824–2831.

Bontempo, L.; Caruso, R.; Fiorillo, M.; Gambino, G.L.; Perini, M.; Simoni, M.; Traulo, P.; Wehrens, R.; Gagliano, G.; Camin, F., 2014. Stable isotope ratios of H, C, N and O in Italian citrus juices. J. Mass Spectrom., 49, 785–791.

Borges, E.M.; Volmer, D.A.; Brandelero, E.; Gelinski, J.M.L.N.; Gallimberti, M.; Barbosa, F., Jr., 2016. Monitoring the authenticity of organic grape juice via chemometric analysis of elemental data. Food Anal. Method, 9, 362–369.

Borges, G.; Crozier, A., 2012. HPLC-PDA-MS fingerprinting to assess the authenticity of pomegranate beverages. Food Chem., 135, 1863–1867.

Borges, G.; Mullen, W.; Crozier, A., 2010. Comparison of the polyphenolic composition and antioxidant activity of European commercial fruit juices. Food Funct., 1, 73–83.

Borràs, E.; Ferré, J.; Boqué, R.; Mestres, M.; Aceña, L.; Busto, O., 2015. Data fusion methodologies for food and beverage authentication and quality assessment—A review. Anal. Chim. Acta, 891, 1–14.

Carter, J.F.; Chesson, L.A. (Eds.). (2017). Food Forensics: Stable Isotopes as a Guide to Authenticity and Origin (1st ed.). CRC Press. https://doi.org/10.1201/9781315151649

Castro-Puyana, M.; Pérez-Míguez, R.; Montero, L.; Herrero, M., 2017. Application of mass spectrometry-based metabolomics approaches for food safety, quality and traceability. Trac-Trend Anal. Chem., 93, 102–118.

Cautela, D.; Laratta, B.; Santelli, F.; Trifirò, A.; Servillo, L.; Castaldo, D., 2008. Estimating bergamot juice adulteration of lemon juice by high-performance liquid chromatography (HPLC) analysis of flavanone glycosides. J. Agric. Food Chem., 56, 5407–5414.

Chang, J.-D.; Zheng, H.; Mantri, N.; Xu, L.; Jiang, Z.; Zhang, J.; Song, Z.; Lu, H., 2016. Chemometrics coupled with ultraviolet spectroscopy: A tool for the analysis of variety, adulteration, quality and ageing of apple juices. Int. J. Sci. Technol., 51, 2474–2484.

Clausen, M.R.; Pedersen, B.H.; Bertram, H.C.; Kidmose, U., 2011. Quality of sour cherry juice of different clones and cultivars (Prunus cerasus L.) determined by a combined sensory and NMR spectroscopic approach. J. Agric. Food Chem., 59, 12124–12130.

Codex Alimentarius, 2005. Codex general standard for fruit juices and nectars; CODEX STAN 247–2005; Codex Alimentarius Commission: Rome, Italy.

Cozzolino, D., 2012. Recent trends on the use of infrared spectroscopy to trace and authenticate natural and agricultural food products. Appl. Spectrosc. Rev., 47, 518–530.

Cozzolino, D.; Cynkar, W.; Shah, N.; Smith, P., 2011. Varietal differentiation of grape juice based on the analysis of near- and mid-infrared spectral data. Food Anal. Method, 5, 381–387.

Cubero-Leon, E.; Peñalver, R.; Maquet, A., 2014. Review on metabolomics for food authentication. Food Res. Int., 60, 95–107.

Cuevas, F.J.; Pereira-Caro, G.; Moreno-Rojas, J.M.; Muñoz-Redondo, J.M.; Ruiz-Moreno, M.J., 2017. Assessment of premium organic orange juices authenticity using HPLC-HR-MS and HS-SPME-GC-MS combining data fusion and chemometrics. Food Control, 82, 203–211.
D'Orazio, G.; Cifuentes, A.; Fanali, S., 2008. Chiral nano-liquid chromatography-mass spectrometry applied to amino acids analysis for orange juice profiling. Food Chem., 108, 1114–1121.
Dasenaki, M.E.; Thomaidis, N.S., 2019. Quality and Authenticity Control of Fruit Juices-A Review. Molecules; 24(6):1014. https://doi.org/10.3390/molecules24061014
De Oliveira, C.R.; Carneiro, R.L.; Ferreira, A.G., 2014. Tracking the degradation of fresh orange juice and discrimination of orange varieties: An example of NMR in coordination with chemometrics analyses. Food Chem., 164, 446–453.
Dugo, P.; Giuffrida, D.; Herrero, M.; Donato, P.; Mondello, L., 2009. Epoxycarotenoids esters analysis in intact orange juices using two-dimensional comprehensive liquid chromatography. J. Sep. Sci., 32, 973–980.
Dzugan, M.; Wesołowska, M.; Zaguła, G.; Puchalski, C., 2018. The comparison of the physicochemical parameters and antioxidant activity of homemade and commercial pomegranate juices. Acta Sci. Pol Technol. Aliment., 17, 59–68.
Ellis, D.I.; Ellis, J.; Muhamadali, H.; Xu, Y.; Horn, A.B.; Goodacre, R., 2016. Rapid, high-throughput, and quantitative determination of orange juice adulteration by Fourier-transform infrared spectroscopy. Anal. Method, 8, 5581–5586.
EMR, 2020. Global fruit juice market outlook. www.expertmarketresearch.com/reports/fruit-juice-market
Esslinger, S.; Riedl, J.; Fauhl-Hassek, C., 2014. Potential and limitations of non-targeted fingerprinting for authentication of food in official control. Food Res. Int., 60, 189–204.
Faria, M.A.; Magalhães, A.; Nunes, M.E.; Oliveira, M.B.P.P., 2013. High resolution melting of trnL amplicons in fruit juices authentication. Food Control, 33, 136–141.
Filip, M.; Vlassa, M.; Copaciu, F.; Coman, V., 2012. Identification of anthocyanins and anthocyanidins from berry fruits by chromatographic and spectroscopic techniques to establish the juice authenticity from market. J. Planar Chromatogr., 25, 534–541.
Foodakai, 2019. Industry report series: The case of non-alcoholic beverages. https://agroknow.com/wp-content/uploads/2020/07/Foodakai_Non-Alcoholic_Beverages_Report.pdf
Gaiad, J.E.; Hidalgo, M.J.; Villafañe, R.N.; Marchevsky, E.J.; Pellerano, R.G., 2016. Tracing the geographical origin of Argentinean lemon juices based on trace element profiles using advanced chemometric techniques. Microchem. J., 129, 243–248.
Godin, J.P.; McCullagh, J.S., 2011. Review: Current applications and challenges for liquid chromatography coupled to isotope ratio mass spectrometry (LC/IRMS). Rapid Commun. Mass Spectrom., 25, 3019–3028.
Grandizoli, C.S.; Campos, F.R.; Simonelli, F.; Barison, A., 2014. Grape juice quality control by means of ^{1}H-NMR spectroscopy and chemometric analyses. Quim Nova, 37, 1227–1232.
Guo, J.; Yue, T.; Yuan, Y., 2012. Feature selection and recognition from nonspecific volatile profiles for discrimination of apple juices according to variety and geographical origin. J. Food Sci., 77, C1090–1096.
Guo, J.; Yue, T.; Yuan, Y.; Wang, Y., 2013. Chemometric classification of apple juices according to variety and geographical origin based on polyphenolic profiles. J. Agric. Food Chem., 61, 6949–6963.
Guyon, F.; Auberger, P.; Gaillard, L.; Loublanches, C.; Viateau, M.; Sabathie, N.; Salagoity, M.H.; Medina, B., 2014. (13)C/(12)C isotope ratios of organic acids, glucose and fructose determined by HPLC-co-IRMS for lemon juices authenticity. Food Chem., 146, 36–40.
Han, J.; Wu, Y.; Huang, W.; Wang, B.; Sun, C.; Ge, Y.; Chen, Y., 2012. PCR and DHPLC methods used to detect juice ingredient from 7 fruits. Food Control, 25, 696–703.
He, J.; Rodriguez-Saona, L.E.; Giusti, M.M., 2007. Midinfrared Spectroscopy for Juice Authentication—Rapid Differentiation of Commercial Juices. J. Agric. Food Chem., 55, 4443–4452.
Hosu, A.; Cimpoiu, C., 2017. Thin-layer chromatography applied in quality assessment of beverages derived from fruits. J. Liq. Chromatogr. Relat. Technol., 40, 239–246.
Ibáñez, C.; García-Cañas, V.; Valdés, A.; Simó, C., 2013. Novel MS-based approaches and applications in food metabolomics. Trac-Trend Anal. Chem., 52, 100–111.
Imarc Group, 2020. Fruit juice market: Global industry trends, share, size, growth, opportunity and forecast 2021–2026. www.imarcgroup.com/fruit-juice-manufacturing-plant
Jandric, Z.; Cannavan, A., 2017. An investigative study on differentiation of citrus fruit/fruit juices by UPLC-QToF MS and chemometrics. Food Control, 72, 173–180.

Jandric, Z.; Islam, M.; Singh, D.K.; Cannavan, A., 2017. Authentication of Indian citrus fruit/fruit juices by untargeted and targeted metabolomics. Food Control, 72, 181–188.

Jandric, Z.; Roberts, D.; Rathor, M.N.; Abrahim, A.; Islam, M.; Cannavan, A., 2014. Assessment of fruit juice authenticity using UPLC-QToF MS: A metabolomics approach. Food Chem., 148, 7–17.

Jha, S.N.; Gunasekaran, S., 2010. Authentication of sweetness of mango juice using Fourier transform infrared-attenuated total reflection spectroscopy. J. Food Eng., 101, 337–342.

Jha, S.N.; Jaiswal, P.; Grewal, M.K.; Gupta, M.; Bhardwaj, R., 2016. Detection of adulterants and contaminants in liquid foods-a review. Crit. Rev. Food Sci. Nutr., 56, 1662–1684.

Karoui, R.; Blecker, C., 2010. Fluorescence spectroscopy measurement for quality assessment of food systems— A review. Food Bioprocess Technol., 4, 364–386.

Kelly, S.D.; Rhodes, C.; Lofthouse, J.H.; Anderson, D.; Burwood, C.E.; Dennis, M. J.; Brereton, P., 2003. Detection of sugar syrups in apple juice by delta(2)H per thousand and delta(13)C per thousand analysis of hexamethylenetetramine prepared from fructose. J. Agricult. Food Chem., 51 (7):1801–1806. doi: 10.1021/jf021044p.

Knolhoff, A.M.; Croley, T.R., 2016. Non-targeted screening approaches for contaminants and adulterants in food using liquid chromatography hyphenated to high resolution mass spectrometry. J. Chromatogr. A, 1428, 86–96.

Koda, M.; Furihata, K.; Wei, F.; Miyakawa, T.; Tanokura, M., 2012. Metabolic discrimination of mango juice from various cultivars by band-selective NMR spectroscopy. J. Agric. Food Chem., 60, 1158–1166

Krüger, S.; Urmann, O.; Morlock, G.E., 2013. Development of a planar chromatographic method for quantitation of anthocyanes in pomace, feed, juice and wine. J. Chromatogr. A, 1289, 105–118.

Longobardi, F.; Ventrella, A.; Bianco, A.; Catucci, L.; Cafagna, I.; Gallo, V.; Mastrorilli, P.; Agostiano, A., 2013. Non-targeted ^{1}H NMR fingerprinting and multivariate statistical analyses for the characterisation of the geographical origin of Italian sweet cherries. Food Chem., 141, 3028–3033.

Lord, N.; Flores Elizondo, C.; Spencer, J., 2017. The dynamics of food fraud: the interactions between criminal opportunity and market (dys)functionality in legitimate business. Criminol. Crim. Justice 17 (5).

Lorente, J.; Vegara, S.; Marti, N.; Ibarz, A.; Coll, L.; Hernandez, J.; Valero, M.; Saura, D., 2014. Chemical guide parameters for Spanish lemon (Citrus limon (L.) Burm.) juices. Food Chem., 162, 186–191.

Lubinska-Szczygeł, M.; Pudlak, D.; Dymerski, T; Namieśnik, J., 2018. Rapid assessment of the authenticity of limequat fruit using the electronic nose and gas chromatography coupled with mass spectrometry. Monatshefte fur Chemie 149 (9):1605–1614. doi: 10.1007/s00706-018-2242-7.

Ludwig C.; Viant M.R., 2010. Two-dimensional J-resolved NMR spectroscopy: review of a key methodology in the metabolomics toolbox. Phytochem. Anal., 21 (1), 22–32. doi:10.1002/pca.1186.

Madesis, P.; Ganopoulos, I.; Sakaridis, I.; Argiriou, A.; Tsaftaris, A., 2014. Advances of DNA-based methods for tracing the botanical origin of food products. Food Res. Int., 60, 163–172.

Maione, C.; de Paula, E.S.; Gallimberti, M.; Batista, B.L.; Campiglia, A.D.; Barbosa, F., Jr.; Barbosa, R.M., 2016. Comparative study of data mining techniques for the authentication of organic grape juice based on ICP-MS analysis. Expert Syst. Appl., 49, 60–73.

Maireva, S.; Usai, T.; Manhokwe, S., 2013. The determination of adulteration in orange based fruit juices. Int. J. Sci. Technol., 2, 365–372.

Marieschi, M.; Torelli, A.; Beghe, D.; Bruni, R., 2016. Authentication of Punica granatum L.: Development of SCAR markers for the detection of 10 fruits potentially used in economically motivated adulteration. Food Chem., 202, 438–444.

Marketline, 2018. Carbonated Soft drinks in Europe. https://store.marketline.com/report/carbonated-soft-drinks-in-europe-3/

Meerza, S.; Gustafson, C.R. (2019). Does prior knowledge of food fraud affect consumer behavior? Evidence from an incentivized economic experiment. PLOS One, 14(12), e0225113. https://doi.org/10.1371/journal.pone.0225113

Mena, P.; Garcia-Viguera, C.; Navarro-Rico, J.; Moreno, D.A.; Bartual, J.; Saura, D.; Marti, N., 2011. Phytochemical characterisation for industrial use of pomegranate (Punica granatum L.) cultivars grown in Spain. J. Sci. Food Agric., 91, 1893–1906.

Moore, J.C.; Spink, J.; Lipp, M., 2012. Development and application of a database of food ingredient fraud and economically motivated adulteration from 1980 to 2010. J. Food Sci., 77, R118–R126.

Muro, C.K.; Doty, K.C.; Bueno, J.; Halamkova, L.; Lednev, I.K., 2015. Vibrational spectroscopy: Recent developments to revolutionize forensic science. Anal. Chem., 87, 306–327.

Navarro-Pascual-Ahuir, M.; Lerma-García, M.J.; Simó-Alfonso, E.F.; Herrero-Martínez, J.M., 2017. Analysis of aliphatic organic acids in commercial fruit juices by capillary electrophoresis with indirect UV detection: Application to differentiation of fruit juices. food anal. Method, 10, 3991–4002.

Nikolaou, C.; Karabagias, I.K.; Gatzias, I.; Kontakos, S.; Badeka, A.; Kontominas, M.G., 2017. Differentiation of fresh Greek orange juice of the merlin cultivar according to geographical origin based on the combination of organic acid and sugar content as well as physicochemical parameters using chemometrics. Food Anal. Method, 10, 2217–2228.

Nuncio-Jáuregui, N.; Calín-Sánchez, Á.; Hernández, F.; Carbonell-Barrachina, Á.A., 2014. Pomegranate juice adulteration by addition of grape or peach juices. J. Sci. Food Agric., 94, 646–655.

Obón, J.M.; Díaz-García, M.C.; Castellar, M.R., 2011. Red fruit juice quality and authenticity control by HPLC. J. Food Compos. Anal., 24, 760–771.

Ogrinc, N.; Bat, K.; Kosir, I.J.; Golob, T.; Kokkinofta, R., 2009. Characterization of commercial Slovenian and Cypriot fruit juices using stable isotopes. J. Agric. Food Chem., 57, 6764–6769.

Pardo, M.A., 2015. Evaluation of a dual-probe real time PCR system for detection of mandarin in commercial orange juice. Food Chem., 172, 377–384.

Primrose, S.; Woolfe, M.; Rollinson, S., 2010. Food forensics: Methods for determining the authenticity of foodstuffs. Trends Food Sci. Technol., 21, 582–590.

Psomiadis, D.; Zisi, N.; Koger, C.; Horvath, B.; Bodiselitsch, B., 2018. Sugar-specific carbon isotope ratio analysis of coconut waters for authentication purposes. J Food Sci Technol., 55(8): 2994–3000. doi: 10.1007/s13197-018-3217-8.

Reid, L.M.; O'Donnell, C.P.; Downey, G., 2006. Recent technological advances for the determination of food authenticity. Trends Food Sci. Technol., 17, 344–353.

Reportlinker, 2019. Global Soft Drinks Industry. Cision PR Newswire.

Rinaldi, M.; Gindro, R.; Barbeni, M.; Allegrone, G., 2009. Pattern recognition and genetic algorithms for discrimination of orange juices and reduction of significant components from headspace solid-phase microextraction. Phytochem. Anal., 20, 402–407.

Rinke, P.; Jamin, E., 2018. Fruit juices. In: FoodIntegrity Handbook. Ed.: Jean-François Morin and Michèle Lees. doi.org/10.32741/fihb

Rinke, P.; Moitrier, S.; Humpfer, E.; Keller, S.; Moertter, M.; Godejohann, M.; Hoffmann, G.; Schaefer, H.; Spraul, M., 2007. An ^{1}H-NMR technique for high throughput screening in quality and authenticity control of fruit juice and fruit juice raw materials—SGF-profiling. Fruit Process., 1, 10–18.

Roussel, S.; Bellon-Maurel, V.; Roger, J.-M.; Grenier, P., 2003. Authenticating white grape must variety with classification models based on aroma sensors, FT-IR and UV spectrometry. J Food Engineer 60 (4):407–419. doi: 10.1016/S0260–8774(03)00064-5.

Rozanska, A.; Dymerski, T.; Namieśnik, J., 2018. Novel analytical method for detection of orange juice adulteration based on ultra-fast gas chromatography. Monatshefte für Chemie 149 (9):1615–1621. doi:10.1007/s00706-018-2233-8.

Rummel, S.; Hoelzl, S.; Horn, P.; Rossmann, A.; Schlicht, C., 2010. The combination of stable isotope abundance ratios of H, C, N and S with 87Sr/86Sr for geographical origin assignment of orange juices. Food Chem., 118, 890–900.

Schmutzer, G.R.; Dehelean, A.; Magdas, D.A.; Cristea, G.; Voica, C., 2016. Determination of stable isotopes, minerals, and volatile organic compounds in Romanian orange juice. Anal. Lett., 49, 2644–2658.

Scott, M.; Knight, A., 2009. Quantitative PCR analysis for fruit juice authentication using PCR and laboratory-on-a-chip capillary electrophoresis according to the Hardy-Weinberg law. J. Agric. Food Chem., 57, 4545–4551.

Shen, F.; Wu, Q.; Su, A.; Tang, P.; Shao, X.; Liu, B., 2016. Detection of adulteration in freshly squeezed orange juice by electronic nose and infrared spectroscopy. Czech J. Food Sci., 34, 224–232.

Sherma, J.; Rabel, F., 2018. A review of thin layer chromatography methods for determination of authenticity of foods and dietary supplements. J. Liq. Chromatogr. Relat. Technol., 41, 645–657.

Snyder, A.B.; Sweeney, C.F.; Rodriguez-Saona, L.E.; Giusti, M.M., 2014. Rapid authentication of concord juice concentration in a grape juice blend using Fourier-transform infrared spectroscopy and chemometric analysis. Food Chem., 147, 295–301.

Sobolev, A.P.; Mannina, L.; Proietti, N.; Carradori, S.; Daglia, M.; Giusti, A.M.; Antiochia, R.; Capitani, D., 2015. Untargeted NMR-based methodology in the study of fruit metabolites. Molecules, 20, 4088–4108.

Spraul, M.; Schütz, B.; Humpfer, E.; Mörtter, M.; Schäfer, H.; Koswig, S.; Rinke, P., 2009b. Mixture analysis by NMR as applied to fruit juice quality control. Magn. Reson. Chem., 47 (S1), S130–S137. doi:10.1002/mrc.2528.

Spraul, M.; Schütz, B.; Rinke, P.; Koswig, S.; Humpfer, E.; Schäfer, H.; Mörtter, M.; Fang, F.; Marx, U.; Minoja, A., 2009a. NMR-based multi parametric quality control of fruit juices: SGF profiling. Nutrients, 1, 148–155.

Syposs, Z.; Knight, G., 2021. Fraud in nonalcoholic ready to drink products. In: Food Fraud. Ed.: Rosalee S. Hellberg, Karen Everstine, Steven A. Sklare. Academic Press. https://doi.org/10.1016/B978-0-12-817242-1.00017-8

Tolle, S.; Lander, V.; Winterhalter, P., 2011. Authentication of Fruit Juices Derived from Morinda citrifolia (Noni). In Progress in Authentication of Food and Wine; Ebeler, S.E., Takeoka, G.R., Winterhalter, P., Eds.; American Chemical Society; Volume 1081, pp. 249–258.

Türkyılmaz, M., 2013. Anthocyanin and organic acid profiles of pomegranate (Punica granatum L.) juices from registered varieties in Turkey. Int. J. Sci. Technol., 48, 2086–2095.

Twohig, M.; Krueger, D.A.; Gledhill, A.; Yang, J.; Burgess, J., 2011. Super fruit juice authenticity using multivariate data analysis, high resolution chromatography, UV and Time of Flight MS detection. Agro Food Ind. Hi Tech., 22, 23–26.

Vaclavik, L.; Schreiber, A.; Lacina, O.; Cajka, T.; Hajslova, J., 2011. Liquid chromatography—mass spectrometry-based metabolomics for authenticity assessment of fruit juices. Metabolomics, 8, 793–803.

Valdés, A.; Beltrán, A.; Mellinas, C.; Jiménez, A.; Garrigós, M.C., 2018. Analytical methods combined with multivariate analysis for authentication of animal and vegetable food products with high fat content. Trends Food Sci. Technol., 77, 120–130.

Vardin, H.; Tay, A.; Ozen, B.; Mauer, L., 2008. Authentication of pomegranate juice concentrate using FTIR spectroscopy and chemometrics. Food Chem., 108, 742–748.

Vigneau, E.; Thomas, F., 2012. Model calibration and feature selection for orange juice authentication by ^{1}H NMR spectroscopy. Chemom. Intell. Lab., 117, 22–30.

Voldrich, M.; Skálová, P.; Kvasnicka, F.; Cuhra, P.; Kubík, M.; Pyš, P., 2002. Authenticity of 100% Orange Juice in the Czech Market in 1996–2001. Czech J. Food Sci., 20, 83–88.

Wang, Z.; Jablonski, J.E., 2016. Targeted and non-targeted detection of lemon juice adulteration by LC-MS and chemometrics. Food Addit. Contam. Part A Chem. Anal. Control Expo. Risk Assess., 33, 560–573.

Willems, J.L.; Low, N.H., 2014. Authenticity analysis of pear juice employing chromatographic fingerprinting. J. Agric. Food Chem., 62, 11737–11747.

Willems, J.L.; Low, N.H., 2018. Structural identification of compounds for use in the detection of juice-to-juice debasing between apple and pear juices. Food Chem., 241, 346–352.

Włodarska, K.; Khmelinskii, I.; Sikorska, E., 2018. Authentication of apple juice categories based on multivariate analysis of the synchronous fluorescence spectra. Food Control, 86, 42–49.

Wu, H.; Wang, J.; Yue, T.; Yuan, Y., 2017. Variety-based discrimination of apple juices by an electronic nose and gas chromatography-mass spectrometry. Int. J. Sci. Technol., 52, 2324–2333.

Wu, H.; Yue, T; Yuan, Y., 2018. Authenticity tracing of apples according to variety and geographical origin based on electronic nose and electronic tongue. Food Analytical Methods 11 (2), 522–532. doi: 10.1007/s12161–017–1023-y.

Xie, L.J.; Ye, X.Q.; Liu, D.H.; Ying, Y.B., 2008. Application of principal component-radial basis function neural networks (PC-RBFNN) for the detection of water-adulterated bayberry juice by near-infrared spectroscopy. J. Zhejiang Univ. Sci. B, 9, 982–989.

Xu, L.; Xu, Z.; Liao, X., 2021. A review of fruit juice authenticity assessments: Targeted and untargeted analyses. Crit Rev Food Sci Nut., 62, 6081–6102. doi: 10.1080/10408398.2021.1895713.

Yamamoto, K.; Yahada, A.; Sasaki, K.; Funakoshi-Yoshida, A.; Ohta, C.; Koga, N.; Ohta, H., 2013. Detection of adulterated shiikuwasha juice by sensory evaluation, colorimetric value and volatile components. Food Sci. Technol. Res., 19, 843–848.

Yamamoto, K.; Yahada, A.; Sasaki, K.; Ogawa, K.; Koga, N.; Ohta, H., 2012. Chemical markers of shiikuwasha juice adulterated with calamondin juice. J. Agric. Food Chem., 60, 11182–11187.
Zhang, Y.; Krueger, D.; Durst, R.; Lee, R.; Wang, D.; Seeram, N.; Heber, D., 2009. International multidimensional authenticity specification (IMAS) algorithm for detection of commercial pomegranate juice adulteration. J. Agric. Food Chem., 57, 2550–2557.
Zielinski, A.A.F.; Haminiuk, C.W.I.; Nunes, C.A.; Schnitzler, E.; van Ruth, S.M.; Granato, D., 2014. Chemical composition, sensory properties, provenance, and bioactivity of fruit juices as assessed by chemometrics: a critical review and guideline. Compr. Rev. Food Sci. Saf., 13, 300–316.

8 Authenticity and Chemometrics of Cereals and Grains and Their Derived Products

Jorgelina Zaldarriaga Heredia, Antonella Evelyn Montemerlo, Marcelo Wagner, Silvana Mariela Azcarate, and José Manuel Camiña

CONTENTS

8.1 WHEAT

Wheat (*Triticum aestivum* L.) is one of the most important cereals produced around the world and it is considered a basic food in diets. Due to its high consumption, several topics are considered necessary to perform quality analysis of this cereal. First, wheat protein adulteration is a growing concern for food manufacturers. On the other hand, wheat provenance is a relevant area studied in recent years, especially in those cases when wheat quality depends on its origin. In this sense, wheat consumers are increasingly demanding assurances including geographical origin labeling. Furthermore, wheat variety identification also has been given great importance in recent years. Finally, the growing demand for organic food emphasizes the need for organic wheat authenticity from other agricultural practices (Liu et al. 2021a).

In order to acquire proper analytical information for wheat analysis, stable isotope analysis and multi-element analysis have been the most implemented techniques to authenticate wheat and its derivatives (Wadood et al. 2019; Wadood et al. 2019; Zhao et al. 2013a). Also, gas and liquid chromatography techniques, as well as DNA-based methods, have proven to be extremely valuable for studying issues related to wheat authenticity and adulteration (Liu et al. 2021a). Currently, spectroscopic techniques such as infrared, fluorescence and nuclear magnetic resonance (NMR) techniques, among others, have made significant advances in these fields contributing with several advantages since they are fast, non-destructive, cost-effective, easy to operate, and environmentally friendly (Wadood et al. 2019).

A plethora of information can be obtained from those analytical platforms and, to carry out the data analysis, chemometric applications have been keys to fulfill with authenticity aims. Thus, several works applying chemometric methods for studies of wheat grains and their derivatives, such as flour and dough made with flour, have been widely reported.

8.1.1 Analysis of Wheat Grains

A large number of works using chemometric techniques for wheat grain analysis have been published, the objectives of these works being very assorted. Foschi et al. developed a non-destructive and routine-compatible method to discriminate three Italian local breeds and to lay

DOI: 10.1201/9781003266167-8

the groundwork for a possible authentication method for Emmer wheats (a traditional Italian wheat species of high nutritional and dietary value). Spectroscopic techniques such as mid-infrared spectroscopy (MIRS) and near-infrared spectroscopy (NIRS) combined with chemometric algorithms as partial least squares–discriminant analysis (PLS-DA) and sequentially orthogonalized–partial least squares–discriminant analysis (SO-PLS-DA) were used as different data fusion (DF) strategies to classify the samples according to geographic origin. The most accurate predictions were provided by the SO-PLS-DA model, which misclassified only one test sample out of 44 (Foschi et al. 2021). In another work, González-Martín et al. analyzed the spectral data obtained from NIRS by discriminant partial least squares analysis (DPLS), correctly classifying 76% of the wheat grain samples and between 90% and 96% of the flour samples. These results demonstrated that NIRS together with chemometric methods provided a rapid tool for classifying wheat grain and flour samples according to their geographic origin (González-Martín et al. 2014). On the other hand, Wadood et al. studied the variation of the stable isotopic proportions of carbon, nitrogen, hydrogen and oxygen in wheat grains along with different processed fractions of three geographical origins over five years, using isotope-ratio mass spectrometry (IRMS). By applying linear discriminant analysis (LDA) to the entire data set, the generated model correctly classified more than 91% of the samples according to geographic origin. The application of these parameters will help to develop an analytical control procedure that can be used to control mislabeling regarding geographic origin of the wheat grain and its products (Wadood et al. 2019). In another study, Zhao et al. investigated the feasibility of using NIRS coupled with chemometrics to discriminate the geographical origins of wheat. After pretreatment of spectral data in wheat grain and whole wheat flour, they applied LDA and DPLS to classify the origins of the samples. They concluded that NIRS combined with chemometrics is a promising method for identifying wheat based on geographic origin (Zhao et al. 2013b).

De Flaviis et al. analyzed the volatile organic compounds (VOC) profile of common and durum wheat grains grown in different fields located at different altitudes for two years. A total of 149 compounds using gas chromatography with mass spectrometry (GC-MS) were identified. The wheat origin, altitude and species effects were investigated using PLS-DA, which allowed the wheat classification from different origin based on its VOC profile. They concluded that environmental conditions are more important than species in determining VOC variation in wheat (De Flaviis et al. 2022a). The same authors analyzed the VOC profile of common wheat and durum wheat grown in different fields located at three different elevations for two years and used chemometric tools such as PLS to describe physiological parameters in wheat. The selected VOCs, related to the conditions of biotic, abiotic and oxidative stress, could describe the changes in the VOC profile of wheat induced by the increase in altitude. Furthermore, common wheat and durum wheat showed different responses to stress at high altitude. These results could be considerably useful for the classification and authentication of wheat products (De Flaviis et al. 2022b).

Arslan et al. evaluated the feasibility of Fourier transform infrared (FTIR) spectroscopy together with multivariate data analysis methods for adulteration detection in high-quality wheat, with barley as an adulterant. They used various chemometric methods such as principal component analysis (PCA), LDA, hierarchical cluster analysis (HCA), soft independent modeling class analogy (SIMCA) and partial least squares regression (PLSR), which were able to classify between adulterated and unadulterated samples, including the quantification of adulteration degree (Arslan et al. 2020).

De Girolamo et al. investigated the infrared spectroscopy application in the analysis of raw durum wheat samples naturally contaminated with ochratoxin A (OTA). The samples were analyzed by Fourier transform near- and mid-infrared spectroscopies (FT-NIR, FT-MIR) using chemometric classification models such as PLS-DA and principal component analysis with linear discriminant analysis (PCA-LDA) to differentiate highly contaminated durum wheat samples from lightly contaminated ones. The results indicated that FT-NIR and FT-MIR analysis, in conjunction with chemometric tools, could be promising, inexpensive and easy-to-use alternatives to rapidly discriminate raw wheat samples from OTA content (De Girolamo et al. 2019).

He et al. combined Vis and NIR spectroscopies with PCA and LDA to discriminate between normal wheat grains and those contaminated with deoxynivalenol mycotoxins (DON). This research showed that Vis and NIR spectroscopies combined with chemometrics have the potential to be used for a non-destructive online detection system of wheat grains contaminated with DON (He et al. 2021). On the other hand, Shi et al. developed a rapid and non-destructive method for the DON content determination in wheat and for the discrimination of the contamination degree using a multispectral imaging system. Genetic algorithms (GAs) and PCA were used as preprocessing methods to obtain the best spectral characteristics. Then, PLS, support vector machines (SVM), and backpropagation neural networks (BPNN) were assessed on the reduced spectral data set. The results exposed that the multispectral imaging technology and chemometrics combination was an efficient and non-destructive method for the DON determination in wheat (Shi et al. 2022).

Jiao et al. employed surface-enhanced Raman spectroscopy (SERS) using active silver nanoparticle nanosensor and coated zinc oxide nanoparticles for rapid and sensitive quantitation of deltamethrin in wheat. Mean centering–successive projection algorithm–partial least squares regression (MC-SPA-PLS) provided an optimal predictive performance. The overall results suggested that SERS coupled with MC-SPA-PLS could be an easy and efficient method to quantify deltamethrin residues levels in wheat (Jiao et al. 2021).

Jaillais et al. evaluated the wheat grain safety quality, analyzing wheat infection with *Fusarium culmorum*. A novel non-destructive methodology based on multispectral imaging and chemometrics to detect wheat contamination was developed. Regarding the acquired results, it was possible to detect infected grains and identify the affected areas by PCA application. Furthermore, the regression analysis performed on the images allowed estimating the contamination degree of each pixel and therefore mapping the contamination of each grain (Jaillais et al. 2015). In another work, Serranti et al. investigated the possibility of using hyperspectral imaging techniques to classify different types of wheat grains infected with *Fusarium*. Thus, PCA, PLS-DA and interval partial least squares–discriminant analysis (iPLS-DA)—this last for the selection of a reduced set of effective wavelength intervals—were implemented for the obtained data analysis, achieving good classification results. The developed procedures could be useful in industries or laboratories for quality control purposes (Serranti et al. 2013). Mishra et al. used e-nose sensor to evaluate wheat infested with *Sitophilus granarius* during four different storage periods with various degrees of infestation. Multiple linear regression (MLR) models were developed to predict uric acid and protein content as infestation indexes. PCA and HCA efficiently sorted the most stored wheat grain samples. The findings of this study open up a convenient, rapid and non-destructive approach to determine insect-infested wheat grain quality at various stages during storage (Mishra et al. 2018).

Unuvar et al. used several spectroscopic methods for the discrimination of durum and common wheat samples, since they are fast, reliable, easy to use, low cost, ecological and non-destructive. Common and durum wheat samples from different genotypes were collected and analyzed by Raman spectroscopy, NIRS, synchronous fluorescence spectroscopy (SFS) and attenuated total reflectance–Fourier transform infrared spectroscopy (ATR-FTIR). Data analysis was performed using PCA and PLS-DA being the chemometric tools successful to distinguish between samples of common and durum wheat flours (Unuvar et al. 2021). On the other hand, Vermeulen et al. evaluated NIR hyperspectral imaging as a tool to discriminate between soft wheat and durum wheat at individual grain and bulk sample levels. For that, four approaches based on morphological criteria, NIR spectral profile, protein content criteria and glass/non-glass grain ratio were studied. PLS-DA was used as classification method for the discrimination model build. The results revealed that a combination of morphological and spectral NIR approaches could detect sample classification fraud with 99% accuracy (Vermeulen et al. 2018).

Kamboj et al. predicted the protein and carbohydrate content of wheat grain stored for one year at 4°C and 21°C using NIRS and chemometrics. Reference data and NIRS data were analyzed by PCA, PLSR, MLR and SVM regression (SVMR). All developed models showed good precision to successfully predict the protein and carbohydrate content of stored wheat grains, achieving the best performance with PLSR and SVMR. This study exhibited that NIRS has the potential to distinguish wheat grains stored under different temperature conditions (Kamboj et al. 2021).

Fortini Spoladore et al. used NIRS data and chemometric techniques to discriminate five genotypes of wheat grown from different cities and years. The spectral analysis by PCA indicated that samples could be grouped according to genotypes. Furthermore, the PLS-DA model using NIRS spectra was able to correctly classify 100% of the samples from the prediction set. In this case, NIRS together with chemometrics can be an alternative for genotype and hardness classification, even when the cereal was grown in different regions and years (Fortini Spoladore et al. 2021).

Williams determined the effectiveness of NIRS and chemometrics to predict physical quality parameters in wheat grains such as texture and specific gravity. These factors could be predicted with acceptable reliability using NIRS combined with artificial neural network (ANN) and least squares support vector machine (LSSVM; Williams 2020).

Wu et al. investigated the possibility of using NIR hyperspectral spectroscopy to differentiate waxy wheat and three partial waxy wheats from wild-type wheat. SVM, PLS-DA and BPNN methods were applied to build discrimination models. The results showed that the combination of NIR hyperspectral spectroscopy with SVM was useful in the identification of waxy and partial waxy wheats of wild-type wheat (Wu et al. 2021).

8.1.2 Analysis of Wheat Flours

Ahmad et al. investigated the influence of flour hydration on spectral signals, classification of farinographic phases and the separation of wheat flours based on their performance in baking. Using fluorescence spectroscopy and PCA, four farinographic phases were clearly separated in hydration, dough, stability and softening. The results demonstrate that fluorescence spectroscopy together with chemometric modeling can be used to characterize and categorize the farinographic kneading process, which is important in the baking industry (Ahmad et al. 2016a). Also, the same authors investigated the potential of fluorescence spectroscopy in conjunction with PLSR to predict analytical, rheological and baking parameters of wheat flours. Good linear regression models were obtained for protein, wet gluten and sedimentation value. As conclusion, fluorescence spectroscopy can be used as a rapid method to predict wheat quality and baking characteristics simply by taking the spectra of the flour without sample preparation (Ahmad et al. 2016b).

Boussard et al. assessed the effect and the biochemical characteristics of wheat flour added with yeast, glucose-oxidase, soybean and horse bean flours. The data set was analyzed by PCA concluding that in terms of lipid oxidation, the biochemical characteristics of the dough are not affected by the presence of glucose oxidase in the formulation, whatever the other ingredients (Boussard et al. 2012).

Cebi et al. developed a straightforward method for the L-cysteine detection in wheat flour using Raman microscopy. L-cysteine is deliberately added to various types of flour since it allows favorable baking conditions such as low viscosity, higher elasticity and increase during baking. L-cysteine detection in wheat flour was successfully achieved using chemometrics (PCA and HCA) combined with Raman microscopy (Cebi et al. 2017).

Correia et al. optimized the samples treatment by microwave-assisted digestion and two operating conditions in the inductively coupled plasma (ICP)-based method to determine 17 elements in different flours—including wheat, cassava and corn—through an experimental design. To achieve this goal, a multivariate data analysis using PCA and HCA was performed, achieving good results in the classification of flours according to their botanical origin (Correia et al. 2017).

Grassi et al. evaluated the gluten aggregation properties of whole wheat semolina for the durum wheat cultivars' differentiation containing different amounts of proteins related to gluten quality. The samples were classified into four quality classes based on the total gluten protein content and the glutenin to gliadin ratio. The class distribution was carried out by PCA, and a clear differentiation was achieved between varieties of high and medium-high quality from those of low and medium-low quality. In conclusion, the gluten aggregation properties proved to be good quality indicators for durum wheat semolina and therefore a reliable tool for the mixes preparation in the pasta-making process (Grassi et al. 2021).

Souza de Almeida Duarte et al. proposed a feasibility study using NIRS and analytical systems based on color histograms assisted by chemometrics (CACHAS) to determine and authenticate the content of cassava starch in wheat flour. The data-driven soft independent modeling class analogy (DD-SIMCA) application presented a similar predictive ability using NIRS and CACHAS for certified wheat flour, authenticating all target samples and correctly recognizing samples that could represent frauds. However, satisfactory results were not obtained for commercial wheat flour using CACHAS. Therefore, NIRS is more useful to offer definitive quantitative and qualitative analysis, whereas CACHAS can only provide alternative preliminary analysis (Souza de Almeida Duarte et al. 2022).

Tao et al. explored the low-cost, ultra-compact, portable micro-NIR spectrometer in the spectral range from 1150 to 2150 nm to detect adulteration of wheat flour. Eight varieties of cassava flour were used as adulterants in wheat flour at five adulteration levels of 5%, 10%, 20%, 30% and 40%. Methods such PCA-LDA and PLS-DA were performed to establish discriminant models. In general, the obtained results demonstrated the usefulness of the low-cost spectrometer to detect adulteration of wheat flour using a rapid and non-destructive method (Tao et al. 2021).

Tian et al. presented a NIRS application for total phenolic content (TPC) prediction in whole wheat flour, which provides several health human benefits. Wheat producers and consumers are increasingly interested in wheats with a higher TPC content, and even whole wheat flour with high levels of phenols may have a higher commercial value in the future. The results demonstrated that NIRS together to PCA and PLSR could accurately quantify phenolic compounds, even at low concentrations (less than 0.2%) in the food matrix (Tian et al. 2021).

Wadood et al. analyzed wheat grains and flour from three genotypes harvested over four years obtained from three different geographical areas of China. The combination of NIRS and chemometrics was able to classify wheat according to origin, production year and genotype. For these purposes, authors applied LDA and multivariate analysis of variance (MANOVA). The best classification performances were obtained from the flour matrix for both geographical origin and production years, with correct percentages of 100% and 73%, respectively. Regarding genotypes, the whole wheat grain showed a better classification percentage (98.2%). All samples were validated using an external validation procedure, and the obtained percentages result satisfactory indicated the suitability of the developed model (Wadood et al. 2019).

Zhao et al. investigated the feasibility of combining spectral and imaging information for the prediction of DON contamination in whole wheat flour using. The data acquired from Vis/NIR spectroscopies and hyperspectral imaging was analyzed by PCA and showed a certain clustering trend between acceptable samples and infected samples (Zhao et al. 2020). On the other hand, Li et al. developed a new methodology that integrates multimolecular infrared spectroscopy (MM-IR) with two-dimensional correlation spectroscopy of two traces (2T-2DCOS) for the qualitative and quantitative in situ determination of DON in wheat flour. Wheat flour samples with DON greater than and less than 1.00 mg kg^{-1} were classified by SVM, with an accuracy rate of up to 100%. Furthermore, a robust quantitative prediction model based on PLSR was established to detect low levels of DON contamination (Li et al. 2021).

8.1.3 Analysis of Wheat Doughs

Geng et al. established a fingerprinting method combined with analysis to differentiate breads made with whole wheat flour from those made using refined wheat flour. The chemical compositions of the bread samples were profiled by ultra-high-performance liquid chromatography–high resolution accurate mass–mass spectrometry (UHPLC-HRAM-MS). PCA and SIMCA were used as chemometric methods for samples differentiation and classification concluding that SIMCA can be a powerful and robust tool for the authentication of breads made with whole wheat flour (Geng et al. 2016).

Giannetti et al. used headspace–solid phase microextraction–gas chromatography–mass spectrometry (HS-SPME-GC-MS) to evaluate the volatile fraction of durum wheat pasta produced under different process conditions (artisanal manufacturing methods or industrial processes). At the same time, HPLC analysis was performed to quantify furosine in the same pulp samples. Furosine is commonly used as an index of nutritional damage that occurs during pasta drying. Since the furosine determination requires time- and solvent-consuming methods, a more convenient approach could be the identification of volatile compounds formed during the drying step and detectable by simple and automated methods. PCA was applied to the experimental data to discriminate the pulp samples according to their drying conditions. A limited number of volatile compounds proved to be crucial in the samples differentiation. Therefore, these compounds could be used as pulp quality markers, alternatively to furosine, or as process markers to keep drying treatment under control (Giannetti et al. 2014).

De Girolamo et al. used FT-IR combined with PLS-DA and LDA to classify commercial durum wheat pasta from Italy and Argentina, adulterated with soft wheat. The results indicated that the proposed methodology represents a promising, inexpensive and easy-to-use screening tool to rapidly analyze durum wheat pasta samples to monitor common wheat adulteration (De Girolamo et al. 2020).

8.2 RICE

Rice is one of the most important cereal crops around the world and constitutes the staple diet of almost half of the world's population. In addition, this annual plant is the main source of calories and protein for humans, mainly in Asian countries, the United States and some European countries. It is the grain with the third-highest production in the world after corn and wheat. The main producers and consumers of rice in the world are China, India, Indonesia, Bangladesh, Vietnam and Thailand (FAO 2021).

Rice is the seed of the monocarpic annual plant of the genus *Oryza*, which contains 21 wild species and two cultivated species (Vlachos and Arvanitoyannis 2008). The cultivated species are *Oryza sativa* (Asian rice) and *Oryza glaberrima* (African rice). Currently, the existence of approximately 5000 different varieties of rice in the world is evaluated. However, almost all of the thousands of varieties cultivated and developed have their origin in *Oryza sativa*. Rice grown in Asia can be divided into three main subspecies: indica, japonica and javanica, the first two being the most common species (Vlachos and Arvanitoyannis 2008). Indica varieties are long-grained and not sticky (e.g., basmati rice). Japonica varieties are short-grained and sticky. Javanica varieties are broad-grained and are grown in tropical climates.

Due to the high consumption of the global population for this food, rice is an ever-present subject for study. In the current literature, authentication and identification of rice is one of the most addressed concerns, particularly when it comes to the recognition of its geographical origin. The identification of rice varieties is also a problem widely reported in the literature. In addition, the dynamic growth of the organic industry and the preference of many consumers for organic foods have led to the need to develop tools capable of certifying rice cultivation methods (Maione and Barbosa 2019).

Normally, rice can be classified according to its physical characteristics. In general, grain length, grain shape, texture, color, translucency and aroma constitute some distinctive characteristics of rice. Basmati rice, jasmine rice and sticky rice are some of the most popular rice varieties (Drivelos and Georgiou 2012). However, the geographical locations and the different varieties of rice cannot be distinguished by their appearance. These limitations lead to authenticity problems and can affect the credibility of producers, traders and the rights of consumers (Gonzálvez et al. 2011). From this point of view, it is vitally important to develop new and sophisticated analytical techniques to ensure the traceability of food and thus prevent illegal marketing, since these frauds often include adulteration methods increasingly difficult to detect (Reid et al. 2006).

Rice identification has been extensively investigated in recent literature with the help of multivariate data analyses. These tools offer powerful methods capable of performing statistical and predictive analysis on data sets described by many variables. In the case of rice, the variables can be diverse and numerous (e.g., chemical elements, physicochemical properties). One of the key criteria for the rice authentication study lies in determining which variables to analyze, since different parameters may contain more or less information about the geographical origin, variety and type of crop used. Last, the combination of the multivariate data analysis method to be used with the chosen parameters is greatly influenced by the types of variables, the number of samples available and the quality of the data set (Śliwińska-Bartel et al. 2021).

8.2.1 Rice Analysis

The main constituents of rice are starch, proteins, lipids, vitamins, minerals and some volatile compounds. However, there are factors that can influence the chemical composition of the rice, such as the place and environment of cultivation, the processing stages (white or whole) and the different varieties. Due to these factors and the complex matrix of rice, the correct choice of analytical tools and the development of methodologies for the detection of fraud in the commercialization of rice are essential. At present, there are three different main research approaches to rice authenticity in the literature: botanical origin (variety), geographic origin and cultivation practices (organic rice).

8.2.2 Rice Variety Authentication

In general, the rice aroma profile is directly related to the botanical origin (i.e., to its variety). Due to consumer preferences, low yield and less resistance to weather conditions and diseases, the price of aromatic rice is several times higher than non-aromatic rice.

Basmati rice is the most popular variety due to its special reputation as a premium product. For this reason, over the years it has been one of the most tested and evaluated rice varieties in terms of authenticity.

Akram et al. investigated the profiles of metabolites in rice varieties—basmati and non-basmati—using polymeric hydrophilic interaction chromatography coupled to an Orbitrap mass analyzer. Among the identified compounds, they performed a statistical analysis by ANOVA and found the characteristic metabolites of basmati and non-basmati rice. The classification of the rice varieties classification was carried out by PCA and LDA (Akram et al. 2017).

India and Pakistan are the largest producers of basmati rice. Arif et al. demonstrated the high potential of combining elemental analysis with chemometric modeling for the classification of Pakistani basmati and non-basmati rice. In this study, the key elemental variables were found, and by means of DD-SIMCA it was possible to discriminate between the two varieties of rice (Arif et al. 2021). On the other hand, Shannon et al. developed a two-level analytical system, consisting of a rapid NIRS analysis at the culture site and a laboratory confirmation technique based on headspace–gas chromatography–mass spectrometry (HS-GC-MS). By using DD-SIMCA, PCA and orthogonal partial least squares–discriminant analysis (OPLS-DA) they were able to identify possible adulterations of high-quality basmati rice (Shannon et al. 2021). More recently, Arslan et al. developed a novel methodology to identify varieties of basmati rice. An array system of colorimetric sensors based on smartphones and a GC-MS analysis of the volatile compounds characteristic of basmati rice, together with the use of chemometric algorithms as PCA, HCA and k-nearest neighbor (k-NN), was able to efficiently discriminate the rice varieties (Arslan et al. 2022).

One of the most common long-grain varieties of rice is the so-called jasmine rice and, due to its high price, it is often mixed with low-quality rice. Timsorn et al. developed a low-cost sensor for the quantitative identification of adulteration of raw jasmine rice. An electronic nose based on metal oxide gas sensors, combined with PCA, SVM and BPNN, was successfully employed for the rapid detection of the percentage of adulteration in high-quality rice (Timsorn et al. 2017). Recently, some authors have detected and authenticated adulterations in rice by spectroscopic techniques and chemometric tools. Le Nguyen Doan et al. used NIR spectra and PLS-DA analysis, while Liu et al. analyzed multispectral images using PCA, PLSR, LSSVM and BPNN in order to identify adulterations of the different samples (Le Nguyen Doan et al. 2021; Liu et al. 2021b).

8.2.3 Rice Geographical Origin Authentication

In recent years, a growing number of publications have dealt with the geographical origins of rice, especially rice from Asia, due to the higher production and consumption on that continent. In addition, these investigations provide greater evidence of fraudulent actions related to the rice trade.

Chung et al. proposed a methodology to determine the authenticity of the geographical origin of brown rice grown during the same year in Korea, China and the Philippines using inductively coupled plasma–atomic emission spectroscopy (ICP-AES) by analysis of 25 constituent elements of brown rice. After a detailed chemometric study, they concluded that only 11 of these elements contribute significantly to the ability to discriminate the place of cultivation of the same type of rice. On the other hand, it should be noted that PLS-DA distinguished these samples according to geographical origin (Chung et al. 2015). The same authors also carried out a geographical origin classification of rice by analyzing, by means of IRMS, the isotopic compositions of four common elements (C, N, O and S). Applying PCA to the obtained experimental data, it was possible to distinguish between crops of the same rice grain grown in Korea, China and the Philippines (Chung et al. 2016).

In recent years, numerous authors reported the use of elemental analysis by inductively coupled plasma–mass spectrometry (ICP-MS), combined with chemometric tools, to classify rice from different countries and regions. The identification of rice from Thailand, France, Japan, India, Italy and Pakistan was carried out by Cheajesadagul et al. using PCA and LDA based on the analysis of 21 elements in 2013. Maione et al. (2016) based their study on the analysis of 20 elements and ANN to classify rice grown in different regions of Brazil. Rice from two different regions of Thailand was successfully classified by Promchan et al. (2016) combining multi-elemental analysis and LDA. More recently, Xu et al. (2021) were able to classify rice from different growing regions of China using SVM and random forest (RF) combined with multi-elemental analysis. On the other hand, some authors were able to combine ICP-MS with another instrumental technique in order to perform a multi-element analysis on other elements present in the rice grain. For example, Chung et al. and Liu et al. correctly assigned the origin of rice from countries such as Cambodia, the Philippines, Japan, Korea, Thailand, Malaysia, and different regions of China. In both cases, classification was achieved by combining multi-elemental analysis, stable isotope analysis by IRMS and chemometric tools, such as PCA, LDA, and OPLS-DA (Chung et al. 2018; Liu et al. 2019). Finally, Du et al. (2018) used mathematical algorithms, ANOVA, PCA and LDA to classify rice from various regions of China, combining electronic nose and ICP-MS, capable of detecting the characteristic volatile compounds and the analysis of 30 elements in aromatic rice.

Lim et al. reported the GC-MS profiles of volatile fractions and organic compounds present in white rice. According to the chemometric analyses by PLS-DA, within the volatile fractions, 1-hexanol and hexanal were the factors that contributed to the classification of rice produced in the different countries (Lim et al. 2018a). Among the different organic components, the phospholipid content presented the greatest variability in rice samples and was the key factor for the classification of the geographical origin of the grains (Lim et al. 2018b). Long et al. developed a rapid and precise methodology to authenticate the geographical origins of white rice, based on the multiple reactions monitoring–mass spectrometry (MRM-MS) analysis of the different phospholipids present in rice. The use of PCA for the analysis of the data obtained made possible the correct classification of the rice samples obtained from China and Korea (Long et al. 2017).

Recently, some authors have proposed an interesting approach for the authentication of the geographical origin of rice, based on NIRS. Guo et al. applied chemometric methods to the data obtained by NIRS from high-quality rice samples from China and from the same samples adulterated with different proportions of low-quality rice. The use of PLSR made it possible to quantify the percentage of adulteration in all samples (Guo et al. 2019). In addition, Teye et al. revealed that NIRS spectra together with chemometrics could be used for the rapid detection of the authenticity and quality of rice from different countries. Using a signal preprocessing method, such as multi-signal calibration (MSC) combined with PCA and k-NN, a classification rate greater than 90% for all evaluated rice categories was achieved (Teye et al. 2019).

NMR spectroscopy in conjunction with chemometrics was also successfully applied for grading rice from different regions of China and Korea. It was possible to identify the concentration profiles of different rice metabolites, such as sucrose, fructose, glucose, succinate and polyphenols. The use of PCA for these profiles made it possible to successfully discriminate the origin of the analyzed rice samples (Chae and Kim 2016).

8.2.4 Rice Cultivation Methods

Organic rice has become increasingly popular in recent years because it is considered healthier and more environmentally friendly. For this reason, it usually has a higher cost and is subject to possible fraud and adulterations. Thus, recently a growing number of publications have been reported that attempt to respond to the adulteration problems related to organic rice.

The authentication of organic rice in Brazil was approached by different authors, based on the elemental analysis by ICP-MS of samples of organic and conventional rice combined with various chemometrics strategies. The analysis of the concentrations of 20 chemical elements was carried out by Borges et al. (2015) performing the interpretation of the data using PCA, HCA, SIMCA and k-NN and providing a solid approach for the authenticity evaluation of organic rice. Meanwhile, Barbosa et al. (2016) combined ICP-MS with SVM to be able to discriminate both types of rice—with a precision of 96%—using only the concentration of two elements.

Xiao et al. addressed the problem of adulteration of organic rice by various methodologies. They reported a NIRS study to discriminate conventional rice and organic rice from different producers in China. To evaluate the obtained data, PCA and PLSR to classify 87.5% of the two types of culture were performed (Xiao et al. 2019). Then, the proteomic analysis of the two kinds of rice, using liquid chromatography–quadrupole time of flight–mass spectrometry (LC-QTOF-MS/MS) was developed. A total of 114 proteins were observed to change significantly between the organic and conventional rice samples. The use of PCA to discriminate between cultivation methods by proteomic analysis laid the foundation for future traceability and authentication studies of organic rice products (Xiao et al. 2019).

According to many North American, European and Chinese organizations, organic products are those produced without the use of fertilizers or pesticides, unlike green products in which a proportion of fertilizers can be used (FAO 2021). In relation to these regulations, being able to differentiate the rice produced according to the types of cultivation has become a challenge for the scientific community. Recent research has developed methodologies capable of discriminating the rice produced according to the different production methodologies. Lui et al. conducted a complementary study of stable isotopes and multi-element analysis combined with chemometric tools to differentiate organically grown rice from green and conventional rice. In this case, individual analysis of carbon and nitrogen isotopes was not enough to distinguish organic from green rice. Complementing these data with elemental analysis and PLS-DA, it was possible to correctly identify the types of rice with 100% precision (Liu et al. 2020). Chung et al. studied the variation of stable isotopes from conventional rice and from certified organic rice farms in Korea. The modeling of the data obtained by SVM demonstrated the possibility of differentiating organic rice from conventional rice, with a predictability of 84.5%. In particular, this methodology was also successful in discriminating pesticide-free rice (green rice) from those mentioned above (Chung et al. 2021).

Recent research related to the authentication of rice focuses mainly on the variety classification, the geographical origin and the cultivation method used. Currently, there are numerous developments and new strategies in rice fraud studies, and the number of analytical techniques used to detect it is growing. New methods for analyzing statistical data allow better classification of samples and provide improved tools to detect adulteration and false labeling of products. Fraud cases have negative impacts on industry and consumer confidence and may be related to health risks. In this way, a theoretical framework is established to fight against the incorrect labeling and counterfeiting of rice. At the same time, food safety and the integrity of the labeling must be guaranteed.

8.3 MAIZE

Maize (*Zea mays* L.) is one of the most important cereals because of its numerous applications in processed foods, where it is the major or minor component. Apart from maize authenticity issues related to cultivar and geographical origin (at national and/or international level) there is another important issue related to genetically modified maize. Various parameters such as proteins, fatty acids, phenolic compounds, pigments and heavy metals were determined in order to identify the maize authenticity. However, the implementation of multivariate analysis is of great importance toward reaching valid conclusions on authenticity issues.

8.3.1 Maize Genetically Modified and Varietal Origin Analysis

Maize grain contains all the genetic information of the crop and is essential for the production and improvement of crops, agricultural biotechnology, human nutrition and food security. Grains can be considered a key element for the success of the harvest (Finch-Savage and Bassel 2015). In recent years, technological innovation has led to the growing production of crops. This tendency may continue due to the increase in the use of genetically modified seeds, as well as the use of resources for its production, such as agrochemicals, fertilizers and equipment that offer more yielding security and stability during harvest (FAO 2021).

Premium quality seed varieties have higher prices because a better variety leads to achieve higher production and profits. Consequently, fraud practices in which mislabel poor quality seed varieties as good quality must be detected to protect farmers and manufacturers. Jia et al. achieved a clear differentiation between transgenic and non-transgenic maize cultivars by analyzing the perfusion profiles and monolithic RP-HPLC of the albumin, globulin, prolamin and glutelin maize fractions together with LDA. Significant differences were observed between transgenic and non-transgenic cultivars in the chromatograms obtained from any of the four protein fractions. LDA was successfully employed with a 100% correct classification percentage (Jia et al. 2015). In addition, the elemental analysis in maize seeds combined with the multivariate analysis, PCA and LDA, allowed to classify maize according two genetic technologies (Zaldarriaga Heredia et al. 2020b). In another work, Raman spectroscopy and chemometric tools were combined for the transgenic maize discrimination. Different spectral preprocessing and algorithms for the variable selection were evaluated to fit a classifier model based on LDA. The results showed spectral bands belonging to carbohydrates and carotenoids were responsible for the discrimination (Dib et al. 2021).

Regarding maize varieties, a novel method was developed to discriminate maize seed varieties based on hyperspectral imaging technology in the Vis-NIR region. From obtained images obtained of the maize samples, a variety discrimination model was established using SVM. The results showed that the model promises to be a new approach for the variety classification of maize seed (Zhou et al. 2020). In addition, NIRS has been used to identify the authenticity of the maize seed variety. Jia et al. obtained NIRS spectra using diffuse reflectance mode combined to SVM and SIMCA to establish the identification models for four varieties of maize. The results showed the feasibility of using NIRS and chemometrics as a rapid method for the identification of maize seeds (Jia et al. 2015). Moreover, the hyperspectral image data and the application of LSSVM based on the LDA wavelength selection algorithm showed a good ability to discriminate different maize varieties (Xia et al. 2019).

8.3.2 Maize Geographical Origin Authentication

Maize with specific geographical indications can affect its price. The geological differences according regions can cause discrepancy on mineral composition of soils where crops are grown and further influences on multi-element concentrations of crops (Zhao et al. 2013a). The mineral element fingerprinting technique combined with multivariate statistical analysis could be a helpful method to identify the geographical origin of maize. In the work reported by Wang et al. the mineral content was determined by ICP-MS as a fingerprint technique to highlight the indicators of favorable elements to determine the geographical origin of maize. PCA, k-NN and sequential linear discriminant analysis (SLDA) were applied as chemometric tools. The results obtained from the multivariate analysis showed that the maize kernels from different growing areas (Jilin, Gansu and Shandong provinces in China) had their own fingerprint characteristics of mineral elements (Wang et al. 2020). Other authors applied FTIR to address geographic origin as a characteristic of authenticity to develop classification models based on samples of maize kernels from three countries (Ukraine, the United States and Peru). PCA, PLS-DA and SIMCA applications allowed to classify the maize samples according to their geographical origin (Achten et al. 2019).

The production area of crops can affect the concentrations of many elements in various cereal crops, including maize. The effects of different agronomic practices on maize production have been studied, such as the application of different fertilizers, the variation in the sowing density applied to the crop and the influence of the slope land where the crop is developed (lowlands and highlands). The multielemental profile is a key parameter of crop quality due to its importance in the quality of food products. For this, inductively coupled plasma–optical emission spectroscopy (ICP-OES) was used to quantify the element profiles in maize grain samples. PCA and CA allowed classifying samples according to type of fertilizer, applied density and different slope of the land, according to the mineral profile. The integral analysis of the mineral profile carried out by means of multivariate analysis allowed to determine differences between the agricultural management practices and their possible effects on the seed quality (Zaldarriaga Heredia et al. 2019, 2020a).

8.3.3 Maize Flour and Its Main Derivatives Analysis

Maize flour can cover the nutritional requirements of human diet, acting as a good food supplement due to its richness in important nutrients for proper functioning of human body. Filho et al. evaluated the correlation between the mineral profile, obtained by atomic absorption spectroscopy (AAS), of yellow and white maize flours and the cereal growing in regions from Brazil. PCA-grouped samples with similar elemental profiles suggest that there is a relationship with the place of origin of the sample. Furthermore, other flour samples from the same region had different concentrations of the same metals, concluding that there are other factors that affect the mineral flour profile, additionally, to its geographical location (Filho et al. 2020). In other hands, the discrimination of 26 local varieties of maize flours from southern Brazil was performed through the use of ATR-FTIR and chemometrics. ATR-FTIR and PCA applied to the spectral data set proved to be useful tools for the screening and rapid discrimination of maize flours with different chemical composition (Kuhnen et al. 2010). The concentrations of total aflatoxins; aflatoxins B1, B2, G1 and G2; and heavy metals were determined in 21 samples of maize flour obtained from different cities and towns in Turkey and the United Kingdom. Toxins were determined by HPLC and heavy metals by ICP-MS. PCA showed that there is a strong relationship between heavy metals and the presence of toxins in maize flour samples (Algül and Kara 2014). Additionally, laser-induced breakdown spectrometry (LIBS) has been used to analyze cornmeal samples adulterated with sorghum. The sorghum proportion in maize flour mixtures has been determined based on its elemental composition. LIBS combined with PLS-DA provided successful classification in flour mix samples. In addition, LIBS and PLSR allowed the quantification of sorghum adulteration in maize flour mixtures. LIBS together with PLS-DA and PLSR analysis can be proposed as a very promising technique for the routine analysis of corn flour mixtures with sorghum (Akın et al. 2021).

8.4 OTHER CEREALS

8.4.1 Quinoa

Quinoa is a gluten-free crop that is receiving increasing attention around the world for its nutritional qualities. However, although the properties of quinoa are superior to that of the common cereals, it is considered a pseudo-cereal. Quinoa research has increased in recent years mainly due to its nutritional value and high yield. NIRS is a rapid, non-destructive technique useful for studying chemical properties. Suitable NIRS calibrations have been developed for the routine determination of dietary constituents using PLSR. The models was developed for the quantification of moisture, ash, lipid, protein and carbohydrate content, demonstrating that the proposed methodology achieved adequate results with low root mean square error prediction (RMSEP; Encina-Zelada et al. 2017). The high amino acid content in quinoa, especially essential amino acids, increased its demand by consumers. Again, NIRS combined with chemometric calibrations contributed to the determination of the amino acid profile in quinoa samples (Escudero et al. 2014). Capillary UV absorption–diode array detection electrophoresis (CE-UV-DAD) was used to obtain multi-feature wavelength electrophoretic profiles of soluble protein extracts from different varieties of quinoa grains. Then, multivariate curve resolution–alternated least squares (MCR-ALS) was applied, followed by PCA and PLS-DA, classifying the quinoa varieties according to their differential protein composition (Galindo-Luján et al. 2021).

Quinoa is consumed whole, but it is also ground to produce high-value flour, which is susceptible to adulteration. Quinoa flour has been receiving an increasing attention as a substitute for wheat flour in bread formulations, due to its immuno-nutritional features. This growing interest in quinoa has increased demand and, consequently, prices. For this reason, FT-MIR was used to detect the presence of three adulterants (soybean, maize and wheat flours). PLS-DA and SIMCA models were used to classify pure from adulterated samples (Rodríguez et al. 2019). Xue et al. applied front face synchronous fluorescence spectroscopy (FFSFS) for a rapid and non-invasive for adulterated quinoa flour detection with maize and soybean flours. PCA and LDA were performed for preliminary qualitative discrimination. Then, prediction models based on the combination of FFSFS spectra and PLSR were built. For mixtures of adulterants, they developed a general model PLS2 for the simultaneous determination of maize and soybean flours. The authors concluded that the study not only has important applications for the quinoa flour authentication, but also it can inform the quantitative analysis of components of other cereal flour mixtures from different botanical origins (Xue et al. 2021). Quinoa flour quality is largely determined by the properties of its main component, starch, as well as its non-starch components. Li and Zhu observed wide variation in chemical composition and physicochemical properties in quinoa flour samples. Thus, to analyze the correlations between the quinoa flour properties and starch, a PCA was performed revealing four distinctive groups of flour samples based on their composition and properties (Li and Zhu 2017).

8.4.2 Sorghum

Most of the world's sorghum production comes from the semi-arid regions of Africa and Asia. This cereal is one of the staples of the poorest countries in the world, especially those of the semiarid tropics. It stands out inasmuch as it is used as food, fodder, and in the biofuel and other industries. The mineral and protein content in sorghum varies due to genotypic and environmental influences, as well as its interactions. The mineral content (Ca, Fe, K, Mn, Na, P, Zn and Mg) and protein in sorghum genotypes was determined. PCA was used to emphasize the variation between genotypes and identify patterns in the data set used for the analysis. PCA explained most of the variability between genotypes. The presence of a considerable amount of variability in the mineral and protein composition suggested that genotypes affect strongly the sorghum nutritional quality (Gerrano et al. 2016). The extraction and assessment of fatty acid profiles from two varieties of sorghum grown in two different seasons were evaluated by GC-MS. Fatty acids—considered as variety descriptors—were assessed PCA and LDA. LDA allowed discrimination between Della and Red for Green (RG) varieties. The high predictive power suggests that fatty acid composition is an adequate fingerprint to reveal metabolic variations of these grains (Mengistie et al. 2021). Otherwise, a rapid quantification method for measurement of starch content and ethanol yield in sorghum cultivars using FT-NIR in diffuse reflectance mode was also developed and validated. Processed spectra were correlated with starch content and ethanol yield applying PLSR. The obtained results showed that FT-NIR can be used as a practical method for the approximate selection of sorghum cultivars to determine the desirable starch content and ethanol yield (Li et al. 2015).

8.4.3 Oats

Oats are a rich source of essential nutrients such as protein, lipids, vitamins, minerals and fiber. Compared to others, this cereal contains the highest protein content, with a good balance in the amino acid profile, which makes it very beneficial for the human diet (Menon et al. 2016). According to market requirements, the quantity of shelled oat grains is one of the most important quality characteristics of oats. Hyperspectral images of oat and shelled oat grain samples have been acquired. PCA was applied to the obtained data for exploratory purposes, and PLS-DA was performed to build the classification models and discriminate the two types of grain. Thus, the developed method could be implemented in the quality control of oat grains (Serranti et al. 2013). Moreover, the potential of the NIR hyperspectral imaging technique was studied in combination with multivariate data analysis to classify oats from gluten-rich grains. PCA showed the possibility of separating oat samples from wheat, barley and rye. Furthermore, the PLS-DA application resulted in good precision in classifying the grains into certain classes based on the obtained spectra. The results showed the potential of hyperspectral NIR images for their application in the control and inspection of the oats quality in different industrial processing stages (Erkinbaev et al. 2017). This cereal has a relatively higher lipid concentration among cereals and the lipid components are especially important for the functional characteristics. Using GC, the fatty acid profile was determined in oat samples. The performed analyses produced a total of 26 types of fatty acids that were contained in 46 oat samples, in which 11 fatty acids were common to all samples. Based on this information, a fingerprint based on the similarity of fatty acid information in different areas and varieties of oat samples could be established. The results of this study provided a basis for quality control of oat and oat products (Qi et al. 2017).

Furthermore, oats are widely utilized for human consumption and in food industries. Due to its high nutritional value and characteristic flavor, oat flour plays an important role in the breakfast cereals group and other processed foods as an alternative or supplement to the ordinary wheat flour. Wang et al. developed a rapid method for the quantitative analysis of wheat flour—used as an adulterant in oat flour—by NIRS and chemometric methods. PLSR was used to develop calibration models to predict wheat flour levels. The results indicated that NIRS and chemometric analysis can provide a rapid method for the quantitative analysis of wheat flour as an adulterant in oats (Wang et al. 2014). LIBS was also combined with multivariate analysis as a rapid method to classify flour types—including oatmeal—and to detect the addition of refined wheat flour to oatmeal samples. The methodology based on the combination of the element profile—determined by LIBS—and PCA provided a successful classification of the several types of flour, differentiating oatmeal from the rest. In addition, PLSR provided a successful quantification of the not allowed amount of refined wheat flour, which is added to bread and oat flour samples (Akin et al. 2020).

8.4.4 Barley

Barley is used as human food, but over time it was mainly combined in animal feed, malt and cereals for brewing, due in part to the increased prominence of wheat and rice over the years. The main advantage to incorporate barley in food products lay in its potential health benefits. For instance, PLSR was used for the recalibration of a prediction model for the protein content to obtain quality traits of malt barley with data from NIRS. In this stage, advantages of developing a recalibration model rather than developing a new calibration model to replace the calibration previously used in the NIRS application system were demonstrated (Abeshu 2021).

Phenolic compounds have been associated with the prevention of cardiovascular diseases, certain types of cancer, and other diseases related to aging, so the consumption of foods rich in polyphenols is becoming of great interest. The phenolic profile of 18 different varieties of waxy and non-waxy barleys was performed by high performance liquid chromatography–diode array detection–mass spectrometry (HPLC-DAD-MS). The waxy genotype of barley presented higher contents of free and bound phenolic compounds. PCA was developed based on free and bound phenolic compounds from barley samples, being able to be grouped by genotype (waxy and non-waxy). This data was then tested by PLS-DA and ANN analysis to discriminate genotypes, achieving a correct classification of 100% in both cases (Gómez-Caravaca et al. 2014).

In Table 8.1 are included the chemometric methods used in cereal analysis and derived: the first three (PCA, PLS-DA and LDA) represent more than 50% of all used chemometric tools cited in this chapter, which are usually the most preferred in food analysis.

Other tools such as SVM, PLSR, ANN and SIMCA are also common in food quality analysis but are less used in cereals. In this sense, the main factors leading to these preferences could be the implemented instrumental method and, related to this, sample pretreatment. Spectroscopic methods as MIRS, NIRS, fluorescence, Raman and others bring spectral information, which must be preprocessed before chemometric analysis and whose results do not always fit adequately. However, in most cases the pretreatment of samples is low or none. On the other hand, spectrometric or chromatographic methods are able to report the concentration of several organic or inorganic analytes, whose information is totally different from a spectrum and more descriptive about composition of samples. But in this case, the pretreatment of sample is an important step in the analysis. In both cases, the behavior of chemometric methods will depend on data quality and differences between classified samples (i.e., differences in composition or spectral characteristic in samples to be classified). In some cases, a spectral analysis can bring correct information to achieve best results; in other ones the differences in composition can be the best choice.

Regarding chemometrics, Figure 8.1 shows the number of reports (in percentages) included in this chapter for the analysis of cereals using chemometric approaches. From the point of view of published works, wheat is the most studied cereal, followed by rice, other cereals, maize and sorghum.

Table 8.1 Cereals: Chemometric Tools, Published Works, and World Production

Chemometric tool	Cites
PCA	46
PLS-DA	21
LDA	16
PLSR	16
SVM	10
HCA	6
SIMCA	6
BPNN	4
k-NN	4
LSSVM	3
DD-SIMCA	3
ANN	3
DPLS	2
PCA-LDA	2
MLR	2
OPLS-DA	2
ANOVA	2
SO-PLS-DA	1
GA	1
MC-SPA-PLS	1
iPLS-DA	1
CACHAS	1
MANOVA	1
RF	1
MSC	1
SLDA	1
CA	1
MCR-ALS	1

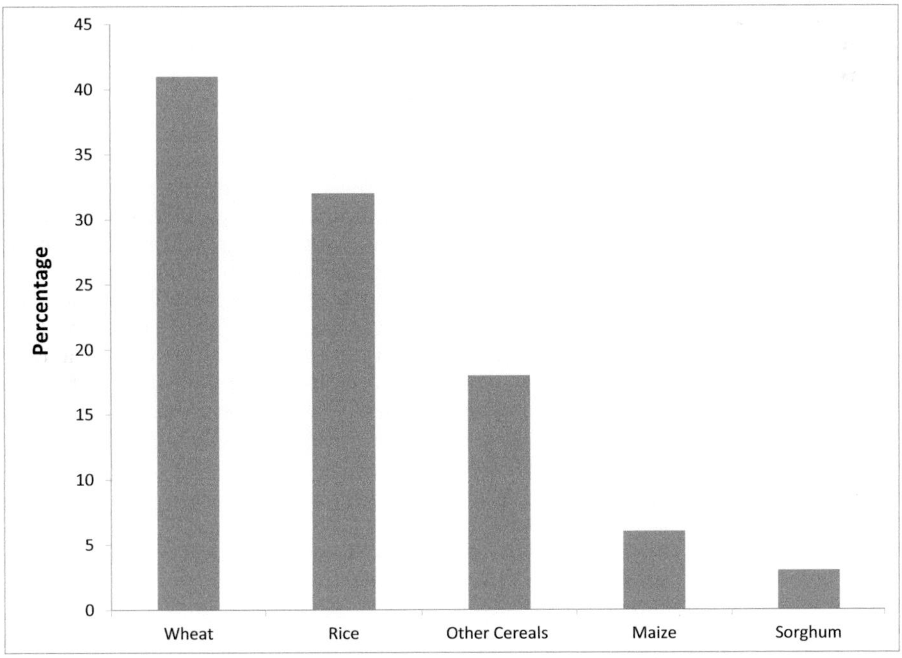

Figure 8.1 Perentage of cited works in this chapter, for the analysis of cereals by chemometrics.

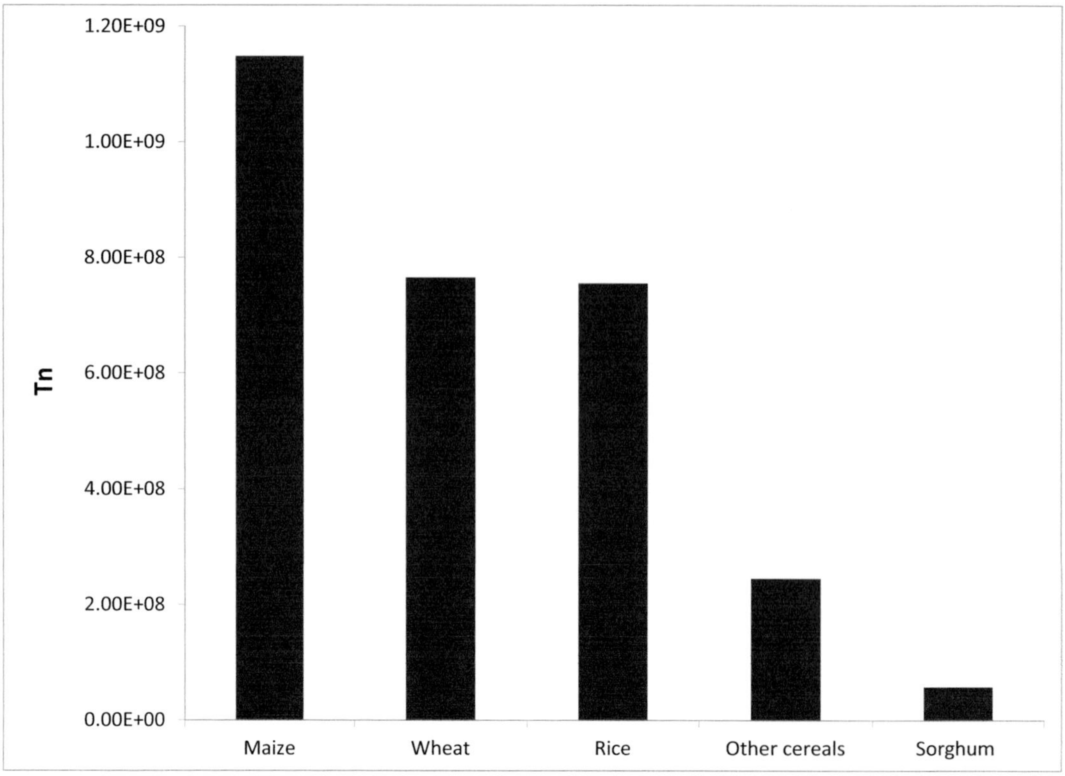

Figure 8.2 World production of cereals in 2020 (FAOSTAT 2021).

However, considering the world production of cereals (FAOSTAT 2021), maize represents the most cultivated cereal, followed by wheat, rice, other cereals and sorghum (Figure 8.2). That means wheat and rice are the most studied cereals by chemometrics, although maize is the most produced and consumed around the world. It is possible that number of works studying maize will increase in the next years, considering the new interest in geographical and varietal origins, as well as other quality indexes.

8.5 CONCLUSIONS

The need for safer and controlled foods requires the implementation of more trustworthy analytical methods to analyze both raw materials and manufactured products. In this sense, the analysis of adulterations and frauds in foods requires ever more sophisticated analytical tools able to detect acute changes in composition of samples. The ability of new and advanced instrumental methods to quantify concentrations on the order of μg L^{-1} or μg kg^{-1} is necessary to assess the obtained data by chemometric techniques to detect those acute changes. In this chapter, the most recent analytical platforms that combine analytical techniques and chemometrics with the objective to perform cereal authentication analysis have been exposed. As was mentioned previously, the list of published works is very extensive to include all of them here; from more than 400 reports published in the last five years, 102 were selected here. However our intention has been to show the most representative reports to facilitate a fast search about the state of the art on this topic.

8.6 ABBREVIATIONS

ANN	artificial neural networks
ATR-FTIR	attenuated total reflectance–Fourier transform infrared spectroscopy
BPNN	backpropagation neural networks
CACHAS	chemometrics-assisted color histogram–based analytical system
CE-UV-DAD	capillary UV absorption diode array detection electrophoresis
DC-MS	diffuse chromatography–mass spectrometry
DD-SIMCA	data-driven soft independent modeling class analogy

DF	data fusion
DON	deoxynivalenol
DPLS	discriminant partial least squares
FFSFS	front face synchronous fluorescence spectroscopy
FT-IR	Fourier transform infrared spectroscopy
FT-MIR	Fourier transform mid-infrared spectroscopy
FT-NIR	Fourier transform near-infrared spectroscopy
GA	genetic algorithms
GC-MS	gas chromatography–mass spectrometry
HCA	hierarchical cluster analysis
HPLC-DAD-MS	high-performance liquid chromatography–diode array detection–mass spectrometry
HS-GC-MS	headspace–gas chromatography–mass spectrometry
HS-SPME-GC-MS	headspace–solid phase microextraction–gas chromatography–mass spectrometry
ICP-AES	inductively coupled plasma–atomic emission spectroscopy
ICP-MS	inductively coupled plasma–mass spectrometry
ICP-OES	inductively coupled plasma–optical emission spectroscopy
iPLS-DA	interval partial least squares–discriminant analysis
IRMS	isotope-ratio mass spectrometry
k-NN	k-nearest neighbor
LC-QTOF-MS/MS	liquid chromatography–quadrupole time of flight–mass spectrometry
LDA	linear discriminant analysis
LIBS	laser-induced breakdown spectrometry
LSSVM	least squares support vector machine
MANOVA	multivariate analysis of variance
MCR-ALS	multivariate curve resolution–alternated least square
MC-SPA-PLS	mean centering–successive projection algorithm–partial least squares
MIRS	mid-infrared spectroscopy
MLR	multiple linear regression
MM-IR	multimolecular–infrared spectroscopy
MRM-MS	multiple reactions monitoring–mass spectrometry
MSC	multi-signal calibration
NIRS	near-infrared spectroscopy
NMR	nuclear magnetic resonance
OPLS-DA	orthogonal partial least squares–discriminant analysis
OTA	ochratoxin A
PCA	principal component analysis
PCA-LDA	principal component analysis–linear discriminant analysis
PLS-DA	partial least squares–discriminant analysis
PLSR	partial least squares regression
RF	random forest
RMSEP	root mean square error prediction
SERS	surface-enhanced Raman spectroscopy
SFS	synchronous fluorescence spectroscopy
SIMCA	soft independent modeling class analogy
SLDA	sequential linear discriminant analysis
SO-PLS-DA	sequentially orthogonalized–partial least squares–discriminant analysis
SVM	support vector machines
SVMR	support vector machine regression
TPC	total phenolic content
UHPLC-HRAM-MS	ultra-high-performance liquid chromatography–high-resolution accurate mass–mass spectrometry
VOC	volatile organic compounds

REFERENCES

Abeshu, Y. 2021. Development of NIRS Re-calibration model for Ethiopian barley (Hordeum Vulgare) lines traits to determine their brewing potential. *Journal of Agriculture and Food Research* 6:100238. doi: 10.1016/J.JAFR.2021.100238.

Achten, E., D. Schütz, M. Fischer, C. Fauhl-Hassek, J. Riedl, and B. Horn. 2019. Classification of grain maize (Zea mays L.) from different geographical origins with FTIR spectroscopy—a suitable analytical tool for feed authentication. *Food Analytical Methods* 12:2172–2184. doi: 10.1007/s12161-019-01558-9.

Ahmad, M. H., M. Nache, S. Waffenschmidt, and B. Hitzmann. 2016a. Characterization of farinographic kneading process for different types of wheat flours using fluorescence spectroscopy and chemometrics. *Food Control* 66:44–52. doi: 10.1016/j.foodcont.2016.01.029.

Ahmad, M. H., M. Nache, S. Waffenschmidt, and B. Hitzmann. 2016b. A fluorescence spectroscopic approach to predict analytical, rheological and baking parameters of wheat flours using chemometrics. *Journal of Food Engineering* 182:65–71. doi: 10.1016/j.jfoodeng.2016.03.006.

Akın, P. A., B. Sezer, S. R. Bean, K. Peiris, M. Tilley, H. Apaydin, and İ. H. Boyacı. 2021. Analysis of corn and sorghum flour mixtures using laser-induced breakdown spectroscopy. *Journal of the Science of Food and Agriculture* 101 (3):1076–1084. doi: 10.1002/jsfa.10717.

Akin, P. A., B. Sezer, T. Sanal, H. Apaydin, H. Koksel, and İ. H. Boyaci. 2020. Multi-elemental analysis of flour types and breads by using laser induced breakdown spectroscopy. *Journal of Cereal Science* 92:102920. doi: 10.1016/J.JCS.2020.102920.

Akram, M. I., I. M. Vincent, A. J. Siddiqui, and S. Ghulam Musharraf. 2017. Polymeric hydrophilic interaction liquid chromatography coupled with orbitrap mass spectrometry and chemometric analysis for untargeted metabolite profiling of natural rice variants. *Journal of Cereal Science* 73:165–173. doi: 10.1016/j.jcs.2017.01.001.

Algül, I. and D. Kara. 2014. Determination and chemometric evaluation of total aflatoxin, aflatoxin B1, ochratoxin A and heavy metals content in corn flours from Turkey. *Food Chemistry* 157:70–76. doi: 10.1016/J.FOODCHEM.2014.02.004.

Arif, M., G. Chilvers, S. Day, S. A. Naveed, M. Woolfe, O. Ye Rodionova, A. L. Pomerantsev, et al. 2021. Differentiating Pakistani long-grain rice grown inside and outside the accepted Basmati Himalayan geographical region using a 'one-class' multi-element chemometric model. *Food Control* 123:107827. doi: 10.1016/j.foodcont.2020.107827.

Arslan, F. N., G. Akin, Ş. N. Karuk Elmas, B. Üner, I. Yilmaz, H. G. Janssen, and A. Kenar. 2020. FT-IR Spectroscopy with chemometrics for rapid detection of wheat flour adulteration with barley flour. *Journal Fur Verbraucherschutz Und Lebensmittelsicherheit* 15 (3):245–261. doi: 10.1007/s00003-019-01267-9.

Arslan, M., M. Zareef, H. Elrasheid Tahir, Z. Guo, A. Rakha, H. Xuetao, J. Shi, Li Zhihua, Z. Xiaobo, and M. Rafiq Khan. 2022. Discrimination of rice varieties using smartphone-based colorimetric sensor arrays and gas chromatography techniques. *Food Chemistry* 368:130783. doi: 10.1016/j.foodchem.2021.130783.

Barbosa, R. M., E. Silva de Paula, A. C. Paulelli, A. F. Moore, J. M. Oliveira Souza, B. Lemos Batista, A. D. Campiglia, and F. Barbosa. 2016. Recognition of organic rice samples based on trace elements and support vector machines. *Journal of Food Composition and Analysis* 45:95–100. doi: 10.1016/j.jfca.2015.09.010.

Borges, E. M., J. M. Lafayette Neves Gelinski, V. C. de Oliveira Souza, F. Barbosa, and B. Lemos Batista. 2015. Monitoring the authenticity of organic rice via chemometric analysis of elemental data. *Food Research International* 77:299–309. doi: 10.1016/j.foodres.2015.06.008.

Boussard, A., C. B.Y. Cordella, L. Rakotozafy, G. Moulin, F. Buche, J. Potus, and J. Nicolas. 2012. Use of chemometric tools to estimate the effects of the addition of yeast, glucose-oxidase, soybean or horse bean flours to wheat flour on biochemical bread dough characteristics. *Chemometrics and Intelligent Laboratory Systems* 113:68–77. doi: 10.1016/j.chemolab.2012.01.006.

Cebi, N., C. E. Dogan, A. Develioglu, M. E. Altuntop Yayla, and O. Sagdic. 2017. Detection of L-cysteine in wheat flour by Raman microspectroscopy combined chemometrics of HCA and PCA. *Food Chemistry* 228:116–124. doi: 10.1016/j.foodchem.2017.01.132.

Chae, Y. K., and S. Hyun Kim. 2016. Discrimination of rice products by geographical origins and cultivars by two-dimensional NMR Spectroscopy. *Bulletin of the Korean Chemical Society* 37 (10):1612–1617. doi: 10.1002/bkcs.10914.

Cheajesadagul, P., C. Arnaudguilhem, J. Shiowatana, A. Siripinyanond, and J. Szpunar. 2013. Discrimination of geographical origin of rice based on multi-element fingerprinting by high resolution inductively coupled plasma mass spectrometry. *Food Chemistry* 141 (4):3504–3509. doi: 10.1016/j.foodchem.2013.06.060.

Chung, I. M., J. K. Kim, J. Keun Lee, and S. Hyun Kim. 2015. discrimination of geographical origin of rice (Oryza sativa L.) by multielement analysis using inductively coupled plasma atomic emission spectroscopy and multivariate analysis. *Journal of Cereal Science* 65:252–259. doi: 10.1016/j.jcs.2015.08.001.

Chung, I. M., J. K. Kim, K. J. Lee, S. K. Park, J. H. Lee, N. Y. Son, Y. I. Jin, and S. Hyun Kim. 2018. Geographic authentication of Asian rice (Oryza sativa L.) using multi-elemental and stable isotopic data combined with multivariate analysis. *Food Chemistry* 240:840–849. doi: 10.1016/j.foodchem.2017.08.023.

Chung, I. M., J. K. Kim, M. Prabakaran, J. H. Yang, and S. Hyun Kim. 2016. Authenticity of rice (Oryza sativa L.) geographical origin based on analysis of c, n, o and s stable isotope ratios: A preliminary case report in Korea, China and Philippine. *Journal of the Science of Food and Agriculture* 96 (7):2433–2439. doi: 10.1002/jsfa.7363.

Chung, I. M., Y. J. Kim, H. S. Moon, H. Y. Chi, and S. Hyun Kim. 2021. Long-term isotopic model study for ecofriendly rice (Oryza Sativa L.) authentication: updating a case study in South Korea. *Food Chemistry* 362:130215. doi: 10.1016/j.foodchem.2021.130215.

Correia, F. O., D. S. Silva, S. S. Lopes Costa, I. Kalini V. Silva, D. Ribeiro da Silva, J. P. H. Alves, C. A. B. Garcia, T. de A. Maranhão, E. A. Passos, and R. G. O. Araujo. 2017. Optimization of microwave digestion and inductively coupled plasma-based methods to characterize cassava, corn and wheat flours using chemometrics. *Microchemical Journal* 135:190–198. doi: 10.1016/j.microc.2017.09.007.

De Flaviis, R., D. Mutarutwa, G. Sacchetti, and D. Mastrocola. 2022a. Could environmental effect overcome genetic? A chemometric study on wheat volatiles fingerprint. *Food Chemistry* 372:131236. doi: 10.1016/j.foodchem.2021.131236.

De Flaviis, R., D. Mutarutwa, G. Sacchetti, and D. Mastrocola. 2022b. Quantitatively unravelling the effect of altitude of cultivation on the volatiles fingerprint of wheat by a chemometric approach. *Food Chemistry* 370:131296. doi: 10.1016/j.foodchem.2021.131296.

De Girolamo, A., C. von Holst, M. Cortese, S. Cervellieri, M. Pascale, F. Longobardi, L. Catucci, A. C. Raffaella Porricelli, and V. Lippolis. 2019. Rapid screening of ochratoxin a in wheat by infrared spectroscopy. *Food Chemistry* 282:95–100. doi: 10.1016/j.foodchem.2019.01.008.

De Girolamo, A., M. C. Arroyo, S. Cervellieri, M. Cortese, M. Pascale, A.F. Logrieco, and V. Lippolis. 2020. Detection of durum wheat pasta adulteration with common wheat by infrared spectroscopy and chemometrics: A case study. *LWT- Food Science and Technology* 127:109368. doi: 10.1016/j.lwt.2020.109368.

Dib, S. R., T. Varão Silva, J. A. Gomes Neto, L. J. Moreira Guimarães, E. J. Ferreira, and E. C. Ferreira. 2021. Raman spectroscopy for discriminating transgenic corns. *Vibrational Spectroscopy* 112:103183. doi: 10.1016/J.VIBSPEC.2020.103183.

Drivelos, S. A., and C. A. Georgiou. 2012. Multi-element and multi-isotope-ratio analysis to determine the geographical origin of foods in the European Union. *TrAC—Trends in Analytical Chemistry* 40:38–51. doi: 10.1016/j.trac.2012.08.003.

Du, M., Y. Fang, F. Shen, B. Mao, Y. Zou, P. Li, F. Pei, and Q. Hu. 2018. Multiangle discrimination of geographical origin of rice based on analysis of mineral elements and characteristic volatile components. *International Journal of Food Science and Technology* 53 (9):2088–2096. doi: 10.1111/ijfs.13795.

Encina-Zelada, C., V. Cadavez, J. Pereda, L. Gómez-Pando, B. Salvá-Ruíz, J. A. Teixeira, M. Ibañez, K. H. Liland, and U. Gonzales-Barron. 2017. Estimation of composition of quinoa (Chenopodium quinoa Willd.) grains by near-infrared transmission spectroscopy. *LWT- Food Science and Technology* 79:126–134. doi: 10.1016/j.lwt.2017.01.026.

Erkinbaev, C., K. Henderson, and J. Paliwal. 2017. Discrimination of gluten-free oats from contaminants using near infrared hyperspectral imaging technique. *Food Control* 80:197–203. doi: 10.1016/J.FOODCONT.2017.04.036.

Escudero, O., M. I. González Martín, G. Wells Moncada, S. Fischer, and J. M. Hernández Hierro. 2014. Amino acid profile of the quinoa (Chenopodium quinoa Willd.) using near infrared spectroscopy and chemometric techniques. *Journal of Cereal Science* 60 (1):67–74. doi: 10.1016/j.jcs.2014.01.016.

FAO. 2021. Food and Agriculture Organization of the United Nations. www.fao.org/home/es

FAOSTAT. 2021. Organización de las Naciones Unidas para la Alimentación y la Agricultura. www.fao.org/faostat/es/

Filho, I. K., A. C. Jaski, M. M. Takayanagui, C. K. De Andradre, M. Butik, and S. P. Quinaia. 2020. Determination and chemometric evaluation of the mineral profile of maize flours. *Journal of Food Composition and Analysis* 92:103579. doi: 10.1016/J.JFCA.2020.103579.

Finch-Savage, W. E., and Bassel, G.W. 2015. Seed vigour and crop establishment: Extending performance beyond adaptation. *Journal of Experimental Botany* 67 (3), 567–591. doi: 10.1093/jxb/erv490

Fortini, S., S. M. B. dos Santos Scholz, and E. Bona. 2021. Genotypic classification of wheat using near-infrared spectroscopy and PLS-DA. *Applied Food Research* 1 (2):100019. doi: 10.1016/j.afres.2021.100019.

Foschi, M., A. Biancolillo, S. Vellozzi, F. Marini, A. A. D'Archivio, and R. Boqué. 2021. spectroscopic fingerprinting and chemometrics for the discrimination of Italian emmer landraces. *Chemometrics and Intelligent Laboratory Systems* 215: 104348. doi: 10.1016/j.chemolab.2021.104348.

Galindo-Luján, R., L. Pont, V. Sanz-Nebot, and F. Benavente. 2021. Classification of quinoa varieties based on protein fingerprinting by capillary electrophoresis with ultraviolet absorption diode array detection and advanced chemometrics. *Food Chemistry* 341:128207. doi: 10.1016/j.foodchem.2020.128207.

Geng, P., J. M. Harnly, and P. Chen. 2016. Differentiation of bread made with whole grain and refined wheat (T. Aestivum) flour using LC/MS-based chromatographic fingerprinting and chemometric approaches. *Journal of Food Composition and Analysis* 47:92–100. doi: 10.1016/j.jfca.2015.12.010.

Gerrano, A. S., M. T. Labuschagne, A. Van Biljon, and N. G. Shargie. 2016. Quantification of mineral composition and total protein content in sorghum [Sorghum Bicolor (L.) Moench] genotypes. *Cereal Research Communications* 44 (2):272–285. doi: 10.1556/0806.43.2015.046.

Giannetti, V., M. Boccacci Mariani, P. Mannino, and E. Testani. 2014. Furosine and flavour compounds in durum wheat pasta produced under different manufacturing conditions: Multivariate chemometric characterization. *LWT—Food Science and Technology* 56 (1):15–20. doi: 10.1016/j.lwt.2013.10.045.

Gómez-Caravaca, A. M., V. Verardo, A. Berardinelli, E. Marconi, and M. F. Caboni. 2014. A chemometric approach to determine the phenolic compounds in different barley samples by two different stationary phases: a comparison between C18 and pentafluorophenyl core shell columns. *Journal of Chromatography A* 1355:134–142. doi: 10.1016/J.CHROMA.2014.06.007.

González-Martín, M. I., G. Wells Moncada, C. González-Pérez, N. Zapata San Martín, F. López-González, I. Lobos Ortega, and J. Miguel Hernández-Hierro. 2014. Chilean flour and wheat grain: tracing their origin using near infrared spectroscopy and chemometrics. *Food Chemistry* 145:802–806. doi: 10.1016/j.foodchem.2013.08.103.

Gonzálvez, A., S. Armenta, and M. De La Guardia. 2011. Geographical traceability of 'arrs de Valencia' rice grain based on mineral element composition. *Food Chemistry* 126 (3):1254–1260. doi: 10.1016/j.foodchem.2010.11.032.

Grassi, S., M. Gullì, G. Visioli, and A. Marti. 2021. Gluten aggregation properties as a tool for durum wheat quality assessment: A chemometric approach. *LWT- Food Science and Technology* 142:111048. doi: 10.1016/j.lwt.2021.111048.

Guo, L., Y. Yu, H. Yu, Y. Tang, J. Li, Y. Du, Y. Chu, S. Ma, Y. Ma, and X. Zeng. 2019. Rapid quantitative analysis of adulterated rice with partial least squares regression using hyperspectral imaging system. *Journal of the Science of Food and Agriculture* 99 (12):5558–5564. doi: 10.1002/jsfa.9824.

He, X., T. Zhao, F. Shen, Q. Liu, Y. Fang, and Q. Hu. 2021. online detection of naturally DON contaminated wheat grains from China using Vis-NIR Spectroscopy and computer vision. *Biosystems Engineering* 201:1–10. doi: 10.1016/j.biosystemseng.2020.11.001.

Jaillais, B., P. Roumet, L. Pinson-Gadais, and D. Bertrand. 2015. Detection of fusarium head blight contamination in wheat kernels by multivariate imaging. *Food Control* 54:250–258. doi: 10.1016/j.foodcont.2015.01.048.

Jiao, T., M. M. Hassan, J. Zhu, S. Ali, W. Ahmad, J. Wang, C. Lv, Q. Chen, and H. Li. 2021. Quantification of deltamethrin residues in wheat by AG@ZNO NFS-based surface-enhanced Raman spectroscopy coupling chemometric models. *Food Chemistry* 337:127652. doi: 10.1016/j.foodchem.2020.127652.

Jia, S., D. An, Z. Liu, J. Gu, S. Li, X. Zhang, D. Zhu, T. Guo, and Y. Yan. 2015. Variety identification method of coated maize seeds based on near-infrared spectroscopy and chemometrics. *Journal of Cereal Science* 63:21–26. doi: 10.1016/J.JCS.2014.07.003.

Kamboj, U., P. Guha, and S. Mishra. 2021. Comparison of PLSR, MLR, SVM regression methods for determination of crude protein and carbohydrate content in stored wheat using near infrared spectroscopy. *Materials Today: Proceedings*. doi: 10.1016/j.matpr.2021.04.540.

Kuhnen, S., J. B. Ogliari, P. F. Dias, E. F. Boffo, I. Correia, A. G. Ferreira, I. Delgadillo, and M. Maraschin. 2010. ATR-FTIR spectroscopy and chemometric analysis applied to discrimination of landrace maize flours produced in southern Brazil. *International Journal of Food Science and Technology* 45 (8):1673–1681. doi: 10.1111/j.1365–2621.2010.02313.

Li, F. L., J. Xie, S. Wang, Y. Wang, and C. Hua Xu. 2021. Direct qualitative and quantitative determination methodology for massive screening of DON in wheat flour based on multi-molecular infrared spectroscopy (MM-IR) with 2T-2DCOS. *Talanta* 234:122653. doi: 10.1016/j.talanta.2021.122653.

Li, G., and F. Zhu. 2017. Physicochemical properties of quinoa flour as affected by starch interactions. *Food Chemistry* 221:1560–1568. doi: 10.1016/j.foodchem.2016.10.137.

Li, J., M. G. C. Danao, S. F. Chen, S. Li, V. Singh, and P. J. Brown. 2015. Prediction of starch content and ethanol yields of sorghum grain using near infrared spectroscopy. *Journal of Near Infrared Spectroscopy* 23 (2):85–92. doi: 10.1255/jnirs.1146.

Lim, D. K., C. Mo, D. K. Lee, N. P. Long, J. Lim, and S. W. Kwon. 2018a. Non-destructive profiling of volatile organic compounds using HS-SPME/GC—MS and its application for the geographical discrimination of white rice. *Journal of Food and Drug Analysis* 26 (1):260–267. doi: 10.1016/j.jfda.2017.04.005.

Lim, D. K., C. Mo, J. H. Lee, N. P. Long, Z. Dong, J. Li, J. Lim, and S. W. Kwon. 2018b. The integration of multi-platform MS-Based metabolomics and multivariate analysis for the geographical origin discrimination of Oryza sativa L. *Journal of Food and Drug Analysis* 26 (2):769–777. doi: 10.1016/j.jfda.2017.09.004.

Liu, H. Y., S. A. Wadood, Y. Xia, Y. Liu, H. Guo, B. L. Guo, and R. Y. Gan. 2021a. Wheat authentication: An overview on different techniques and chemometric methods. *Critical Reviews in Food Science and Nutrition* in press, 1–24. doi: 10.1080/10408398.2021.1942783.

Liu, W., X. Xu, C. Liu, and L. Zheng. 2021b. Nondestructive detection of authenticity of Thai jasmine rice using multispectral imaging. *Journal of Food Quality* 2021:6642220. doi: 10.1155/2021/6642220.

Liu, Z., W. Zhang, Y. Zhang, T. Chen, S. Shao, L. Zhou, Y. Yuan, T. Xie, and K. M. Rogers. 2019. Assuring food safety and traceability of polished rice from different production regions in China and Southeast Asia using chemometric Models. *Food Control* 99:1–10. doi: 10.1016/j.foodcont.2018.12.011.

Liu, Z., Y. Yuan, T. Xie, Y. Zhang, S. Shao, J. Nie, W. Xia, K. M. Rogers, and W. Zhang. 2020. Long-term agricultural effects on the authentication accuracy of organic, green, and conventional rice using isotopic and elemental chemometric analyses. *Journal of Agricultural and Food Chemistry* 68 (5):1213–1225. doi: 10.1021/acs.jafc.9b06847.

Long, N. P., D. K. Lim, C. Mo, G. Kim, and S. W. Kwon. 2017. Development and assessment of a lysophospholipid-based deep learning model to discriminate geographical origins of white rice. *Scientific Reports* 7 (1):1–10. doi: 10.1038/s41598-017-08892-0.

Maione, C., B. L. Batista, A. Dobal Campiglia, F. Barbosa, and R. M. Barbosa. 2016. Classification of geographic origin of rice by data mining and inductively coupled plasma mass spectrometry. *Computers and Electronics in Agriculture* 121:101–107. doi: 10.1016/j.compag.2015.11.009.

Maione, C., and R. M. Barbosa. 2019. Recent applications of multivariate data analysis methods in the authentication of rice and the most analyzed parameters: A review. *Critical Reviews in Food Science and Nutrition* 59 (12):1868–1879. doi. 10.1080/10408398.2018.1431763.

Mengistie, E., A. M. Alayat, F. Sotoudehnia, N. Bokros, S. DeBolt, and A. G. McDonald. 2021. Fatty acid profile-based chemometrics to differentiate metabolic variations in sorghum. *ACS Food Science & Technology* 1 (11):2127–2134 doi: 10.1021/acsfoodscitech.1c00320.

Menon, R., T. Gonzalez, M. Ferruzzi, E. Jackson, D. Winderl, and J. Watson. 2016. Oats-from farm to fork. In *Advances in Food and Nutrition Research*. 1st ed. Vol. 77. Elsevier Inc. doi: 10.1016/bs.afnr.2015.12.001.

Mishra, G., S. Srivastava, B. K. Panda, and H. N. Mishra. 2018. prediction of sitophilus granarius infestation in stored wheat grain using multivariate chemometrics & fuzzy logic-based electronic nose analysis. *Computers and Electronics in Agriculture* 152:324–332. doi: 10.1016/j.compag.2018.07.022.

Nguyen D., D. Le, Q. C. Nguyen, F. Marini, and A. Biancolillo. 2021. Authentication of rice (Oryza sativa l.) using near infrared spectroscopy combined with different chemometric classification strategies. *Applied Sciences* 11 (1):362. doi: 10.3390/app11010362.

Promchan, J., D. Günther, A. Siripinyanond, and J. Shiowatana. 2016. Elemental imaging and classifying rice grains by using laser ablation inductively coupled plasma mass spectrometry and linear discriminant analysis. *Journal of Cereal Science* 71:198–203. doi: 10.1016/j.jcs.2016.08.017.

Qi, X. G., L. Zhu, C. Wang, H. Zhang, L. Wang, and H. Qian. 2017. Development of standard fingerprints of naked oats using chromatography combined with principal component analysis and cluster analysis. *Journal of Cereal Science* 74:224–230. doi: 10.1016/J.JCS.2017.02.009.

Reid, L. M., C. P. O'Donnell, and G. Downey. 2006. Recent technological advances for the determination of food authenticity. *Trends in Food Science and Technology* 17 (7):344–353. doi: 10.1016/j.tifs.2006.01.006.

Rodríguez, S. D., G. Rolandelli, and M. P. Buera. 2019. Detection of quinoa flour adulteration by means of FT-MIR spectroscopy combined with chemometric methods. *Food Chemistry* 274:392–401. doi: 10.1016/j.foodchem.2018.08.140.

Serranti, S., D. Cesare, F. Marini, and G. Bonifazi. 2013. Classification of oat and groat kernels using NIR hyperspectral imaging. *Talanta* 103:276–284. doi: 10.1016/J.TALANTA.2012.10.044.

Serranti, S., D. Cesare, and G. Bonifazi. 2013. The development of a hyperspectral imaging method for the detection of fusarium-damaged, Yellow berry and vitreous Italian durum wheat kernels. *Biosystems Engineering* 115 (1):20–30. doi: 10.1016/j.biosystemseng.2013.01.011.

Shannon, M., C. H. Ratnasekhar, T. F. McGrath, A. P. Kapil, and C. T. Elliott. 2021. A two-tiered system of analysis to tackle rice fraud: The Indian Basmati study. *Talanta* 225:122038. doi: 10.1016/j.talanta.2020.122038.

Shi, Y., W. Liu, P. Zhao, C. Liu, and L. Zheng. 2022. Rapid and nondestructive quantification of deoxynivalenol in individual wheat kernels using near-infrared hyperspectral imaging and chemometrics. *Food Control* 131:108420. doi: 10.1016/j.foodcont.2021.108420.

Śliwińska-Bartel, M., D. T. Burns, and C. Elliott. 2021. Rice fraud a global problem: A review of analytical tools to detect species, country of origin and adulterations. *Trends in Food Science and Technology* 116:36–46. doi: 10.1016/j.tifs.2021.06.042.

Souza de Almeida Duarte, E., V. E. de Almeida, G. Bezerra da Costa, M. C. Ugulino de Araújo, G. Véras, P. H. Gonçalves Dias Diniz, and D. D. de Sousa Fernandes. 2022. Feasibility study on quantification and authentication of the cassava starch content in wheat flour for bread-making using NIR spectroscopy and digital images. *Food Chemistry* 368:130843. doi: 10.1016/j.foodchem.2021.130843.

Tao, F., L. Liu, C. Kucha, and M. Ngadi. 2021. Rapid and non-destructive detection of cassava flour adulterants in wheat flour using a handheld microNIR spectrometer. *Biosystems Engineering* 203:34–43. doi: 10.1016/j.biosystemseng.2020.12.010.

Teye, E., C. L.Y. Amuah, T. McGrath, and C. Elliott. 2019. Innovative and rapid analysis for rice authenticity using hand-held NIR spectrometry and chemometrics. *Spectrochimica Acta—Part A: Molecular and Biomolecular Spectroscopy* 217:147–154. doi: 10.1016/j.saa.2019.03.085.

Tian, W., G. Chen, G. Zhang, D. Wang, M. Tilley, and Y. Li. 2021. Rapid determination of total phenolic content of whole wheat flour using near-infrared spectroscopy and chemometrics. *Food Chemistry* 344:128633. doi: 10.1016/j.foodchem.2020.128633.

Timsorn, K., Y. Lorjaroenphon, and C. Wongchoosuk. 2017. Identification of adulteration in uncooked jasmine rice by a portable low-cost artificial olfactory system. *Measurement: Journal of the International Measurement Confederation* 108:67–76. doi: 10.1016/j.measurement.2017.05.035.

Unuvar, A., I. H. Boyaci, and H. Koksel. 2021. A novel approach for rapid discrimination of common and durum wheat flours using spectroscopic analyses combined with chemometrics. *Journal of Cereal Science* 100:103269. doi: 10.1016/j.jcs.2021.103269.

Vermeulen, P., M. Suman, J. A. Fernández Pierna, and V. Baeten. 2018. Discrimination between durum and common wheat kernels using near infrared hyperspectral imaging. *Journal of Cereal Science* 84:74–82. doi: 10.1016/j.jcs.2018.10.001.

Vlachos, A., and I. S. Arvanitoyannis. 2008. A review of rice authenticity/adulteration methods and results. *Critical Reviews in Food Science and Nutrition* 48 (6):553–598. doi: 10.1080/10408390701558175.

Wadood, S. A., B. Guo, X. Zhang, and Y. Wei. 2019. Geographical origin discrimination of wheat kernel and white flour using near-infrared reflectance spectroscopy fingerprinting coupled with chemometrics. *International Journal of Food Science and Technology* 54 (6):2045–2054. doi: 10.1111/ijfs.14105.

Wadood, S. A., G. Boli, and W. Yimin. 2019. Geographical traceability of wheat and its products using multielement light stable isotopes coupled with chemometrics. *Journal of Mass Spectrometry* 54 (2):178–188. doi: 10.1002/jms.4312.

Wang, F., H. Zhao, C. Yu, J. Tang, W. Wu, and Q. Yang. 2020. Determination of the geographical origin of maize (Zea Mays L.) using mineral element fingerprints. *Journal of the Science of Food and Agriculture* 100 (3):1294–1300. doi: 10.1002/jsfa.10144.

Wang, N., X. Zhang, Z. Yu, G. Li, and B. Zhou. 2014. Quantitative analysis of adulterations in oat flour by FT-NIR spectroscopy, incomplete unbalanced randomized block design, and partial least squares. *Journal of Analytical Methods in Chemistry* 2014:393596. doi: 10.1155/2014/393596.

Williams, P. C. 2020. Application of chemometrics to prediction of some wheat quality factors by near-infrared spectroscopy. *Cereal Chemistry* 97 (5):958–966. doi: 10.1002/cche.10318.

Wu, Y., Y. Yun, J. Chen, and D. Liu. 2021. Discrimination of waxy wheats using near-infrared hyperspectral spectroscopy. *Food Analytical Methods* 14 (8):1704–1713. doi: 10.1007/s12161-021-02008-1.

Xia, C., S. Yang, M. Huang, Q. Zhu, Y. Guo, and J. Qin. 2019. Maize seed classification using hyperspectral image coupled with multi-linear discriminant analysis. *Infrared Physics & Technology* 103:103077. doi: 10.1016/J.INFRARED.2019.103077.

Xiao, R., L. Liu, D. Zhang, Y. Ma, and M. O. Ngadi. 2019. Discrimination of organic and conventional rice by chemometric analysis of NIR spectra: a pilot study. *Journal of Food Measurement and Characterization* 13 (1):238–249. doi: 10.1007/s11694-018-9937-7.

Xiao, R., L. Li, and Y. Ma. 2019. A label-free proteomic approach differentiates between conventional and organic rice. *Journal of Food Composition and Analysis* 80:51–61. doi: 10.1016/j.jfca.2019.04.004.

Xue, S. S., J. Tan, J. Y. Xie, and M. F. Li. 2021. Rapid, simultaneous and non-destructive determination of maize flour and soybean flour adulterated in quinoa flour by front-face synchronous fluorescence spectroscopy. *Food Control* 130:108329. doi: 10.1016/J.FOODCONT.2021.108329.

Xu, F., F. Kong, H. Peng, S. Dong, W. Gao, and G. Zhang. 2021. Combing machine learning and elemental profiling for geographical authentication of Chinese geographical indication (GI) rice. *Npj Science of Food* 5 (1):1–6. doi: 10.1038/s41538-021-00100-8.

Zaldarriaga Heredia, J., C.A. Moldes, M. Savio, S. M. Azcarate, R. A. Gil, and J. M. Camiña. 2019. Assessment of agricultural practices in maize crops (Zea Mays) based on elemental profile and chemometrics analysis. *Asian Journal of Research in Crop Science* 3 (2):1–9. doi: 10.9734/ajrcs/2019/v3i230043.

Zaldarriaga Heredia, J., C. A. Moldes, R. A. Gil, and José M. Camiña. 2020a. Effect of topography on maize grains elemental profile: A chemometric approach. *Current Analytical Chemistry* 16 (8):1079–1087. doi: 10.2174/1573411016666200319095312.

Zaldarriaga Heredia, J., C. A. Moldes, R. A. Gil, and José M. Camiña. 2020b. Elementomic and genetic technology assessment in maize (Zea Mayze) plants based on analysis of leaf and seed tissues. *Microchemical Journal* 159:105569. doi: 10.1016/j.microc.2020.105569.

Zhao, H., B. Guo, Y. Wei, and B. Zhang. 2013a. Multi-element composition of wheat grain and provenance soil and their potentialities as fingerprints of geographical origin. *Journal of Cereal Science* 57 (3):391–397. doi: 10.1016/J.JCS.2013.01.008.

Zhao, H., B. Guo, Y. Wei, and B. Zhang. 2013b. Near infrared reflectance spectroscopy for determination of the geographical origin of wheat. *Food Chemistry* 138 (2–3):1902–1907. doi: 10.1016/j.foodchem.2012.11.037.

Zhao, T., M. Chen, X. Jiang, F. Shen, X. He, Y. Fang, Q. Liu, and Q. Hu. 2020. Integration of spectra and image features of Vis/NIR hyperspectral imaging for prediction of deoxynivalenol contamination in whole wheat flour. *Infrared Physics and Technology* 109:103426. doi: 10.1016/j.infrared.2020.103426.

Zhou, Q., W. Huang, S. Fan, F. Zhao, D. Liang, and X. Tian. 2020. Non-destructive discrimination of the variety of sweet maize seeds based on hyperspectral image coupled with wavelength selection algorithm. *Infrared Physics & Technology* 109:103418. doi: 10.1016/J.INFRARED.2020.103418.

9 Authenticity and Chemometrics of Honey

Dimitrios D. Ntakoulas, Ioannis N. Pasias,
Kalomoira G. Raptopoulou, Charalampos Proestos, and Philipp Weller

CONTENTS

9.1 INTRODUCTION

Honey is a natural product known for its beneficial health properties since ancient times. According to the definition set by European Union Council Directive 2001/110/EC, honey is the natural sweet substance produced by *Apis mellifera* bees from the nectar of plants or from secretions of living parts of plants, or excretions of plant-sucking insects on the living parts of plants, which the bees collect, transform by combining with specific substances of their own, deposit, dehydrate, store and leave in honeycombs to ripen and mature (Council Directive 2001/110/EC). It consists mainly of sugars and other components such as enzymes, amino acids, organic acids,

DOI: 10.1201/9781003266167-9

carotenoids, vitamins, minerals and aromatic substances. The chemical composition, color, odor and taste of honey depend primarily on the botanical and geographical origin and are also influenced by packaging and storage conditions (Tornuk et al., 2013; Escuredo et al., 2014).

Honey is also rich in flavonoids and phenolic acids that have a wide range of organic activities and act as natural antioxidants (Alqarni et al., 2014). Honey also shows antibacterial, antimicrobial and anticancer activity that acts in different ways on cell proliferation. Moreover, honey shows antidiabetic and cardio-protective capacity and is also beneficial for the nervous system and for the respiratory system against bacterial infections and asthma (Cianciosi et al., 2018). Consequently, there is now an increase in the consumption of honey as a basic element of a healthy diet that is adopted by many people. We therefore understand that honey is a valuable product for human health and that it is essential to guarantee its quality and authenticity. The authenticity of honey has two aspects. The first aspect concerns the production procedure, and the prevention of adulteration by the addition of other food ingredients or by the use of appropriate apiculture practices. The second aspect concerns the labeled origin (botany and/or geographic) and the "organic" provenance. Since honey from particular botanical sources and geographic locations has a higher price than others, it is important to make sure that it is authenticated (Ruoff & Bogdanov, 2004).

Honey shall comply with accepted general regulatory standards. In this chapter, the existing legislation on honey and all the different methods of adulteration will be discussed. Furthermore, all modern methods and techniques of analysis, as well as the chemometry used to identify the authenticity of honey, will be presented.

9.2 CHEMICAL COMPOSITION OF HONEY

Honey is a saturated sugar solution and contains about 200 chemical compounds (Escuredo et al., 2013), such as minerals, proteins, vitamins, organic acids, phenolic compounds, minerals and a large amount of volatile compounds (Pontes et al., 2007; Ciulu et al., 2011; Alqarni et al., 2014). Figure 9.1 shows the mean percentage content of the main constituents in honey (Santos & González, 2017).

Some honey ingredients come directly from nectar or honeydew, while others are composite or decomposition products. The type and amount of these ingredients vary according to the botanical origin, the strength of the bee colony, the climatic conditions and the ripening stage of the honey. The chemical composition of honey is also influenced by a number of other parameters associated with its treatment, packaging and storage conditions, age and oxygen exposure.

9.2.1 Sugars

While 25 different sugars have been identified in honey, they do not coexist in all types of honey. Monosaccharides, primarily glucose and fructose, make up approximately 85% of honey sugars.

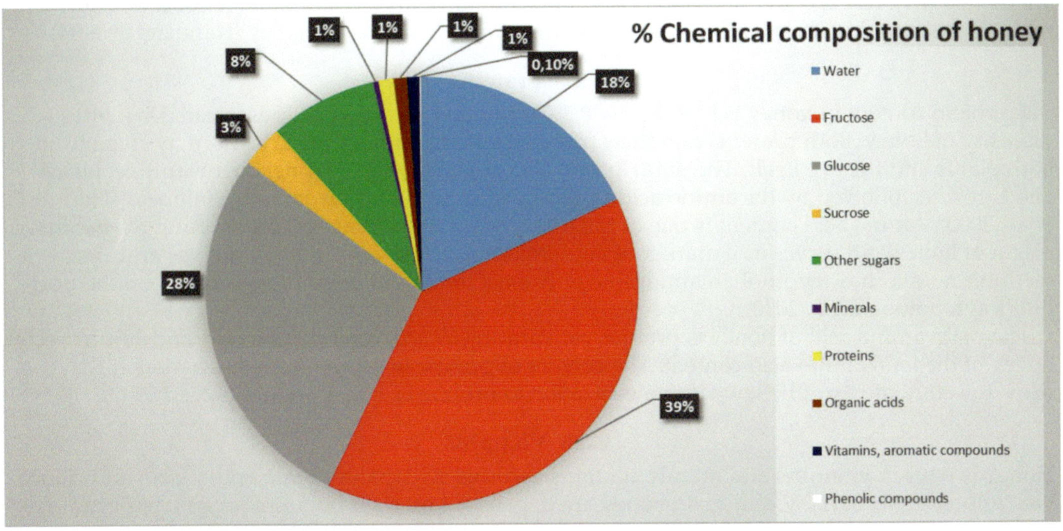

Figure 9.1 Percentage content of the main constituents of honey.

Maltose and sucrose followed at approximately 7.5% and 1.5%, respectively. As it matures, sucrose is broken down into its structural units by the invertase enzyme. Other disaccharides such as maltose, isomaltose, trehalose and turanose have also been identified in honey, as well as the trisaccharides raffinose, melezitose and erlose (Codex Alimentarius, 2001; Tananaki, 2006; Puścion-Jakubik et al., 2020).

The concentration of sugars is related to the phenomenon of granulation, which results in the solidification of honey. The granulation rate is mainly related to the glucose/fructose ratio. When the glucose/fructose ratio is high, the crystallization of honey occurs quickly, and when the glucose/fructose ratio is lower, the crystallization of honey occurs slowly. The glucose/fructose ratio at 2.5:1 is the critical value of honey crystallization, at which honey can quickly form stable crystalline deposition (Ma et al., 2017; Amariei et al., 2020).

9.2.2 Water

The water content of honey is expressed by the term *moisture*. Moisture is a major parameter and affects the quality of honey. Honey's moisture content influences its stability against fermentation and crystallization. Honey with high moisture ratio ferments faster due to the microbial growth, including *Zygosaccharomyces* (Singh I. & Singh S., 2018). The fermentation process leads to the formation of alcohol and in the presence of oxygen, the alcohol will form acetic acid and water, which causes honey to have a sour taste. The moisture content of the honey depends principally on the degree of maturity and on the type of honey and climatic conditions.

9.2.3 Enzymes

Honey contains enzymes that originate in both plants and the bee itself. The main enzymes found in honey are diastase, invertase, catalase, phosphatases, amylases, maltase, ascorbinoxidases, β-fructofuranosidase and glucose oxidase. The glucose oxidase enzyme catalyzes the oxidation of glucose into gluconic acid (which assists in the absorption of calcium), along with the simultaneous production of hydrogen peroxide (which provides antimicrobial properties). Invertase is produced by the hypopharyngeal glands of bee workers and catalyzes the decomposition of disaccharide sucrose in its structural units, glucose and fructose. Diastase converts starch into short-chain sugars and is used for quality criteria. Catalase converts hydrogen peroxide to water and oxygen. As a result of amylase activity, long starch chains are transformed into maltose and dextrin (Tananaki, 2006; Walker, 2009; Eteraf-Oskouei & Najafi, 2013; Prica et al., 2015).

9.2.4 Organic Acids

Organic acids, although constituting a small percentage of the mass of honey, play an important role in determining its taste, color and pH, as well as its antioxidant and antimicrobial activity (Suto et al., 2020). The main organic acids found in honey are acetic, butyric, oxalic, citric, formic, gluconic, lactic, malic, succinic and fumaric acids, while gluconic acid predominates. These acids are related to honey age and deterioration in honey quality under poor storage conditions. In recent years, organic acids have served to confirm the authenticity of honey (Karabagias, 2014).

9.2.5 Proteins

The protein content in honey is low (below 0.2% w/w) and its presence is associated with the viscosity of honey. Both proteins and their free constructive units, amino acids, may come from both plants and the bee itself. The main amino acids in honey are proline, phenylaniline, lysine and tyrosine, followed by the amino acids valine, arginine, histidine and glutamic acid (Hermosín et al., 2003). Amino acid content is one of the most common methods of determining the botanic origin of honey and serves as a marker of authenticity. Certain amino acids such as arginine, methionine, alanine, tryptophan and lysine only exist in certain floral honeys based on their origin (Raptopoulou et al., 2020).

The main amino acid of honey is proline, constituting 59% in honeydew honey and 49% in nectar honey of the total amino acid content. The content of proline in honey is used as a marker that confirms the authenticity of honey. The concentration of proline in honey should exceed 25 mg/100 g.

9.2.6 Vitamins

Honey is poor in vitamins and mainly contains ascorbic acid, pantothenic acid, niacin and riboflavin (Ajibola et al., 2012). Vitamins in honey are dependent on their botanical origin. For example, riboflavin and niacin are mainly found in citrus honey, whereas ascorbic acid is mainly found in eucalyptus honey.

9.2.7 Inorganic Elements

The honey content of inorganic elements depends on their botanic origin, soil composition and environmental factors. So far, 54 chemical elements have been identified in various types of honey and can be divided into three groups: macro-elements (Na, Ca, K, Mg, S, Cl, P), trace elements (Al, Mn, Pb, Zn, Cu, Cd, Tl, Co, Ni, Rb, Be, BA, Bi, U, V, Fe, Pt, Pd, Te, Hf, Mo, Sn, Sb, La, I, Sm, Tb, Dy, Sd, Th, PR, Nd, TM, Yb, Lu, Gd, Ho, ER, Cr, Ce, As, Br, B, Cd, Hg, Sr, Se) and heavy metals (trace elements that have a specific gravity at least five times higher than the specific gravity of water and inorganic elements). According to scientific literature, the most important heavy metals are Pb, Cd, Hg, Cr, Cu, Mn, Ni, Zn and Ag. Heavy metals are harmful and toxic in low concentrations due to their potential to accumulate in living beings (Tananaki, 2006; Solayman et al., 2016).

9.2.8 5-Hydroxymethylfurfural (HMF)

5-Hydroxymethylfurfural (HMF) is a six-carbon heterocyclic organic compound that is present at low concentrations (1–15 mg/kg) in fresh honey. HMF is produced through the Maillard reaction (a non-enzymatic browning reaction) by sugar degradation during the heating process or long storage of honey. The existence of fructose, glucose, minerals and acids in honey stimulates the production of this compound. HMF is used as a quality criterion of honey, as its concentration is low in fresh honey and increases during heat treatment or aging. In addition, studies show that honeys stored at low temperatures have low levels of HMF, whereas honeys stored at medium and higher temperatures have higher levels of HMF. Furthermore, the storage of honey in metal containers increases the value of HMF (Markowicz Bastos et al., 2012; Shapla et al., 2018). The Codex Alimentarius standard has established the upper limit for HMF in honey at 40 mg/kg (with the upper limit at 80 mg/kg for honeys produced in tropical regions) to ensure that the product has not undergone extensive aging or heating process and is safe for the consumers (Codex Alimentarius, 2001).

9.2.9 Phenolic Compounds

Phenolic compounds exist in large amounts in the plant kingdom and are transferred to honey by pollen or nectar from plants visited by honey bees (Olas, 2020). Among phenolic compounds, flavonoids and phenolic acids have anticancer, anti-inflammatory, anti-atherogenic, anti-thrombotic, antibacterial and antioxidant properties, as well as analgesic activity (Tungmunnithum et al., 2018). The quantity of various phenolic compounds and the phenolic profile vary among honeys and depend on the different geographical factors, floral sources, environmental factors, seasonal factors and the collection method of nectar (Cheung et al., 2019).

The main flavonoids present in honey are the following: myricetin (3,3′,4′,5,5′,7-hexahydroxyflavone), tricetin (5,7,3′,4′,5′-pentahydroxyflavone), quercetin [2-(3,4-dihydroxyphenyl)-3,5,7-trihydroxychromen-4-one], kaempherol [3,5,7-trihydroxy-2-(4-hydroxyphenyl)chromen-4-one, luteolin [2-(3,4-dihydroxyphenyl)-5,7-dihydroxychromen-4-one], galangin [3,5,7-trihydroxy-2-phenylchromen-4-one], chrysin (5,7-dihydroxy-2-phenylchromen-4-one), naringenin [5,7-dihydroxy-2-(4-hydroxyphenyl)-2,3-dihydrochromen-4-one], hesperidin ((2S)-5-hydroxy-2-(3-hydroxy-4-methoxyphenyl)-7-[(2S,3R,4S,5S,6R)-3,4,5-trihydroxy-6-[[(2R,3R,4R,5R,6S)-3,4,5-trihydroxy-6-methyloxan-2-yl]oxymethyl]oxan-2-yl]oxy-2,3-dihydrochromen-4-one) and rutin (2-(3,4-dihydroxyphenyl)-5,7-dihydroxy-3-[(2S,3R,4S,5S,6R)-3,4,5-trihydroxy-6-[[(2R,3R,4R,5R,6S)-3,4,5-trihydroxy-6-methyloxan-2-yl]oxymethyl]oxan-2-yl]oxychromen-4-one). The existence of flavonoids in honey depends mainly on its botanical origin. For that reason, some of them are used as markers of botanical origin (Jasicka-Misiak et al., 2012; Kurtagic et al., 2015).

The main phenolic acids present in honey are gallic acid (3,4,5-trihydroxybenzoic acid), salicylic acid (2-hydroxybenzoic acid), p-hydroxybenzoic acid (4-hydroxybenzoic acid), 3-hydroxybenzoic acid, protocatechuic acid (3,4-dihydroxybenzoic acid), syringic acid (4-hydroxy-3,5-dimethoxybenzoic acid), caffeic acid [(E)-3-(3,4-dihydroxyphenyl)prop-2-enoic acid], vanillic acid (4-hydroxy-3-methoxybenzoic acid), p-Coumaric acid [(E)-3-(4-hydroxyphenyl)prop-2-enoic acid], ferulic acid [(E)-3-(4-hydroxy-3-methoxyphenyl)prop-2-enoic acid] and abscisic acid [(2Z,4E)-5-[(1R)-1-hydroxy-2,6,6-trimethyl-4-oxocyclohex-2-en-1-yl]-3-methylpenta -2,4-dienoic acid] (Kurtagic et al., 2015; Chin & Sowndhararajan, 2020).

9.3 HONEY TYPES

9.3.1 Classification of Honey According to Its Origin

Honey may be classified by origin, method of harvest and presentation. Depending on its origin, honey is classified as blossom honey, honeydew honey, monofloral honey and multifloral honey

[35]. According to the Codex Alimentarius, blossom honey or nectar honey comes from nectars of plants, and honeydew honey comes from secretions of living parts of plants or excretions of plant-sucking insects (Codex Alimentarius, 2001). Monofloral honey is the type of honey that has a unique flavor or other characteristic because it originates mainly from the nectar of a plant species. Multifloral honey contains nectar from several botanical sources, none of which is predominant (e.g., meadow blossom honey, forest honey; Bogdanov, 2016).

9.3.1.1 Blossom or Nectar Honey

Nectar is the gentle, viscous secretion of the nectaries or glands of the flowers, stems and leaves of a plant. Nectar in flowers is mostly used to attract pollinators, such as insects, fruit bats, songbirds and hummingbirds. Nectar is collected by bees and is used as a raw material in the production of honey. The quantity and composition of nectar is influenced by climatic conditions such as temperature, moisture and sunshine. Greater quantities of nectar are produced on warm,

Table 9.1 Properties of the Most Known Blossom Honeys in the World

Common Name	Botanical Name of Plant	Place of Harvest	Color	Granulation Speed	Flavor
Eucalyptus	*Eucalyptus* spp.	S. Europe, Oceania, Africa, S. America	Yellow to brown	Rapid to medium	Medium-strong caramel
Acacia	*Robinia pseudoacacia*	Africa, America, Oceania, temperate Europe	Light white	Slow	Sweet delicate with hints of vanilla
Lavender	*Lavandula intermedia*	Temperate Europe, Asia and N. America	Light white to extra light	Rapid	Floral, well-rounded flavor
Heather	*Calluna vulgaris*	Europe	Brown-reddish	Medium	Strong, caramelized
Orange blossom	*Citrus* spp.	Europe, temperate and subtropical, N. America, S. America	Very light white	Rapid	Medium floral, fruity
Lime	*Tilia* spp.	Temperate Europe and Asia	White to yellow	Rapid to medium	Mouthwatering flavor of citrus fruits
Linden	*Tilia* spp.	Temperate and subtropical N. America	White to amber	Medium	Sweet, spicy and slightly bitter
Rosemary	*Rosmarinus officinalis*	Temperate Europe, Asia, Africa	Light white to extra light	Rapid	Floral, fruity
Sunflower	*Helianthus annuus*	Temperate Europe, S. and N. America, Asia, Africa, Oceania; tropical Africa, N. America.	Yellow to golden light	Rapid	Slightly floral
Thyme	*Thymus capitatus*	Mediterranean and temperate Europe, N. America, Oceania	Yellow to light brown	Fast to medium	Resinous, herbal, savory
White clover	*Trifolium repens*	Europe, N. America	Light white to light	Rapid	Sweet and flowery
Chestnut	*Castanea sativa*	Europe	Reddish-brown amber to dark amber	Slow	Caramelized, bitter
Strawberry tree	*Arbutus unedo*	Italy, Greece, coastal Mediterranean countries	Dark amber when young and light; brown when crystallized	Rapid	Bitter taste with aromatic notes of bitter caramel

Source: Bogdanov (2016).

wet days. The quantity and composition of nectar is also influenced by geographic location, soil composition, plant species and age. Nectar is mostly an aqueous solution of glucose, fructose and sucrose sugars, but it also contains traces of salts, proteins, acids and essential oils. Nectar also contains amino acids, lipids, antioxidants, and color- and odor-related compounds. The distinctive odor of honey is associated with the presence of volatile compounds such as monoterpenes, hydrocarbons, aromatic and nitrogenous compounds. Blossom honey forms crystals faster due to its higher content in glucose (Tananaki, 2006). Some honey types, such as linden and black lobster, are very much alike in different places of the world, while other types such as thyme, eucalyptus and orange blossom can vary significantly depending on the place of origin. Table 9.1 shows the properties of the most popular blossom honeys in the world (Bogdanov, 2016).

9.3.1.2 *Honeydew Honey*

Honeydew honey is produced by bees (*Apis mellifera*) from secretions of living parts of plants or excretions of plant-sucking insects and has very different characteristics from blossom honey. These insects belong to the Hemiptera family and especially to their superfamilies Cicadoidea, Aphidoidea, Coccoidea, Psylloidea and Aleyrodoidea. These have specialized oral molecules, with which they pierce the plant tissue. These insects suck the sap of the plant, and much of it is not digested but passes through special filter chambers in insects and is converted to honeydew. The honeydew is discarded from the body of insects in the form of drops that fall on the surface of the leaves and branches and are finally collected by the bees. In this way, in the digestive system of insects the original plant sap is enriched with enzymes, while the range of carbohydrates changes with the breakdown of oligosaccharides and the synthesis of new sugars (Crane & Walker, 1985; Codex Alimentarius, 2001; Tananaki, 2006). The chemical composition of the honeydew honey varies depending on the origin of the plant and the insect of origin. The chemical composition and amount of honeydew also depend on climatic conditions and especially on temperature and wind. In addition, the chemical composition of honeydew is affected by the composition of soil, the age of the plant, and the age of the insect. Honeydew honeys exhibit higher content of bioactive compounds such as amino acids, proteins and phenolic compounds, and for that reason they present higher antioxidant and antimicrobial activity. Honeydew honey also presents higher values of electrical conductivity, ash percentage, pH, net absorbance, higher content of trisaccharides and disaccharides, lower level of monosaccharides, peculiar sensory features and darker color compared to blossom honeys (Codex Alimentarius, 2001; Tananaki, 2006; Bogdanov, 2016; Seraglio et al., 2019).

Honeydew honeys crystallize more slowly due to their higher fructose content than glucose. Fructose has higher solubility than glucose and remains in solution form, whereas glucose crystallizes rapidly and transforms into glucose monohydrate (Seraglio et al., 2019). The largest quantities of honey are produced in forests located far from polluted urban centers and from crops sprayed with plant protection substances, so they are more free from the presence of harmful substances and pesticides than other honeys. Honeydew honey production varies among countries as honey secretions are associated with the host plant and insect (Tananaki, 2006). There are numerous varieties of honeydew honey in the world, but the main ones are fir honey, oak honey, pine honey, *Metcalfa pruinosa* honey and spruce honey.

Fir honeydew honey (*Abies* species) is produced particularly in central Europe. The insects that produce it belong to the genus *Cinara*. In Italy this kind of honey is produced in the Tuscan Alps and Apennines, and in Germany this variety comes from the black and Bavarian forests. In Greece, fir honey is produced from Greek fir (*Abies cephalonica*), mainly by the insect *Physokermes hemicryphus*, which represents 5% of the total honey production in Greece. In Turkey, fir honey is derived from the *Abies bornmuelleriana* trees of the Marmara region.

Oak (*Quercus* species) honey is produced mainly in Italy but also in Greece, France, Croatia, Serbia, Bulgaria and Slovakia. The insects responsible for honey production in oaks are primarily *Tuberculatus annulatus* and *T. borealis*. Fir honey is brown in color, crystallizes very slowly and tastes like toffee (Bogdanov, 2016). Pine honey (*Pinus* species) occurs mainly in Greece, Turkey and the Italian island of Ischia. The insect *Marchalina hellenica* is nourished by the sap of pine trees and produces large amounts of pine honeydew. Pine honey makes up approximately 60% of the annual production of Greek honey. Pine honey is amber brown, crystallizes very slowly and has a moderate flavor of sweetness and iodine, particularly in areas where sea pines grow (Bogdanov, 2016).

Metcalfa pruinosa honey is the type of honeydew honey that is produced from the sap-sucking insect *Metcalfa pruinosa*, which was introduced to Europe in the early 1980s initially in Italy and extended to France, Slovenia and Greece. It nourishes over 200 species of cultivated, wild and ornamental plants and thus spreads rapidly (Giovani et al., 1998). Spruce (*Picea abies*) honeydew honey comes mainly from the Alps. The sap-sucking insects involved in the production of these honeydews belong to *Physokermes hemicryphus, Cinara piceae* and *Cinara costata*. Spruce honey is light to dark amber, sometimes with reddish veins, crystallizes very slowly and has an average intensity and a slightly sweet and persistent taste (Bogdanov, 2016). Beech honey (*Nothofagus* species) is produced exclusively in New Zealand from two species: black beech (*Nothofagus solandri*) and red beech (*Nothofagus fusca*). The sap-sucking insects that produce the honeydew are *Ultracoelostoma brittini* and *Ultracoelostoma assimile*. Beech honeydew honey is dark in color and has a silky texture with hints of flavor like caramel nectar and berries.

9.3.2 Classification of Honey According to Its Production Method

According to the production method, honey may be classified under the following categories (Codex Alimentarius, 2001; Council Directive 2001/110/EC):

1. Extracted honey comes from the centrifugation of honeycombs and represents the highest percentage of honey produced in the world.
2. Pressed honey is obtained by pressing combs without brood with or without the application of an average heat of not more than 45°C.
3. Drained honey is obtained by draining decapped, broodless honeycombs.
4. Filtered honey is obtained by removing foreign organic or inorganic matter in such a way as to significantly remove pollen.

9.3.3 Classification of Honey According to Its Presentation

According to its presentation, honey can be classified into the following categories (Codex Alimentarius, 2001; Council Directive 2001/110/EC; Bogdanov, 2016):

1. Normal honey is in liquid or crystalline form or a mixture of them.
2. Comb honey is stored by bees in the cells of freshly built, broodless combs or thin comb foundation sheets, made solely of beeswax and sold in sealed whole combs or sections of such combs.
3. Cut comb contains one or more pieces or chunks of comb honey.
4. Finally, according to European Directive 2001/110/EC, another type is baker's honey, which is suitable for industrial uses, or as an ingredient in other foodstuffs that are then processed. It may have a foreign taste or odor, may have begun to ferment or have fermented, or may have been overheated (Council Directive 2001/110/EC).

Figure 9.2 summarizes the primary types of honey.

9.4 REGULATION LIMITS

Honey is a natural product with a particular composition that is influenced by various factors such as geographical and botanical origin, climatic conditions, beekeepers' manipulations, and storage, processing and packaging conditions. It is therefore necessary to have legislation that guarantees the quality of the product and the health of consumers. The Codex standard for honey adopted in 1981 by the Codex Alimentarius Commission, revised in 1987, 2001 and amended in 2019, is the basis of honey legislation. The European Union followed the Codex recommendations issued the Directive 2001/110/EC amended by Directive 2014/113/EU in order to determine the requirements need to be met in the production process and marketing of honey in EU Member States (Codex Alimentarius, 2001; Council Directive 2001/110/EC; Thrasyvoulou et al., 2018). The Codex and the Directive set regulatory limits for some physicochemical properties of honey that are directly related to its quality. These physicochemical properties include sugar content, moisture content, water insoluble content, electrical conductivity, acidity, diastase activity and hydroxymethylfurfural (HMF) content determined after processing and blending (Codex Alimentarius, 2001; Council Directive 2001/110/EC). The Codex and the Directive dictate that honey can be labeled according to its botanical origin, in the case that the product comes exclusively or mainly from the labeled source and possesses the microscopic, physicochemical and organoleptic attributes of the botanical source. Baker's honey and filtered honey cannot be labeled by their botanical origin. Both the Codex and the Directive do not define the aforementioned characteristics of monofloral honeys, enabling the trade of this honey type problematic and uncertain (Thrasyvoulou et al., 2018).

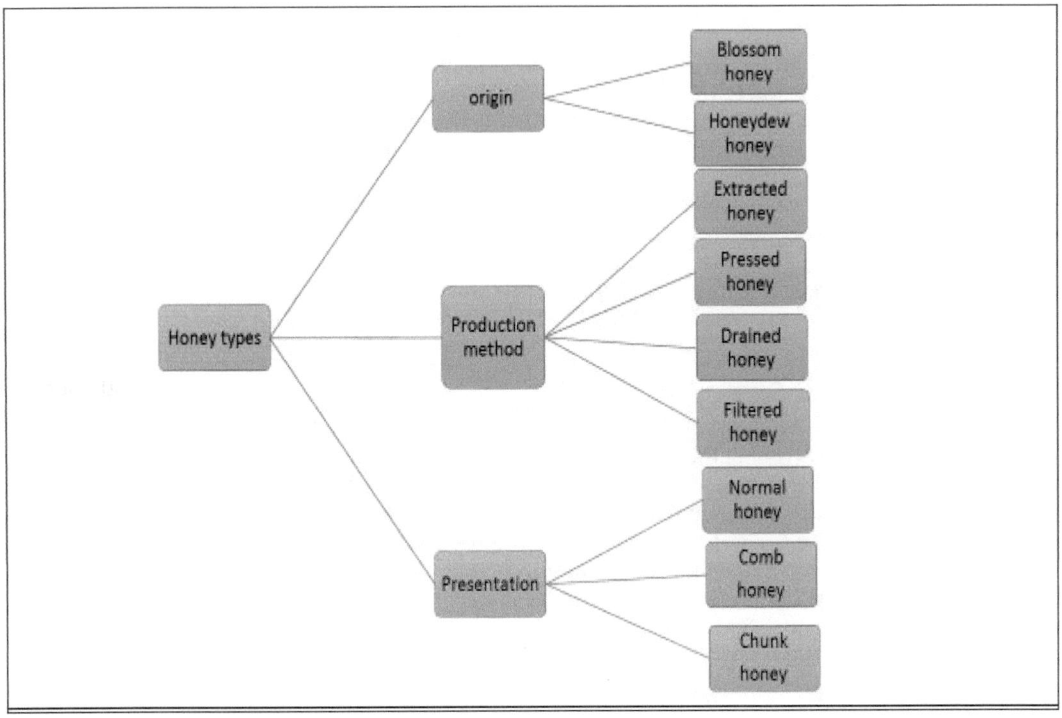

Figure 9.2 Honey types.

To fill this gap, several countries have set their own limits on the characteristics of their monofloral honeys (Thrasyvoulou et al., 2018). The botanical origin of the honey is determined with the help of melissopalynology, where is the examination of the type of pollen and its honey content. The identification of different pollen types is performed by microscopic analysis. However, there are some problems in the application of this method, such as that different proportions of pollen are produced by different plant species, the nectar yield may be different in female and male blossoms, the produced pollen may vary seasonally, bees can collect pollen without collecting nectar and the pollen may be collected by beekeepers at the entrance of the beehive. In addition, pollen may be added to honey fraudulently after harvest, further limiting the method. In the case of citrus honey, melissopalynology is not an appropriate method for identifying honey of floral origin because the quantity of pollen is small and highly variable (Thrasyvoulou et al., 2018). Although pollen analysis presents the aforementioned limitations, when used in combination with physicochemical and organoleptic characteristics it can give reliable results (Anklam, 1998). Thus, many countries have established the minimum percentage of pollen required for the characterization of monofloral honeys (Von der Ohe et al., 2004).

Regarding the geographical origin of honey, the Codex specifies that the country of origin must be indicated on the packaging of honey. The European Directive also requires the country of origin of honey to be included on the honey label. In addition, the Directive states that if honey comes from more than one Member State or from a third country, the indication of geographical origin must be substituted by the terms "blend of EU honey", "blend of non-EU honey", or "blend of EU and non-EU honey" (Codex Alimentarius, 2001; Council Directive 2001/110/EC; Thrasyvoulou et al., 2018).

Moisture is one of the most important parameters of honey quality. Honey produced by bees has a moisture content that depends on its botanical origin, climatic conditions and beekeeping practices. The moisture content of honey defines its stability against granulation and fermentation. Both the Codex and the European Directive set the limit for moisture up to 20% for most honeys and no more than 23% for heather (*Calluna*) honey. Moreover, the European Directive sets the limit for baker's honey from heather (*Calluna*) up to 25%. The term *baker's honey* is not mentioned in the Codex Alimentarius (Codex Alimentarius, 2001; Council Directive 2001/110/EC).

Usually the honey that comes from well-sealed honeycombs has a moisture content less than 18% (Thrasyvoulou et al., 2018).

Regarding the sugar content in glucose and fructose (sum), both the Codex and the European Directive require that the lower limit for blossom honey is 60 g/100 g, while for honeydew honey and blends of blossom honey and honeydew honey, the lower limit is 45 g/100 g (Codex Alimentarius, 2001; Council Directive 2001/110/EC). The reason for this difference is that honeydew honeys contain less sugars.

Both the Codex Alimentarius and the European Directive set the upper limit for sucrose content at 5 g/100 g except honeys derived from species alfalfa (*Medicago sativa*), *Citrus* spp., false acacia (*Robinia pseudoacacia*), French honeysuckle (*Hedysarum*), Menzies banksia (*Banksia menziesii*), red gum (*Eucalyptus camaldulensis*) and leatherwood (*Eucryphia lucida* and *Eucryphia milliganii*), where the upper limit for sucrose content is 10 g/100 g; and honeys derived from species lavender (*Lavandula* spp.) and borage (*Borago officinalis*), where the upper limit for sucrose content is 15 g/100 g (Codex Alimentarius, 2001; Council Directive 2001/110/EC).

The measurement of water insoluble content was introduced at a time when a large percentage of honey was harvested by pressing the combs, and it is an important way to detect honey impurities that exceed the permissible limits. Both the Codex and the European Directive set the upper limit for water insoluble solids content at 0.1 g/100 g except for pressed honey, for which the upper limit is 0.5 g/100 g (Codex Alimentarius, 2001; Council Directive 2001/110/EC).

Electrical conductivity is a good criterion to determine the botanical origin of honey. Honey contains mineral salts and organic acids, compounds that are called "ionizable", and have the property to conduct electric current when they are in solution (Yadata, 2014). Both the Codex and the European Directive set the following limits for electrical conductivity in honey (Codex Alimentarius, 2001; Council Directive 2001/110/EC).

1. For honeys not listed under (2) or (3) and blends of these, the electrical conductivity must not be more than 0.8 mS/cm;
2. For honeydew and chestnut honey and blends of these except with those listed under (3), the electrical conductivity must not be less than 0.8 mS/cm;
3. Exceptions: Strawberry tree (*Arbutus unedo*), bell heather (*Erica*), eucalyptus, lime (*Tilia* spp.), ling heather (*Calluna vulgaris*), manuka or jelly bush (*Leptospermum*), and tea tree (*Melaleuca* spp.).

So the general rule is that honey with electrical conductivity greater than 0.8 mS/cm is honeydew or chestnut honey and honey with electrical conductivity less than 0.8 mS/cm is blossom honey.

Acidity is a result of the organic acids (e.g., acetic, citric, oxalic, tartaric) that exist in honey, either from bee secretions or from plant nectar. Both the Codex and the Directive set the upper limit for free acidity in honey at 50 milliequivalents acid per kg. In addition, the European Directive refers to baker's honey where the free acidity must not be more than 80 milliequivalents acid per 1000 g (Codex Alimentarius, 2001; Council Directive 2001/110/EC).

Diastase activity is a honey quality criterion and is used to determine if honey has undergone an intensive heating process. The heating process motivates the damage of thermosensitive aromatic compounds. This damage is dependent upon temperature and heating time (Tosi et al., 2008). Moreover, diastase activity is affected by storage conditions, the age of the bees, the amount of nectar flow and the honey collecting period (Raptopoulou et al., 2020). According to the Codex Alimentarius, the diastase activity of honey, determined after processing and/or blending must be not less than 8 Schade units, except for honeys with a low natural enzyme content, in which diastase activity must not be less than 3 Schade units. The European Directive defines that diastase activity in honey (except baker's honey) must not be less than 8 Schade units, except for honeys with low natural enzyme content (e.g., citrus honeys) and an HMF content of not more than 15 mg/kg, where the diastase activity must not be less than 3 Schade units (Codex Alimentarius, 2001; Council Directive 2001/110/EC).

HMF (5-hydroxymethylfurfural) is a six-carbon heterocyclic organic compound that is present in low concentrations in fresh honey. HMF is produced by the Maillard reaction (a non-enzymatic browning reaction) by the breakdown of sugar during the heating process or prolonged storage of honey. HMF is used as a quality criterion because it is present in low concentrations in fresh honey and increases its concentration during heat treatment or aging (Raptopoulou et al., 2020). HMF has negative effects on human health, such as carcinogenicity, cytotoxicity toward mucous membranes, mutagenicity and chromosomal abnormalities (Shapla et al., 2018). According to the Codex

Alimentarius and the European Directive, the HMF content of honey shall not be more than 40 mg/kg. However, in the case of honey of declared origin from countries or regions with tropical ambient temperatures and blends of these honeys, the HMF content must not be more than 80 mg/kg (Codex Alimentarius, 2001; Council Directive 2001/110/EC). In the European Union, the mandatory indications for labeling honey must be adapted to the general rules on food identification (Directive 2011/91/EU on indications or marks identifying the lot to which a foodstuff belongs) and on the provision of food information to consumers (Regulation (EU) No 1169/2011 on the provision of food information to consumers). Honey is considered a non-perishable, single-ingredient product and therefore the list of required indications, in addition to geographical origin, includes:

1. The name of the food ("honey", "baker's honey" or "filtered honey");
2. The date of minimum durability, "best before" date and any special storage conditions and/or conditions of use;
3. The net quantity expressed in units of mass;
4. If relevant, instructions for use where it would be difficult to make appropriate use of the food in the absence of such instructions;
5. The name and the address of the food business operator responsible for the food information (the operator under whose name or business name the honey is marketed, or the importer into the European Union if that operator is not established in the European Union);
6. The identification of the lot to which the product belongs.

Finally, it is important to emphasize that there are countries (e.g., non-EU countries) that have set their own limits to ensure the quality of honey. These limits may differ from the European Directive for honey and as a result do not meet European Union regulation (Thrasyvoulou et al., 2018).

9.5 HONEY FRAUD SYSTEM TO IDENTIFY ADULTERATION

9.5.1 Honey Fraud History and Legislation

Honey fraud is a planned and illegal action performed to acquire an unfair financial gain by handling money and selling a product that does not meet the existing, accepted regulations for honey. These intentional violations of existing regulations may hinder the proper functioning of the internal market and may also constitute a risk to public health.

Honey fraud is not a new phenomenon, as it has been officially recorded in the United States since 1889, when honey was deemed to be the most adulterated product. In recent years, the international traffic of adulterated honey has risen to unbearable levels. In addition, in recent years, average honey production per hive has decreased as a result of a number of factors that affect plants and bees. The reduction in honey production per hive, in combination with honey fraud, has caused a huge reduction in beekeepers' incomes, who have to compete on prices with sellers of low-priced fraudulent products. According to the World Congress of Beekeepers, adulteration of honey constitutes the greatest threat to beekeepers in the history of beekeeping. Moreover, according to the US Pharmacopeia's Food Fraud Database, honey ranks as the third "favourite" food target for adulteration, behind only milk and olive oil (US Pharmacopeia, 2018). Similarly, the European Union considers honey to be at high risk of being fraudulent (European Parliament, 2013).

The European Commission (Regulation (EU) 2017/625) considers that four key operational criteria should be present for food fraud:

1. *Violation of EU rules*: This criterion refers to the rules that are codified in the EU agri-food chain legislation (Article 1(2) of Regulation (EU) 2017/625).
2. *Deception of customers*: This criterion requires some form of deception of the consumers/customers, and the deceptive parameter should constitute a threat for the public health.
3. *Undue advantage*: This criterion refers to direct or indirect economic advantage for the perpetrator.
4. *Intentionality*: This criterion can be controlled in a number of ways that prove that specific non-compliance is not happening inadvertently (e.g., the substitution of a high quality component with another of lower quality).

Apimondia, the International Federation of Beekeepers' Associations, issued the Apimondia Statement on honey fraud in January 2020, and proposes the use of honey production methods, which allow bees to fully do their job in order to maintain the integrity and quality of honey. According to the Codex Alimentarius and the European Directive, honey is a natural product and no other ingredients should be added to it. Honey fraud includes mainly the following types (Apimondia, 2020):

1. Adulteration with different substances such us sugars and water; the mixing of honey with different sugar syrups that are manufactured mainly from corn, wheat, rice, beet sugar and cane sugar results in an increase in the mass of the product;
2. Feeding of bees with sugars during the main nectar flow period;
3. Incorrect labeling of botanical source, geographical origin or other compositional characteristics of honey;
4. The use of ion-exchange resins to reduce or remove residues (e.g., of veterinary drugs), offensive aromas or other components of honey such as HMF or to lighten the color of honey.
5. The harvest of immature honey, before the bees manage to turn the nectar or honeydew into a product that will have the composition of the original honey. In addition, the removal of water from immature honey that shows a high percentage of moisture changes the natural composition of honey, as heat treatment is required.

9.5.2 Ensuring the Authenticity of Honey

According to the Apimondia Statement, the strategy to fight honey fraud must include the following measures (Apimondia, 2020):

1. Informing beekeepers about the current situation and informing them about the negative effects of honey adulteration, as well as educating them on how to avoid it;
2. Informing packers and retailers about the need to carry out testing on honeys from countries where the legislation for honey does not meet the requirements of generally accepted standards;
3. Informing consumers through the media;
4. Periodic review of honey standards and use of advanced techniques and analytical methods (formal and informal) for the detection of honey fraud;
5. Collaboration between national and international authorities and institutions to ensure the authenticity and quality of honey;
6. Faithful implementation of existing food fraud legislation to prevent such illegal activities;
7. Use of third-party audits to ensure the compliance to internationally accepted standards, the food safety of the product and the company's fraud management system, which must include a fraud vulnerability assessment and a mitigation strategy, as well as the traceability of honey back to the beekeeper and the apiary.

In order to ensure the authenticity and the quality of honey as a natural product, a series of actions must be followed. First, beekeepers shall use good beekeeping practices in order to avoid adulteration of honey with products used for artificial feeding of bees. Feeding of bees with sugar products may be necessary some periods of the year. This often happens when bees are short of stores and there is a risk of starvation (e.g., during winter). Beekeepers should keep a record of the beekeeping practices they have applied to their beehives, which must include the type of food and the period the bees were fed. In any case, the feeding of bees with syrups and other products during the main nectar flow period must be avoided.

In addition, the companies engaged in the packaging, trading and processing of honey must maintain a documented honey fraud management system, which should involve a vulnerability characterization to fraud, risk assessment and severity analysis, preventive actions evaluation, a risk mitigation strategy, critical points identification in order to control the origin of the honey fraud, monitoring the critical limits, corrective actions and a program for implementation and review (Apimondia, 2020; García & Schwarzinger, 2021). According to the Apimondia Statement on honey fraud, the measures used to prevent the honey fraud include traceability of honey, laboratory testing and auditing systems.

Honey should be able to be traced back to beekeepers, to its botanical source from which bees collected the nectar or honeydew and to its geographical origin. Computer applications with the use of barcodes on honey packages shall ensure the traceability of honey, and all product information, from production to consumption, should be displayed to an online program (García & Schwarzinger, 2021). Traceability also means transparency of beekeeper's practices. According to Hazard Analysis Critical Control Point (HACCP) requirements, beekeepers should keep a record of the beekeeping practices followed during honey production, during honey harvesting and processing, as well as of the storage conditions. Traceability of honey must be aligned with the main food safety certification standards that are recognized by the Global Food Safety Initiative (GFSI), the Global Standard for Food Safety (BRC) and the IFS Food Fraud Guidelines. Both of these standards require the implementation of Vulnerability Assessment and Critical Control Points (VACCP), including organizational and analytical measures (Apimondia, 2020).

In order to assess the risk of fraud, the vulnerability of occurrence and the severity of fraud have to be measured. The fraud vulnerability prevention must be based on the assessment of possible opportunities, motivations and control measures in place. The main parameters and criteria used to evaluate the mitigation plan and the risk assessment are the historical incidents concerning honey fraud, the ease of fraudulent activity, the supply chain complexity and their confidence, the economic factors, the current control measures and the speed of the information provided (Tromp and Séchet, 2018). Vulnerability should be lowered to an acceptable level, and the whole process must be inspected periodically.

Testing of honey is also an important action to fight honey fraud. Honey fraud is a dynamic phenomenon, and the effectiveness of the methods used to detect it usually decreases over time. This is happening because the fraudster's side is constantly inventing ways to deceive testing (Dübecke et al., 2011). National and international authorities and institutions engaged in honey always have to go a step forward by developing new methods and using the appropriate techniques in order to detect immediately honey fraud.

In recent years a number of different syrup types are used to adulterate honey, and some of them are specially designed to cheat the existing testing methods, since these contain trace compounds and minor components that are used as analytical markers. It is practically impossible for a method to detect all the different types of honey fraud. The methods and techniques used to control honey adulteration must be constantly evolving and adapting to the new requirements. The significance of implementing the suitable technique or method and not only the technique or method proposed by authorities (official methods), has to be emphasized due to the limitations of official methods and the dynamic nature of honey fraud (Apimondia, 2020).

It is important to conduct risk assessment (including VACCP) before choosing the method we will use to detect fraud in honey. The selection of best testing method shall be in the frame of a comprehensive and detailed food fraud management system that must take into consideration the supply chain of honey, the history of honey adulteration occasions from that supplier or origin, the supplier's profile and the relationship with them, financial abnormalities of the territory related to honey and the most usual adulteration methods used to the region of honey origin (Apimondia, 2020). Another important element that the honey fraud management system should include is the use of third-party audits. The audits shall be carried out independently by specially trained personnel. So these audits should be conducted on site, during the honey production period, by personnel who have sufficient knowledge of good beekeeping practices and of the parameters that affect the quality of honey. These personnel must have the ability to recognize and identify deviations and errors both during the production stage (e.g., feeding during the main nectar flow period, harvesting immature honey), and during the honey processing stage (e.g., vacuum dehumidifiers, ion-exchange resins, syrups). In addition, auditors must be present and collect samples from different stages of the honey production and processing chain for laboratory testing (Apimondia, 2020).

9.6 ANALYTICAL METHODS AND CHEMOMETRICS

9.6.1 Overview

Methods for the authentication analysis of honeys have significantly evolved over the last two decades, driven by increasingly sophisticated adulteration procedures, as described in section 9.5.1. The main two factors influencing the market price are geographical origin and botanical origin, with the latter particularly relevant for monofloral or unifloral honeys. This leads to a complex

Table 9.2 Cumulative Results from the EU Food Fraud Report

Non-compliance	Physicochemical parameters	Botanical Source	Geographical Origin	Sugars Added	Other Labeling Issues	Total
% of non-compliant samples	2%	7%	2%	6%	2%	19%

Source: Aries et al. (2016).

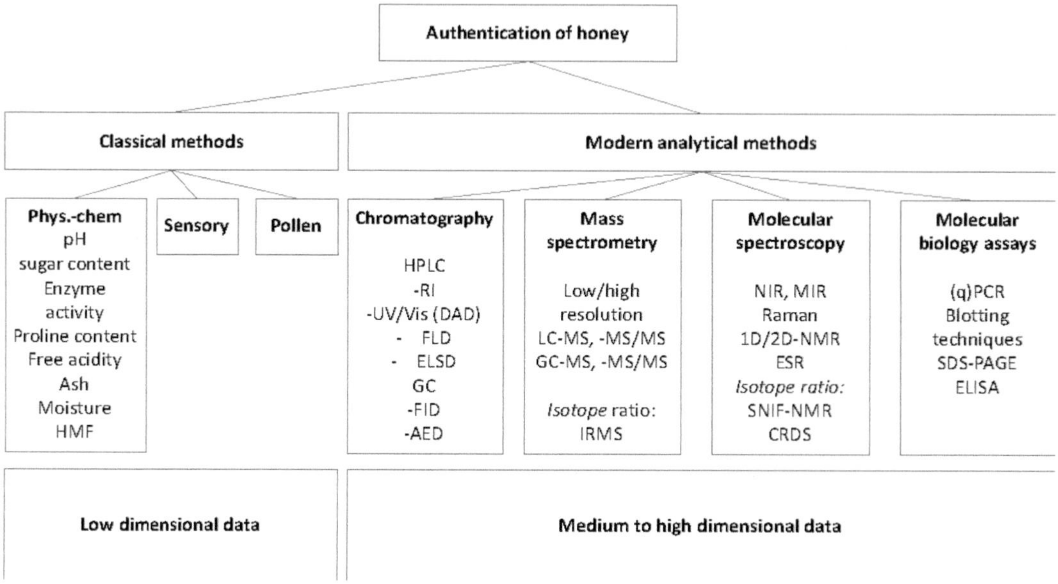

Figure 9.3 Schematics of honey authentication methods.

Source: Adopted from Ling Chin and Sowndhararajan (2020).

landscape of analytical tasks to be solved by the analyst, which is only feasible by maximizing the potential of reliable, robust and modern analytical methods with the use of chemometric tools.

This is in particular true for honey, as the statistics of adulterated honey samples show. The food fraud report of the EU from 2015–2017 (Table 9.2) gives an overview of the most common incidents, and in principle, these are mainly addition of sugar syrups from various (cheaper) sources and/or mislabeling of the botanical/geographical origin (Aries et al., 2016) Roughly one-fifth of all honey samples analyzed were at least "suspicious".

Beyond these main and often clear cases, there are a large number of other possible fraudulent actions, such as non-declared removal of pollen by filtration, undeclared heat treatment by pasteurization or the forceful maturation of unripe honey by water removal (Bogdanov, 2002)

As the Codex Alimentarius standards for honey (CXS 12–1981) are more general criteria than a collection of detailed methods for the authentication of honey, expert groups all over the world have worked intensively over the last decades to develop and define suitable methods. The International Honey Commission (IHC) has gathered such methods in a compendium (www.ihc-platform.net/ihcmethods2009.pdf), which is a valuable basis for routine laboratories to confirm marketing standards in the framework of food authenticity control. Still, they are not always sufficiently effective enough to detect adulterations. Figure 9.3 gives an overview on commonly applied methods for the authentication of honey and a preliminary idea of the dimensionality of the data, which is key for chemometric approaches.

The analytical techniques for the authentication of honeys described in the following sections are focused on those with a chemometric component; thus, the list of possible methods will not be complete.

9.6.2 Targeted versus Non-Targeted Methods

In the field of food authentication, a plethora of publications are available on the application of different analytical techniques on various food matrices (e.g., food integrity handbook by Elliot et al., 2018). Still, the majority of techniques are so-called targeted approaches, which aim for a predefined set of potentially characteristic substances for an adulteration (Ballin and Laursen, 2019; Cavanna et al., 2018; Creydt and Fischer, 2020).

As in any field of food authentication, the idea of finding specific authentication marker compounds is tempting but, depending on the analytical question, difficult if not impossible. One of the reasons is the enormous variance of factors influencing honey in its composition as a natural product, factors that are often difficult to control.

The general composition of honey varies significantly, mainly depending on the botanical origin (Table 9.3), but in particular, it is the low abundant species, such as proteins, peptides, amino and carboxylic acids, volatiles, phenolic substances, trace elements in their concentrations and ratios that show strong variation, not only among the different types of honeys but also within one distinct type.

These minor constituents vary significantly depending on the composition of the nectar harvested by the bees. Floral nectar composition is mainly influenced by the plant species but also by geography and climate and the soil properties. Furthermore, bees foraging the nectar are also exposed to environmental pollution of the area, and finally honey harvest, processing and storage leave their chemical footprints.

This leaves the analyst with a complex, multidimensional pattern with a plethora of factors, which makes it challenging to define markers or sets of markers that allow a reliable and robust differentiation of honeys from unknown sources, in particular if the analytical question is diffuse (e.g., classification of geographical origin).

While targeted methods are principally not limited to the detection of only one compound at a time, but can do so for a larger number (e.g., in pesticide multi-methods), one restriction still remains: the information gathered is limited to the scope of predefined substances and, therefore, often provides only limited information to safely protect against food fraud and adulteration. This is a downfall, as there are potentially dozens or hundreds of substances that may be added to foods in general—the scope for honey is surely more limited—yet a targeted approach still lacks the flexibility of finding new and unknown fraudulent strategies that are no longer based on known substances.

In contrast, non-targeted approaches go a different way: all amenable data is collected and retrospectively interpreted, without the need for selection prior to analysis. This is a common approach in metabolomics and has gained ground in the food sector both in mass spectrometry

Table 9.3 Average Composition of Honeys

	Blossom Honey		Honeydew Honey	
Parameter	**Average**	**Min-Max**	**Average**	**Min-Max**
Water content [%]	17.2	15–20	16.3	15–20
Fructose [%]	38.2	30–45	31.8	28–40
Glucose [%]	31.3	24–40	26.1	19–32
Sucrose [%]	0.7	0.1–4.8	0.5	0.1–4.7
Other disaccharides [%]	5.0	28	4.0	16
Melezitose [%]	<0.1	–	4.0	0.3–22.0
Erlose [%]	0.8	0.56	1.0	0.16
Other oligosaccharides [%]	3.8	0.5–1	13.1	0.1–6
Total sugars [%]	79.7	–	80.5	–
Minerals [%]	0.2	0.1–0.5	0.9	0.6–2
Amino acids, proteins [%]	0.3	0.2–0.4	0.6	0.4–0.7
Acids [%]	0.5	0.2–0.8	1.1	0.8–1.5
pH	3.9	3.5–4.5	5.2	4.5–6.5

Source: White (1975).

and spectroscopy-based strategies. A multitude of derivatives have emerged and were named according to their field of application (e.g., "foodomics", "lipidomics"), as well as by the type of analysis (e.g., "volatilomics"). In the last years, this demand for new technologies has been driven massively by more and more complex authentication questions, such as production origin (geography), production type (conventional, organic) or the issue of illegal blending at lower concentration levels, where targeted methods often fail.

However, it has to be considered that a strategy based on a single method, such as LC-ESI-MS is limited in its detection capabilities to the substances that are accessible via ESI. With regard to a holistic description of the sample characteristics (e.g., for a geographic differentiation), this might not be sufficient and can be overcome by adding additional, complementary methods into the authentication toolbox.

While non-targeted methods seem superior in their flexibility and the overview on the sample, it still has to be considered that it may be reasonable to additionally add specific targets to the analytical portfolio for particular issues. For example, the adulteration of honeys by adding sugar syrups may easily be identified via targeted analysis of specific markers without the need for a profiling method, while for detection of a mislabeling of geographical and botanical honey origin, a combination of both targeted and untargeted profiling methods can be used. It is important to stress that various botanical and geographical origins of a honey can have a combined influence on the sample at the same time, as characteristic parameters, such as trace elements and/or organic impurities or artifacts and cannot be unequivocally separated. Here, profiling methods can serve both purposes.

Most hardware manufacturers meanwhile offer such functionality in their software packages, but there are other, independent tools available, such as XCMS (https://xcmsonline.scripps.edu) or MetaboAnalyst (www.metaboanalyst.ca), which were fundamentally designed for metabolomic studies, but since metabolomic principles are applicable as well to food authentication, it has been described in the literature also for honey authentication (e.g., Koulis et al., 2021; Lippolis et al., 2020). Some authors legitimately differentiate between "target screening", "suspect screening" and "untargeted screening", which demonstrates that the terms "targeted" and "untargeted" are not unambiguous. In this context, it seems more reasonable to differentiate between "expected knowns", "unexpected knowns", "expected unknowns" and "unexpected unknowns", as suggested by Righetti et al. (2016) or Rychlik et al. (2017) in the field of mycotoxin analysis.

9.6.3 Analytical Data in View of Chemometrics

The main difference of targeted and untargeted approaches with view on chemometric processing is that targeted methods typically generate only a limited number of so-called features, a term that is commonly used for the signals extracted from the analysis. Features should be relevant data, in an optimal case characteristic for a specific botanical origin (e.g., abscisic acid for lime honey) or a geographical provenance. This could be a set of physicochemical parameters, values for a set of trace elements, a NIR or MIR spectrum, or a high-dimensional LC-HRMS analysis.

In chemometric data analysis, most algorithms require a 2D matrix ($n \times m$, with n objects and m variables); for example, PCA, LDA or PLS-DA (here, it is rather two 2D matrices, **X** and a depending **Y** matrix). In a targeted analysis, this is quite straightforward: the objects are the samples measured and the features are, for example, signal intensities of a certain compound. As the target compounds are usually refined and confirmed via reference standard (if available), there is usually no need to use a third dimension, such as the retention time in chromatography. Thus, it is likely that the number of objects (i.e., samples measured) is higher or at least equal to the number of variables (i.e., features of the analysis) and not vice versa. This is an important perquisite for at least some of the linear discriminant analysis algorithms, such as LDA. The situation changes completely when full-scan MS data is used in combination with retention time data, as it is done in fully untargeted approaches. Here, high-dimensional data is generated in form of a 3D matrix: m/z × retention time × intensity of signal. This is data that most algorithms cannot work directly with, apart from so-called multi-way PCA (PARAFAC) or TUCKER3 (Bro, 1997). This leaves the analyst with two issues: very large amounts of data and 3D (or even higher dimensional data if MS/MS data is generated, for example). Prior to data analysis, these 3D data have to be unfolded into a 2D data matrix, which is obviously even larger than the data set before. This is where dimensionality reduction algorithms such as PCA or PLS jump in. Apart from specific differences in the way these algorithms reduce dimensionality and explain the variance of the data (or group the covariant data), the principle is similar: the algorithms look for covariant information that can be aggregated. The matrix decomposition generates a much smaller model of the original data space ($n \times o$ with $o << m$), which can be used for further analysis.

9.6.4 Classical Analytical Parameters for Honey Authentication

Classical physicochemical parameters are the first pillar of classical honey authentication and can provide valuable information on the origin of monofloral honeys. They comprise typically pH, sugar content, the amino acid proline, enzyme and specifically, diastase activity, moisture and ash content, free acidity and the content of hydroxy methyl furfural (HMF). These parameters are mainly defined by country-specific legislation (e.g., for the EU by Council Directive 2001/110/EC). The basis for these analyses methods has been defined by the IHC.

While these parameters vary distinctly depending on the type of flowers, as well as on the season, climate and finally, geography, it is clear that they are not bulletproof in terms of their reliability. Yet, they can be first indications of adulteration (Table 9.4).

9.6.5 Extension with Sugar Syrups

The extension of honeys with cheaper sugar syrups is a common way to maximize an economic gain. As honey basically is an oversaturated solution of a limited number of monosaccharides at high concentrations (glucose, fructose), together with lower concentrated disaccharides (sucrose, maltose, isomaltose, maltulose, turanose and kojibiose), trisaccharides (melezitose, maltotriose) and some oligosaccharides (erlose, theanderose, panose), the latter are the result of the incomplete breakdown of higher saccharides found in nectars and honeydew.

Typically, the ratio of fructose and glucose is approximately 1.2:1.0, however, these values vary distinctly among the floral origins (da Silva et al., 2016; de la Fuente et al., 2011; Escuredo et al., 2014).

The fructose to glucose ratio indicates the crystallization tendency of honeys (Manikis and Thrasivoulou, 2001) and generally, honey samples with levels of below 30% m/m of glucose have a lower tendency to crystallize (e.g., for honeydew or chestnut), while higher glucose honeys, such as canola or sunflower have a higher tendency.

Consequently, the use of pure glucose syrup would not only be clearly detectable, but also would lead to atypic crystallization behavior. In a similar way, addition of sucrose would be easy to detect by the IHC methods (Table 9.4), even though sucrose is enzymatically converted at least partially by the contained invertase.

An adulterant much closer to honey is invert sugar syrup (IS), a mixture of glucose and fructose. The easiest way to produce IS is the acidic hydrolysis of sucrose, which however leaves detectable traces of the condensation product HMF, and thus it is seldomly used for adulteration (Anklam, 1998). Since HMF is toxicologically relevant, modern processes are employing invertase enzymes at lower temperatures (60°C) and moderate acidic conditions, which reduces HMF formation.

The most common adulterant for honey, however, remains high-fructose corn syrup (HFSC), in particular due to the fact that it is cheap to produce and easily available on the commodity market. HFSC 55, which refers to its dry-matter weight of fructose, is mostly produced as sweetener for

Table 9.4 Classical Analytical Parameters and Their Importance for Adulteration

Analytical Parameter	Assessment of Parameter
Free acidity (pH titration)	Spoilage by fermentation/botanical origin
Ash content (gravimetry)	Botanical origin
Conductivity	Botanical origin
Moisture (refractive index)	Addition of water
HMF content (colorimetry/HPLC)	Age, heat treatment, also source-specific pattern of invert sugar syrups
Diastase (enzyme assay)	Heat treatment
Invertase (enzyme assay)	Age/freshness
Sugars (mainly glucose, fructose, maltose, sucrose)	Addition of sugars, invert sugar syrups, also botanical origin
Proline (colorimetry)	Maturity

soft drinks, but also features a composition fairly similar to honey: 55% fructose, 41% glucose and approximately 4% oligosaccharides (White, 1975). If only the sugar profile is monitored (e.g., by HPLC), even significant additions of up to 50% might not be detected as Codex specification violations. It is self-explanatory that this approach, however, is principally detectable via the dilution of the remaining characteristic sugars, minerals, volatiles and so on that are not present in HFCS syrups, yet the large variation of the latter in nature might still make room for an optimized adulteration. To overcome this issue, target-based approaches focusing on atypical trace artifacts from sugar syrups are more appropriate and furthermore, detection of isotope ratios of the sugars quickly reveal at least lower sophisticated attempts of fraud.

9.6.6 Analytical Strategies

9.6.6.1 Sensory Analysis

Sensory evaluation is the second pillar of classical honey authenticity evaluation, as a particular monofloral honey may depict unique organoleptic characteristics, such as olfactory profile and taste or color. The IHC has published a harmonized framework for the characterization of European monofloral honeys (von der Ohe et al., 2004). This comprises a unified description of terms used in sensory description of unifloral honey.

In particular when combined with commonly applied physicochemical parameters, such as pH, free acidity, proline content or conductivity, sensory assessment can be a helpful addition for the characterization of unifloral honeys from different geographical origins, in particular, when the information is used in a multivariate way (Bertoncelj et al., 2011; Moumeh et al., 2020; Patrignani et al., 2018; Rodríguez García et al., 2006).

However, its reliability in the authentication of honey is strongly dependent on the availability of specifically trained panelists. Even though such expert panels are reported to be stunningly precise, in particular in combination with the support of pollen analysis data (Rodríguez García et al., 2006) another limiting factor is the lack of standardization, which is inherent to sensory assessments.

This led to new approaches, which attempted to support classical sensory knowledge with electronic noses (e-noses) or electronic tongues (e-tongues). Most of the systems are based on metal oxide sensor (MOS) arrays with typically a broad range of specificity to a number of substances, such as alcohols, sulfur compounds or aromatic substances.

Modern sensory-based approaches include the use of e-nose and/or e-tongue sensors to detect adulterations (e.g., acacia and linden honey adulterated with sugar syrups; Bodor et al., 2020).

9.6.6.2 Pollen Analysis

Melissopalynology, or pollen analysis, is the third pillar of classical honey authenticity analysis. Sensory and physicochemical analyses are used to determine the origin of monofloral honey, while melissopalynological analysis is commonly used to identify floral pollen grains found in honey through microscopic examination (Louveaux et al., 1978). Pollen grains of different plant types have a distinct morphology that can be identified by microscopy. Monofloral honey should contain at least 45% of pollen grains from a single species to be labeled as such, compared to multifloral honey that produced pollen grains from several plant species, none of which is considered predominant.

The melissopalynological approach, however, may not be suitable for some types of honey such as citrus or cotton, as pollen content is variable and generally low. The amount of pollen depends on the plant species, the harvest season and the nectar yield of the male and female flowers. In some cases, pollen may be filtered through the bee's honey bag and added fraudulently to the honey (Ling Chin and Sowndhararajan, 2020).

This significant natural variety of pollen content makes melissopalynology complex, even for trained experts, therefore it is commonly accompanied by sensory analysis and determination of certain physicochemical characteristics (Ling Chin and Sowndhararajan, 2020; Louveaux et al., 1978)

While none of the three methodologies (physicochemical, sensory and pollen analysis) generates high-dimensional data, such as obtained from mass spectrometric or spectroscopic techniques that force the analyst to use chemometric methods, these still can bring significant benefits also into this field. This is mainly due to the fact that classical analytical approaches are usually univariate (i.e., individual compounds are treated independently and not in a holistic approach).

However, this often leads to a failure to see trends or tendencies hidden in the combined information gathered. As an example, the e-tongue sensor arrays mentioned above was paired with

Table 9.5 Selected Examples for Chemometric Physicochemical, Sensory, and Melissopalynological Analyses

Authentication Issue	Parameter	Analytical Method	Chemometric Method	References
Botanical origin	Color	CIELAB data by UV-Vis spectroscopy	HCA, PCA	(Tuberoso et al., 2014)
Botanical origin	Color	Colorimeter	PCA, classification/ regression tree	(Bergamo et al., 2019)
Geographical origin	Color + antioxidant capacity	UV-Vis, ABTS/DPPH assays	PCA, HCA	(Castiglioni et al., 2018)
Botanical and geographical origin	Physicochemical properties	Physicochemical analysis	Classification/ regression tree	(Fechner et al., 2016)
			PCA, LDA	(Popek et al., 2017)
Botanical origin	Physicochemical properties and pollen	Microscopy and physicochemical analysis	PCA, cluster analysis (CA)	(Combarros-Fuertes et al., 2019)
Botanical origin	Sensory and pollen analysis	e-tongue, pollen analysis	SIMCA	(Rodríguez et al., 2015)
Botanical origin	Voltammetric properties, physicochemical parameters	Voltammetric e-tongue, physicochemical analysis	PCA, LDA, PLS-DA	(di Rosa et al., 2018)
Botanical origin	Sensory, physicochemical, antioxidant, phenols, sugars, colorimetry	e-tongue, physicochemical analysis, TPC, FRAP, CUPRAC, CIELAB	PCA, LDA	(Oroian and Ropciuc, 2019)
Botanical origin	Pollen analysis	Semi-automated microscopy/image recognition	CNN	(Olsson et al., 2021)

physicochemical data (e.g., conductivity, sugar content, acidity) and modeled and classified by an artificial neural network (ANN) with nearly perfect separation quality (Major et al., 2011).

A number of further studies have described a successful application of MOS-based e-noses and e-tongues to differentiate floral origins in combination with physicochemical parameters (e.g., Dias et al., 2008; Romano et al., 2016).

Table 9.5 shows selected examples for authenticity studies based on physicochemical parameters, sensory assessments and melissopalynology in conjunction with chemometrics.

This demonstrates that chemometric strategies can not only optimize the use of data from individual classical methods, such as physicochemical, sensory or pollen analysis, but furthermore can maximize the knowledge by intelligent combination. Nearly all of the studies published demonstrate that unsupervised chemometric methods (i.e., algorithms such as PCA, which are based on unlabeled data) are often not able to differentiate in complex or "diffuse" cases, where the data were overlapping. This was particular the case for sample sets with a larger number of "classes" (e.g., geographical origins) and is attributed to the fact that PCA maximizes the total variance of all data—which not necessarily means that all sample classes will be separated. In this context—and this is virtually a general rule independent from the analytical techniques used—supervised techniques, such as PLS-DA, are significantly more powerful due to the fact that these maximize the inter-class variance. These approaches, however, strongly depend on a thoughtful validation (e.g., by k-fold cross-validation, bootstrapping or comparable techniques to avoid overfitting).

9.6.6.3 Hyphenated Separation Techniques

Hyphenated separation techniques are surely among the most commonly used analytical tools in laboratories all over the world and in particular, in the field of food authentication. Their strengths are mainly the high separation power for various analytes in complex matrices, such as foodstuffs, and their flexibility to adapt to a variety of different detector systems.

The main part of the studies employs either liquid chromatography (LC) or gas chromatography (GC), with a clear majority of the studies focusing on LC. This is mainly due to the fact that the scope of analytes is commonly considered to be wider and the sample preparation usually is less complex, while GC is limited to the (thermally stable) volatile organic carbon fraction (VOCs).

Capillary electrophoresis (CE), while being a high-performance separation technique with a broad scope of analytes and not only limited to proteins or peptides, but also to small molecules, in particular ionic species, plays an inferior role in this context. This might be due to the fact that CE is by far less common in routine laboratories than LC or GC systems. Nevertheless, a number of studies that evaluated the use of CE in the authentication of monofloral honeys were published (Ramón-Sierra et al., 2015; Yue et al., 2021).

9.6.7. Liquid Chromatography

LC is usually understood in the context of HPLC, but it should be stressed that even simpler LC-based techniques such as high-performance thin layer chromatography (HPTLC) can be used in the field of authenticity control (e.g., for phenolic compounds as markers for lime and acacia honeys; Stanek et al., 2019), although the use is limited to targeted approaches with a lower number of analytes due to the limited selectivity.

The majority of LC methods used are reverse phase (RP) HPLC methods, as they are optimally suitable for water-based samples with analytes of low to medium polarity. A smaller share of the methods use high-performance anion-exchange chromatography (HPAEC), such as for specific sugars, (e.g., the disaccharide turanose; Gašić et al., 2015) Most HPLC methods published are targeted methods, with a variety of different detectors used: UV-Vis detectors, pulsed amperometric detectors (PAD), ICP-OES, and low- and high-resolution mass spectrometry.

Commonly employed marker compounds are the phenol derivatives, mainly phenolic acids and the more complex flavonoids. However, the scope of analytes in honey ranges from apolar (e.g., monoterpenes) to highly polar (e.g., phenolic acids), which makes analyses challenging, mainly due to the low retention of the polar compounds. As a countermeasure, one could use hydrophilic interaction chromatography (HILIC) phases, which are partially complementary to RP phase material.

Apart from the phenolic compounds, a number of other target compounds are commonly analyzed: protoalkaloids and alkaloids, such as synephrine (Tette et al., 2017), pyrrolizidines (Hungerford et al., 2019), sugars (Gašić et al., 2015) and other organic acids (Jandric et al., 2017). The scope of target compounds also defines the sample preparation demands (e.g., SPE; Michalkiewicz et al., 2008).

When used in conjunction with chemometric approaches, as most of the studies nowadays do, the analyst should keep in mind that all sample preparation steps are potentially discriminating individual compounds of different chemistry. This may lead to shifts in chemometric analyses, such as PCA or supervised methods (e.g., PLS-DA) due to additional variance that is not sample-related but an artifact from the sample preparation.

More and more, an increasing number of non-targeted, metabolomic approaches have been described, where the sample preparation has to be kept at an absolute minimum in order to avoid discrimination of individual substances. In particular, the vast substance group of phenolic compounds are at least partially highly susceptible to discriminative effects. In this context, generic liquid-liquid-extraction strategies, such as the use of ethyl acetate, which also can extract less polar compounds, such as the flavonoids or terpenes. While it is generally possible to perform non-targeted approaches also with less selective detectors, such as DADs, it is clear that metabolomic approaches significantly profit from the much higher selectivity of mass spectrometric detection. With increasingly improving sensitivity and selectivity of these systems, in particular for the high-resolution mass spectrometric detectors (HRMS), such as Orbitraps or QTOFs, also the need for sophisticated sample preparation has decreased, therefore so-called dilute-and-shoot approaches have gained importance (Li et al., 2017; Seraglio et al., 2016) Here, samples are diluted, centrifuged and filtered and then subjected to analysis. This approach is advisable for metabolomics-based strategies, in particular for fully untargeted ones using spectral fingerprints, as these are highly susceptible to artifacts from preparation.

Table 9.6 shows selected examples for targeted and untargeted analyses authenticity studies based on liquid chromatography methods, partially combined with other parameters.

9.6.7.1 Gas Chromatography

Gas chromatography (GC) based methods are widely used mainly for the volatile organic carbon (VOC) fraction of honey samples (see Table 9.7). VOC-based analyses are very attractive since the volatiles are directly linked to the aroma profile of honeys and thus to the botanical origin. The majority of these molecules are esters, aldehydes, ketones, alcohols, carboxylic acids, terpenes and terpenoids, furans and alkanes with a large number of other lower abundant species. GC analysis

Table 9.6 Selected Examples for Targeted and Untargeted LC-Based Methods

Authentication Issue	Parameter	Analytical Method	Chemometric Method	Reference
Geographical origin of Serbian polyfloral honeys	Targeted: phenolic compounds (e.g., chlorogenic acid, quercetin)	UHPLC-HRMS (Orbitrap)	PCA, PLS-DA	(Gašić et al., 2014)
Geographic origin of Greek thyme honey	Targeted: phenolic compounds (e.g., chrysin, kaempferol) + physicochemical parameters	HPLC-UV, physicochemical parameters	M-ANOVA, LDA	(Karabagias et al., 2014c)
Botanical origin (chaste, canola)	Targeted: kaempferol, morin, ferulic acid	HPLC-DAD-LRMS/MS	PCA, PLS-DA, SIMCA	(Zhou et al., 2014)
Botanical origin (sage honey)	Targeted: boron, potassium, kaempferol, turanose	UHPLC-DAD-MS/MS, Orbitrap-MS Total phenols, HPAEC-PAD, ICP-OES	PCA	(Gašić et al., 2015)
Botanical origin of floral honeys	Targeted: phenolic compounds, flavonoids, sugars, organic acids, terpenes	UHPLC-Orbitrap-MS	PCA, OPLS-DA	(Li et al., 2017)
Botanical origin of New Zealand honeys	Untargeted analysis with potential structures	UPLC-QToF-MS, IRMS, ICP-MS, IR	PCA, OPLS-DA	(Jandric et al., 2017)
Botanical and geographical origin of floral honeys	Untargeted analysis, suspect screening and targeted	UPLC-QToF-MS	PCA, PLS-DA	(Koulis et al., 2021)

Table 9.7 Selected Examples for Targeted and Untargeted GC-Based Methods

Authentication Issue	Parameter	Analytical Method	Chemometric Method	Reference
Geographic origin (Spain, Romania, Czech Republic)	Targeted: aldehydes, alcohols, terpenes	DHS-GC-MS, sugar analysis, physicochemical parameters	ANOVA, PCA, SLDA	(Juan-Borrás et al., 2014)
Geographic origin (Greece)	Targeted: organic acids, aldehydes, alcohols, terpenes	SHS-SPME-GC-MS, physicochemical parameters	MANOVA, LDA	(Karabagias et al., 2014a, 2014b)
Geographical origin (Galicia, Malaysia, Bangladesh)	Targeted: toluene	SHS-SPME-GC-QToF-MS	PCA	(Moniruzzaman et al., 2014)
Botanical origin (heather, alder buckthorn, raspberry, canola)	Targeted: organic acids, aldehydes, alcohols, terpenes	SHS-SPME-GC-QToF-MS	HCA, CA	(Seisonen et al., 2015)
Geographical and entomological origin (regions in Brazil, 8 species of stingless bees)	Targeted: organic acids, esters, TDN, aldehydes, alcohols, terpenes	SHS-GC-MS	PCA, LDA	(de Lima Morais da Silva et al., 2017)
Botanical origin of European honeys (canola, acacia, honeydew)	Untargeted plus marker identification	SHS-GC-IMS	PCA, LDA, kNN	(Gerhardt et al., 2018)
Entomological origin	Targeted: aldehydes, terpenes, terpene oxides, ketones, alcohols	SHS-GC-IMS, SHS-SPME-GC-MS	PCA, OPLS-DA	(Wang et al., 2019a)
Botanical origin and adulteration with sugar syrups	Untargeted plus marker identification	SHS-GC-IMS	PCA, OPLS-DA	(Wang et al., 2019b)

of the VOC fraction is typically carried out via indirect injection methods, such as headspace (either static, SHS or dynamic/purge and trap, DHS) or by additional (online) extraction steps, such as headspace solid phase microextraction (HS-SPME) or in-tube extraction (ITEX). Most of the GC methods published are MS-based methods, either low-resolution (LRMS) or high-resolution (HRMS) methods, where the majority is still based on quadrupole LRMS detectors. This is because in comparison to liquid chromatography, the number of available HRMS detectors is much lower. While analyzing a plethora of VOC compounds, still the majority of methods are targeted approaches; only a smaller number (Ruiz-Matute et al., 2007; Xagoraris et al., 2021) are untargeted methods.

Nearly all of the GC-MS-based systems are using electron ionization, which is mainly due to the high reproducible spectra and the availability of established spectral databases. A smaller number of articles were published based on proton transfer reaction (PTR) ionization (e.g., Ballabio et al., 2018; Schuhfried et al., 2016).

With regard to chemometric data analysis, it has to be stressed that every technology used for enrichment is somehow based on adsorption/desorption principles (e.g., on an SPME fiber, an ITEX tube or an TENAX tube). Due to the repeated thermal stress of heating and cooling, these fibers will degrade after a certain time, which will lead to a gradually reduced sorption behavior—which will add additional variance into the data that is not associated with a biological or geographical origin.

An emerging technology that quickly gained popularity is ion mobility spectrometry, hyphenated to gas chromatography (GC-IMS). The majority of these systems are based on ^{3}H or (less commonly) ^{63}Ni ionization sources, while a smaller number employ corona discharge (CD). The radioactive sources feature high sensitivities in the lower ppbv-level (Arroyo-Manzanares et al., 2019; Gerhardt et al., 2018) in particular for proton-affine substances, which often allows to measure without pre-enrichments, such as SPME. The advantage over electron ionization is that the ionization is much softer, hence leaving analyte ions intact. These systems are highly attractive due to their performance and robustness and at the same time, simplicity and lower cost of ownership, as they do not require vacuum technology. GC-IMS ionizes analytes to produce so-called protomers, which are basically proton-water clusters that are then separated according to their drift time in a buffer gas, which allows to approximate the collision cross section (CCS). These drift time IMS (DTIMS) systems are the most common principle in GC-IMS. The resulting 3D fingerprint is highly complex and due to the complementary character of drift time versus retention time, the data is truly orthogonal, which makes the data optimal for chemometric analysis. As GC-IMS systems inherently lack m/z information, there are still no publicly available CCS-databases, which is why the chemometric analysis is typically performed as fully untargeted. A number of recent studies has evaluated GC-IMS for the authentication of the botanical origin or the detection of adulterants (Arroyo-Manzanares et al., 2019; Gerhardt et al., 2018; Wang et al., 2019b, 2019a)

9.6.7.2 Ambient Desorption Techniques for Mass Spectrometry

The use of ambient desorption or ionization techniques (AIMS) is a fast-evolving practice in modern mass spectrometry and is gaining popularity mainly due to its character: samples may be analyzed in their native state, or at least with minimalistic sample preparation. In conjunction with the enormous development of mass spectrometry, in particular HRMS systems, AIMS has a high potential to overcome at least a number of limitations that come with conventional hyphenation techniques. An excellent overview on the progress over recent years is given by (Arrizabalaga-Larrañaga et al., 2021).

There are a number of studies related to the use of direct analysis in real time (DART) for the authentication of monofloral honeys (Lippolis et al., 2020), easy ambient sonic-spray ionization mass spectrometry (EASI-MS; Porcari et al., 2016) or rapid evaporative ionization mass spectrometry (REIMS; H. Wang et al., 2019), to name some of the most prominent examples.

With regard to chemometric analyses, ambient desorption techniques are challenging in terms of reproducibility and desorption behavior, which significantly depends on the plethora of factors, such as particle size, water content and overall matrix composition.

9.6.7.3 Matrix-Assisted Laser Ionization Desorption Mass Spectrometry

Together with the ambient desorption techniques, MALDI-TOF analysis depicts a special case of mass spectrometry–based honey authentication.

The sample preparation is not as minimalistic (e.g., as in AIMS), as it requires co-crystallization with a matrix (e.g., 2,5-dihydroxybenzoic acid [DHB]). However this step is principally automatable, which would increase the duty cycle of such instruments drastically. In general, MALDI-TOF data is complex due to its high dimensionality and thus optimally suitable for chemometric data analysis.

A number of studies have been published in recent years, focusing on the protein fingerprint (e.g., Wang et al., 2009) the metabolomic fingerprint (Brendel et al., 2021) or the fingerprint of bacteria that are characteristic for specific regions or plants (Pomastowski et al., 2019). Furthermore, also the detection of an adulteration with sugar syrups based on the oligosaccharide and polysaccharide profiles was evaluated (Qu et al., 2019).

9.6.7.4 Spectroscopy

Spectroscopy-based methods are highly attractive in the context of honey authentication as they are typically fast, non-destructive and, depending on the type, they may be used directly with no or virtually no sample preparation (see Table 9.8). While it is possible to allocate individual signals or bands to substances in NMR spectroscopy, most other techniques, such as NIR, MIR, Raman, UV-Vis or fluorescence, do not allow the direct determination of specific analytes from the spectrum. Thus, the majority of approaches are fingerprint or profiling based, taking into account the complete spectrum or selected parts. In conjunction with appropriate and technology-specific preprocessing, spectral data are highly interesting for use with powerful chemometric tools. Applying the correct preprocessing steps is vital, as physical phenomena may introduce a significant amount of variance to the data that—if not corrected for—may compromise the differentiation of botanical or geographical origin. As a simple rule of thumb, it is obvious that when variance generated by the measurement procedure is higher than the inherent sample variance due to its provenance, it may lead to false interpretations.

Common preprocessing steps for IR data are typically correction of scattering effects due to varying particle sizes of the sample, normalization to correct for differences in path lengths and derivatization to reduce noise. On the other hand, NMR data commonly features non-linear shifts due to pH value variations or the need for precise phasing.

Most spectroscopic data are of 2D tensor type (i.e., a common 2D data matrix) and thus ideal for nearly all types of algorithms. It is relatively scarce that 3D data is generated, but also these approaches exist (e.g., by 2D fluorescence; Lenhardt et al., 2015), where parallel factor analysis (PARAFAC) was used for the differentiation of the botanical origin of Estonian honeys or 2D-NMR (e.g., unfolded HMBC data) for the classification of Italian honeys (Lolli et al., 2008).

Table 9.8 Selected Examples for Spectroscopy-Based Methods

Authentication Issue	Parameter	Analytical Method	Chemometric Method	Reference
Botanical origin of Estonian honeys	Spectral fingerprint	Front-face fluorescence spectroscopy	PCA, PARAFAC	(Kivima et al., 2014)
Botanical origin, authentication	Spectral fingerprint	Front-face fluorescence spectroscopy	PCA, PLS-DA, PARAFAC	(Lenhardt et al., 2015)
Botanical origin and detection of adulterations	NIR and MIR spectral fingerprints	NIR, MIR	PLS-DA	(Gan et al., 2016)
Botanical origin of monofloral and polyfloral European honeys	MIR fingerprints	µMIR	SIMCA	(Brendel et al., 2021)
Origin and composition of European acacia honeys	NMR fingerprints	^{1}H-NMR	PLS-DA	(Schievano et al., 2020)
Botanical origin of European monofloral	NMR fingerprints	^{1}H-NMR	PCA, PCA-LDA	(Gerhardt et al., 2016)
honeys (acacia, canola, honeydew)	NBCSA assay	NBCSA colorimetric assay	PLS-DA, LDA	(Chaharlangi et al., 2021)

IR spectroscopy is used in form of both near (NIR) and middle (MIR) infrared, where both principles have their advantages and disadvantages. NIR can be used in the form of transmission (cuvette or glass vial) in diffuse reflection (fiber or Ulbricht sphere) and has longer path lengths, which allow to penetrate the sample deeper and is generally easier to be handled in the field. However, since NIR mainly is based on less intense overtones with lower spectral resolution and is highly susceptible to water, it can only be used in combination with chemometric tools. MIR, on the other hand, has intrinsically a higher spectral resolution and sensitivity but is basically restricted to attenuated reflectance measurements (ATR), which have a very limited penetration depth into the sample of approximately 1l, or alternatively, to diffuse reflection (DRIFT), which is seldom used, however, due to the fact that it still requires sample preparation with KBr, thus its use is limited to NIR (Mendes and Duarte, 2021) An interesting alternative is specialized MIR transmission approaches with ultra-low-profile cuvettes and cooled MCT detectors, which allow for a significantly higher reproducibility than conventional FT-MIR systems, which allow a good differentiation quality for unifloral and also polyfloral honeys in combination with SIMCA (Brendel et al., 2021)

Raman spectroscopy is a spectroscopic technique used in condensed matter physics and chemistry to study vibration, rotational and other low-frequency modes. One of the main advantages is the absence of interference by water. A number of studies describe the use of Raman spectra in combination with chemometric processing, such as detection of adulteration (Oroian et al., 2018); differentiation of botanical origin (Corvucci et al., 2015); or geographical origin (Anguebes-Franseschi et al., 2019).

NMR spectroscopy and, in particular, proton NMR has evolved from a organic chemist's tool to a reliable and accepted profiling platform in the last decades (e.g., Sobolev et al., 2016). High-field NMR systems are demanding in terms of their needs to infrastructure and thus are bound to specialized laboratories, therefore there are increasingly more low-field Halbach magnet FT-NMR systems found in the market, which were already demonstrated to be applicable for the detection of HFCS adulteration of honey (e.g., Ribeiro et al., 2014). Despite the high demand to infrastructure, high-field NMR spectroscopy is omnipresent in the field of food authentication and also in honey authentication. One of the reasons is the high robustness and the ability to detect numerous compounds within one measurement in complex matrices—such as honey—without complicated sample preparation. The interesting aspect is that while it is a fingerprinting-based spectroscopy and can be used to predict geographical, botanical or entomological origin in combination with appropriate modeling and classification techniques, it also can be used for targeted approaches not only to identify but also to quantify dozens of substances. This is also reflected in the number of publications on the botanical origin (Gerhardt et al., 2016; Kuballa et al., 2018; Schievano et al., 2020); geographical or production origin (Consonni et al., 2019; Schievano et al., 2019; Spiteri et al., 2020); or the adulteration with HFCS (e.g., (Bertelli et al., 2010; Ribeiro et al., 2014).

9.6.7.5 Isotope Analysis

One of the most commonly applied techniques for authentication of honeys is the analysis of the stable isotopes of hydrogen, oxygen, nitrogen and carbon. As the ratios of these isotopes ($^{2}H/^{1}H$, $^{18}O/^{16}O$, 15N/14N and $^{13}C/^{12}C$) are significantly influenced by plant type, climatic conditions, geography, production and other factors, the analysis—and moreover the relations—of the stable isotopes are a very interesting field of analysis. It is complex and thus financially less attractive to fraudsters to adjust isotope ratios to specific values, which is in particular the case for cheap extensions with sugar syrups, such as HFCS. While the vast majority of plants used for honey production belong to the C3 plants, the main sources for cheap sugar or starch belong to the C4 plants (e.g., sugar cane or corn). The different pathways in photosynthesis lead to different enrichments of the ^{13}C isotope, as $^{13}CO_2$ reacts slower during fixation and leads to higher depletions in C3 plants compared to C4 plants. This phenomenon is called kinetic isotope effect. This means that an addition of sugar syrups from C4 sources to honey will shift the so-called delta values ($\delta^{13}C$ in ‰) in a way that adulteration can be visible. For honeys from C3 plants, the $\delta^{13}C$ values typically range from –22‰ to –33‰, for C4 plant–derived honeys, a range of –10‰ to –20‰ is normal and for CAM (crassulacean acid metabolism) plants, such as pineapple and cactus, $\delta^{13}C$ values are reported to range between –11‰ and –13.5‰. An addition of C4 sugar syrups will typically lead to a shift; values less than –23.5 ‰ are at least subject to potential adulteration (González Martín et al., 1998; Padovan, 2003).

The protein in the honey may be used as an internal reference, as the $\delta^{13}C$ of protein will remain constant upon adulteration. The majority of the studies on adulteration are based on isotope ratio

Table 9.9 Selected Methods Based on Isotope Ratio Analysis

Authentication Issue	Parameter	Analytical Method	Chemometric Method	Reference
Geographical and botanical origin of acacia, sunflower, canola linden, honeydew and other honey	$\delta^{13}C$, $\delta^{18}O$, δ^2H and d^2H from ethanol	IRMS, SNIF-NMR	PCA	(Dinca et al., 2015)
Botanical origin (e.g., citrus, rhododendron, eucalyptus, acacia)	$\delta^{13}C$ of honey, ethanol and protein, $\delta^{15}N$ of protein, elements	IRMS, SNIF-NMR, ICP-OES	PCA	(Bontempo et al., 2017)
Botanical origin (4 species)	$\delta^{13}C$, color, antioxidant capacity, P, Sn	IRMS, ICP-OES	MANOVA, LDA	(Karabagias et al., 2016)
Geographical origin	$\delta^{13}C$, oligosaccharides, polyphenols	IRMS, GC-MS	PCA, PLS-DA	(She et al., 2019)
Geographical origin (6 Chinese regions, acacia)	$\delta^{13}C$, polyphenols, oligosaccharides,	IRMS, GC-MS	Logistic regression	(Berriel and Perdomo, 2019)
Botanical origin (eucalyptus, pasture from Uruguay)	$\delta^{13}C$ of honey and protein	IRMS	ANOVA, HCA, PCA	(Geană et al., 2020)

mass spectrometry, either in form of EA-IRMS with combustion (C, N, H, S) or pyrolysis (O) or in hyphenation with LC for $\delta^{13}C$ or GC (mainly for $\delta^{13}C$). LC hyphenation is typically used for the pre-separation of the sugar fraction to specifically detect the $\delta^{13}C$ ratios of the individual sugars, but it has also been described for the use with organic acids, mainly gluconic acid, for the differentiation of the geographic origin (Kawashima et al., 2019; Suto et al., 2019).

In terms of the botanical and geographical origin, a number of studies evaluated the applicability of stable isotopes (see Table 9.9).

Vasić et al., 2020) published a study on the discrimination of five varieties of honey by $\delta^{13}C$ and additionally with 20 elements and 14 carbohydrates. Interestingly, they discussed the $\delta^{13}C$ value of the protein fraction also a potential indicator for the botanical origin. Another study by (Magdas et al., 2021) analyzed a number of Romanian and French honey from different botanical origins. They used isotope ratios together with rare earth elements (La, Ce, Pr) and demonstrated that by chemometric combination of these parameters, they were able to differentiate the geographies, but mainly due to the δ^2H and $\delta^{18}O$ values attributed to the different climatic conditions.

While in general, isotope analysis generates important and valuable information, most of the studies used additional markers, such as elements, phenols or oligosaccharides to increase the differentiation quality.

9.6.7.6 *Elemental Analysis*

The elemental fingerprint of honeys is dependent on a variety of factors, from geographical origin to botanical origin, as well as environmental conditions (Pohl et al., 2017).

In principle, the elements in honey stem from the pollen grains and the nectar or the honeydew the bees forage on (Jovetić et al., 2017; Madejczyk and Barałkiewicz, 2008). As the soil mineral content varies with geography due to geochemical and geological characteristics, it becomes clear that this is an influencing factor for the geographical differentiation. Conversely, different plants feature different characteristics of uptake of minerals form the soils and thus, the combination of soil chemistry and plant biology will influence the elemental fingerprint (Đogo Mračević et al., 2020; Squadrone et al., 2020a). Furthermore, environmental factors, industry, agriculture, the resulting pollution levels, as well as climatic conditions affect the pattern of elements in the nectar and therefore in the honey (Kováčik et al., 2016; POHL, 2009).

The most common techniques for elemental fingerprinting are ICP-OES, ICP-MS, ED-XRF and ICP-HRMS to overcome isobaric interferences. As all of these techniques generate a plethora of data, these are ideally suitable for the use with chemometric methods, also in conjunction with other data, such as chemical or isotope data. Apart from the common elements, often also rare earth elements (REEs) are incorporated. A study compared 40 elements together with the REEs by

ICP-MS in mono- and multifloral honeys from Italy (Squadrone et al., 2020b). The trace elements showed differences in different botanical origins and the REEs varied within mono- and polyfloral honeys. A similar result was found in a further study focusing on honeys from different continents (Squadrone et al., 2020a). Interestingly, the lighter fraction of the REEs, the so-called LREEs, such as La, Ce, Pr, Sm, Eu and Gd, showed higher mobilities in the plants compared to the heavier fraction, the HREEs and thus, feature higher potential for the geographical origin.

More recent approaches described the use of laser-induced breakdown spectroscopy (LIBS) for the geographical origin of acacia and polyfloral honeys (Zhao et al., 2020) and energy-dispersive X-ray fluorescence (ED-XRF) to classify honeys according to botanical and geographical origin (Fiamegos et al., 2020).

Finally, the availability of high resolution mass spectrometry for ICP systems, ICP-HRMS opened up further ways to optimize elemental fingerprinting, mainly due to the fact that isobaric interferences are reduced. This increases the number of available elements from approximately 70 to 270 elements, which allows a new approach called "elemental metabolomics", which significantly could improve the differentiation quality in honey authentication (Zhang et al., 2017).

9.6.7.7 Molecular Biological Approaches

Analytical procedures targeting the genetic information in honey are attractive in multiple ways: honey carries the genetic information from the plant, the bee foraging on the nectar and, potentially, on the environmental signatures resulting from the geography.

One option is quantitative PCR (qPCR), which is considered one of the most sensitive tools for the detection even of traces of plant-specific DNA (e.g., by TaqMan™ assays) for the detection of a number of biological origins, such as acacia, citrus, clover, heather, eucalyptus and a number of others to deduce also the geographical origin (Laube et al., 2010). The authors were successful in the authentication of Corsican honey; the prerequisite, however, is a profound knowledge of the endemic flora.

LAMP (loop mediated isothermal amplification) is reported to be superior to PCR approaches in terms of specificity, sensitivity, speed and cost-per-analysis and was applied for the authentication of monofloral canola honey (Wang et al., 2020).

The interpretation of traces of DNA via metabarcoding was applied (Bruni et al., 2015; Galimberti et al., 2014; Hawkins et al., 2015; Richardson et al., 2015; Valentini et al., 2010) to describe the provenance of pollen found either in the honey or in beehives. Furthermore, traces of DNA from other sources, such as bacteria or fungi may be used as authentication markers (Wirta et al., 2021).

Shotgun sequencing of DNA in honey is reported to be suitable to characterize the so-called hologenome of the honeybees and thus, also the geographic origin (Bovo et al., 2020).

Next-generation sequencing (NGS) was also described a as tool for the authentication of honeys (e.g., Haynes et al., 2019). A multiplex next-generation sequencing approach was also described by Keller et al. (2015).

9.6.7.8 Data Fusion

In nearly all of the cases discussed in the literature, it became clear that for complex matrices such as honey, paired with evenly complex tasks, such as geographical differentiation, there is no single "perfect method". Consequently, it seems logical not only to make use of all complex information from one single technique but also to combine data from different analytical, multimodal sources. This is commonly known as "data fusion" and at least in theory should be able to provide more accurate information on a sample. There are three principal techniques in data fusion: low-, mid-, and high-level data fusion (Borràs et al., 2015). From a principal perspective, all types of data can be combined with each other, provided these are appropriately preprocessed.

A number of studies exist in the field of honey authentication. Ballabio et al. (2018) demonstrated that the combination of data from multiple spectroscopic sources, such as NIR, MIR, Raman and mass spectrometric data from PTR-MS and e-nose improved the discrimination quality for the botanical origin. A similar approach was described by for NMR and MS data (Spiteri et al., 2016), NIR and Raman data (Tahir et al., 2019), e-tongue and spectroscopy (di Rosa et al., 2018; Ulloa et al., 2013) and GC-IMS and MIR data (Schwolow et al., 2019). Most of these studies proved that data fusion often improves discriminative power, however, there is still no general rule in which cases data fusion makes sense and in which cases it biases the results. Depending on the analytical technologies used, suitable weighting functions have to be considered.

9.7 CONCLUSIONS

Honey is probably one of the most complex food products that we consume; thus it is incredibly complex to identify *the* perfect method for authentication, if there is one. Analytical methods have undergone an exponential degree of development in the last decades to help answering the continuously more complex issues that the globalization has brought upon trade with honey.

While it is clear that conventional analytical parameters as discussed in the preceding chapters before will still be important and relevant for producers, distributors and routine laboratories, a clear trend toward technologies with higher selectivity and sensitivity is observable. High-end spectroscopy, spectrometry, and multiplexed biochemical assays have long made their way into honey authentication. As these systems generate significantly more and more complex data, it is virtually impossible to extract all relevant information by univariate approaches, but more and more sophisticated multivariate data analysis is required; chemometrics, machine learning, deep learning and artificial intelligence are just some of the buzzwords that are found not only in research articles but also increasingly in the field of routine applications. This however leads—or will lead—to a change of paradigms: analytical chemists have the inherent ambition to explain chemical composition differences by individual, quantifiable "marker" substances. For diffuse questions, such as geography or low-level blending, this might be wishful thinking rather than reality. Such issues can only be resolved by a consequent development and use of multivariate data analysis procedures, which logically will also shift the qualification profile of analytical chemists toward a more profound understanding of data processing. In particular, advanced strategies, such as convolutional neural networks or data fusion, are double-edged swords, which may cut careless users as well. A solid understanding of such techniques will be a requirement for the future and should find its way into the curricula of analytical chemistry students.

REFERENCES

Ajibola, A., Chamunorwa, J.P., Erlwanger, K.H., 2012. Nutraceutical values of natural honey and its contribution to human health and wealth. Nutrition & Metabolism (Lond) 9, 61. doi:10.1186/1743-7075-9-61

Alqarni A.S., Owayss A.A., Mahmoud A.A., Hannan M.A., 2014. Mineral content and physical properties of local and imported honeys in Saudi Arabia. Journal of Saudi Chemical Society 18, 618–625. https://doi.org/10.1016/j.jscs.2012.11.009

Amariei S., Norocel L., Scripcă L.A., 2020. An innovative method for preventing honey crystallization. Innovative Food Science & Emerging Technologies 66, 102481. https://doi.org/10.1016/j.ifset.2020.102481.

Anguebes-Franseschi, F., Abatal, M., Pat, L., Flores, A., Córdova Quiroz, A. v., Ramírez-Elias, M.A., San Pedro, L., May Tzuc, O., Bassam, A., 2019. Raman spectroscopy and chemometric modeling to predict physical-chemical honey properties from Campeche, Mexico. Molecules 24, 4091. https://doi.org/10.3390/molecules24224091

Anklam, E., 1998. A review of the analytical methods to determine the geographical and botanical origin of honey. Food Chemistry 63, 549–562. https://doi.org/10.1016/S0308-8146(98)00057-0

Apimondia Statement on honey fraud, Version 2, January 2020. https://www.apiservices.biz/documents/articles-en/apimondia_statement_on_honey_fraud_v_2.pdf (accessed on August 24, 2022).

Aries, E., Burton, J., Carrasco, L., de Rudder, O., Maquet, A., 2016. Scientific support to the implementation of a Coordinated Control Plan with a view to establishing the prevalence of fraudulent practices in the marketing of honey N° SANTE/2015/E3/JRC/SI2.706828. https://www.google.com/search?q=Aries%2C+E.%2C+Burton%2C+J.%2C+Carrasco%2C+L.%2C+de+Rudder%2C+O.%2C+Maquet%2C+A.%2C+2016.+Scientific+support+to+the+implementation+of+a+Coordinated+Control+Plan+with+a+view+to+establishing+the+prevalence+of+fraudulent+practices+in+the+marketing+of+honey+N%C2%B0+SANTE%2F2015%2FE3%2FJRC%2FSI2.706828.&sxsrf=ALiCzsae3g0Dpwc62-2oFhu8v0uDyx-yIg%3A1663946990936&ei=7tAtY5bSOLi-xc8PwPC3EA&ved=0ahUKEwiWgcfKnav6AhU4X_EDHUD4DQIQ4dUDCA4&oq=Aries%2C+E.%2C+Burton%2C+J.%2C+Carrasco%2C+L.%2C+de+Rudder%2C+O.%2C+Maquet%2C+A.%2C+2016.+Scientific+support+to+the+implementation+of+a+Coordinated+Control+Plan+with+a+view+to+establishing+the+prevalence+of+fraudulent+practices+in+the+marketing+of+honey+N%C2%B0+SANTE%2F2015%2FE3%2FJRC%2FSI2.706828.&gs_lp=Egdnd3Mtd2l6uAEM-AEB-AECMgcQIxjqAhgn-MgcQIxjqAhgnMgcQIxjqAhgnMgcQIxjqAhgnMgcQIxjqAhgnMgcQIxjqAhgnMgcQIxjqAhgn-MgcQIxjqAhgnMgcQIxjqAhgnMgcQIxjqAhgnqAIKSIoiUMgTWMgTcAF4AMgBAJABA-JgBAK-ABAKoBAOIDBCBBGADiAwQgRhgAiAYB&sclient=gws-wiz (accessed on September 23, 2022)

Arrizabalaga-Larrañaga, A., Ayala-Cabrera, J.F., Seró, R., Santos, J.F., Moyano, E., 2021. Ambient ionization mass spectrometry in food analysis, in: Food Toxicology and Forensics. Elsevier, pp. 271–312. https://doi.org/10.1016/B978-0-12-822360-4.00006-6
Arroyo-Manzanares, N., García-Nicolás, M., Castell, A., Campillo, N., Viñas, P., López-García, I., Hernández-Córdoba, M., 2019. Untargeted headspace gas chromatography—Ion mobility spectrometry analysis for detection of adulterated honey. Talanta 205, 120123. https://doi.org/10.1016/j.talanta.2019.120123
Ballabio, D., Robotti, E., Grisoni, F., Quasso, F., Bobba, M., Vercelli, S., Gosetti, F., Calabrese, G., Sangiorgi, E., Orlandi, M., Marengo, E., 2018. Chemical profiling and multivariate data fusion methods for the identification of the botanical origin of honey. Food Chemistry 266, 79–89. https://doi.org/10.1016/j.foodchem.2018.05.084
Ballin, N.Z., Laursen, K.H., 2019. To target or not to target? Definitions and nomenclature for targeted versus non-targeted analytical food authentication. Trends in Food Science & Technology 86, 537–543. https://doi.org/10.1016/j.tifs.2018.09.025
Bergamo, G., Seraglio, S.K.T., Gonzaga, L.V., Fett, R., de Mello Castanho Amboni, R.D., Dias, C.O., Costa, A.C.O., 2019. Differentiation of honeydew honeys and blossom honeys: a new model based on colour parameters. Journal of Food Science and Technology 56, 2771–2777. https://doi.org/10.1007/s13197-019-03737-2
Berriel, V., Perdomo, C.H., 2019. Differentiating pasture honey from eucalyptus honey based on carbon isotopic data in Uruguay. Heliyon 5, e01228. https://doi.org/10.1016/j.heliyon.2019.e01228
Bertelli, D., Lolli, M., Papotti, G., Bortolotti, L., Serra, G., Plessi, M., 2010. Detection of Honey Adulteration by Sugar Syrups Using One-Dimensional and Two-Dimensional High-Resolution Nuclear Magnetic Resonance. Journal of Agricultural and Food Chemistry 58, 8495–8501. https://doi.org/10.1021/jf101460t
Bertoncelj, J., Golob, T., Kropf, U., Korošec, M., 2011. Characterisation of Slovenian honeys on the basis of sensory and physicochemical analysis with a chemometric approach. International Journal of Food Science & Technology 46, 1661–1671. https://doi.org/10.1111/j.1365-2621.2011.02664.x
Bodor, Z., Kovacs, Z., Rashed, M.S., Kókai, Z., Dalmadi, I., Benedek, C., 2020. Sensory and Physicochemical Evaluation of Acacia and Linden Honey Adulterated with Sugar Syrup. Sensors 20, 4845. https://doi.org/10.3390/s20174845
Bogdanov, S., 2002. Harmonised Methods of the International Honey Commission. Liebefeld.
Bogdanov, S., 2016. Honey types. *Bee Product Science*. https://www.bee-hexagon.net/english/bee-products/ (accessed on September 2022).
Bontempo, L., Camin, F., Ziller, L., Perini, M., Nicolini, G., Larcher, R., 2017. Isotopic and elemental composition of selected types of Italian honey. Measurement 98, 283–289. https://doi.org/10.1016/j.measurement.2015.11.022
Bovo, S., Utzeri, V.J., Ribani, A., Cabbri, R., Fontanesi, L., 2020. Shotgun sequencing of honey DNA can describe honey bee derived environmental signatures and the honey bee hologenome complexity. Scientific Reports 10, 9279. https://doi.org/10.1038/s41598-020-66127-1
Brendel, R., Schwolow, S., Gerhardt, N., Schwab, J., Rau, P., Oest, M., Rohn, S., Weller, P., 2021. MIR spectroscopy versus MALDI-ToF-MS for authenticity control of honeys from different botanical origins based on soft independent modelling by class analogy (SIMCA)—A clash of techniques? Spectrochimica Acta Part A: Molecular and Biomolecular Spectroscopy 263, 120225. https://doi.org/10.1016/j.saa.2021.120225
Bro, R., 1997. PARAFAC. Tutorial and applications. Chemometrics and Intelligent Laboratory Systems 38, 149–171. https://doi.org/10.1016/S0169-7439(97)00032-4
Bruni, I., Galimberti, A., Caridi, L., Scaccabarozzi, D., de Mattia, F., Casiraghi, M., Labra, M., 2015. A DNA barcoding approach to identify plant species in multiflower honey. Food Chemistry 170, 308–315. https://doi.org/10.1016/j.foodchem.2014.08.060
Castiglioni, S., Stefano, M., Pisani, M., Carloni, P., 2018. Geographical characterisation of multifloral honeys from the Marche region (Italy) according to their antioxidant activity and colour using a chemometric approach. International Journal of Food Science & Technology 53, 571–581. https://doi.org/10.1111/ijfs.13630
Cavanna, D., Righetti, L., Elliott, C., Suman, M., 2018. The scientific challenges in moving from targeted to non-targeted mass spectrometric methods for food fraud analysis: A proposed validation workflow to bring about a harmonized approach. Trends in Food Science & Technology 80, 223–241. https://doi.org/10.1016/j.tifs.2018.08.007

Chaharlangi, M., Tashkhourian, J., Bordbar, M.M., Brendel, R., Weller, P., Hemmateenejad, B., 2021. A paper-based colorimetric sensor array for discrimination of monofloral European honeys based on gold nanoparticles and chemometrics data analysis. Spectrochimica Acta Part A: Molecular and Biomolecular Spectroscopy 247, 119076. https://doi.org/10.1016/j.saa.2020.119076

Cheung, Y., Meenu, M., Yu, X. & Xu, B., 2019. Phenolic acids and flavonoids profiles of commercial honey from different floral sources and geographic sources. International Journal of Food Properties 22, 290–308. https://doi.org/10.1080/10942912.2019.1579835

Chin, N.L., & Sowndhararajan, K., 2020. A Review on Analytical Methods for Honey Classification, Identification and Authentication, Honey Analysis—New Advances and Challenges, Vagner de Alencar Arnaut de Toledo and Emerson Dechechi Chambó, IntechOpen. https://www.intechopen.com/chapters/71122

Cianciosi D., Forbes-Hernández T.Y., Afrin S., Gasparrini M., Reboredo-Rodriguez P., Manna P.P., Zhang J., Bravo Lamas L., Martínez Flórez S., Agudo Toyos P., Quiles J.L., Giampieri F., & Battino M., 2018. Phenolic Compounds in Honey and Their Associated Health Benefits: A Review. Molecules 11, 2322. doi: 10.3390/molecules23092322

Ciulu M., Solinas S., Floris I., Panzanellia A., Pilo M.I., Piu P.C., et al., 2011. RP-HPLC determination of water-soluble vitamins in honey. Talanta 83, 924–929. https://doi.org/10.1016/j.talanta.2010.10.059

Codex Alimentarius—Revised Codex Standard for honey—2001: Codex stan. 12–1981, Rev.1 (1987), Rev. 2 (2001), Amended 2019, 8 p. https://www.fao.org/fao-who-codexalimentarius/sh-proxy/fr/?lnk=1&url=https%253A%252F%252Fworkspace.fao.org%252Fsites%252Fcodex%252FStandards%252FCXS%2B12-1981%252FCXS_012e.pdf

Combarros-Fuertes, P., Valencia-Barrera, R.M., Estevinho, L.M., Dias, L.G., Castro, J.M., Tornadijo, M.E., Fresno, J.M., 2019. Spanish honeys with quality brand: a multivariate approach to physicochemical parameters, microbiological quality, and floral origin. Journal of Apicultural Research 58, 92–103. https://doi.org/10.1080/00218839.2018.1494918

Consonni, R., Bernareggi, F., Cagliani, L.R., 2019. NMR-based metabolomic approach to differentiate organic and conventional Italian honey. Food Control 98, 133–140. https://doi.org/10.1016/j.foodcont.2018.11.007

Corvucci, F., Nobili, L., Melucci, D., Grillenzoni, F.-V., 2015. The discrimination of honey origin using melissopalynology and Raman spectroscopy techniques coupled with multivariate analysis. Food Chemistry 169, 297–304. https://doi.org/10.1016/j.foodchem.2014.07.122

Council Directive 2001/110/EC of 20 December 2001 relating to honey. 2001. Available online: https://eur-lex.europa.eu/legal-content/EN/ALL/?uri=CELEX:32001L0110 (accessed on November 2, 2021). https://www.fao.org/fao-who-codexalimentarius/sh-proxy/fr/?lnk=1&url=https%253A%252F%252Fworkspace.fao.org%252Fsites%252Fcodex%252FStandards%252FCXS%2B12-1981%252FCXS_012e.pdf

Crane, E., Walker, P., 1985. Important Honeydew Sources and their Honeys, Bee World 66, 105–112. https://doi.org/10.1080/0005772X.1985.11098832

Creydt, M., Fischer, M., 2020. Food authentication in real life: How to link nontargeted approaches with routine analytics? ELECTROPHORESIS 41, 1665–1679. https://doi.org/10.1002/elps.202000030

da Silva, P.M., Gauche, C., Gonzaga, L.V., Costa, A.C.O., Fett, R., 2016. Honey: Chemical composition, stability and authenticity. Food Chemistry 196, 309–323. https://doi.org/10.1016/j.foodchem.2015.09.051

de la Fuente, E., Ruiz-Matute, A.I., Valencia-Barrera, R.M., Sanz, J., Martínez Castro, I., 2011. Carbohydrate composition of Spanish unifloral honeys. Food Chemistry 129, 1483–1489. https://doi.org/10.1016/j.foodchem.2011.05.121

de Lima Morais da Silva, P., de Lima, L.S., Caetano, Í.K., Torres, Y.R., 2017. Comparative analysis of the volatile composition of honeys from Brazilian stingless bees by static headspace GC—MS. Food Research International 102, 536–543. https://doi.org/10.1016/j.foodres.2017.09.036

Dias, L.A., Peres, A.M., Vilas-Boas, M., Rocha, M.A., Estevinho, L., Machado, A.A.S.C., 2008. An electronic tongue for honey classification. Microchimica Acta 163, 97–102. https://doi.org/10.1007/s00604-007-0923-8

Dinca, O.-R., Ionete, R.E., Popescu, R., Costinel, D., Radu, G.-L., 2015. Geographical and botanical origin discrimination of Romanian honey using complex stable isotope data and chemometrics. Food Analytical Methods 8, 401–412. https://doi.org/10.1007/s12161-014-9903-x

di Rosa, A., Marino, A., Leone, F., Corpina, G., Giunta, R., Chiofalo, V., 2018. Characterization of Sicilian honeys pollen profiles using a commercial e-tongue and melissopalynological analysis for rapid screening: A pilot study. Sensors 18, 4065. https://doi.org/10.3390/s18114065

Đogo Mračević, S., Krstić, M., Lolić, A., Ražić, S., 2020. Comparative study of the chemical composition and biological potential of honey from different regions of Serbia. Microchemical Journal 152, 104420. https://doi.org/10.1016/j.microc.2019.104420

Dübecke, A., Beckh, G., Lüllmann, C., 2011. Pyrrolizidine alkaloids in honey and bee pollen. Food additives & contaminants. Part A, Chemistry, analysis, control, exposure & risk assessment 28, 348–358. https://doi.org/10.1080/19440049.2010.541594

Elliot, C., van Goethem, M., Amaral, J.S., Suman, M., Raezke, K.P., Valli, E., Maestri, E., 2018. FoodIntegrity Handbook. Eurofins Analytics France. https://doi.org/10.32741/fihb

Escuredo O., Míguez M., Fernández-González M., Seijo M.C., 2013. Nutritional value and antioxidant activity of honeys produced in a European Atlantic area. Food Chemistry 138, 851–856. https://doi.org/10.1016/j.foodchem.2012.11.015

Escuredo, O., Dobre, I., Fernández-González, M., Seijo, M.C., 2014. Contribution of botanical origin and sugar composition of honeys on the crystallization phenomenon. Food Chemistry 149, 84–90. https://doi.org/10.1016/j.foodchem.2013.10.097

Eteraf-Oskouei, T., Najafi, M., 2013. Traditional and modern uses of natural honey in human diseases: a review. Iranian Journal of Basic Medical Sciences, 16, 731–742. Available online: www.ncbi.nlm.nih.gov/pmc/articles/PMC3758027/

European Commission (2018). Available online: https://eur-lex.europa.eu/legal-content/EN/TXT/?uri=celex%3A32017R0625.

European Parliament (2013). Draft report on the food crisis, fraud in the food chain and the control thereof.(2013/2091(INI). Available online: www.europarl.europa.eu/doceo/document/A-7-2013-0434_EN.html.

Fechner, D.C., Moresi, A.L., Ruiz Díaz, J.D., Pellerano, R.G., Vazquez, F.A., 2016. Multivariate classification of honeys from Corrientes (Argentina) according to geographical origin based on physicochemical properties. Food Bioscience 15, 49–54. https://doi.org/10.1016/j.fbio.2016.05.002

Fiamegos, Y., Dumitrascu, C., Ghidotti, M., de la Calle Guntiñas, M.B., 2020. Use of energy-dispersive X-ray fluorescence combined with chemometric modelling to classify honey according to botanical variety and geographical origin. Analytical and Bioanalytical Chemistry 412, 463–472. https://doi.org/10.1007/s00216-019-02255-6

Galimberti, A., de Mattia, F., Bruni, I., Scaccabarozzi, D., Sandionigi, A., Barbuto, M., Casiraghi, M., Labra, M., 2014. A DNA barcoding approach to characterize pollen collected by honeybees. PLOS One 9, e109363. https://doi.org/10.1371/journal.pone.0109363

Gan, Z., Yang, Y., Li, J., Wen, X., Zhu, M., Jiang, Y., Ni, Y., 2016. Using sensor and spectral analysis to classify botanical origin and determine adulteration of raw honey. Journal of Food Engineering 178, 151–158. https://doi.org/10.1016/j.jfoodeng.2016.01.016

García, N., Schwarzinger, S., 2021. Honey fraud, Editor(s): Rosalee S. Hellberg, Karen Everstine, Steven A. Sklare, Food Fraud, Academic Press 2021; Chapter 15, 309–334. https://doi.org/10.1016/B978-0-12-817242-1.00019-1

Gašić, U., Kečkeš, S., Dabić, D., Trifković, J., Milojković-Opsenica, D., Natić, M., Tešić, Ž., 2014. Phenolic profile and antioxidant activity of Serbian polyfloral honeys. Food Chemistry 145, 599–607. https://doi.org/10.1016/j.foodchem.2013.08.088

Gašić, U.M., Natić, M.M., Mišić, D.M., Lušić, Darija v., Milojković-Opsenica, D.M., Tešić, Ž. Lj., Lušić, Dražen, 2015. Chemical markers for the authentication of unifloral Salvia officinalis L. honey. Journal of Food Composition and Analysis 44, 128–138. https://doi.org/10.1016/j.jfca.2015.08.008

Geană, E.-I., Ciucure, C.T., Costinel, D., Ionete, R.E., 2020. Evaluation of honey in terms of quality and authenticity based on the general physicochemical pattern, major sugar composition and δ13C signature. Food Control 109, 106919. https://doi.org/10.1016/j.foodcont.2019.106919

Gerhardt, N., Birkenmeier, M., Kuballa, T., Ohmenhaeuser, M., Rohn, S., Weller, P., 2016. Differentiation of the botanical origin of honeys by fast, non-targeted ^{1}H-NMR profiling and chemometric tools as alternative authenticity screening tool, in: Proceedings of the XIII International Conference on the Applications of Magnetic Resonance in Food Science. IM Publications, p. 33. https://doi.org/10.1255/mrfs.7

Gerhardt, N., Birkenmeier, M., Schwolow, S., Rohn, S., Weller, P., 2018. Volatile-compound fingerprinting by headspace-gas-chromatography ion-mobility spectrometry (HS-GC-IMS) as a benchtop alternative to ^{1} H NMR profiling for assessment of the authenticity of honey. Analytical Chemistry 90, 1777–1785. https://doi.org/10.1021/acs.analchem.7b03748

Giovani, C., Padovani, R., Frilli, F., Barbattini, R. Greatti, M., 1998. Potassium Evaluation in Metcalfa Pruinosa Honeydew Honey by Means of 40K Measurement. Journal of Apicultural Research 37, 47–49. https://doi.org/10.1080/00218839.1998.11100954

González Martín, I., Marqués Macías, E., Sánchez Sánchez, J., González Rivera, B., 1998. Detection of honey adulteration with beet sugar using stable isotope methodology. Food Chemistry 61, 281–286. https://doi.org/10.1016/S0308-8146(97)00101-5

Hawkins, J., de Vere, N., Griffith, A., Ford, C.R., Allainguillaume, J., Hegarty, M.J., Baillie, L., Adams-Groom, B., 2015. Using DNA metabarcoding to identify the floral composition of honey: A new tool for investigating honey bee foraging preferences. PLOS One 10, e0134735. https://doi.org/10.1371/journal.pone.0134735

Haynes, E., Jimenez, E., Pardo, M.A., Helyar, S.J., 2019. The future of NGS (next generation sequencing) analysis in testing food authenticity. Food Control 101, 134–143. https://doi.org/10.1016/j.foodcont.2019.02.010

Hermosín I., Chicón R.M., Dolores Cabezudo M., 2003. Free amino acid composition and botanical origin of honey. Food Chemistry 83, 263–268. https://doi.org/10.1016/S0308-8146(03)00089-X

Hungerford, N., Carter, S., Anuj, S., Tan, B., Hnatko, D., Martin, C., Sharma, E., Yin, M., Nguyen, T., Melksham, K., Fletcher, M., 2019. Analysis of pyrrolizidine alkaloids in Queensland honey: Using low temperature chromatography to resolve stereoisomers and identify botanical sources by UHPLC-MS/MS. Toxins 11, 726. https://doi.org/10.3390/toxins11120726

Jandric, Z., Frew, R.D., Fernandez-Cedi, L.N., Cannavan, A., 2017. An investigative study on discrimination of honey of various floral and geographical origins using UPLC-QToF MS and multivariate data analysis. Food Control 72, 189–197. https://doi.org/10.1016/j.foodcont.2015.10.010

Jasicka-Misiak, I., Poliwoda, A., Dereń, M., Kafarski, P., 2012. Phenolic compounds and abscisic acid as potential markers for the floral origin of two Polish unifloral honeys. Food Chemistry 131, 1149–1156. https://doi.org/10.1016/j.foodchem.2011.09.083

Jovetić, M., Trifković, J., Stanković, D., Manojlović, D., Milojković-Opsenica, D., 2017. Mineral content as a tool for the assessment of honey authenticity. Journal of AOAC International 100, 862–870. https://doi.org/10.5740/jaoacint.17-0145

Juan-Borrás, M., Domenech, E., Hellebrandova, M., Escriche, I., 2014. Effect of country origin on physicochemical, sugar and volatile composition of acacia, sunflower and Tilia honeys. Food Research International 60, 86–94. https://doi.org/10.1016/j.foodres.2013.11.045

Karabagias, I., 2014. Differentiation of the geographical and botanical origin of Greek honey samples with the help of instrumental chemical and statistical analysis, Doctoral thesis, University of Ioannina.

Karabagias, I.K., Badeka, A., Kontakos, S., Karabournioti, S., Kontominas, M.G., 2014a. Characterization and classification of Thymus capitatus (L.) honey according to geographical origin based on volatile compounds, physicochemical parameters and chemometrics. Food Research International 55, 363–372. https://doi.org/10.1016/j.foodres.2013.11.032

Karabagias, I.K., Badeka, A., Kontakos, S., Karabournioti, S., Kontominas, M.G., 2014b. Characterisation and classification of Greek pine honeys according to their geographical origin based on volatiles, physicochemical parameters and chemometrics. Food Chemistry 146, 548–557. https://doi.org/10.1016/j.foodchem.2013.09.105

Karabagias, I.K., Casiello, G., Kontakos, S., Louppis, A.P., Longobardi, F., Kontominas, M.G., 2016. Investigating the impact of botanical origin and harvesting period on carbon stable isotope ratio values ($^{13}C/^{12}C$) and different parameter analysis of Greek unifloral honeys: A chemometric approach for correct botanical discrimination. International Journal of Food Science & Technology 51, 2460–2467. https://doi.org/10.1111/ijfs.13227

Karabagias, I.K., Vavoura, M. v., Badeka, A., Kontakos, S., Kontominas, M.G., 2014c. Differentiation of Greek thyme honeys according to geographical origin based on the combination of phenolic compounds and conventional quality parameters using chemometrics. Food Analytical Methods 7, 2113–2121. https://doi.org/10.1007/s12161-014-9851-5

Kawashima, H., Suto, M., Suto, N., 2019. Stable carbon isotope ratios for organic acids in commercial honey samples. Food Chemistry 289, 49–55. https://doi.org/10.1016/j.foodchem.2019.03.053
Kivima, E., Seiman, A., Pall, R., Sarapuu, E., Martverk, K., Laos, K., 2014. Characterization of Estonian honeys by botanical origin. Proceedings of the Estonian Academy of Sciences 63, 183. https://doi.org/10.3176/proc.2014.2.08
Koulis, G.A., Tsagkaris, A.S., Aalizadeh, R., Dasenaki, M.E., Panagopoulou, E.I., Drivelos, S., Halagarda, M., Georgiou, C.A., Proestos, C., Thomaidis, N.S., 2021. Honey phenolic compound profiling and authenticity assessment using HRMS targeted and untargeted metabolomics. Molecules 26, 2769. https://doi.org/10.3390/molecules26092769
Kováčik, J., Grúz, J., Biba, O., Hedbavny, J., 2016. Content of metals and metabolites in honey originated from the vicinity of industrial town Košice (eastern Slovakia). Environmental Science and Pollution Research 23, 4531–4540. https://doi.org/10.1007/s11356-015-5627-8
Kuballa, T., Brunner, T.S., Thongpanchang, T., Walch, S.G., Lachenmeier, D.W., 2018. Application of NMR for authentication of honey, beer and spices. Current Opinion in Food Science 19, 57–62. https://doi.org/10.1016/j.cofs.2018.01.007
Kurtagic, H, Barudanovic, S., Durmić, V., 2015. Determination of Rutin, Quercetin, Naringenin and Hesperetin in the Honey from Bosnia and Herzegovina (B & H) in Relation to the Composition of Pollen. Journal of Environmental Science and Engineering A 4, 615–622. http://dx.doi.org/10.17265/2162-5298/2015.12.001
Laube, I., Hird, H., Brodmann, P., Ullmann, S., Schöne-Michling, M., Chisholm, J., Broll, H., 2010. Development of primer and probe sets for the detection of plant species in honey. Food Chemistry 118, 979–986. https://doi.org/10.1016/j.foodchem.2008.09.063
Lenhardt, L., Bro, R., Zeković, I., Dramićanin, T., Dramićanin, M.D., 2015. Fluorescence spectroscopy coupled with PARAFAC and PLS DA for characterization and classification of honey. Food Chemistry 175, 284–291. https://doi.org/10.1016/j.foodchem.2014.11.162
Ling Chin, N., Sowndhararajan, K., 2020. A review on analytical methods for honey classification, identification and authentication, in: honey analysis—New advances and challenges. IntechOpen. https://doi.org/10.5772/intechopen.90232
Lippolis, V., de Angelis, E., Fiorino, G.M., di Gioia, A., Arlorio, M., Logrieco, A.F., Monaci, L., 2020. Geographical origin discrimination of monofloral honeys by direct analysis in real time ionization-high resolution mass spectrometry (DART-HRMS). Foods 9, 1205. https://doi.org/10.3390/foods9091205
Li, Y., Jin, Y., Yang, Shupeng, Zhang, W., Zhang, J., Zhao, W., Chen, L., Wen, Y., Zhang, Yongxin, Lu, K., Zhang, Yaping, Zhou, J., Yang, Shuming, 2017. Strategy for comparative untargeted metabolomics reveals honey markers of different floral and geographic origins using ultrahigh-performance liquid chromatography-hybrid quadrupole-orbitrap mass spectrometry. Journal of Chromatography A 1499, 78–89. https://doi.org/10.1016/j.chroma.2017.03.071
Lolli, M., Bertelli, D., Plessi, M., Sabatini, A.G., Restani, C., 2008. Classification of Italian honeys by 2D HR-NMR. Journal of Agricultural and Food Chemistry 56, 1298–1304. https://doi.org/10.1021/jf072763c
Louveaux, J., Maurizio, A., Vorwohl, G., 1978. Methods of melissopalynology. Bee World 59, 139–157. https://doi.org/10.1080/0005772X.1978.11097714
Ma, Y., Zhang, B., Li, H., Li, Y., Hu, J., Li, J., Wang, H., Deng, Z., 2017. Chemical and molecular dynamics analysis of crystallization properties of honey. International Journal of Food Properties 20, 725–733. https://doi.org/10.1080/10942912.2016.1178282
Madejczyk, M., Barałkiewicz, D., 2008. Characterization of Polish rape and honeydew honey according to their mineral contents using ICP-MS and F-AAS/AES. Analytica Chimica Acta 617, 11–17. https://doi.org/10.1016/j.aca.2008.01.038
Magdas, D.A., Guyon, F., Puscas, R., Vigouroux, A., Gaillard, L., Dehelean, A., Feher, I., Cristea, G., 2021. Applications of emerging stable isotopes and elemental markers for geographical and varietal recognition of Romanian and French honeys. Food Chemistry 334, 127599. https://doi.org/10.1016/j.foodchem.2020.127599
Major, N., Marković, K., Krpan, M., Šarić, G., Hruškar, M., Vahčić, N., 2011. Rapid honey characterization and botanical classification by an electronic tongue. Talanta 85, 569–574. https://doi.org/10.1016/j.talanta.2011.04.025
Manikis, I., Thrasivoulou, A., 2001. The relation of physicochemical characteristics of honey and the crystallization sensitive parameters (This project was funded by the Commission of the European Communities under the craft action of the Brite-Euram programme).

Markowicz Bastos, D., Monaro, É., Siguemoto, E., Séfora, M., 2012. Maillard reaction products in processed foods: pros and cons. in: Valdez B (ed) Food Industrial Processes-Methods and Equipment, 1st edn. InTech, Rijeka, pp. 281–300.

Mendes, E., Duarte, N., 2021. Mid-infrared spectroscopy as a valuable tool to tackle food analysis: A literature review on coffee, dairies, honey, olive oil and wine. Foods 10, 477. https://doi.org/10.3390/foods10020477

Michalkiewicz, A., Biesaga, M., Pyrzynska, K., 2008. Solid-phase extraction procedure for determination of phenolic acids and some flavonols in honey. Journal of Chromatography A 1187, 18–24. https://doi.org/10.1016/j.chroma.2008.02.001

Moniruzzaman, M., Rodríguez, I., Ramil, M., Cela, R., Sulaiman, S.A., Gan, S.H., 2014. Assessment of gas chromatography time-of-flight accurate mass spectrometry for identification of volatile and semi-volatile compounds in honey. Talanta 129, 505–515. https://doi.org/10.1016/j.talanta.2014.06.019

Moumeh, B., Dolores Garrido, M., Diaz, P., Peñaranda, I., Linares, M.B., 2020. Chemical analysis and sensory evaluation of honey produced by honeybee colonies fed with different sugar pastes. Food Science & Nutrition 8, 5823–5831. https://doi.org/10.1002/fsn3.1843

Olas, B., 2020. Honey and Its Phenolic Compounds as an Effective Natural Medicine for Cardiovascular Diseases in Humans? Nutrients 12, 283. https://doi.org/10.3390/nu12020283

Olsson, O., Karlsson, M., Persson, A.S., Smith, H.G., Varadarajan, V., Yourstone, J., Stjernman, M., 2021. Efficient, automated and robust pollen analysis using deep learning. Methods in Ecology and Evolution 12, 850–862. https://doi.org/10.1111/2041-210X.13575

Oroian, M., Ropciuc, S., 2019. Romanian honey authentication using voltammetric electronic tongue. Correlation of voltammetric data with physico-chemical parameters and phenolic compounds. Computers and Electronics in Agriculture 157, 371–379. https://doi.org/10.1016/j.compag.2019.01.008

Oroian, M., Ropciuc, S., Paduret, S., 2018. Honey adulteration detection using Raman spectroscopy. Food Analytical Methods 11, 959–968. https://doi.org/10.1007/s12161-017-1072-2

Padovan, G., 2003. Detection of adulteration of commercial honey samples by the 13C/12C isotopic ratio. Food Chemistry 82, 633–636. https://doi.org/10.1016/S0308-8146(02)00504-6

Patrignani, M., Ciappini, M.C., Tananaki, C., Fagúndez, G.A., Thrasyvoulou, A., Lupano, C.E., 2018. Correlations of sensory parameters with physicochemical characteristics of Argentinean honeys by multivariate statistical techniques. International Journal of Food Science & Technology 53, 1176–1184. https://doi.org/10.1111/ijfs.13694

Pohl, P., 2009. Determination of metal content in honey by atomic absorption and emission spectrometries. TrAC Trends in Analytical Chemistry 28, 117–128. https://doi.org/10.1016/j.trac.2008.09.015

Pohl, P., Bielawska-Pohl, A., Dzimitrowicz, A., Jamroz, P., Welna, M., Lesniewicz, A., Szymczycha-Madeja, A., 2017. Recent achievements in element analysis of bee honeys by atomic and mass spectrometry methods. TrAC Trends in Analytical Chemistry 93, 67–77. https://doi.org/10.1016/j.trac.2017.05.009

Pomastowski, P., Złoch, M., Rodzik, A., Ligor, M., Kostrzewa, M., Buszewski, B., 2019. Analysis of bacteria associated with honeys of different geographical and botanical origin using two different identification approaches: MALDI-TOF MS and 16S rDNA PCR technique. PLOS One 14, e0217078. https://doi.org/10.1371/journal.pone.0217078

Pontes, M., Marques, J.C., Câmara, J.S., 2007. Screening of volatile composition from Portuguese multifloral honeys using headspace solid-phase microextraction-gas chromatography–quadrupole mass spectrometry. Talanta 74, 91–103. https://doi.org/10.1016/j.talanta.2007.05.037

Popek, S., Halagarda, M., Kursa, K., 2017. A new model to identify botanical origin of Polish honeys based on the physicochemical parameters and chemometric analysis. LWT 77, 482–487. https://doi.org/10.1016/j.lwt.2016.12.003

Porcari, A.M., Fernandes, G.D., Barrera-Arellano, D., Eberlin, M.N., Alberici, R.M., 2016. Food quality and authenticity screening via easy ambient sonic-spray ionization mass spectrometry. The Analyst 141, 1172–1184. https://doi.org/10.1039/C5AN01415H

Prica, N., Živkov Baloš, M., Jakšić, S., Mihaljev, Ž., Kartalović, B., Babić, J., Savić, S., 2015. Moisture and acidity as indicators of the quality of honey originating from Vojvodina region. Archives of Veterinary Medicine 7, 99–109. https://doi.org/10.46784/e-avm.v7i2.135

Puścion-Jakubik, A., Borawska, M.H., Socha, K., 2020. Modern Methods for Assessing the Quality of Bee Honey and Botanical Origin Identification. Foods 9, 1028. https://doi.org/10.3390/foods9081028

Qu, L., Jiang, Y., Huang, X., Cui, M., Ning, F., Liu, T., Gao, Y., Wu, D., Nie, Z., Luo, L., 2019. High-Throughput Monitoring of Multiclass Syrup Adulterants in Honey Based on the Oligosaccharide and Polysaccharide Profiles by MALDI Mass Spectrometry. Journal of Agricultural and Food Chemistry 67, 11256–11261. https://doi.org/10.1021/acs.jafc.9b05317

Ramón-Sierra, J.M., Ruiz-Ruiz, J.C., de la Luz Ortiz-Vázquez, E., 2015. Electrophoresis characterisation of protein as a method to establish the entomological origin of stingless bee honeys. Food Chemistry 183, 43–48. https://doi.org/10.1016/j.foodchem.2015.03.015

Raptopoulou, K.G., Pasias, I.N., Proestos, C., 2020. Recent Advances in Analytical Techniques for the Determination of Authenticity and Adulteration of Honey and its Products, 125–145. https://doi.org/10.2174/9789811405112120040007

Ribeiro, R. de O.R., Mársico, E.T., Carneiro, C. da S., Monteiro, M.L.G., Júnior, C.C., Jesus, E.F.O. de, 2014. Detection of honey adulteration of high fructose corn syrup by Low Field Nuclear Magnetic Resonance (LF ^{1}H NMR). Journal of Food Engineering 135, 39–43. https://doi.org/10.1016/j.jfoodeng.2014.03.009

Richardson, R.T., Lin, C.-H., Sponsler, D.B., Quijia, J.O., Goodell, K., Johnson, R.M., 2015. Application of ITS2 Metabarcoding to Determine the Provenance of Pollen Collected by Honey Bees in an Agroecosystem. Applications in Plant Sciences 3, 1400066. https://doi.org/10.3732/apps.1400066

Righetti, L., Paglia, G., Galaverna, G., Dall'Asta, C., 2016. Recent advances and future challenges in modified mycotoxin analysis: Why HRMS has become a key instrument in food contaminant research. Toxins 8, 361. https://doi.org/10.3390/toxins8120361

Rodríguez García, J.C., Iglesias Rodríguez, R., Peña Crecente, R.M., Barciela García, J., García Martín, S., Herrero Latorre, C., 2006. Preliminary Chemometric Study on the Use of Honey as an Environmental Marker in Galicia (Northwestern Spain). Journal of Agricultural and Food Chemistry 54, 7206–7212. https://doi.org/10.1021/jf060823t

Rodríguez, I., Serrano, S., Galán-Soldevilla, H., Piva, G., Ubera, J.L., 2015. Sensory analysis integrated by palynological and physicochemical determinations plays a key role in differentiating unifloral honeys of similar botanical origins (*Myrtaceae* honeys from southern Spain). International Journal of Food Science & Technology 50, 1545–1551. https://doi.org/10.1111/ijfs.12802

Romano, A., Cuenca, M., Makhoul, S., Biasioli, F., Martinello, L., Fugatti, A., Scampicchio, M., 2016. Comparison of e-noses: The case study of honey. Italian Journal of Food Science 28, 326–337.

Ruiz-Matute, A.I., Soria, A.C., Martínez-Castro, I., Sanz, M.L., 2007. A new methodology based on GC–MS to detect honey adulteration with commercial syrups. Journal of Agricultural and Food Chemistry 55, 7264–7269. https://doi.org/10.1021/jf070559j

Ruoff K., Bogdanov S. (2004) Authenticity of honey and other bee products. Apiacta, 38 317–327. Available online: www.apimondia.com/docs/apiacta/2003/ruoff_1.pdf.

Rychlik, M., Kanawati, B., Schmitt-Kopplin, P., 2017. Foodomics as a promising tool to investigate the mycobolome. TrAC Trends in Analytical Chemistry 96, 22–30. https://doi.org/10.1016/j.trac.2017.05.006

Santos-Buelga, C., González-Paramás, A.M., 2017. Chemical Composition of Honey. In: Alvarez-Suarez J. (eds) Bee Products – Chemical and Biological Properties. Springer, Cham 2017. https://doi.org/10.1007/978-3-319-59689-1_3

Schievano, E., Sbrizza, M., Zuccato, V., Piana, L., Tessari, M., 2020. NMR carbohydrate profile in tracing acacia honey authenticity. Food Chemistry 309, 125788. https://doi.org/10.1016/j.foodchem.2019.125788

Schievano, E., Stocchero, M., Zuccato, V., Conti, I., Piana, L., 2019. NMR assessment of European acacia honey origin and composition of EU-blend based on geographical floral markers. Food Chemistry 288, 96–101. https://doi.org/10.1016/j.foodchem.2019.02.062

Schuhfried, E., Sánchez del Pulgar, J., Bobba, M., Piro, R., Cappellin, L., Märk, T.D., Biasioli, F., 2016. Classification of 7 monofloral honey varieties by PTR-ToF-MS direct headspace analysis and chemometrics. Talanta 147, 213–219. https://doi.org/10.1016/j.talanta.2015.09.062

Seisonen, S., Kivima, E., Vene, K., 2015. Characterisation of the aroma profiles of different honeys and corresponding flowers using solid-phase microextraction and gas chromatography—mass spectrometry/olfactometry. Food Chemistry 169, 34–40. https://doi.org/10.1016/j.foodchem.2014.07.125

Seraglio, S.K.T., Valese, A.C., Daguer, H., Bergamo, G., Azevedo, M.S., Gonzaga, L.V., Fett, R., Costa, A.C.O., 2016. Development and validation of a LC-ESI-MS/MS method for the determination of phenolic compounds in honeydew honeys with the diluted-and-shoot approach. Food Research International 87, 60–67. https://doi.org/10.1016/j.foodres.2016.06.019

Seraglio, K.S.T., Silva, B., Bergamo, G., Brugnerotto, P., Gonzaga, L.V., Fett, R., Costa, A.C.O., 2019. An overview of physicochemical characteristics and health-promoting properties of honeydew honey. Food Research International 119, 44–66. https://doi.org/10.1016/j.foodres.2019.01.028

Shapla, U.M., Solayman, M., Alam, N., et al., 2018. 5-Hydroxymethylfurfural (HMF) levels in honey and other food products: effects on bees and human health. Chemistry Central Journal 12, 35. https://doi.org/10.1186/s13065-018-0408-3

She, S., Chen, L., Song, H., Lin, G., Li, Y., Zhou, J., Liu, C., 2019. Discrimination of geographical origins of Chinese acacia honey using complex 13C/12C, oligosaccharides and polyphenols. Food Chemistry 272, 580–585. https://doi.org/10.1016/j.foodchem.2018.07.227

Singh, I., Singh, S., 2018. Honey moisture reduction and its quality. Journal of food science and technology 55, 3861–3871. https://doi.org/10.1007/s13197-018-3341-5

Sobolev, A.P., Circi, S., Mannina, L., 2016. Advances in Nuclear Magnetic Resonance Spectroscopy for Food Authenticity Testing, in: Advances in Food Authenticity Testing. Elsevier, pp. 147–170. https://doi.org/10.1016/B978-0-08-100220-9.00006-0

Solayman, Md., Asiful Islam, Md., Sudip, P., Yousuf, A., Ibrahim Khalil, Md., Alam, N., Gan, S.H., 2016. Physicochemical Properties, Minerals, Trace Elements, and Heavy Metals in Honey of Different Origins: A Comprehensive Review. Comprehensive Reviews In Food Science And Food Safety 15, 219–233. https://doi.org/10.1111/1541-4337.12182

Spiteri, C., Lia, F., Farrugia, C., 2020. Determination of the Geographical Origin of Maltese Honey Using ^{1}H NMR Fingerprinting. Foods 9, 1455. https://doi.org/10.3390/foods9101455

Squadrone, S., Brizio, P., Stella, C., Mantia, M., Pederiva, S., Brusa, F., Mogliotti, P., Garrone, A., Abete, M.C., 2020a. Trace elements and rare earth elements in honeys from the Balkans, Kazakhstan, Italy, South America, and Tanzania. Environmental Science and Pollution Research 27, 12646–12657. https://doi.org/10.1007/s11356-020-07792-7

Squadrone, S., Brizio, P., Stella, C., Pederiva, S., Brusa, F., Mogliotti, P., Garrone, A., Abete, M.C., 2020b. Trace and rare earth elements in monofloral and multifloral honeys from Northwestern Italy; A first attempt of characterization by a multi-elemental profile. Journal of Trace Elements in Medicine and Biology 61, 126556. https://doi.org/10.1016/j.jtemb.2020.126556

Stanek, N., Kafarski, P., Jasicka-Misiak, I., 2019. Development of a high performance thin layer chromatography method for the rapid qualification and quantification of phenolic compounds and abscisic acid in honeys. Journal of Chromatography A 1598, 209–215. https://doi.org/10.1016/j.chroma.2019.04.052

Suto, M., Kawashima, H., Suto, N., 2019. Heart-cutting two-dimensional liquid chromatography combined with isotope ratio mass spectrometry for the determination of stable carbon isotope ratios of gluconic acid in honey. Journal of Chromatography A 1608, 460421. https://doi.org/10.1016/j.chroma.2019.460421

Suto, M., Kawashima, H., Nakamura, Y., 2020. Determination of Organic Acids in Honey by Liquid Chromatography with Tandem Mass Spectrometry. Food Analytical Methods 13, 2249–2257. https://doi.org/10.1007/s12161-020-01845-w

Tananaki Chrysousa, 2006. Study of the factors that affect the volatile components of honeys derived from honeydew, Doctoral thesis, Aristotle University of Thessaloniki.

Tette, P.A.S., Guidi, L.R., Bastos, E.M.A.F., Fernandes, C., Gloria, M.B.A., 2017. Synephrine—A potential biomarker for orange honey authenticity. Food Chemistry 229, 527–533. https://doi.org/10.1016/j.foodchem.2017.02.108

Thrasyvoulou, A., Tananaki, C., Goras, G., Karazafiris, E., Dimou, M., Liolios, V., Kanelis, D. Gounari, S. (2018). Legislation of honey criteria and standards. Journal of Apicultural Research 57, 88–96. https://doi.org/10.1080/00218839.2017.1411181

Tornuk, F., Karaman, S., Ozturk, I., Toker, O.S, Tastemur, B., Sagdic, O., et al., 2013. Quality characterization of artisanal and retail Turkish blossom honeys: Determination of physicochemical, microbiological, bioactive properties and aroma profile. Industrial Crops and Products 46, 124–131. https://doi.org/10.1016/j.indcrop.2012.12.042

Tosi, E., Martinet, R., Ortega, M., Lucero, H., Ré E., 2008. Honey diastase activity modified by heating. Food Chemistry 106, 883–887. https://doi.org/10.1016/j.foodchem.2007.04.025

Tromp, S., Séchet, B., 2018. IFS Food Fraud Guidelines, 1–50. https://ifs-productintegrity.com/wp-content/uploads/2020/09/IFS_Guideline_Product_Fraud_Mitigation_V2_EN.pdf

Tuberoso, C.I.G., Jerković, I., Sarais, G., Congiu, F., Marijanović, Z., Kuś, P.M., 2014. Color evaluation of seventeen European unifloral honey types by means of spectrophotometrically determined CIE L*Cab*h(ab)° chromaticity coordinates. Food Chemistry 145, 284–291. https://doi.org/10.1016/j.foodchem.2013.08.032

Tungmunnithum, D., Thongboonyou, A., Pholboon, A., Yangsabai, A., 2018. Flavonoids and Other Phenolic Compounds from Medicinal Plants for Pharmaceutical and Medical Aspects: An Overview. Medicines (Basel), 5, 93. https://doi.org/10.3390/medicines5030093

US Pharmacopeia, 2018. Food fraud database. Available online: www.foodfraud.org/.

Valentini, A., Miquel, C., Taberlet, P., 2010. DNA barcoding for honey biodiversity. Diversity 2, 610–617. https://doi.org/10.3390/d2040610

Vasić, V., Đurđić, S., Tosti, T., Radoičić, A., Lušić, D., Milojković-Opsenica, D., Tešić, Ž., Trifković, J., 2020. Two aspects of honeydew honey authenticity: Application of advance analytical methods and chemometrics. Food Chemistry 305, 125457. https://doi.org/10.1016/j.foodchem.2019.125457

von der Ohe, W., Persano Oddo, L., Piana, M.L., Morlot, M., Martin, P., 2004. Harmonized methods of melissopalynology. Apidologie 35, S18—S25. https://doi.org/10.1051/apido:2004050

Walker, G.P., 2009. Chapter 228 – Salivary Glands, Editor(s): Vincent H. Resh, Ring T. Cardé, Encyclopedia of Insects (Second Edition), Academic Press, Pages 897–901. https://doi.org/10.1016/B978-0-12-374144-8.00237-X

Wang, H., Cao, X., Han, T., Pei, H., Ren, H., Stead, S., 2019. A novel methodology for real-time identification of the botanical origins and adulteration of honey by rapid evaporative ionization mass spectrometry. Food Control 106, 106753. https://doi.org/10.1016/j.foodcont.2019.106753

Wang, J., Kliks, M.M., Qu, W., Jun, S., Shi, G., Li, Q.X., 2009. Rapid determination of the geographical origin of honey based on protein fingerprinting and barcoding using MALDI TOF MS. Journal of Agricultural and Food Chemistry 57, 10081–10088. https://doi.org/10.1021/jf902286p

Wang, X., Rogers, K.M., Li, Y., Yang, S., Chen, L., Zhou, J., 2019a. Untargeted and targeted discrimination of honey collected by *Apis cerana* and *Apis mellifera* based on volatiles using HS-GC-IMS and HS-SPME-GC—MS. Journal of Agricultural and Food Chemistry 67, 12144–12152. https://doi.org/10.1021/acs.jafc.9b04438

Wang, X., Yang, S., He, J., Chen, L., Zhang, J., Jin, Y., Zhou, J., Zhang, Y., 2019b. A green triple-locked strategy based on volatile-compound imaging, chemometrics, and markers to discriminate winter honey and sapium honey using headspace gas chromatography-ion mobility spectrometry. Food Research International 119, 960–967. https://doi.org/10.1016/j.foodres.2019.01.004

Wang, Y., Zhang, M., Wang, D., Zhang, Y., Jiao, X., Liu, Y., 2020. Development of a real-time LAMP assay for monofloral honey authentication using rape honey. CyTA—Journal of Food 18, 309–314. https://doi.org/10.1080/19476337.2020.1749135

White, J.W., 1975. Composition of honey, in: Crane E (ed) Honey, a Comprehensive Survey. London, pp. 157–206.

Wirta, H., Abrego, N., Miller, K., Roslin, T., Vesterinen, E., 2021. DNA traces the origin of honey by identifying plants, bacteria and fungi. Scientific Reports 11, 4798. https://doi.org/10.1038/s41598-021-84174-0

Xagoraris, M., Revelou, P.-K., Alissandrakis, E., Tarantilis, P.A., Pappas, C.S., 2021. Greek honey authentication: Botanical approach. Encyclopedia 1, 1322–1333. https://doi.org/10.3390/encyclopedia1040099

Yadata, D., 2014. Detection of the Electrical Conductivity and Acidity of Honey from Different Areas of Tepi. Food Science and Technology 2, 59–63. https://doi.org/10.13189/FST.2014.020501

Yue, J., Fan, J., Li, Y., Ren, H., 2021. Rapid authentication of mono-floral honeys by capillary zone electrophoresis. Journal of Food Measurement and Characterization 15, 3553–3560. https://doi.org/10.1007/s11694-021-00914-w

Zhang, P., Georgiou, C.A., Brusic, V., 2017. Elemental metabolomics. Briefings in Bioinformatics bbw131. https://doi.org/10.1093/bib/bbw131

Zhou, J., Yao, L., Li, Y., Chen, L., Wu, L., Zhao, J., 2014. Floral classification of honey using liquid chromatography—diode array detection—tandem mass spectrometry and chemometric analysis. Food Chemistry 145, 941–949. https://doi.org/10.1016/j.foodchem.2013.08.117

10 Authenticity and Chemometrics of Spices, Herbs, and Seasonings

Araceli Rivera-Pérez, Roberto Romero-González, and Antonia Garrido Frenich

CONTENTS

10.1 INTRODUCTION

Spices and herbs are aromatic plant parts with valuable organoleptic properties that are widely used as condiments to improve the flavor of foods, as well as preservative agents in food storage (Vallverdú-Queralt et al., 2014; Rivera-Pérez et al., 2020). Spices are non-leafy plant parts including seeds, roots, bark, rhizomes, berries, or flowers (e.g., black pepper, ginger, saffron, among others). In contrast, the term *herb* is used for leafy parts further used as seasonings such as coriander or dill (Kabak and Dobson, 2017). The Food and Drug Administration (FDA) provides a general definition for *spice* as "aromatic vegetable substances, in the whole, broken, or ground form, whose significant function in food is seasoning rather than nutrition" (US Food and Drug Administration, 2018). Their use as seasonings is well-known since ancient times (i.e., since about 2000 BC), in ancient India, China and Egypt (Iha and Trucksess, 2019). Nowadays, there is growing consumption of plant-derived condiments, most likely due to the popular consumer belief that "natural is the healthiest", highly related to their nutritional benefits (Lopez et al., 2015). In addition, many condiments produced in the Mediterranean area (e.g., oregano, thyme, rosemary) are reported to

DOI: 10.1201/9781003266167-10

contain different compounds such as polyphenols, and flavonoids (among others) that are responsible for their antimicrobial, antioxidant and anti-inflammatory properties (Christaki et al., 2012; Di Bella et al., 2019). Nevertheless, herbs and spices have become targeted matrices for food fraud since supply chains in the condiment industry are generally complex, long and pass through different countries (Galvin-King, Haughey and Elliott, 2018). Besides, spices and herbs are commonly provided in powder form, which makes them more vulnerable to food fraud (Modupalli et al., 2021). Some adulterant practices included the adulteration of high-value condiments such as saffron with safflower and gardenia, oregano adulterated with olive leaves, or black pepper adulteration with papaya seeds (Negi, Pare and Meenatchi, 2021).

In the last decade, metabolomics tools have become a powerful approach to address the authenticity and quality control of foods, enabling the comprehensive characterization of food matrices to be further used for their authenticity assessment and identification of markers (Kalogiouri et al., 2020). There are two main metabolomics approaches applied in the food authenticity field: targeted and untargeted (Jimenez-Carvelo and Cuadros-Rodríguez, 2021). Targeted approaches refer to the determination of already known analytes of the sample using the corresponding reference standards (Kalogiouri et al., 2020) or assessing the compositional analysis of profiles, known as metabolic profiling (Jimenez-Carvelo and Cuadros-Rodríguez, 2021). Untargeted approaches involve the obtention of unspecific instrument signals, also known as fingerprinting methodology, which is characterized by the detection of as many metabolites as possible without their mandatory quantification or identification (Cuadros-Rodríguez et al., 2016). The general scheme of metabolomics studies includes the establishment of the aim of the study, sample preparation and extraction, data acquisition, data processing and statistical analysis by chemometrics (Figure 10.1).

Concerning extraction, the sample preparation is a primordial step that provided an overview of metabolites from a complex sample and greatly influences the metabolite coverage (Oms-Oliu,

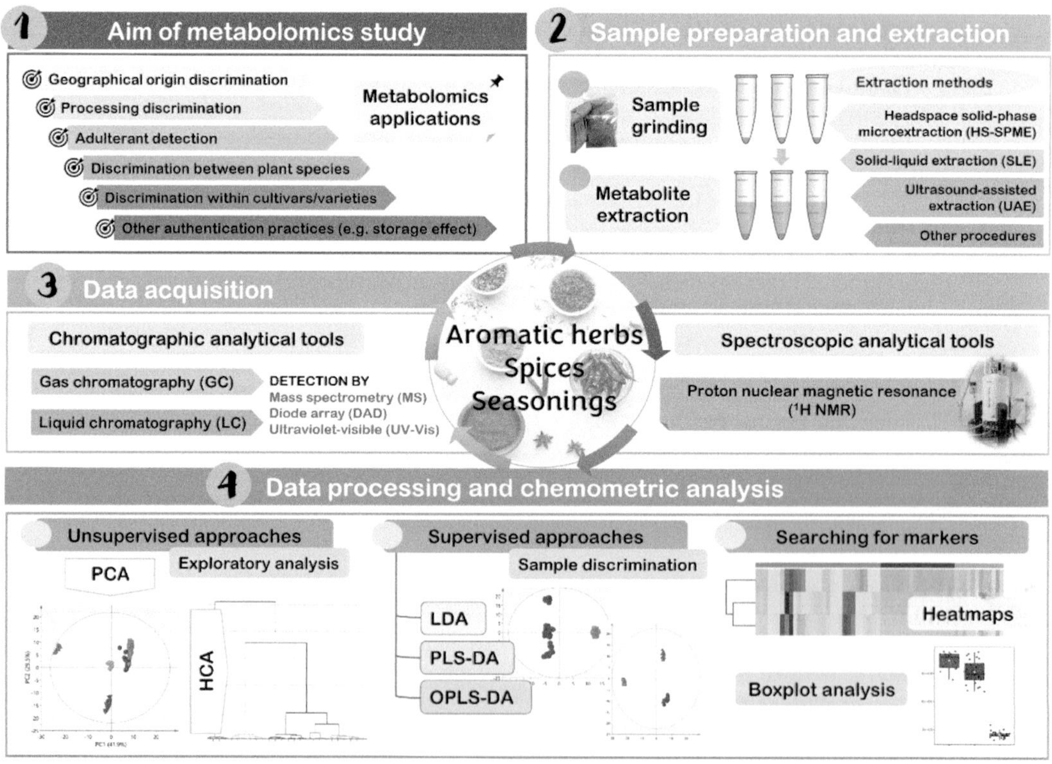

Figure 10.1 General scheme of metabolomics workflow applied for authentication of aromatic herbs, spices, and seasonings. Abbreviations: HCA, hierarchical cluster analysis; LDA, linear discriminant analysis; OPLS-DA, orthogonal partial least squares–discriminant analysis; PCA, principal component analysis; PLS-DA, partial least squares–discriminant analysis.

Source: Original figure by the authors.

Odriozola-Serrano and Martín-Belloso, 2013). It is noteworthy that the selected extraction protocol must be compatible with the analytical platform to be used (Medina et al., 2019). Currently, novel microextraction techniques such as headspace–solid phase microextraction (HS-SPME) have been reported as the "golden standard" for the extraction of volatile analytes from condiments (Figure 10.1; Farag et al., 2020; Di Donato et al., 2021; Reale et al., 2021; Feng et al., 2022). Other time-effective alternatives such as ultrasound-assisted extraction (UAE) or solid-liquid extraction (SLE) are also among the most common options for multi-extraction of polar and non-polar metabolites (Chemat et al., 2017; Rivera-Pérez, Romero-González and Garrido Frenich, 2021d; Hegazi et al., 2022; Rivera-Pérez, Romero-González and Garrido Frenich, 2022; Suzuki et al., 2022). Detailed and exhaustive information about the most used sample extraction procedures for condiment authentication is described in this chapter.

Currently, chromatographic techniques, such as gas chromatography (GC) and liquid chromatography (LC), are being employed as powerful analytical platforms in metabolomics approaches applied for quality control of condiments due to their remarkable efficiency to separate compounds from a complex mixture (Figure 10.1; Negi, Pare and Meenatchi, 2021). Due to the complexity of plant-derived matrices, mass spectrometry (MS) is the main detection technique coupled to chromatographic techniques, given its high sensitivity and high coverage of metabolites, allowing their detection at sub-microgram concentrations (Ellis et al., 2015). Nevertheless, the application of classical detection by ultraviolet-visible (UV-Vis) or diode array (DAD) has also been combined with LC-based approaches (Tabtabaei et al., 2019; Vidal-Casanella et al., 2020). An advanced version of MS consisting of high-resolution mass spectrometry (HRMS; especially using Orbitrap or quadrupole-time-of-flight, TOF, analyzers) is being more used since accurate mass acquisition considerably reduces the number of candidates per metabolite, leading to the reliable identification of markers (Kalogiouri et al., 2020; Rivera-Pérez, Romero-González and Garrido Frenich, 2021a, 2021d). In recent years, proton nuclear magnetic resonance spectroscopy (^{1}H NMR) is receiving special attention in condiment fingerprinting since it is fast, non-destructive, non-invasive and especially indicated for the detection of a wide range of non-polar and polar metabolites by high-throughput analysis (Petrakis et al., 2015; Abd Ghafar et al., 2020). However, the limitation of NMR spectroscopy is its less sensitivity compared to MS-based approaches (Kim and Verpoorte, 2010; Maulidiani et al., 2013).

Metabolomics studies of condiments usually lead to complex and high-dimensional analytical data and thus, they require the use of chemometric tools to obtain valuable data (Rivera-Pérez, Romero-González and Garrido Frenich, 2021d). Multivariate data analysis (MVDA) is a statistical approach commonly used to handle the challenging data processing in metabolomics studies (Abd Ghafar et al., 2020). The most widely used MVDA techniques include unsupervised approaches for exploratory analysis such as principal component analysis (PCA), and hierarchical cluster analysis (HCA), as well as supervised analyses for discrimination/classification purposes, including linear discriminant analysis (LDA) and partial or orthogonal partial least squares–discriminant analysis, as indicated in Figure 10.1 (PLS-DA and OPLS-DA, respectively; Kucharska-Ambrożej and Karpinska, 2020; Chaji et al., 2021).

This chapter aims to provide an overview of the most promising applications of extraction protocols and analytical techniques in combination with chemometric and metabolomics tools to assess the authenticity and quality control of spices, aromatic herbs, and plant-derived condiments. It comprises the relevant studies of seasonings published in the last decade (2012–2022). Emphasis will be put on the most widely applied chromatographic techniques (GC and LC) in combination with MS, HRMS or classical detection, as well as the application of promising ^{1}H NMR-based metabolomics as a non-destructive approach for quality assessment of spices, herbs and seasonings.

10.2 SAMPLE PREPARATION AND METABOLITE EXTRACTION METHODS FROM AROMATIC HERBS, SPICES, AND SEASONINGS

The sample extraction is a crucial step in the food authentication field since it is directly linked to the obtention of maximum metabolite information from the sample, which may be further employed for authentication purposes. In most cases, extraction methods are firstly applied in authentication workflows for the obtention of the sample profile showing its characteristic metabolic composition. Pretreatments are often chosen on the basis of the analytical technique further used for data acquisition. In this context, the extraction methods more often reported for spice/herb/seasoning fingerprinting include hydrodistillation, maceration, SLE and so on, while more "green" alternatives such as UAE and HS-SPME are becoming highly demanded due to their several advantages including automatization, reduced sample and solvent consumption, among others (Rivera-Pérez, Romero-González and Garrido Frenich, 2021b). Figure 10.2 shows an overview of the most used extraction methods in metabolomics approaches applied for authentication of herbs and spices.

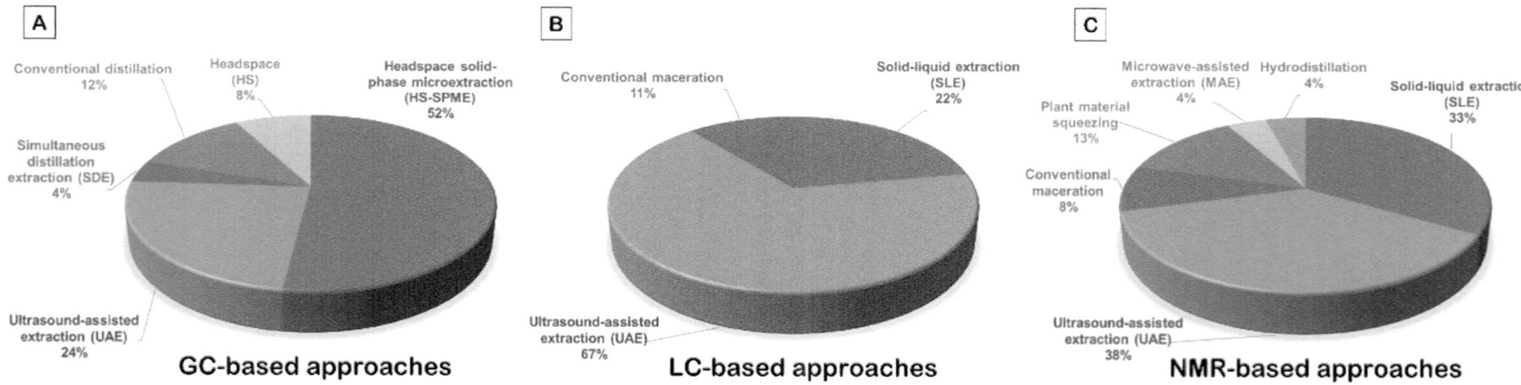

Figure 10.2 Overview of the main extraction methods used to assess the authenticity of aromatic herbs, spices, and seasonings according to the analytical tool used for data acquisition. (a) Gas chromatography (GC); (b) liquid chromatography (LC); and (c) nuclear magnetic resonance (NMR).

Source: Original figure by the authors.

10.2.1 Extraction Procedures Suitable for Chromatographic Methods: GC-Based Approaches

The obtention of the comprehensive volatile composition of condiments for further authenticity and food traceability purposes via GC analyses and application of chemometric tools is highly influenced by the extraction method applied for metabolite profiling. Conventional extraction methods such as Clevenger distillation (Liu et al., 2018; Li et al., 2020), steam distillation (Satyal et al., 2016), and simultaneous distillation-extraction (SDE; Barata et al., 2021) have been utilized for the isolation of essential oils of condiments to further assess their authenticity by GC-based fingerprint analysis (Table 10.1; Figure 10.2). Nevertheless, these classical extraction methods have some disadvantages since they require large solvent volumes and long extraction times (Ávila et al., 2009).

Table 10.1 Summary of GC-Based Metabolomics Approaches Used for Authenticity Assessment of Culinary Herbs, Spices, and Seasonings

Matrix	Purpose of Study	Extraction Method	Analytical Method	Chemometric Analysis	References
Saffron (*Crocus sativus* L.)	Adulterant detection (calendula, safflower, and turmeric)	HS-SPME (CW/DVB, 50°C, 30 min)	GC-MS (IT)	PCA, PLS-DA, SIMCA	(Di Donato et al., 2021)
Saffron (*Crocus sativus* L.)	Adulterant detection (calendula and safflower) Geographical origin discrimination (Spain and Iran) Freshness effect	HS-SPME (DVD/CAR/PDMS, 50°C, 30 min)	GC-MS	PCA, HCA, OPLS-DA	(Farag et al., 2020)
Saffron (*Crocus sativus* L.)	Adulterant detection (turmeric and marigold)	HS	GC-FID	LDA, PLS-DA	(Morozzi et al., 2019)
Pepper (*Piper nigrum* L.)	Adulterant detection (*Carica papaya* L.)	UAE (MeOH, 15 min)	GC-MS	PCA, PLS-DA	(Gul et al., 2018)
11 spices (allspice, anise, clove, etc.)	Quality control	HS-SPME (PDMS, CAR/PDMS, DVB/CAR/PDMS, 60°C, 30 min)	GC-MS	HCA, PCA	(Matsushita et al., 2018)
Nutmeg (*Myristica fragrans*)	Quality control (seed and mace differentiation) Processing practices (roasting)	HS-SPME (DVD/CAR/PDMS, 50°C, 30 min)	GC-MS	PCA, HCA, OPLS-DA	(Farag, Mohsen and El-Gendy, 2018)
6 spices and herbs (rosemary, sage, thyme, etc.)	Quality control	HS-SPME (PDMS, DVB/CAR/PDMS, 60°C, 30 min)	GC-MS	PCA	(Sgorbini et al., 2015)
Black pepper (*Piper nigrum* L.)	Geographical origin discrimination (Sri Lanka, Vietnam, Brazil) Processing practices (sterilized vs. non-sterilized)	UAE (*n*-hexane, 30 min)	GC-HRMS (Orbitrap)	PCA, HCA, OPLS-DA	(Rivera-Pérez, Romero-González and Garrido Frenich, 2021d)
Black pepper (*Piper nigrum* L.)	Geographical origin discrimination (Brazil, India, Indonesia, Vietnam)	UAE (MeOH, 1 h)	GC-MS	PLS-DA	(Liang et al., 2021)

Table 10.1 Summary of GC-Based Metabolomics Approaches Used for Authenticity Assessment of Culinary Herbs, Spices and Seasonings (Continued)

Matrix	Purpose of Study	Extraction Method	Analytical Method	Chemometric Analysis	References
Saffron (*Crocus sativus* L.)	Geographical origin discrimination (Italy, Argentina, Greece, Iran, Lebanon, India, Spain, Turkey) Processing practices (drying) Age effect	HS-SPME (DVB/CAR/PDMS, 40°C, 5 min)	GC-MS	PLS-DA	(Rocchi et al., 2019)
Saffron (*Crocus sativus* L.)	Geographical origin discrimination (Italy, Iran, Morocco)	HS-SPME (PDMS, 45°C, 30 min)	GC-MS (IT)	PCA, LDA	(D'Archivio et al., 2018)
Saffron (*Crocus sativus* L.)	Geographical origin discrimination (Iran, China)	Clevenger distillation (AcOEt, 1 h)	GC-MS	PCA, OPLS-DA	(Liu et al., 2018)
Sumac (*Rhus coriaria*)	Geographical origin discrimination (Palestine, Jordan, Egypt) Processing practices (roasting)	HS-SPME (DVD/CAR/PDMS, 50°C, 30 min)	GC-MS	PCA, OPLS-DA	(Farag, Fayek and Reidah, 2018)
Saffron (*Crocus sativus* L.)	Geographical origin discrimination (Greece, Spain, Iran, Morocco)	HS-SPME (DVB/CAR/PDMS, 35°C, 20 min)	GC-MS	LDA	(Karabagias et al., 2017)
Black cumin (*Nigella sativa*)	Geographical origin discrimination (Germany, Italy, Romania, Czech Republic, Turkey, Egypt, Syria, India) Processing practices (roasting)	HS-SPME (DVD/CAR/PDMS, 50°C, 30 min)	GC-MS	PCA, HCA, OPLS-DA	(Farag et al., 2017)
Thyme (*Thymus vulgaris* L.)	Geographical origin discrimination (France, Serbia)	Steam distillation (3 h)	GC-MS	HCA	(Satyal et al., 2016)
Red and green huajiao	Species discrimination (China)	HS-SPME (DVB/CAR/PDMS, 60°C, 30 min)	GC-MS	PCA, PLS-DA	(Feng et al., 2022)
Bell pepper (*Capsicum annuum*)	Discrimination within varieties (Italy)	HS-SPME (DVB/CAR/PDMS, 90°C, 10 min)	GC-MS (IT)	PLS-DA	(Reale et al., 2021)
Black pepper (*Piper nigrum* L.)	Discrimination within varieties (Brazil)	SDE (Likens-Nickerson, *n*-pentane, 2 h)	GC-MS	HCA	(Barata et al., 2021)
Saffron (*Crocus sativus* L.)	Discrimination within varieties (Iran)	1. UAE (MeOH:ACN, 38:62 v/v, 22 min) 2. DLLME ($CHCl_3$)	GC-FID	PCA, PLS-DA	(Aliakbarzadeh, Parastar and Sereshti, 2016)

Matrix	Purpose of Study	Extraction Method	Analytical Method	Chemometric Analysis	References
Saffron (*Crocus sativus* L.)	Discrimination within varieties (Iran)	1. UAE (MeOH:ACN, 38:62 v/v, 22 min) 2. DLLME ($CHCl_3$)	GC-MS	PCA, MCR-ALS	(Aliakbarzadeh, Sereshti and Parastar, 2016)
Black and white pepper (*Piper nigrum* L.)	Discrimination within varieties (China)	Clevenger hydrodistillation (5 h)	GC-FID	PCA, HCA	(Li et al., 2020)
Saffron (*Crocus sativus* L.)	Discrimination within varieties (Italy) Processing practices (drying)	HS-SPME (PDMS, 20 min)	GC-MS (IT)	ANOVA	(Macchia et al., 2013)
Saffron (*Crocus sativus* L.)	Storage time effect	1. UAE (MeOH:ACN, 38:62 v/v, 22 min) 2. DLLME ($CHCl_3$)	GC-FID	PCA, PLS-DA	(Sereshti et al., 2018)
Oregano (*Origanum vulgare*)	Organic vs. conventional	HS	GC-FID	PCA, OPLS-DA	(Gao et al., 2014)

Abbreviations: ACN, acetonitrile; AcOEt, ethyl acetate; ANOVA, analysis of variance; CAR, carboxen; $CHCl_3$, chloroform; CW, carbowax; DART, direct analysis real time; DLLME, dispersive liquid-liquid microextraction; DVB, divinylbenzene; FID, flame ionization detection; GC, gas chromatography; HCA, hierarchical cluster analysis; HRMS, high-resolution mass spectrometry; HS, headspace; HS-SPME, headspace–solid phase microextraction; IT, ion trap; LDA, linear discriminant analysis; MCR-ALS, multivariate curve resolution–alternating least squares; MeOH, methanol; MS, mass spectrometry; OPLS-DA, orthogonal partial least squares–discriminant analysis; PCA, principal component analysis; PDMS, polydimethylsiloxane; PLS-DA, partial least squares–discriminant analysis; SDE, simultaneous distillation extraction; SIMCA, soft independent model class analogy; UAE, ultrasound-assisted extraction.

Thus, increasing interest in developing extraction procedures without these drawbacks has led to the current application of more environmental-friendly extraction alternatives, such as extraction methods assisted by ultrasound (UAE; Gul et al., 2018; Liang et al., 2021; Rivera-Pérez, Romero-González and Garrido Frenich, 2021d); automatized HS-SPME (Macchia et al., 2013; Sgorbini et al., 2015; Farag et al., 2017, 2020; Karabagias et al., 2017; D'Archivio et al., 2018; Farag, Fayek and Reidah, 2018; Farag, Mohsen and El-Gendy, 2018; Matsushita et al., 2018; Rocchi et al., 2019; Di Donato et al., 2021; Reale et al., 2021; Feng et al., 2022), and HS (Gao et al., 2014; Morozzi et al., 2019) (Table 10.1). Reducing the time and solvent consumption in the metabolite extraction step is particularly advantageous considering the high number of samples that must be analyzed in metabolomics approaches to obtain significant statistical results. The combination of two extraction methods has also been tested by combining UAE and dispersive liquid-liquid microextraction (DLLME; Aliakbarzadeh, Parastar and Sereshti, 2016; Aliakbarzadeh, Sereshti and Parastar, 2016; Sereshti et al., 2018).

Essential oils (EOs) represent a volatile mixture of secondary metabolites that are usually concentrated in aromatic plant parts such as bark, leaves or fruits (Bakkali et al., 2008). Essential oil isolation by classical Clevenger-hydrodistillation (for 5 h) was successfully applied to assess variations in constituents of *Piper nigrum* L. (black pepper) from different Chinese origins (Li et al., 2020; Table 10.1). Furthermore, this extraction procedure was also highlighted as the most appropriate extraction procedure for further geographical authentication of *Crocus sativa* L. (saffron; Liu et al., 2018). In this study (Liu et al., 2018), three extraction methods were compared (UAE, DLLME and distillation), and the findings revealed the suitability of the distillation among the tested extraction steps to obtain a higher number of metabolites from saffron samples, including enhanced safranal and other safranal-derived contents. The essential oil composition of *Thymus vulgaris* L. (thyme) from France and Serbia was also assessed for thyme chemotype discrimination by steam distillation for 3 h (Satyal et al., 2016). Simultaneous distillation extraction (SDE) was reported as an adequate extraction method for volatiles from fruits and leaves of black pepper (Barata et al., 2021) using a Likens-Nickerson apparatus and *n*-pentane as the extraction solvent. Although the SDE extraction was fit for EO extraction to be coupled to GC-MS fingerprinting, it is not especially recommended for routine analysis and it is not so environmental-friendly (Lopez et al., 2015) compared with other alternatives in line with green chemistry principles. Besides, high temperatures reached during the SDE process have been reported as responsible for the thermal degradation of some volatiles and the formation of some reaction products that were not initially present in related matrices, such as tea (Yang, Baldermann and Watanabe, 2013).

In this regard, despite the application of conventional SDE and distillation extraction methods coupled to GC-based analysis (Figure 10.2), there is an increasing trend toward the application of less laborious, time and solvent-consuming extraction methods. Therefore, simpler and more environmental-friendly extraction methods have been recently introduced especially for metabolite extraction from black pepper based on UAE (Gul et al., 2018; Liang et al., 2021; Rivera-Pérez, Romero-González and Garrido Frenich, 2021d) using organic solvents suitable for further GC-based fingerprinting, such as *n*-hexane and methanol (MeOH; Table 10.1). In particular, UAE is a SLE method that uses ultrasonic waves formed in an ultrasonic bath to produce cavitation effects on the solid resulting in an enhanced metabolite extraction from plant matrices to the extraction solvent (Palmieri et al., 2020; Rivera-Pérez et al., 2020). UAE avoids the large use of extraction solvent volumes and reduces environmental impact in terms of energy and time (Chemat et al., 2017). Rivera-Pérez et al. evaluated the influence of the extraction solvent and the extraction time on the overall extracted metabolites from black pepper to assess its geographical origin and processing (Rivera-Pérez, Romero-González and Garrido Frenich, 2021d). Various solvents, including *n*-hexane, ethyl acetate (AcOEt), and dichloromethane (CH_2Cl_2), as well as the corresponding 50:50 (v/v) solvent mixtures, were assessed, and the findings revealed that the highest number of metabolites were obtained using *n*-hexane (Rivera-Pérez, Romero-González and Garrido Frenich, 2021d). Besides, increasing the UAE extraction time from 30 to 45 min improved the metabolite extraction. Nevertheless, the higher extraction time, the higher sample overheating, leading to the degradation of thermolabile metabolites (Rivera-Pérez, Romero-González and Garrido Frenich, 2021d). Therefore, the high reproducibility, reduced solvent consumption, and simpler sample manipulation are some of the main advantages of innovative UAE-based extraction methods compared with conventional ones (Chemat et al., 2017). Furthermore, UAE time effectiveness makes it one of the method of choice for multi-sample extraction required in metabolomics approaches (Rivera-Pérez, Romero-González and Garrido Frenich, 2021c).

Other promising combinations, such as UAE-DLLME, have been successfully applied as the sample preparation step for Iranian variety discrimination, and storage time effect assessment of saffron coupled to GC with MS or flame ionization detection (FID; Aliakbarzadeh, Parastar and Sereshti, 2016; Aliakbarzadeh, Sereshti and Parastar, 2016; Sereshti et al., 2018; Table 10.1). DLLME is a liquid-liquid microextraction (LLME) in which the mixture made of the extraction and the dispersing solvents, are rapidly injected into an aqueous solution to produce a cloudy solution. Then, dispersion is removed by centrifugation or low temperatures, and the phase of interest is collected using a microsyringe and further analyzed (Zgoła-Grześkowiak and Grześkowiak, 2011). In saffron studies (Aliakbarzadeh, Parastar and Sereshti, 2016; Aliakbarzadeh, Sereshti and Parastar, 2016; Sereshti et al., 2018), the volatile profile of this condiment was concentrated using a UAE-DLLME in which the sample was firstly sonicated using a MeOH:acetonitrile (ACN) (38:62 v/v) solvent mixture for 22 min (UAE step), and then the supernatant was mixed and preconcentrated with a low volume of chloroform ($CHCl_3$). In the next step, a 7% (w/v) sodium chloride (NaCl) aqueous solution was injected into the solvent mixture to form a cloudy solution of $CHCl_3$ further disrupted by centrifugation and injected into the GC system (Aliakbarzadeh, Parastar and Sereshti, 2016; Aliakbarzadeh, Sereshti and Parastar, 2016; Sereshti et al., 2018). DLLME is an effective and low-sample, time- and solvent-consuming extraction approach, but an important drawback is the use of organic and highly toxic extraction solvents (Rajabi et al., 2014).

According to the summarized information presented in Table 10.1, sample pretreatment by HS-SPME has received considerable attention in recent years. HS-SPME has turned into the most popular sample extraction for the characterization of the volatile profile of a wide diversity of condiments by GC-based fingerprinting (huajiao, bell pepper, saffron, sumac, nutmeg and black cumin, among others; Macchia et al., 2013; Sgorbini et al., 2015; Farag et al., 2017, 2020; Karabagias et al., 2017; D'Archivio et al., 2018; Farag, Fayek and Reidah, 2018; Farag, Mohsen and El-Gendy,

2018; Matsushita et al., 2018; Rocchi et al., 2019; Di Donato et al., 2021; Reale et al., 2021; Feng et al., 2022; Figure 10.2). In HS-SPME methods, a fused silica fiber coated with different types of polar or non-polar adsorbents is placed in the vapor phase of the headspace vial containing the sample. The fiber is exposed to the sample during the adsorption time (at a certain adsorption temperature) and compounds are adsorbed into the fiber. Then, the fiber is injected into the GC injector and metabolites are desorbed for separation and identification (Sporkert and Pragst, 2000). The effectiveness of HS-SPME extraction is highly influenced by several factors such as the adsorption time, temperature, sample amount, and especially, by the type of fiber sorbent. In this regard, the most commonly used fiber to assess the aroma profile of saffron is the divinylbenzene/carboxen/polydimethylsiloxane combination (DVB/CAR/PDMS; Sgorbini et al., 2015; Farag et al., 2017, 2020; Karabagias et al., 2017; Farag, Fayek and Reidah, 2018; Farag, Mohsen and El-Gendy, 2018; Matsushita et al., 2018; Rocchi et al., 2019; Reale et al., 2021; Feng et al., 2022), which is especially suitable for the extraction of both volatile and semi-volatile compounds (Yang, Baldermann and Watanabe, 2013). Another type of fiber, carbowax (CW)/DVB, was reported as especially suitable for volatile compound extraction to detect adulteration of Italian saffron with plant-derived adulterants such as calendula, safflower and turmeric (Di Donato et al., 2021). Non-polar semi-volatile and volatile compounds may be extracted using non-polar coating as PDMS, successfully applied for saffron characterization for geographical, processing (drying), and variety differentiation (Macchia et al., 2013; D'Archivio et al., 2018). As can be seen, HS-SPME is a solvent-free, simple and automatized procedure that has become the prevalent sample pretreatment in GC metabolomics approaches applied for quality control of condiments, instead of the use of more laborious and less environmental-friendly conventional methods (Di Donato et al., 2021).

Untargeted GC-based approaches for condiment authenticity based on FID detection are usually combined with HS samplers (Figure 10.2). This approach based on HS-GC-FID fingerprinting was performed to detect adulterants in saffron (Morozzi et al., 2019), as well as to discriminate among conventional and organic oregano samples (Gao et al., 2014; Table 10.1).

10.2.2 Extraction Procedures Suitable for Chromatographic Methods: LC-Based Approaches

Currently, several conventional extraction methods such as SLE (D'Archivio et al., 2016; Maldini et al., 2016; Senizza et al., 2019; Tabtabaei et al., 2019; Biancolillo et al., 2020; Hegazi et al., 2022) and

Table 10.2 Summary of LC-Based Metabolomics Approaches Used for Authenticity Assessment of Culinary Herbs, Spices, and Seasonings

Matrix	Purpose of Study	Extraction Method	Analytical Method	Chemometric Analysis	References
Saffron (*Crocus sativus* L.)	Adulterant detection (calendula and safflower) Geographical origin discrimination (Iran and Spain)	SLE (MeOH, 6 min)	UHPLC-HRMS	HCA, PCA, OPLS-DA	(Hegazi et al., 2022)
Oregano (*Origanum vulgare*)	Adulterant detection (olive, myrtle, sumac, cistus, hazelnut leaves)	UAE (H_2O:MeOH, 1:1 v/v, 15 min)	UHPLC-HRMS (QTOF)	PCA, OPLS-DA	(Black et al., 2016)
Turmeric and curry	Quality control	UAE (MeOH, DMSO, 40 min)	HPLC-DAD	PCA	(Vidal-Casanella et al., 2020)
Turmeric and curry	Quality control	UAE (DMSO, 15 min)	UHPLC-HRMS (Q-Orbitrap)	PCA, PLS-DA	(Núñez et al., 2020)
Black pepper (*Piper nigrum* L.)	Quality control	Maceration (70% MeOH, 4°C, overnight)	UHPLC-MS/MS (QTRAP)	PCA	(Gu et al., 2018)

(Continued)

Table 10.2 Summary of LC-Based Metabolomics Approaches Used for Authenticity Assessment of Culinary Herbs, Spices and Seasonings (Continued)

Matrix	Purpose of Study	Extraction Method	Analytical Method	Chemometric Analysis	References
Rosemary, thyme, oregano, cinnamon, cumin, bay	Quality control	UAE (50% EtOH containing 0.1% FA, 5 min)	LC-HRMS (LTQ-Orbitrap)	PCA	(Vallverdú-Queralt et al., 2014)
Ginger (*Zingiber officinale*)	Quality control	UAE (MeOH, 30 min)	HPLC-DAD	HCA, PCA	(Feng et al., 2014)
Black pepper (*Piper nigrum* L.)	Geographical origin discrimination (Brazil, India, Indonesia, Vietnam)	UAE (MeOH, 1 h)	HPLC-MS (LTQ)	PLS-DA	(Liang et al., 2021)
Black pepper (*Piper nigrum* L.)	Geographical origin discrimination (Sri Lanka, Vietnam, Brazil) Processing practices (sterilized vs. non-sterilized)	UAE (H_2O:ACN, 50:50 v/v, 30 min)	UHPLC-HRMS (Q-Orbitrap)	PCA, OPLS-DA	(Rivera-Pérez, Romero-González and Garrido Frenich, 2021a)
Paprika (*Capsicum annuum* L.)	Geographical origin discrimination (Spain, Hungary, Czech Republic) Discrimination within flavor varieties (spicy, sweet, bittersweet)	UAE (MeOH:acetone, 1:1 v/v, 10 min)	UHPLC-HRMS (Q-Orbitrap)	PCA, PLS-DA	(Arrizabalaga-Larrañaga et al., 2021)
Paprika (*Capsicum annuum* L.)	Geographical origin discrimination (Spain, Czech Republic)	UAE (H_2O:ACN, 20:80 v/v, 15 min)	UHPLC-HRMS (Q-Orbitrap)	PCA, PLS-DA	(Barbosa, Saurina et al., 2020b)
Paprika (*Capsicum annuum* L.)	Geographical origin discrimination (Spain, Czech Republic) Discrimination within flavor varieties (spicy, sweet, bittersweet)	UAE (H_2O:ACN, 20:80 v/v, 15 min)	UHPLC-MS/MS (QqQ)	PCA, PLS-DA	(Barbosa, Campmajó et al., 2020)

Matrix	Purpose of Study	Extraction Method	Analytical Method	Chemometric Analysis	References
Paprika (*Capsicum annuum* L.)	Geographical origin discrimination (Spain, Czech Republic) Discrimination within flavor varieties (spicy, sweet, bittersweet)	UAE (H_2O:ACN, 20:80 v/v, 15 min)	UHPLC-HRMS (Q-Orbitrap)	PCA, PLS-DA	(Barbosa, Saurina et al., 2020a)
Paprika (*Capsicum annuum* L.)	Geographical origin discrimination (Spain, Czech Republic) Discrimination within flavor varieties (spicy, sweet, bittersweet)	UAE (H_2O:ACN, 20:80 v/v, 15 min)	HPLC-UV-Vis	LDA	(Cetó et al., 2020)
Saffron (*Crocus sativus* L.)	Geographical origin discrimination (Iran, Italy)	SLE (ultrapure H_2O, 1 h)	HPLC-UV-Vis	PCA, LDA	(Tabtabaei et al., 2019)
Saffron (*Crocus sativus* L.)	Geographical origin discrimination (Italy, Iran) Adulterant detection (stamens, tepals)	SLE (80% MeOH containing 1% FA, 5 min)	UHPLC-HRMS (QTOF)	HCA, OPLS-DA	(Senizza et al., 2019)
Ginger (*Zingiber officinale*)	Geographical origin discrimination (China, Ghana)	1. Maceration (MeOH, 48 h) 2. UAE (1 h 30 min)	UHPLC-HRMS (QTOF)	PCA, HCA, OPLS-DA	(Mais et al., 2018)
Saffron (*Crocus sativus* L.)	Geographical origin discrimination (Iran, China)	UAE (50% EtOH, 20 min)	HPLC-DAD	PCA, OPLS-DA	(Liu et al., 2018)
Saffron (*Crocus sativus* L.)	Geographical origin discrimination (Spain, Czech Republic, Tukey)	UAE (EtOH:H_2O, 70:30 v/v, 1 h)	UHPLC-HRMS (QTOF)	PCA, OPLS-DA	(Rubert et al., 2016)
Saffron (*Crocus sativus* L.)	Geographical origin discrimination (Iran, Italy)	Maceration (70% EtOH at pH 2.0, overnight)	HPLC-DAD	PCA	(Masi et al., 2016)
Saffron (*Crocus sativus* L.)	Geographical origin discrimination (Spain, Iran)	UAE (EtOH:borate buffer at pH 9.0, 50:50 v/v, 15 min)	HPLC-HRMS (QTOF)	PCA, PLS-DA, OPLS-DA	(Guijarro-Díez et al., 2015)

(Continued)

Table 10.2 Summary of LC-Based Metabolomics Approaches Used for Authenticity Assessment of Culinary Herbs, Spices and Seasonings (Continued)

Matrix	Purpose of Study	Extraction Method	Analytical Method	Chemometric Analysis	References
Ginger (*Zingiber officinale*)	Geographical origin discrimination (China, India, Malaysia, Thailand, Vietnam)	UAE (MeOH, 1 h)	HPLC-DAD	HCA, PCA, LDA	(Yudthavorasit, Wongravee and Leepipatpiboon, 2014)
Cinnamomum species	Species discrimination	UAE (60% MeOH, 20 min)	UHPLC-HRMS (Orbitrap)	PCA, PLS-DA	(Wang, Harrington and Chen, 2020)
Saffron (*Crocus sativus* L.)	Discrimination within varieties (Italy)	SLE (H_2O:MeOH, 1:1 v/v, 1 h)	HPLC-DAD	PLS-DA, LDA	(Biancolillo et al., 2020)
Paprika (*Capsicum annuum* L.)	Discrimination within Spanish varieties (La Vera, Murcia PDO)	UAE (H_2O:ACN, 20:80 v/v, 15 min)	HPLC-UV-Vis	PCA, LDA	(Cetó et al., 2018)
Rosemary (*Rosmarinus officinalis* L.)	Discrimination within varieties (Italy)	SLE (MeOH)	HPLC-MS/MS	PCA	(Maldini et al., 2016)
Saffron (*Crocus sativus* L.)	Discrimination within varieties (Italy)	SLE (H_2O:MeOH, 1:1 v/v, 1 h)	HPLC-DAD	LDA	(D'Archivio et al., 2016)

Abbreviations: ACN, acetonitrile; DAD, diode array detection; DMSO, dimethylsulfoxide; EtOH, ethanol; FA, formic acid; HCA, hierarchical cluster analysis; HPLC, high-performance liquid chromatography; HRMS, high-resolution mass spectrometry; LC, liquid chromatography; LDA, linear discriminant analysis; LTQ, linear ion trap quadrupole analyzer; MeOH, methanol; MS, mass spectrometry; MS/MS, tandem mass spectrometry; OPLS-DA, orthogonal partial least squares–discriminant analysis; PCA, principal component analysis; PLS-DA, partial least squares–discriminant analysis; Q, single quadrupole; QqQ, triple quadrupole; QTRAP, triple quadrupole–linear ion trap analyzer; SLE, solid-liquid extraction; TOF, time-of-flight analyzer; UAE, ultrasound-assisted extraction; UHPLC, ultra-high-performance liquid chromatography; UV-Vis, ultraviolet-visible detection.

maceration (Masi et al., 2016; Gu et al., 2018; Mais et al., 2018; Figure 10.2) are being used as sample preparation in LC-based metabolomics approaches for the authenticity of spices and aromatic herbs (Table 10.2).

On the one hand, conventional maceration, using MeOH as the extraction solvent, was applied to obtain ginger (*Zingiber officinale*) fingerprints to further trace their geographical origin (China vs. Ghana; Mais et al., 2018; Table 10.2). Besides, this conventional extraction method was suitable to extract phenolic and polyhydroxy metabolites for quality control of black pepper samples (Gu et al., 2018). However, this traditional method required long extraction times and larger volume consumption, and it can lead to the decomposition of thermolabile bioactive compounds (Azmir et al., 2013). On the other one, SLE has been described as an effective extraction method for untargeted analysis of saffron (D'Archivio et al., 2016; Tabtabaei et al., 2019; Biancolillo et al., 2020; Hegazi et al., 2022), using different LC-compatible solvents such as MeOH (Hegazi et al., 2022), ultrapure water (Tabtabaei et al., 2019), and the corresponding H_2O:MeOH (1:1 v/v) mixture (D'Archivio et al., 2016; Biancolillo et al., 2020).

Particularly, the study of Senizza et al. performed polyphenolic profiling from saffron samples applying a simple SLE method using an alcoholic solution consisting of 80% MeOH acidified with 1% formic acid (Senizza et al., 2019; Table 10.2). In another study, phenolic fingerprints of rosemary (*Rosmarinus officinalis* L.) extracts were obtained by methanolic SLE to differentiate between rosemary cultivars in Sardinia (Italy; Maldini et al., 2016).

Despite the use of traditional SLE methods, nowadays, UAE has become the extraction method of choice in LC-metabolomics approaches performed for seasoning authenticity (Table 10.2; Figure 10.2). Different extraction solvents and solvent mixtures are being used such as MeOH (Feng et al., 2014; Yudthavorasit, Wongravee and Leepipatpiboon, 2014; Vidal-Casanella et al., 2020; Wang, Harrington and Chen, 2020; Liang et al., 2021); ethanol (EtOH; 50%; Liu et al., 2018); 50% EtOH containing 0.1% formic acid (Vallverdú-Queralt et al., 2014); dimethylsulfoxide (DMSO; Núñez et al., 2020; Vidal-Casanella et al., 2020); MeOH:acetone (1:1 v/v; Arrizabalaga-Larrañaga et al., 2021); H_2O:MeOH (1:1 v/v; Black et al., 2016); EtOH:H_2O (70:30 v/v; Rubert et al., 2016); EtOH:borate buffer (pH 9.0, 50:50 v/v; Guijarro-Díez et al., 2015); H_2O:ACN (50:50 v/v; Rivera-Pérez, Romero-González and Garrido Frenich, 2021a); and especially the mixture consisting of H_2O:ACN (20:80 v/v; Cetó et al., 2018, 2020; Barbosa, Campmajó et al., 2020; Barbosa, Saurina et al., 2020a, 2020b). UAE extraction times range from shorter times such as 10 min to more than 1 hour of extraction, with the predominant use of 15 min.

Barbosa et al. research described a simple UAE extraction method using an H_2O:ACN (20:80 v/v) solution as the extracting agent in paprika (*Capsicum annuum* L.) authentication studies (Barbosa, Campmajó et al., 2020; Barbosa, Saurina et al., 2020a, 2020b; Table 10.2). This sample pretreatment allowed the obtention of untargeted ultra-high-performance liquid chromatography (UHPLC)-HRMS paprika fingerprints to trace their geographical origin (Spain and the Czech Republic) and flavor varieties (spicy, sweet and bittersweet paprika; Barbosa, Saurina et al., 2020a). Furthermore, this sample pretreatment was pointed out as a suitable extraction method for targeted extraction of 36 phenolic and polyphenolic compounds from paprika for geographical and flavor variety authentication of the condiment (Barbosa, Campmajó et al., 2020). Besides, UHPLC-HRMS (Q-Orbitrap) polyphenolic and capsaicinoid profiling was performed using this UAE (H_2O:ACN, 20:80 v/v) sample preparation (Barbosa, Saurina et al., 2020b). This UAE extraction procedure was also adequate to obtain fingerprints of paprika samples based on phenolic compounds further combined with high-performance liquid chromatography (HPLC)-UV-Vis analysis (Cetó et al., 2018, 2020). Another study described the use of UAE and another extraction solvent mixture (MeOH:acetone, 1:1 v/v) for the simultaneous extraction of 10 capsaicinoid and carotenoid compounds from paprika for its geographical origin authentication and flavor discrimination (spice, sweet and bittersweet; Arrizabalaga-Larrañaga et al., 2021). Thus, these studies revealed the potential of this simple sample treatment to obtain paprika fingerprints or quantify reliable chemical descriptors to be further used for the sample classification by UHPLC-MS or HPLC-UV-Vis metabolomics approaches.

Particularly, saffron fingerprints have been obtained by UAE methods using different extracting solvents including EtOH (50%; Liu et al., 2018); EtOH:H_2O (70:30 v/v; Rubert et al., 2016); and EtOH:borate buffer at pH 9.0 (50:50 v/v; Guijarro-Díez et al., 2015; Table 10.2). Guijarro-Díez et al. research demonstrated that enhanced extraction of saffron components was achieved by the synergetic EtOH and borate buffer solvent mixture (pH 9.0) compared to EtOH and aqueous buffer solutions (tested at pH 2.5 and 9.0; Guijarro-Díez et al., 2015). Rubert et al. research also optimized the UAE extraction procedure applied for saffron UHPLC-HRMS (QTOF) fingerprinting by assessing the selection of the solvent to extract as maximum components from saffron as possible (Rubert et al., 2016). Initially, different solvents including water, MeOH, ACN, EtOH, MeOH:H_2O (50:50 v/v), ACN:H_2O (75:25 v/v), and EtOH:H_2O (70:30 v/v) were tested. The findings revealed that the largest number of saffron features was extracted using the EtOH:H_2O (70:30 v/v) mixture, isolating both highly polar and non-polar compounds (Rubert et al., 2016).

UAE methods have been described for the untargeted LC-MS analysis of black pepper using pure MeOH (Liang et al., 2021) or aqueous/organic solvent mixtures such as H_2O:ACN (50:50 v/v; Rivera-Pérez, Romero-González and Garrido Frenich, 2021a; Table 10.2). As a dried matrix, the use of aqueous and organic solvent mixtures led to the hydration of the sample and favors the extraction of polar metabolites, leading to an enhanced extraction performance (Rubert et al., 2016). In this regard, UAE with the mixture H_2O:MeOH (1:1 v/v) was highlighted as a valuable extraction procedure to obtain oregano (*Origanum vulgare*) fingerprints by UHPLC-HRMS analysis to be further discriminated from its common adulterants such as olive, myrtle, or sumac leaves (Black et al., 2016).

Polyphenolic and curcuminoid profiling of turmeric and curry samples were performed using DMSO and MeOH, respectively, and the extracts were further analyzed by HPLC-DAD (Vidal-Casanella et al., 2020; Table 10.2). This extraction method based on UAE with DMSO was also suitable for both turmeric and curry polyphenolic and curcuminoid profiling, in this case, performed by UHPLC-HRMS (Q-Orbitrap; Núñez et al., 2020). This organic solvent was selected due to its suitability to extract both curcuminoids and polyphenol profiles (Núñez et al., 2020). Besides, the phenolic profile of six widely consumed aromatic herbs and spices (rosemary, thyme, oregano, cinnamon, cumin and bay) was assessed by UAE using an ethanolic solution (50%) acidified with 0.1% formic acid for their quality control by LC-HRMS analysis (Vallverdú-Queralt et al., 2014).

Untargeted metabolomics approaches based on HPLC-DAD fingerprinting were usually combined with UAE (MeOH) for quality control and origin traceability of ginger samples (Feng et al., 2014; Yudthavorasit, Wongravee and Leepipatpiboon, 2014; Table 10.2). As can be observed, the findings of all these investigations support that UAE is a simpler, more cost-effective, and more efficient alternative compared with more time-consuming traditional methods (Danlami et al., 2014).

10.2.3 Extraction Procedures Suitable for Spectroscopic Methods: NMR-Based Approaches

NMR fingerprints of condiments have been obtained in combination with conventional extraction methods, such as maceration (Pieri et al., 2012; Liang et al., 2015); by plant material squeezing (Becerra-Martínez et al., 2017; Florentino-Ramos et al., 2019; Villa-Ruano et al., 2019); hydrodistillation for essential oil extraction (Freitas et al., 2018); or SLE (Cagliani et al., 2015; Ordoudi et al., 2015; Petrakis et al., 2015; Consonni et al., 2016; Dowlatabadi et al., 2017; Farag et al., 2018; Saviano et al., 2019; Suzuki et al., 2022). Despite the use of alternative techniques, such as microwave-assisted extraction (MAE, Sobolev et al., 2014), UAE (Jung et al., 2012; Awin et al., 2016; Gad and Bouzabata, 2017; Windarsih, Rohman and Swasono, 2018, 2019; Jo et al., 2020; Horn et al., 2021; Mandrone et al., 2021; Rivera-Pérez, Romero-González and Garrido Frenich, 2022) has become prevalent for metabolite extraction from spices and herbs for further ^{1}H NMR fingerprinting (Table 10.3; Figure 10.2).

Table 10.3 Summary of ^{1}H NMR-Based Metabolomics Approaches Used for Authenticity Assessment of Culinary Herbs, Spices, and Seasonings

Matrix	Purpose of Study	Extraction Method	NMR Solvent	Chemometric Analysis	References
Paprika (*Capsicum annuum* L.)	Adulterant detection (Azorubine, Ponceau 4R, beetroot, sumac)	UAE (MeOH-*d4* and 180 mM D_2O phosphate buffer mixture at pH 7.0, 1:1 v/v, 10 min)	MeOH-*d4*:D_2O (1:1 v/v)	DD-SIMCA	(Horn et al., 2021)
Turmeric (*Curcuma longa*)	Adulterant detection (*Curcuma heyneana*)	UAE (MeOH-*d4* and D_2O phosphate buffer mixture at pH 6.0, 1:1 v/v, 20 min)	MeOH-*d4*:D_2O (1:1 v/v)	PCA, OPLS-DA	(Windarsih, Rohman and Swasono, 2019)
Turmeric (*Curcuma longa*)	Adulterant detection (*Curcuma manga*)	UAE (MeOH-*d4* and D_2O phosphate buffer mixture at pH 6.0, 1:1 v/v, 20 min)	MeOH-*d4*:D_2O (1:1 v/v)	PCA, OPLS-DA	(Windarsih, Rohman and Swasono, 2018)
Saffron (*Crocus sativus* L.)	Adulterant detection (safflower and synthetic dyes)	SLE (100 mM D_2O phosphate buffer at pH 7.0, 5 min)	D_2O	PCA, PLS-DA	(Dowlatabadi et al., 2017)
Saffron (*Crocus sativus* L.)	Adulterant detection (*Crocus sativus* stamens, safflower, turmeric, gardenia)	SLE (DMSO-d_6, 3 min)	DMSO-d_6	OPLS-DA, O2PLS-DA	(Petrakis et al., 2015)
Saffron (*Crocus sativus* L.)	Quality control (Italian PDO vs. commercial saffron)	SLE (DMSO-d_6, 3 min)	DMSO-d_6	OPLS-DA	(Cagliani et al., 2015)

Matrix	Purpose of Study	Extraction Method	NMR Solvent	Chemometric Analysis	References
Thyme (*Thymus vulgaris* L.)	Quality control	Maceration (DMSO-d_6, 24 h)	DMSO-d_6	PCA, HCA	(Pieri et al., 2012)
Cinnamon barks (*Cinnamomum cassia, C. verum/dubium*)	Geographical origin discrimination (Vietnam, China, Sri Lanka)	SLE (MeOH, 1 h)	DMSO-d_6	PCA, HCA	(Suzuki et al., 2022)
Black pepper (*Piper nigrum* L.)	Geographical origin discrimination (Sri Lanka, Vietnam, Brazil) Processing practices (sterilized vs. non-sterilized)	UAE ($CDCl_3$ or MeOH-*d4*, 10 min)	$CDCl_3$, MeOH-*d4*	PCA, OPLS-DA	(Rivera-Pérez, Romero-González and Garrido Frenich, 2022)
Allium species (garlic, onion)	Geographical origin discrimination (South Korea, China)	UAE (10 mM D_2O phosphate buffer at pH 7.2, 5 min)	D_2O	PCA	(Jo et al., 2020)
Turmeric (*Curcuma longa*)	Geographical origin discrimination (Algeria, Egypt)	UAE (DMSO-d_6, 15 min)	DMSO-d_6	PCA, HCA	(Gad and Bouzabata, 2017)
Saffron (*Crocus sativus* L.)	Geographical origin discrimination (Greece, Spain, Hungary, Turkey, Italy)	MAE (MeOH-*d4*, 40 °C, 10–30 min)	MeOH-*d4*	PCA	(Sobolev et al., 2014)
Curcuma species	Processing practices (drying)	UAE (MeOH-*d4* and D_2O phosphate buffer mixture at pH 6.0, 1:1 v/v, 30 min)	MeOH-*d4*:D_2O (1:1 v/v)	PCA	(Awin et al., 2016)
Black garlic (*Allium sativum* L.)	Processing practices	Maceration (10% DMSO-d_6/90% D_2O, 1 h)	DMSO-d_6:D_2O	PCA	(Liang et al., 2015)
Origanum species	Species discrimination (*O. vulgare* vs. *O. onites*)	UAE (MeOH-*d4* and 90 mM D_2O phosphate buffer mixture at pH 6.0, 1:1 v/v, 20 min)	MeOH-*d4*:D_2O (1:1 v/v)	PCA, OPLS-DA, PLS-DA	(Mandrone et al., 2021)
Chili pepper (*Capsicum annuum*)	Species discrimination (Mexico)	Plant material squeezing	D_2O	PCA, HCA, LDA	(Florentino-Ramos et al., 2019)
Cinnamomum species	Species discrimination (*C. cassia* vs. *C. verum*)	SLE (MeOH)	MeOH-*d4*	PCA, OPLS-DA	(Farag et al., 2018)

Table 10.3 Summary of ^{1}H NMR-Based Metabolomics Approaches Used for Authenticity Assessment of Culinary Herbs, Spices and Seasonings (Continued)

Matrix	Purpose of Study	Extraction Method	NMR Solvent	Chemometric Analysis	References
Ocimum species	Species discrimination	Essential oil extraction by hydrodistillation (Clevenger, 2 h)	$CDCl_3$	HCA, PCA, PLS-DA	(Freitas et al., 2018)
Curcuma species	Species discrimination (*C. aromatica* vs. *C. longa*)	UAE (MeOH-*d4* and 0.2 M D_2O phosphate buffer mixture at pH 7.0, 40°C, 20 min)	MeOH-*d4*:D_2O	PCA, OPLS-DA	(Jung et al., 2012)
Serrano pepper (*Capsicum annuum* 'Serrano')	Discrimination within varieties (Mexico)	Plant material squeezing	D_2O	PCA, HCA, PLS-DA	(Villa-Ruano et al., 2019)
Onion (*Allium cepa* L.)	Discrimination within varieties	SLE (MeOH, H_2O, $CHCl_3$, 10 min)	$CDCl_3$, D_2O	PCA, PLS-DA	(Saviano et al., 2019)
Serrano pepper (*Capsicum annuum* L.)	Discrimination within varieties (Mexico)	Plant material squeezing	D_2O	PCA, OPLS-DA	(Becerra-Martínez et al., 2017)
Saffron (*Crocus sativus* L.)	Storage time effect	SLE (DMSO-d_6, 3 min)	DMSO-d_6	PLS-DA, OPLS-DA	(Consonni et al., 2016)
Saffron (*Crocus sativus* L.)	Storage time effect	SLE (DMSO-d_6, 3 min)	DMSO-d_6	OPLS-DA	(Ordoudi et al., 2015)

Abbreviations: ^{1}H NMR, proton nuclear magnetic resonance spectroscopy; $CDCl_3$, deuterated chloroform; $CHCl_3$, chloroform; D_2O, deuterium oxide; DD-SIMCA, data-driven soft independent model class analogy; DMSO, dimethylsulfoxide; DMSO-*d6*, deuterated dimethylsulfoxide; HCA, hierarchical cluster analysis; LDA, linear discriminant analysis; MAE, microwave-assisted extraction; MeOH, methanol; MeOH-*d4*, deuterated methanol; O2PLS-DA, bidirectional orthogonal partial least squares–discriminant analysis; OPLS-DA, orthogonal partial least squares–discriminant analysis; PCA, principal component analysis; PLS-DA, partial least squares–discriminant analysis; SLE, solid-liquid extraction; UAE, ultrasound-assisted extraction.

Concerning extracting solvents, condiment lipophilic extracts are widely obtained using non-polar deuterated solvents such as deuterated chloroform ($CDCl_3$; Freitas et al., 2018; Saviano et al., 2019; Rivera-Pérez, Romero-González and Garrido Frenich, 2022) or deuterated dimethylsulfoxide (DMSO-*d6*; Pieri et al., 2012; Cagliani et al., 2015; Ordoudi et al., 2015; Petrakis et al., 2015; Consonni et al., 2016; Gad and Bouzabata, 2017; Suzuki et al., 2022; Table 10.3). Rivera-Pérez et al. applied a ^{1}H NMR fingerprinting approach for black pepper authentication and thus, both polar and lipophilic UAE-obtained extracts were investigated to obtain maximum compositional information about major compounds of black pepper (Rivera-Pérez, Romero-González and Garrido Frenich, 2022). In this study, $CDCl_3$ was selected as the best option for the assessment of lipophilic extracts instead of DMSO-*d6*, since the last one is more viscous, causing severe difficulties during the sample filtration, and longer ^{1}H NMR experiment times are needed as a consequence of suppression water signal required in ^{1}H NMR (DMSO-*d6*) measurements (Rivera-Pérez, Romero-González and Garrido Frenich, 2022). $CDCl_3$-based ^{1}H NMR analysis has also been used to discriminate between onion varieties (*Allium cepa* L.; Saviano et al., 2019) and *Ocimum* species (Freitas et al.,

2018). Tetramethylsilane (TMS) is commonly used in $CDCl_3$-based ^{1}H NMR fingerprint analysis as the reference standard for NMR signal referencing at δ 0.0 ppm (Freitas et al., 2018).

Other studies highlighted the suitability of DMSO-*d6* as the extraction solvent for quality control of cinnamon barks (Suzuki et al., 2022); turmeric (Gad and Bouzabata, 2017); saffron (Cagliani et al., 2015; Ordoudi et al., 2015; Petrakis et al., 2015; Consonni et al., 2016); and thyme (Pieri et al., 2012; Table 10.3). DMSO-*d6* allowed the simultaneous extraction of both lipophilic and hydrophilic compounds, as well as the obtention of NMR signals with sharp line width (Cagliani et al., 2015; Petrakis et al., 2015). Different reference material has been described for quantitative ^{1}H NMR or chemical shift referencing in ^{1}H NMR (DMSO-*d6*) measurements such as sodium 3-(trimethylsilyl)-1-propane-1,1,2,2,3,3-*d6*-sulfonate (DSS-*d6*) (Suzuki et al., 2022); 3-(trimethylsilyl) propionic-2,2,3,3-*d4* acid sodium salt (TMSP-2,2,3,3-*d4*) (Gad and Bouzabata, 2017); or anthracene (Pieri et al., 2012).

Regarding the obtention of polar black pepper extracts, deuterated methanol (MeOH-*d4*) allowed the extraction of a higher number of ^{1}H NMR signals compared with other polar solvents such as deuterium oxide (D_2O)-based phosphate buffer solution (pH 7.0), and a mixture of D_2O phosphate buffer:MeOH-*d4* (1:1 v/v; Rivera-Pérez, Romero-González and Garrido Frenich, 2022; Table 10.3). Furthermore, MeOH-*d4* has been utilized as the extracting solvent of choice to obtain polar extracts of *Cinnamomum* species by SLE (Farag et al., 2018) or saffron by MAE (Sobolev et al., 2014). It is noteworthy that hexamethyldisiloxane (HMDS) and especially TMSP-2,2,3,3-*d4*, are reference materials widely used for chemical shift referencing in ^{1}H NMR (MeOH-*d4*) measurements (Sobolev et al., 2014; Farag et al., 2018; Rivera-Pérez, Romero-González and Garrido Frenich, 2022).

Merging the extraction power of a two-polar solvent mixture consisting of MeOH-*d4* and D_2O-based phosphate buffer has been especially reported for the assessment of polar extracts of spices and herbs such as paprika (Horn et al., 2021; *Origanum* species (Mandrone et al., 2021; turmeric (Windarsih, Rohman and Swasono, 2018, 2019); and *Curcuma* species (Jung et al., 2012; Awin et al., 2016; Table 10.3). Horn et al. described the preparation of the buffer solution by solving disodium hydrogen phosphate dihydrate ($Na_2HPO_4{\cdot}2H_2O$) and potassium dihydrogen phosphate (KH_2PO_4) in D_2O (Horn et al., 2021). Then, the apparent pH of aqueous extracts was usually adjusted to pH = 6.0 (Awin et al., 2016; Windarsih, Rohman and Swasono, 2018, 2019; Mandrone et al., 2021) or pH = 7.0 (Jung et al., 2012; Horn et al., 2021) using 0.1 M hydrochloric acid (HCl) or sodium hydroxide (NaOH) solutions. The use of phosphate buffer salts, as well as the pH adjustment, is highly recommended to deal with the acidity of plant extracts and control their negative impact in variations of chemical shifts (Deborde et al., 2019).

Aqueous extracts of *Allium* species (garlic and onion; Jo et al., 2020) and saffron (Dowlatabadi et al., 2017) were assessed using a D_2O phosphate buffer solution (10–100 mM, pH 7.0–7.2; Table 10.3). Aqueous extracts of chili pepper (Florentino-Ramos et al., 2019) and serrano pepper (Becerra-Martínez et al., 2017; Villa-Ruano et al., 2019) were also obtained using D_2O as the extraction solvent. In this regard, extraction with non-buffered D_2O included the addition of ethylenediaminetetraacetic acid (EDTA) to achieve high resolution and optimize line widths of ^{1}H NMR signals, as well as to limit the effects of the paramagnetic ions found in plant issues (Deborde et al., 2019; Florentino-Ramos et al., 2019). Besides, sodium azide (NaN_3) is used as an antibacterial agent to inhibit bacterial growth (Dona et al., 2016).

10.3 ANALYTICAL AND CHEMOMETRIC TOOLS USED FOR ADULTERANT DETECTION AND QUALITY CONTROL OF CONDIMENTS

10.3.1 Detection of Adulteration and Quality Control by GC-Based Metabolomics Approaches

According to the revised bibliography (Table 10.1), volatile GC-based fingerprinting coupled to MS detection has become the preferred analytical technique for adulterant detection in condiments (Gul et al., 2018; Farag et al., 2020; Di Donato et al., 2021), although the use of GC combined with FID has also been reported (Morozzi et al., 2019).

GC-MS fingerprinting was used to differentiate methanolic extracts of black pepper from its common adulterant (papaya seeds; Gul et al., 2018; Table 10.1). Metabolic profiling revealed that black pepper and papaya composition consisted of fatty acids, organic acids, amino acids, and sugars, among others, at different ratios. Fingerprints were log-transformed for further multivariate data analysis by PCA and PLS-DA, allowing the discrimination between black pepper and its adulterant. The variable importance in projection (VIP) approach was used to reveal piperine and cholest-4-en-3-one as reliable markers. Higher concentrations of piperine were characteristic of black pepper samples, whereas cholest-4-en-3-one was a significant metabolite of papaya composition (Gul et al., 2018).

Untargeted GC-MS profiling in combination with chemometrics allowed the discrimination of saffron from its common adulterants: *Calendula officinalis* L. (calendula) and *Carthamus tinctorius* L. (safflower; Farag et al., 2020; Table 10.1). OPLS-DA was used for successful saffron vs. adulterants differentiation, showing high model goodness-of-fit (R^2 = 0.96–0.99) and predictability (Q^2 = 0.91–0.99). In this study, safranal was revealed as an exclusive marker to distinguish saffron from safflower, whereas calendula was characterized by the enrichment in monoterpenes, such as sabinene, α-phellandrene, α/β-pinene, and α-thujene (Farag et al., 2020). In another study, HS-SPME-GC-MS volatile profiling was combined with PLS-DA for distinguishing Italian saffron samples from adulterated samples with calendula, safflower, and turmeric (Di Donato et al., 2021). This metabolomics approach was able to detect saffron adulteration with turmeric or calendula at 2% (w/w) and safflower 3% (w/w) adulteration levels, respectively, in comparison with 5%–20% (w/w) adulteration levels detected by untargeted spectroscopic methods (Di Donato et al., 2021).

GC-MS fingerprinting has been widely applied to the quality control of different seasonings including nutmeg (*Myristica fragrans*; Farag, Mohsen and El-Gendy, 2018) and other different spices and herbs (e.g., anise, clove, rosemary; Sgorbini et al., 2015; Matsushita et al., 2018; Table 10.1). Farag et al. developed a method for quality control of nutmeg by HS-SPME-GC-MS volatile profiling to differentiate between most commercially nutmeg form (seeds) from other plant parts (arillus and fruits; Farag, Mohsen and El-Gendy, 2018). HS-SPME-GC-MS-based methods were proposed for routine quality control of six spices and herbs including clove (*Syzygium aromaticum* L.), American peppertree (*Schinus molle* L.), black pepper and white pepper (*Piper nigrum* L.), rosemary (*Rosmarinus officinalis* L.), sage (*Salvia officinalis* L.) and thyme (*Thymus vulgaris* L.; Sgorbini et al., 2015). In this study, two GC approaches for chemotype differentiation were assessed in combination with PCA. The first one corresponded to untargeted HS-SPME-GC-MS fingerprinting, whereas the second one was a targeted approach based on quantification of key markers selected for each plant matrix (Sgorbini et al., 2015). As an example, thymol and carvacrol were selected as significant markers for thyme, eugenol was for clove, and linalool and bornyl acetate (among others) were selected for rosemary characterization (Sgorbini et al., 2015). Similarly, authentication of other commercial condiments (e.g., allspice, coriander, star anise) was achieved by the obtention of HS-SPME-GC-MS fingerprints (Matsushita et al., 2018). In this case, unsupervised chemometric techniques (HCA and PCA) provided enough discrimination potential to assess differences among the fingerprints, showing well-differentiated clusters among the tested samples according to their distinctive major components (Matsushita et al., 2018).

10.3.2 Detection of Adulteration and Quality Control by LC-Based Metabolomics Approaches

UHPLC is a reliable separation technique that leads to shorter analysis times, better resolution, and sensitivity compared with conventional LC, especially in complex food matrices such as plant-derived condiments (Rivera-Pérez, Romero-González and Garrido Frenich, 2021c). UHPLC in combination with HRMS detection analyzers has been used for adulterant detection in saffron (Senizza et al., 2019; Hegazi et al., 2022), and oregano samples (Black et al., 2016; Table 10.2).

Regarding saffron adulteration, UHPLC-HRMS fingerprinting and supervised OPLS-DA modeling allowed the discrimination of saffron from adulterated samples with its allied flowers (safflower and calendula; Hegazi et al., 2022). The chemometric analysis resulted in two OPLS-DA models to distinguish saffron from these two adulterants with high goodness-of-fit and predictability (R^2 values of 0.995–0.998 and Q^2 values of 0.972–0.976, respectively; Hegazi et al., 2022; Figure 10.3A–C; Table 10.2). Oxo-hydroxy-undecenoic acid-*O*-hexoside was identified as a unique marker for authentic saffron (Figure 10.3B), whereas safflower was enriched in kaempferol-*O*-neohesperidoside and N,N,N,tri-*p*-coumaroyl spermidine. Furthermore, calendulaglycoside C and di-*O*-caffeoyl quinic acid isomer were pointed out as markers indicative of saffron adulterated with calendula (Hegazi et al., 2022; Figure 10.3D).

Another UHPLC-HRMS (QTOF) metabolomics-based approach was used to assess the discrimination between adulterated samples (with different other flower parts, i.e., stamens and tepals) and authentic saffron (Senizza et al., 2019; Table 10.2). Supervised OPLS-DA led to a statistical model characterized with excellent R^2 and Q^2 parameters of 0.99 that allowed discriminating authentic saffron from stamen or tepal-adulterated samples. In this study, the VIP approach revealed 77 metabolites able to differentiate the authentic saffron from the adulterated one: 36% of the

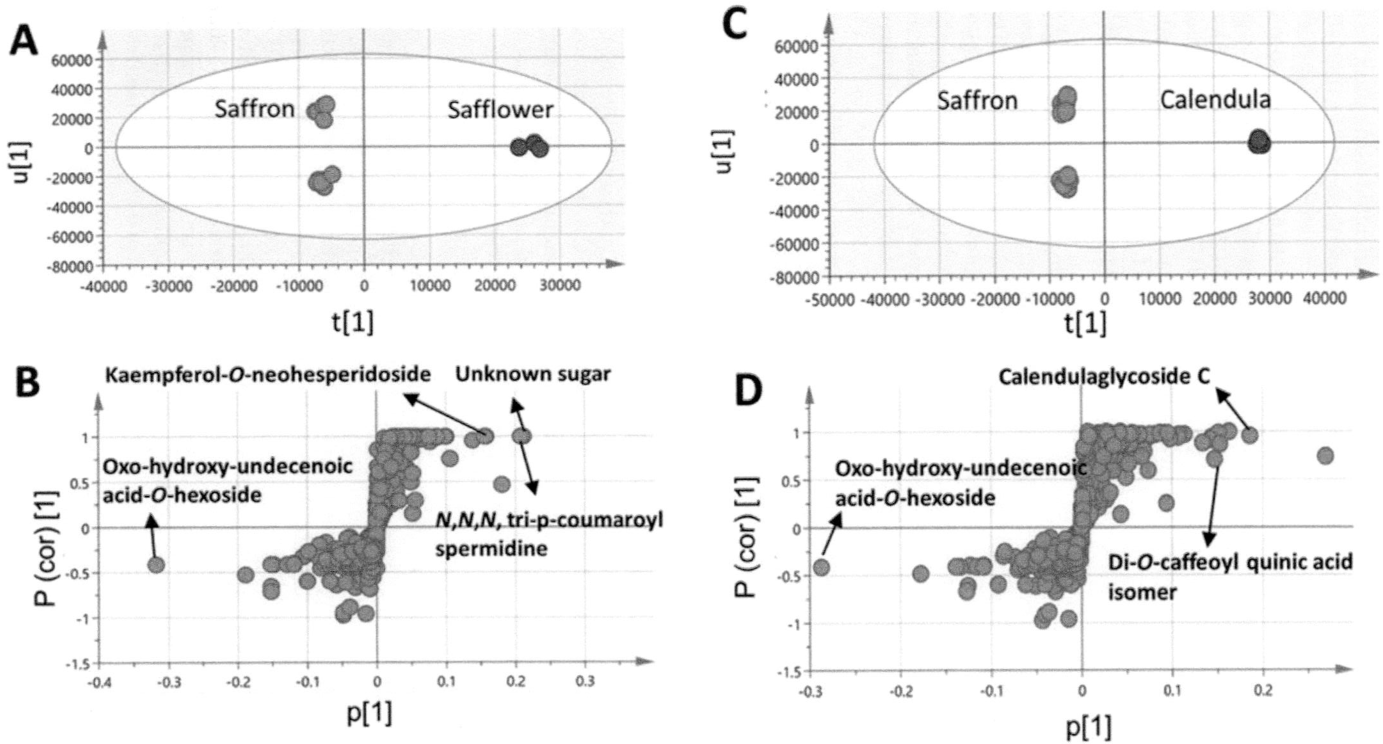

Figure 10.3 OPLS-DA score plots derived from UHPLC-HRMS data of saffron samples vs. common adulterants. (a) Safflower and (c) calendula; S-plots from safflower (b) and calendula (d), showing the covariance p[1] against the correlation p(cor)[1] of the most discriminating variables.

Source: Reprinted with permission from Hegazi et al. (2022). Copyright 2022 Elsevier.

annotated markers were flavonoids, 16% were lignans, phenolic acids accounted for 13% and the remaining VIP markers were other miscellaneous compounds (Senizza et al., 2019).

Profiling analysis of oregano and adulterants (olive, myrtle, hazelnut, sumac and cistus leaves) was performed using untargeted UHPLC-HRMS (QTOF) analysis (Black et al., 2016; Table 10.2). HRMS data were acquired using electrospray ionization (ESI) operating in positive and negative ion modes. OPLS-DA was then carried out and two models were generated using positive and negative-acquired data for reliable oregano versus adulterants differentiation ($R^2 = 0.993–0.997$ and $Q^2 = 0.944–0.949$; Black et al., 2016).

Besides, LC-based strategies have been applied for the quality control of different seasonings including turmeric and curry (Núñez et al., 2020; Vidal-Casanella et al., 2020); ginger (Feng et al., 2014); black pepper (Gu et al., 2018); and other widely used culinary herbs and spices such as rosemary, thyme and cumin (Vallverdú-Queralt et al., 2014; Table 10.2). Untargeted metabolomics focused on phenolic and polyhydroxy profiling was applied for quality control of black pepper by UHPLC and tandem mass spectrometry (MS/MS) analysis using a hybrid triple quadrupole-linear ion trap (QTRAP) analyzer for metabolite identification, and data were further treated performing PCA (Gu et al., 2018).

Vidal-Casanella et al. exploited polyphenolic and curcuminoid fingerprinting obtained by classical HPLC-DAD in combination with unsupervised PCA patterning for quality control of turmeric and curry, respectively (Vidal-Casanella et al., 2020; Table 10.2). Fingerprints were recorded at $\lambda = 280$ nm for phenolic compounds and $\lambda = 420$ nm for curcuminoids. HPLC-DAD fingerprinting combined with chemometrics (PCA, HCA) has also been successfully applied for quality control of methanolic extracts of ginger (Feng et al., 2014). In this study, 280 nm was selected as the detection wavelength (Feng et al., 2014). Recently, a new analytical strategy based on polyphenolic and curcuminoid profiling by UHPLC separation coupled to HRMS detection (using a hybrid quadrupole (Q)-Orbitrap analyzer) and PLS-DA chemometric treatment was applied to authenticate turmeric and curry samples (Núñez et al., 2020). Well-differentiated clusters were observed within the curry and turmeric samples considering the first two latent variables (LV1 vs. LV2). Loading plots revealed that 4-*O*-caffeoylquinic acid, chlorogenic acid, nepetin-7-glucoside, rutin and *D*-(-)-quinic acid were relevant markers of curry samples, whereas some polyphenolic compounds such as *trans*-cinnamic, ferulic and sinapic acids help on the differentiation of turmeric samples (Núñez et al., 2020). Other HRMS analyzer, such as the hybrid linear ion trap quadrupole-Orbitrap one (LTQ-Orbitrap-HRMS), was applied to study the phenolic profile of widely used culinary herbs and spices (rosemary, thyme, oregano, cinnamon, cumin and bay; Vallverdú-Queralt et al., 2014). This analytical tool in combination with unsupervised PCA showed that catechin, epicatechin and quercetin are highly correlated with cumin and cinnamon samples. Higher levels of *p*-coumaric and ferulic acids were characteristic of bay, thyme and oregano, while rosmarinic, caffeic, syringic and *p*-hydroxybenzoic acids were characteristic of rosemary samples (Vallverdú-Queralt et al., 2014).

10.3.3 Detection of Adulteration and Quality Control by ^{1}H NMR-Based Metabolomics Approaches

^{1}H NMR fingerprinting has been used for adulterant detection and quality control of different aromatic herbs and species including thyme (Pieri et al., 2012); paprika (Horn et al., 2021), turmeric (Windarsih, Rohman and Swasono, 2018, 2019); and saffron (Cagliani et al., 2015; Petrakis et al., 2015; Dowlatabadi et al., 2017; Table 10.3).

The suitability of ^{1}H NMR metabolic fingerprinting to assess adulteration of paprika has been recently described (Horn et al., 2021; Table 10.3). This study demonstrated the potential of data-driven soft independent model class analogy (DD-SIMCA) to discriminate authentic paprika from adulterated one with Azorubine, Ponceau 4R, beetroot and sumac, achieving reliable discrimination from Azorubine and Ponceau 4R azo dyes at adulteration levels of 0.1%, from beetroot at 10% and from sumac powder at 20% (Horn et al., 2021). ^{1}H NMR profiling combined with PCA was a suitable strategy for quality control of *Thymus vulgaris* L. samples in terms of chemotype discrimination (thymol, carvacrol and linalool chemotypes; Pieri et al., 2012). Besides, a targeted approach based on quantitative ^{1}H NMR (q ^{1}H NMR) was developed for the determination of minor thyme significant metabolites, such as *p*-cymene, γ-terpinene, rosmarinic acid, and 3,4,3′,4′-tetrahydroxy-5,5′-diisopropyl-2,2′-dimethylbiphenyl (Pieri et al., 2012).

Polar turmeric (*Curcuma longa*) extracts were analyzed by ^{1}H NMR fingerprinting and combined with chemometric tools (PCA and OPLS-DA) to detect turmeric adulteration with other *Curcuma* species, namely *C. heyneana* (Windarsih, Rohman and Swasono, 2019) and *C. manga* (Windarsih,

Rohman and Swasono, 2018; Table 10.3). The model established to discriminate turmeric (*C. longa*) from *C. heyneana* was characterized by good goodness-of-fit and predictability parameters (R^2 = 0.992 and Q^2 = 0.899, respectively; Windarsih, Rohman and Swasono, 2019). Besides, ^{1}H NMR fingerprinting showed specific signals in the aromatic region that may be used to discriminate between *C. longa* and *C. heyneana*. In this regard, the intensity of curcuminoid signals at δ 6.93 and δ 7.57 ppm (aromatic region) decreased in turmeric extracts with the increasing adulteration with *C. heyneana*, whereas an increased signal at δ 8.46 ppm (formic acid) is observed as *C. heyneana* adulteration increases (Windarsih, Rohman and Swasono, 2019; Figure 10.4). These results were in line with previous findings focused on turmeric adulteration with *C. manga* since adulteration with *C. manga* was also revealed by lower signal intensities in the aromatic region (δ 6.0–8.0 ppm) in comparison with authentic *C. longa* extracts (Windarsih, Rohman and Swasono, 2018).

^{1}H NMR spectroscopy together with chemometric multivariate data analysis has also been described for reliable adulterant detection (Petrakis et al., 2015; Dowlatabadi et al., 2017) and quality control (Cagliani et al., 2015) of saffron (Table 10.3). PLS-DA modeling showed clear clustering to discriminate synthetic dyes (tartrazine and naphthol yellow) and safflower from authentic Iranian saffron (Dowlatabadi et al., 2017). Authentic Greek saffron and four typical plant-derived adulterants (i.e., *Crocus sativus* stamens, safflower, turmeric, gardenia) were also investigated by ^{1}H NMR fingerprinting in combination with supervised OPLS-DA (R^2 = 0.945 and Q^2 = 0.923) and bidirectional O2PLS-DA (R^2 = 0.976 and Q^2 = 0.960; Petrakis et al., 2015). The S-plots generated from the OPLS-DA model highlighted higher picrocrocin (δ 1.12, 1.16, 2.08, 4.28, and 10.04 ppm) and crocin (δ 1.96, 4.16, 5.40, 6.52, 6.64, 6.84, and 7.32 ppm) levels in authentic Greek saffron compared with adulterated samples, which contained higher concentrations of fatty acids (δ 1.20 and 1.24 ppm;

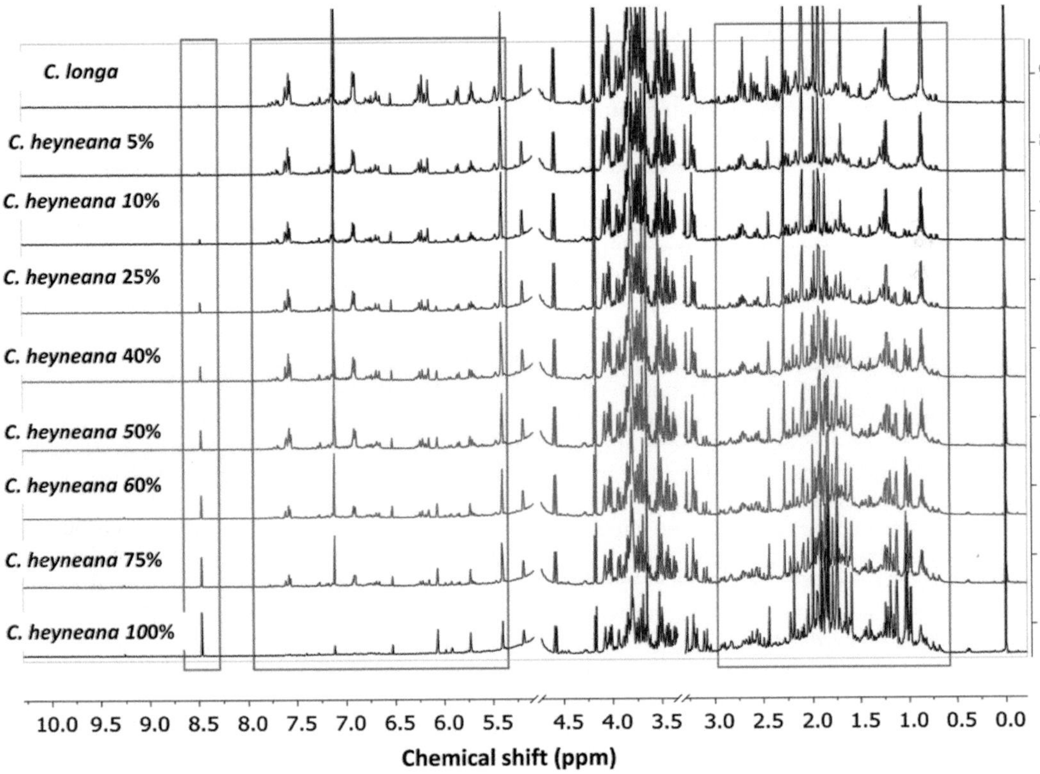

Figure 10.4 ^{1}H NMR spectra of *Curcuma longa* (turmeric) adulterated with *C. heyneana*.

Source: Reprinted with permission from Windarsih, Rohman and Swasono (2019). Copyright 2019 Elsevier.

Petrakis et al., 2015). Italian Protected Designation of Origin (PDO) saffron from L'Aquila, San Gimignano, and Sardinia were successfully differentiated by commercial saffron samples using OPLS-DA modeling ($R^2 = 0.988$ and $Q^2 = 0.846$; Cagliani et al., 2015). Accordingly, the enrichment in picrocrocin and crocins were observed in Italian PDO saffron compared with higher fatty acid content characteristic of commercial samples (Cagliani et al., 2015), in good agreement with other saffron studies (Petrakis et al., 2015).

10.4 ANALYTICAL AND CHEMOMETRIC TOOLS USED FOR GEOGRAPHICAL ORIGIN AND/OR PROCESSING DISCRIMINATION OF CONDIMENTS

10.4.1 Origin and/or Processing Discrimination by GC-Based Metabolomics Approaches

It is well-known that plant primary and secondary metabolism are highly influenced by environmental conditions resulting in a characteristic metabolic composition that can be further assessed for geographical traceability of the seasoning (Manukyan, 2019). Metabolomics approaches based on GC-MS have been widely applied for comprehensive characterization of plant-derived condiments such as thyme (Satyal et al., 2016); black cumin (Farag et al., 2017); saffron (Karabagias et al., 2017; D'Archivio et al., 2018; Liu et al., 2018; Rocchi et al., 2019; Farag et al., 2020); sumac (Farag, Fayek and Reidah, 2018); and black pepper (Liang et al., 2021; Rivera-Pérez, Romero-González and Garrido Frenich, 2021d), allowing the reliable differentiation of samples based on their region of production (Table 10.1). Besides, post-harvest treatments applied to ensure the highest quality of condiments (e.g., drying, sterilization, or grinding processes) lead to metabolic variations in spice composition. In this regard, the combination of GC-MS and chemometrics was capable of distinguishing metabolic changes in spices and herbs produced after different post-harvest processes including sterilization (Rivera-Pérez, Romero-González and Garrido Frenich, 2021d); drying (Macchia et al., 2013; Rocchi et al., 2019); and roasting (Farag et al., 2017; Farag, Fayek and Reidah, 2018; Farag, Mohsen and El-Gendy, 2018).

The application of GC-MS was found useful, for example, in the determination of the origin of saffron samples from Greece, Spain, Iran and Morocco by HS-SPME-GC-MS characterization combined with LDA (Table 10.1), resulting in a correct prediction rate of 95.5% (Karabagias et al., 2017). In this study (Karabagias et al., 2017), 1,3-dimethyl-benzene, 2,4-dimethyl-benzaldehyde, and safranal (among others) were highlighted as the metabolites most important to geographical discrimination between origins. In another study (D'Archivio et al., 2018), geographical classification of saffron samples from L'Aquila (Italy), Iran and Morocco by LDA led to a 91% correct discrimination rate using aroma profiling obtained by HS-SPME-GC-MS. Other chemometric supervised methods such as OPLS-DA have been successfully used for Iranian-Chinese (Liu et al., 2018) or Iranian-Spanish (Farag et al., 2020) geographical differentiation of saffron. In the study of Rocchi et al., HS-SPME-GC-MS fingerprinting and PLS-DA revealed that the aldehyde pattern could be used to differentiate the origin of saffron within the following tested countries: Italy, Argentina, Greece, Iran, Lebanon, India, Spain and Turkey (Rocchi et al., 2019).

Regarding other high-value condiments, black pepper samples were successfully classified according to their geographical origin (Brazil, India, Indonesia and Vietnam) by GC-MS fingerprinting and multivariate data analysis by cubit-root data transformation and PLS-DA, but the identification of markers was not attempted (Liang et al., 2021; Table 10.1). In contrast, an innovative and high-throughput chemometric approach was recently applied for the geographical traceability and processing discrimination of black pepper, as well as for the identification of origin volatile markers (Rivera-Pérez, Romero-González and Garrido Frenich, 2021d). The research of Rivera-Pérez et al. showed the feasibility of a novel metabolomics approach based on GC-HRMS (Orbitrap analyzer) fingerprinting combined with supervised and unsupervised chemometric techniques (PCA, HCA and OPLS-DA) for the authentication of black pepper in terms of origin (Brazil, Sri Lanka and Vietnam) and processing (sterilized and non-sterilized black pepper). The PCA score plot (Figure 10.5A) and HCA dendrogram (Figure 10.5B) of black pepper GC-HRMS fingerprints showed a well-differentiated cluster for black pepper from Sri Lanka on the first two components and a tight cluster for Vietnamese samples, whereas two Brazilian subgroups were observed most likely due to the influence of different micro-climates found in the large Brazil country. Sterilized samples were highly associated with their corresponding non-sterilized ones according to PCA and HCA group patterns (Figure 10.5A–B). Further supervised methods (OPLS-DA) pointed out excellent discrimination of the samples considering the three geographical origins investigated (Figure 10.5C), and non-sterilized versus sterilized black

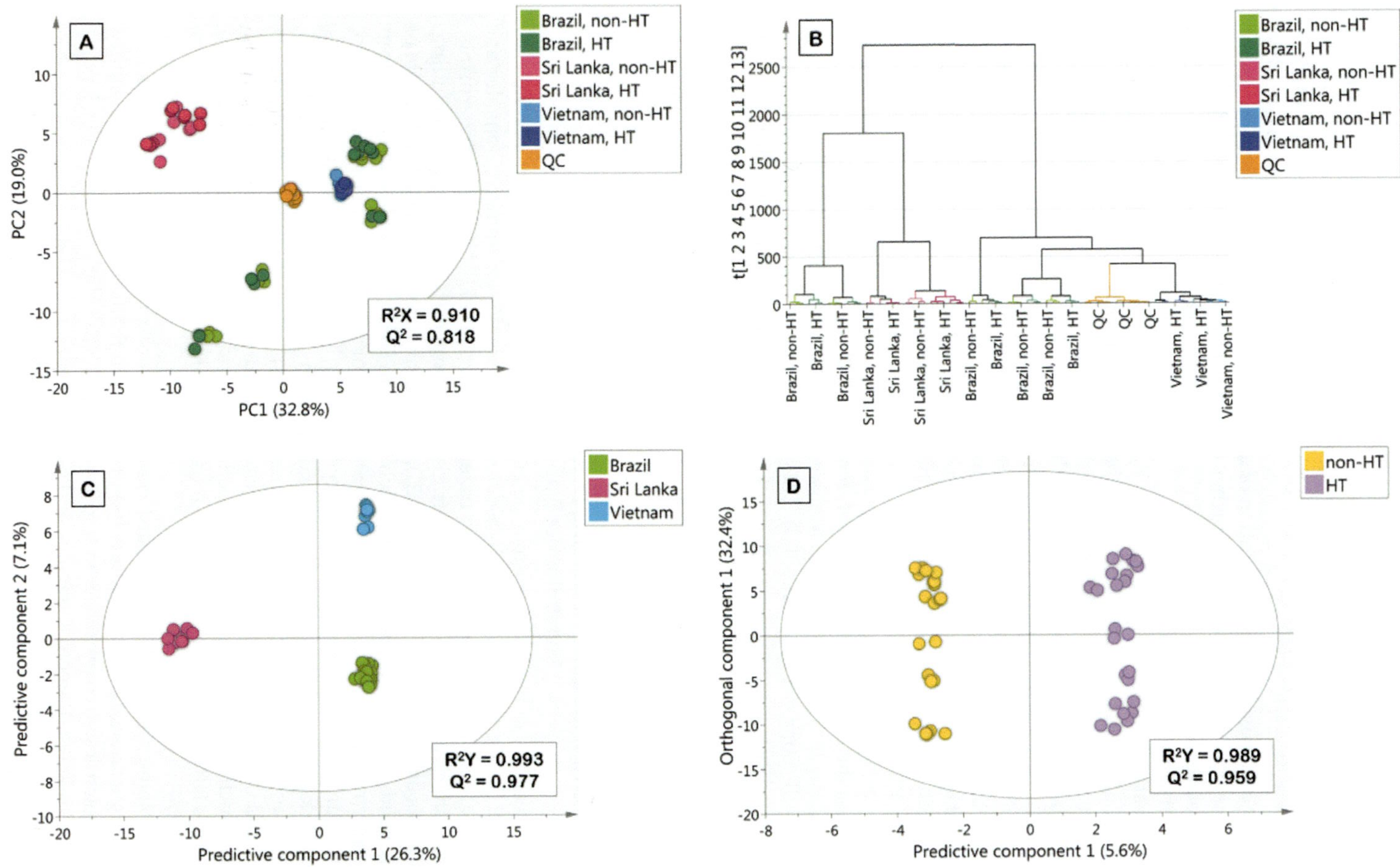

Figure 10.5 (a) PCA score plot of GC-Orbitrap-HRMS fingerprints of black pepper; (b) dendrogram obtained by HCA; (c) OPLS-DA score plot for black pepper samples according to the geographical origin (Brazil, Sri Lanka and Vietnam); and (d) OPLS-DA score plot for non-sterilized (non-HT) vs. sterilized (HT) black pepper samples.

Source: Reprinted with permission from Rivera-Pérez, Romero-González and Garrido Frenich (2021d). Copyright 2021 American Chemical Society.

pepper comparison (Figure 10.5D), with high model goodness-of-fit and predictability (R^2 and Q^2 parameters >0.9).

Furthermore, the VIP approach revealed 12 key markers having the highest discrimination potential, including sesquiterpenes (α-cubebene), alkenylbenzenes (eugenol and *trans*-anethole), diterpenes (dehydroabietal), monoterpenes (cuminaldehyde), and amides, among others. In this regard, the use of an HRMS analyzer has the advantage of providing accurate mass acquisition to reduce the number of candidates per discriminant compound, and thus, reliable putative identification of markers was achieved in comparison with LRMS-based approaches (Luca et al., 2021). It is worth noting that this research (Rivera-Pérez, Romero-González and Garrido Frenich, 2021d) is the only recent GC study focused on the application of an HRMS-based metabolomics approach (Table 10.1) for condiment authentication. Thus, further investigations about the feasibility of HRMS-based methods are demanding to trace the origin and processing of seasonings, as well as for the identification of reliable markers.

The post-harvest treatment based on the drying of the stigmas has a critical influence on the final quality of saffron (Macchia et al., 2013). The metabolic effects produced as a result of the drying of saffron were assessed by HS-SPME-GC-MS in combination with analysis of variance (ANOVA) (Macchia et al., 2013) and PLS-DA (Rocchi et al., 2019; Table 10.1). ANOVA results revealed that safranal content in saffron increased with the drying exposure times, while at the same time, longer drying exposure led to overall decreased identified compounds (Macchia et al., 2013). In contrast, Rocchi et al. study reported no variations in safranal levels during drying, whereas it pointed out some saffron aldehyde volatile compounds (1,3,4-trimethyl-3-cyclohexene-1-carboxaldehyde, octanal, and 2,6,6-trimethyl-1-cyclohexene-1-carboxaldehyde) as markers for drastic drying conditions (Rocchi et al., 2019).

Untargeted GC-MS metabolomics and multivariate data analysis (OPLS-DA) was widely applied in recent years to assess the impact of roasting treatment (120–180°C for 15–120 min) on sumac (Farag, Fayek and Reidah, 2018); nutmeg (Farag, Mohsen and El-Gendy, 2018); and black cumin (Farag et al., 2017) volatile composition (Table 10.1). This heat treatment led to higher furfural and nonanal levels in roasted sumac samples (Farag, Fayek and Reidah, 2018). In nutmeg, this metabolomics approach revealed (*E*)-isoeugenol, methoxyeugenol, and myristicin as key markers indicative of the roasting process (Farag, Mohsen and El-Gendy, 2018). Remarkable metabolic variations were observed in the black cumin aroma profile, leading to a significant decrease in thymoquinone levels and a marked increase of thymol and estragole as a consequence of heat treatment-based processing (Farag et al., 2017).

10.4.2 Origin and/or Processing Discrimination by LC-Based Metabolomics Approaches

The geographical traceability of black pepper may be assessed by fingerprinting based on LC separation, and LRMS (Liang et al., 2021) or HRMS detection (Rivera-Pérez, Romero-González and Garrido Frenich, 2021a) (Table 10.2). An analytical strategy based on HPLC-MS (LTQ) fingerprinting and PLS-DA was investigated for black pepper traceability from different countries including Brazil, India, Indonesia and Vietnam (Liang et al., 2021), leading to an average classification rate of 87.7%. In contrast, an advanced metabolomics approach based on UHPLC and HRMS (Q-Orbitrap analyzer) was recently revealed as a powerful strategy not only for geographical origin (Brazilian, Vietnamese and Sri Lankan samples) but also for processing (sterilized vs. non-sterilized samples) discrimination of black pepper, leading to OPLS-DA models with excellent performance parameters (R^2 = 0.990–0.991 and Q^2 = 0.899–0.970) and fully correct classification rate (100%; Rivera-Pérez, Romero-González and Garrido Frenich, 2021a). Besides, eight high discriminant metabolites were putatively identified including reynosin, artabsinolide D and tatridin B as markers for Vietnamese black pepper, whereas fatty acid derivates were the most frequent markers for processing discrimination (e.g., 10,16-dihydroxyhexadecanoic acid). Up to now, this research is the only study that applied an LC-metabolomics approach to evaluate the effect of sterilization processing on black pepper metabolic composition, and furthermore, the first research that investigated the feasibility of a promising metabolomics strategy based on mid-level GC-LC-NMR data fusion for black pepper authenticity (Rivera-Pérez, Romero-González and Garrido Frenich, 2021a).

Regarding origin discrimination of ginger, an analytical method based on classical HPLC-DAD gingerol-related fingerprint analysis (detection performed at λ = 230 nm) was combined with PCA for reliable clustering of samples from China, India, Malaysia, Thailand and Vietnam (Yudthavorasit, Wongravee and Leepipatpiboon, 2014; Table 10.2). [8]-Gingerol was characteristic of Indian samples; methyl [6]-gingerol was highlighted as a reliable marker for ginger from Thailand;

[10]-gingerol, diacetoxy-[8]-gingerdiol, and methyl diacetoxy-[8]-gingerdiol were indicative of Chinese ginger (Yudthavorasit, Wongravee and Leepipatpiboon, 2014). Recently, an untargeted UHPLC-HRMS (QTOF) metabolomics approach has been described to discriminate between ginger from two origins (Ghana and China; Mais et al., 2018). OPLS-DA modeling and the VIP analysis led to the identification of six compounds with VIP scores >1.5 as markers: *L*-glutamine, 5-hydroxy-1,7-bis(4-hydroxy-3-methoxyphenyl)-3-heptanone, 4-gingerol, 3-acetoxy-5-hydroxy-1,7-bis(4-hydroxy-3-methoxyphenyl)heptane, 3,5-diacetoxy-1,7-bis(4-hydroxy-3-methoxyphenyl) heptane, and 6-shogaol (Mais et al., 2018).

Several targeted UHPLC-MS metabolomics approaches have been widely applied to trace geographical origin (Spain, Hungary and the Czech Republic) of paprika (Barbosa, Campmajó et al., 2020; Arrizabalaga-Larrañaga et al., 2021; Table 10.2). An UHPLC-Q-Orbitrap targeted method was developed for the determination of ten capsaicinoids and carotenoids in paprika (Arrizabalaga-Larrañaga et al., 2021), showing good linearity (R^2 >0.998), and limits of quantification (LOQs) were between 0.21 and 51 mg/kg. A tandem mass spectrometry (MS/MS) approach using a triple quadrupole (QqQ) analyzer was used for the discrimination of paprika from two production origins: La Vera PDO and Murcia PDO (Spain) and the Czech Republic (Barbosa, Campmajó et al., 2020). The method was optimized for the determination of 36 phenolic compounds by UHPLC-MS/MS-QqQ (Figure 10.6A) and targeted data was further submitted to PCA and PLS-DA modeling. The method showed good linearity R^2 >0.995 and high precision (RSD <19.2%), with LOQs ranging from 0.03 to 1000 µg/L (Barbosa, Campmajó et al., 2020). Figure 10.6A shows the total ion chromatogram in negative ion mode, whereas Figure 10.6B illustrates the PCA score plot (PC1 vs. PC2) illustrating sample clustering of 111 paprika samples. As can be seen, good differentiation between La Vera PDO, Murcia PDO and the Czech Republic samples was obtained using quantified concentrations of 36 phenolic compounds as chemical descriptors to achieve sample discrimination regarding the paprika region of production (Barbosa, Campmajó et al., 2020).

Regarding untargeted strategies for paprika origin traceability, paprika polyphenolic and capsaicinoid profiling obtained by UHPLC-HRMS (Q-Orbitrap) in combination with PLS-DA was proposed as a source of analytical data to establish sample authenticity according to the region of production (Spain and the Czech Republic; Barbosa, Saurina et al., 2020b; Table 10.2). In another study, phenolic fingerprints of paprika were obtained by classical HPLC-UV-Vis and LDA chemometric treatment that allowed the authentication of paprika samples from Spain (La Vera and Murcia PDO) and the Czech Republic (Cetó et al., 2020). A comprehensive fingerprinting method based on UHPLC-HRMS (Q-Orbitrap) has been applied to trace paprika region of production (Barbosa, Saurina et al., 2020a). In this case, untargeted UHPLC-HRMS fingerprints demonstrated to be excellent chemical descriptors to discriminate the production regions under study (La Vera PDO, Murcia PDO and the Czech Republic paprika), with 100% correct sample classification rates by PLS-DA (Barbosa, Saurina et al., 2020a).

Geographical traceability of saffron has been investigated by classical strategies based on DAD or UV-Vis detection (Masi et al., 2016; Liu et al., 2018; Tabtabaei et al., 2019), with the predominant application of advanced HRMS detection using TOF or hybrid QTOF analyzers (Rubert et al., 2016; Senizza et al., 2019; Hegazi et al., 2022; Table 10.2). As an example, Liu et al. research combined HPLC-DAD fingerprinting with PCA and OPLS-DA to differentiate between Iran and Chinese saffron (Liu et al., 2018). The two regions of production were well-separated by a reliable OPLS-DA model with R^2 and Q^2 values of 0.848 and 0.816, respectively. LDA treatment on HPLC-UV-Vis fingerprints of Iranian versus Italian saffron led to the identification of four variables: (1) safranal; (2) *cis*-crocetin (β-*D*-gentiobiosyl, β-*D*-glucosyl) ester (C3Gg); (3) *trans*-crocetin (β-*D*-neapolitanosyl, β-*D*-glucosyl) ester (T4ng); and (4) *trans*-crocetin (β-*D*-gentiobiosyl, β-*D*-glucosyl) ester (T3Gg) as the most discriminant metabolites (Tabtabaei et al., 2019). Regarding the application of untargeted UHPLC-TOF-based approaches for saffron geographical origin purposes, a metabolomics study focused on Spanish, Czech and Turkish saffron samples revealed that saffron produced in La Mancha and Aragón (Spain) showed glycerophospholipids and their oxidized lipids as the most significant markers (Rubert et al., 2016). Comparison of Italian PDO saffron versus Iranian saffron fingerprints by OPLS-DA (R^2 = 0.93 and Q^2 = 0.81) highlighted the presence of flavonoids, phenolic acids, lignans and other polyphenols as markers to discriminate PDO versus non-PDO saffron samples (Senizza et al., 2019). An OPLS-DA model with excellent performance (R^2 of 0.999 and Q^2 of 0.989; Hegazi et al., 2022) was established for clear discrimination between Iranian and Spanish saffron. Besides, the corresponding S-plot revealed the applicability of oxo-hydroxy-undecenoic acid-*O*-hexoside, crocetin di-*O*-gentiobiosyl ester, and kaempferol-*O*-sophoroside as markers indicative of high-quality saffron from Spain (Hegazi et al., 2022).

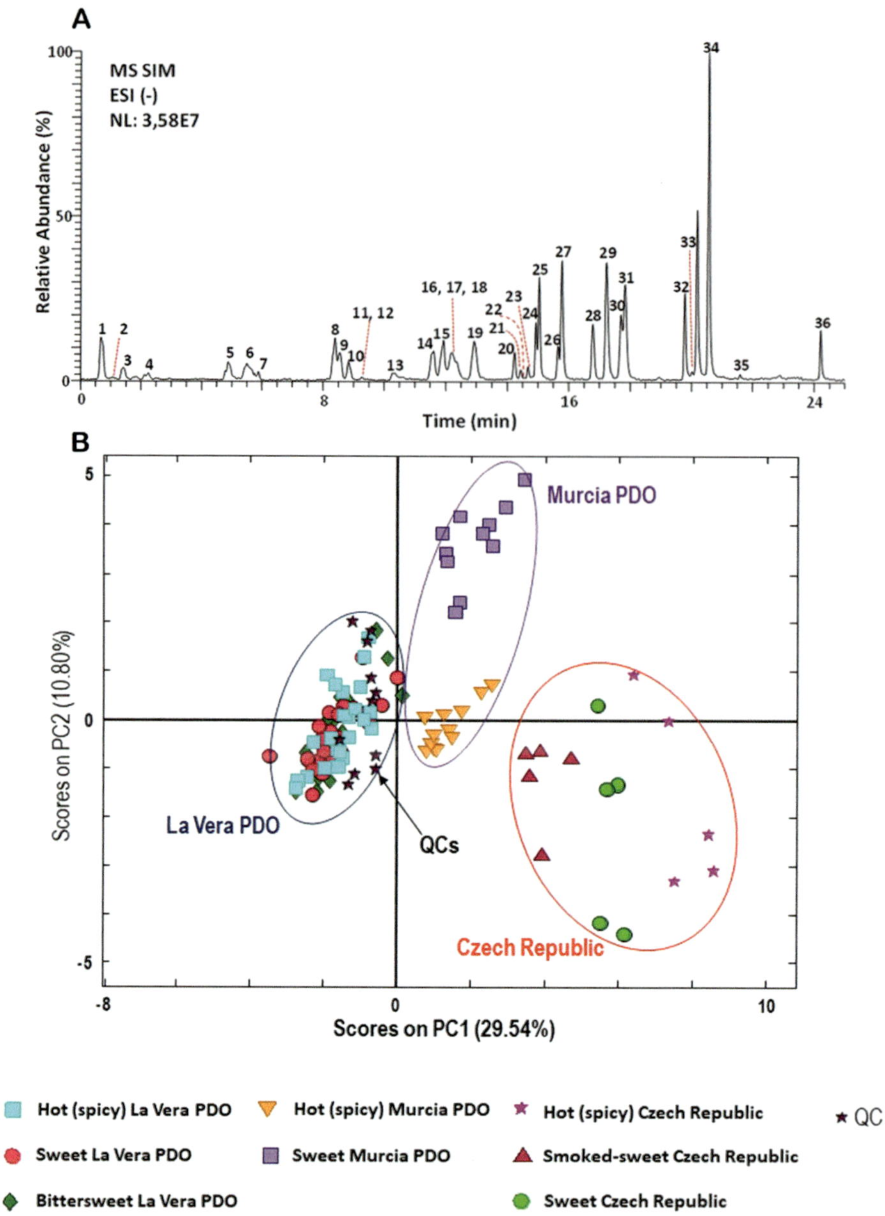

Figure 10.6 (a) Total ion chromatogram (TIC) obtained by UHPLC-MS analysis of the 36 targeted phenolic compounds (standard solution of 500 µg/L in water); (B) PCA score plot (PC1 vs. PC2) of paprika samples from Spain (La Vera PDO and Murcia DPO) and the Czech Republic, showing sample clustering by using the 36 phenolic compound concentrations as chemical descriptors.

Source: Reprinted with permission from Barbosa, Campmajó et al. (2020) (adapted). Copyright 2020 American Chemical Society.

10.4.3 Origin and/or Processing Discrimination by ^{1}H NMR-Based Metabolomics Approaches

The ^{1}H NMR-based fingerprint analysis coupled to multivariate analysis has been demonstrated to be an efficient method to discriminate within different geographical origins of seasonings such as cinnamon (Suzuki et al., 2022); black pepper (Rivera-Pérez, Romero-González and Garrido Frenich,

2022); garlic and onion (Jo et al., 2020); turmeric (Gad and Bouzabata, 2017); and saffron (Sobolev et al., 2014; Table 10.3).

PCA and HCA of the ^{1}H NMR spectra of cinnamon barks showed clear discrimination between cinnamon from Vietnam, China and Sri Lanka. In particular, Vietnamese cinnamon was found to differ significantly from the other tested origins due to their higher coumarin levels, which were further quantified using a q ^{1}H NMR-based method (characteristic signals at δ 6.46 and δ 8.02 ppm; Suzuki et al., 2022; Table 10.3). The geographical origin of black pepper (Brazil, Vietnam and Sri Lanka) was traced by combining ^{1}H NMR fingerprinting and OPLS-DA ($R^2 = 0.962$–0.971 and $Q^2 = 0.787$–0.928) using both polar and lipophilic compositional information obtained from MeOH-*d4* and $CDCl_3$ extracts, respectively (Rivera-Pérez, Romero-González and Garrido Frenich, 2022). Black pepper has demonstrated a characteristic ^{1}H NMR metabolic profile dominated by piperine, which was revealed as a key marker found at significantly higher levels in Sri Lankan black pepper (Rivera-Pérez, Romero-González and Garrido Frenich, 2022). Garlic and onion (*Allium* species) from two regions of production (South Korea and China) were differentiated using unsupervised PCA, which showed that the significant differences for reliable discrimination were in the amino acid content and the aromatic region (Jo et al., 2020).

Algerian and Egyptian *Curcuma longa* ^{1}H NMR fingerprints and unsupervised chemometric techniques (PCA and HCA) were utilized to assess the differences between tested origins (Gad and Bouzabata, 2017; Table 10.3). The findings revealed that origin differentiation was mainly due to the region of essential oils/fatty acids (δ 0.4–3 ppm; Gad and Bouzabata, 2017). Chemometric treatment by PCA of ^{1}H NMR metabolic profiles of saffron samples from different regions of production (Greece, Spain, Hungary, Turkey and Italy) showed significant differences in the content of some metabolites (picrocrocin, crocetin esters, glucose, acetic acid and formic acid) responsible for origin differentiation (Sobolev et al., 2014).

On the other hand, untargeted ^{1}H NMR-based metabolomics and multivariate data analysis (PCA and OPLS-DA) have been applied to assess the impact of processing practices on the metabolic composition of black pepper (Rivera-Pérez, Romero-González and Garrido Frenich, 2022); *Curcuma* species (Awin et al., 2016); and black garlic (Liang et al., 2015; Table 10.3). This strategy revealed that lipophilic ($CDCl_3$) black pepper subjected to heat treatment processing (sterilization) led to significantly higher fatty acid content in sterilized black pepper (Rivera-Pérez, Romero-González and Garrido Frenich, 2022). Drying (air, freeze and oven) of *Curcuma* samples led to higher amino acid content (leucine, lysine, methionine, phenylalanine, valine and alanine) in oven-dried samples, whereas higher curcuminoid content was found in freeze- and air-dried samples since curcuminoids are thermolabile compounds degraded during the oven drying process (Awin et al., 2016). Multivariate data analysis (PCA) applied to black garlic (*Allium sativum* L.) fingerprints subjected to thermal processing pointed out metabolic variations in 38 metabolites compared with raw garlic, including fructose, glucose, acetic acid, formic acid, pyroglutamic acid, cycloalliin and 5-(hydroxymethyl)furfural (5-HMF; Liang et al., 2015).

10.5 ANALYTICAL AND CHEMOMETRIC TOOLS USED FOR OTHER AUTHENTICATION PURPOSES: Differentiation of Plant Species, Discrimination within Varieties/Cultivars, and Other Relevant Effects on Condiments

10.5.1 Other Authentication Practices by GC-Based Metabolomics Approaches

GC-based metabolomics and multivariate data analysis have been employed for other authentication purposes, including differentiation of cultivar/variety of condiments samples belonging to the same origin of production (Macchia et al., 2013; Aliakbarzadeh, Parastar and Sereshti, 2016; Aliakbarzadeh, Sereshti and Parastar, 2016; Li et al., 2020; Barata et al., 2021; Reale et al., 2021) and to differentiate harvesting practices (organic vs. conventional grown; Gao et al., 2014; Table 10.1).

Characterization of eight kinds of huajiao, a high-value Chinese spice, was performed using HS-SPME-GC-MS fingerprinting (Table 10.1). Further multivariate data analysis (PLS-DA and VIP analysis) highlighted four key volatile compounds: linalool, linalyl acetate, *D*-limonene, and sabinene (Feng et al., 2022). Dried bell pepper (*Capsicum annuum*) is a worldwide consumed spice used as a condiment, whose aroma profiling was carried out by HS-SPME-GC-MS analysis and chemometrics (PLS-DA; Reale et al., 2021). This approach was suitable to point out the volatile changes between two different bell peppers harvested in diverse Italian areas, achieving 100% of correct classification and the identification of 16 marker volatile compounds with VIP scores > 1 (e.g., 3-carene, *p*-cymene, (-)-limonene, (+)-carvone; Reale et al., 2021). Unsupervised techniques including HCA and PCA in combination with GC volatile analysis coupled to MS (Barata et al., 2021) or FID (Li et al., 2020) detection were applied to differentiate pepper (*Piper nigrum* L.) cultivated within Brazilian or Chinese production areas, respectively.

GC-MS (Aliakbarzadeh, Sereshti and Parastar, 2016) and GC-FID (Aliakbarzadeh, Parastar and Sereshti, 2016) fingerprinting approaches combined with PCA and PLS-DA, respectively, were described to obtain information on metabolic differences in Iranian saffron cultivars (Table 10.1). The first approach based on GC-MS analysis identified 11 markers including secondary saffron metabolites, such as linoleic acid, and nanocosanol to discriminate between Iranian saffron varieties (Aliakbarzadeh, Sereshti and Parastar, 2016). Assessing the influence of the cultivation area on Iranian saffron by GC-FID fingerprinting led to the identification of nine metabolites (e.g., β-isophoren, α-isophoron, ketoisophorone, dihydrooxophorone and eucarvone) for variety differentiation (Aliakbarzadeh, Parastar and Sereshti, 2016).

Currently, organic food has gained attention due to the higher demand by consumers. GC-FID fingerprinting and OPLS-DA was shown as an effective approach in differentiating organically and conventionally grown oregano (Gao et al., 2014; Table 10.1). The following metabolites were of significance to reveal organically grown oregano: 3,7- dimethyl-1,6-octadien-3-ol, thymoquinone, and γ- terpinene (Gao et al., 2014).

Besides, the GC-chemometrics approach was able to reveal freshness/aging (Rocchi et al., 2019; Farag et al., 2020) and storage time (Sereshti et al., 2018) effects on the volatile composition of saffron (Table 10.1). Untargeted aroma profiling based on GC-MS (Farag et al., 2020) or GC-FID (Sereshti et al., 2018) combined with unsupervised PCA revealed ketoisophorone as a freshness marker, whereas safranal was indicative of aging (Sereshti et al., 2018; Farag et al., 2020). Other research revealed some main volatile compounds of saffron (2-methyl-5-(1-methylethyl)-2,5-cyclohexadiene-1,4-dione and 1,3,4-trimethyl-3-cyclohexene-1-carboxaldehyde) to discriminate older stored saffron by GC-MS fingerprinting and subsequent PLS-DA modeling (Rocchi et al., 2019).

10.5.2 Other Authentication Practices by LC-Based Metabolomics Approaches

UHPLC-HRMS (Q-Orbitrap) targeted methods have been applied to discriminate between different flavor varieties (spicy, sweet and bittersweet) of paprika samples from Spain (La Vera and Murcia PDO), Hungary and the Czech Republic (Barbosa, Saurina et al., 2020a; Arrizabalaga-Larrañaga et al., 2021; Table 10.2). Particularly, targeted UHPLC-HRMS (Q-Orbitrap) analysis of 10 capsaicinoids and carotenoids in paprika samples in combination with PLS-DA treatment revealed that the capsaicinoid content was highly related to the flavor paprika type (Arrizabalaga-Larrañaga et al., 2021). Other research showed that full classification among flavor varieties (sweet, spicy and smoked-sweet) of paprika from Murcia PDO (Spain) and the Czech Republic was achieved by targeted UHPLC-MS/MS (QqQ) determination of 36 phenolic compounds further combined with multivariate data analysis (PLS-DA; Barbosa, Campmajó et al., 2020). In contrast, this strategy was not enough to discriminate between spicy, sweet and bittersweet La Vera PDO paprika samples (Barbosa, Campmajó et al., 2020). This targeted UHPLC-MS/MS metabolomics approach combined with PLS-DA and VIP analysis highlighted some paprika metabolites such as vanillin, kaempferol and *p*-coumaric acid as most important to discriminate within Murcia DOP varieties, whereas in the case of Czech Republic samples, nepetin-7-glucoside was the marker with the highest discrimination potential within Czech flavor paprika varieties (Barbosa, Campmajó et al., 2020).

Untargeted methodologies based on UHPLC-HRMS (Q-Orbitrap; Barbosa, Saurina et al., 2020a) or HPLC-UV-Vis (Cetó et al., 2020) fingerprinting was combined with supervised chemometric techniques (LDA or PLS-DA) to classify paprika from Spain and the Czech Republic regarding their flavor variety (spicy, sweet and bittersweet; Table 10.2). UHPLC-HRMS fingerprinting and PLS-DA was an excellent strategy to fully discriminate within taste varieties for each paprika PDO or production region including La Vera PDO, Murcia PDO and the Czech Republic (Barbosa, Saurina et al., 2020a). Another fingerprinting method based on classical HPLC-UV-Vis detection was applied to differentiate paprika varieties (Cetó et al., 2018). In this case, phenolic profiling by HPLC-UV-Vis (combined with PCA and LDA) allowed differentiating within La Vera and Murcia PDO paprika varieties (spicy, sweet and bittersweet), all produced in Spain (Cetó et al., 2018).

Other untargeted metabolomics approaches based on classical HPLC-DAD fingerprinting were evaluated to classify different saffron varieties cultivated in Italy (D'Archivio et al., 2016; Biancolillo et al., 2020; Table 10.2). D'Archivio et al. research concluded that saffron from Sardinia may be discriminated from saffron produced in central Italy (Abruzzo, Tuscany and Umbria) due to significant differences in crocin levels (D'Archivio et al., 2016).

A targeted approach was performed by developing a HPLC-MS/MS method for the determination of 12 flavonoids and phenolic acids from Italian rosemary (three regions of Sardinia were considered: La Maddalena, Tinnari and Asinara; Maldini et al., 2016; Table 10.2). Unsupervised PCA highlighted that syringic acid, quinic acid and kaempferol were important variables for the differentiation of rosemary samples from Asinara and Tinnari (Maldini et al., 2016). Four different *Cinnamomum* species (*C. burmannii, C. loureiroi, C. cassia,* and *C. verum*) were characterized by UHPLC-HRMS (Orbitrap) fingerprinting (Wang, Harrington and Chen, 2020). PCA and PLS-DA were able to identify several key compounds for cinnamon differentiation such as methoxycinnamaldehyde, cinnamaldehyde and coumarin, among others (Wang, Harrington and Chen, 2020).

10.5.3 Other Authentication Practices by ^{1}H NMR-Based Metabolomics Approaches

Metabolomics approaches based on ^{1}H NMR fingerprinting and chemometrics have been used to discriminate samples from different species (Jung et al., 2012; Farag et al., 2018; Freitas et al., 2018; Florentino-Ramos et al., 2019; Mandrone et al., 2021), and also to differentiate between varieties of the same plant species (Becerra-Martínez et al., 2017; Saviano et al., 2019; Villa-Ruano et al., 2019; Table 10.3).

Regarding the discrimination between plant species, combining ^{1}H NMR fingerprinting of *Origanum* samples and PCA (98.6% of the variance between samples was explained with good predictability of $Q^2 = 0.896$) was an effective strategy to discriminate two different species: *O. vulgare* and *O. onites* (Mandrone et al., 2021; Table 10.3). OPLS-DA modeling ($R^2 = 1$ and $Q^2 = 0.985$) was further performed to find distinctive markers, revealing apigenin (δ 6.51 ppm) and *p*-cymene (δ 0.98 ppm) as markers for *O. onites,* and salvianolic acid B (δ 6.35 ppm) for *O. vulgare* (Mandrone et al., 2021).

A q ^{1}H NMR approach combined with multivariate data analysis by OPLS-DA was developed for successful differentiation between *Cinnamomum cassia* and *C. verum* (Farag et al., 2018; Table 10.3; Figure 10.7A). In this case, eugenol (δ 3.84 ppm and δ 7.68 ppm) was described as a marker for *C. verum* samples, whereas *C. cassia* was characterized by the enrichment in ω-2 and ω-1 fatty acids (δ 1.28 and δ 1.32 ppm; Farag et al., 2018; Figure 10.7B).

The essential oils of seven different *Ocimum* species were investigated by ^{1}H NMR profiling and unsupervised chemometrics (PCA and HCA; Freitas et al., 2018; Table 10.3). The findings revealed three *Ocimum* chemotypes including eugenol as the distinctive marker for *O. gratissimum, O. micranthum* and *O. tenuiflorum;* estragole for *O. basilicum, O. basilicum* var. *purpuracens* and *O. selloi;* and methyl cinnamate for *O. americanum* (Freitas et al., 2018). The polar extracts of two *Curcuma* species (*C. aromatica* and *C. longa*) from two geographical origins (Jeju-do and Jin-do) were analyzed by ^{1}H NMR and data were submitted to PCA and OPLS-DA (Jung et al., 2012). In this study, both statistical models showed clear differentiation between *Curcuma* species with good model goodness-of-fit ($R^2 = 0.796–0.911$) and predictive ability ($Q^2 = 0.851–0.984$). The *Curcuma* species were mainly differentiated due to the levels of sugars (glucose, fructose and sucrose) and essential oils (eucalyptol, curdione and germacrone; Jung et al., 2012). Florentino-Ramos et al. investigated metabolic differences between 11 different species of chili pepper (*Capsicum annuum*) cultivated in Mexico using untargeted ^{1}H NMR-based metabolomics combined with PCA, HCA and LDA (Florentino-Ramos et al., 2019; Table 10.3). Remarkably higher amino acid content was characteristic of *C. annuum* 'Mirasol' and *C. annuum* 'Agua' compared with the rest of the chili peppers. The significantly higher levels of malic acid and fumaric acid could be used as markers of *C. annuum* 'Húngaro' and *C. annuum* 'Árbol', respectively. Besides, sucrose and galactose were markers of *C. annuum* 'Mirasol' due to their high concentrations (Florentino-Ramos et al., 2019).

Concerning the use of untargeted ^{1}H NMR-based approaches to discriminate within varieties of the same plant species, this strategy was applied to discriminate ten varieties of serrano pepper (*Capsicum annuum* 'Serrano'; Villa-Ruano et al., 2019) produced in Mexico, as well as to discriminate within two Mexican cultivars (Veracruz vs. Oaxaca) of serrano pepper (Becerra-Martínez et al., 2017; Table 10.3). In Villa-Ruano et al. research, 48 metabolites were identified by ^{1}H NMR, and PCA and PLS-DA led to serrano pepper variety clustering. Furthermore, PLS-DA score plots were investigated and citric acid, fumaric acid, malic acid, glucose, fructose and sucrose were found as differential metabolites (Villa-Ruano et al., 2019). Discrimination of Veracruz versus Oaxaca Mexican varieties of serrano pepper by OPLS-DA ($R^2 = 0.999$ and $Q^2 = 0.998$ values) pointed out that higher levels of sucrose, aspartate citrate, lactate and leucine were found in Veracruz serrano pepper, whereas serrano pepper from Oaxaca contained enhanced concentrations of fumarate, acetate, formate, malonate, phosphocholine, pyruvate and succinate (Becerra-Martínez et al., 2017). The untargeted ^{1}H NMR analysis of three *Allium cepa* L. varieties (White San Pietro onion, Red Tropea, and Coppery Montoro) showed that Red Tropea onion displayed the highest polyphenol and flavonoid contents. On the other hand, the yellow Coppery Montoro variety was characterized by a complex sulfur compound profile (Saviano et al., 2019).

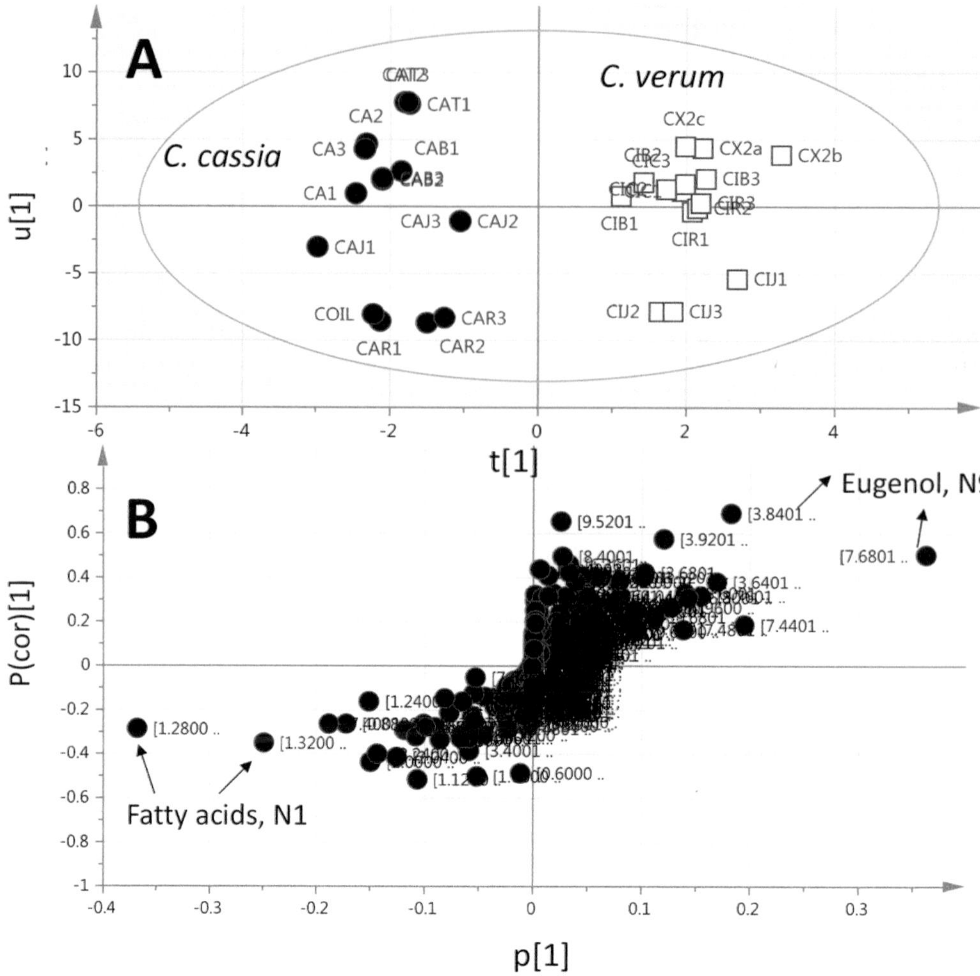

Figure 10.7 (a) OPLS-DA score plot showing discrimination between *Cinnamomum cassia* vs. *C. verum* ¹H NMR fingerprints; and (b) loading plot showing the covariance p[1] against the correlation p(cor)[1] of the discriminating variables of the OPLS-DA model.

Source: Reprinted with permission from Farag et al. (2018). Copyright 2018 Elsevier.

Finally, untargeted NMR approaches coupled to multivariate data analysis were also reported for other authentication purposes including assessing the effect of storage time (Ordoudi et al., 2015; Consonni et al., 2016) on the metabolic composition of saffron (Table 10.3). A two-model OPLS-DA model ($R^2 = 0.946$ and $Q^2 = 0.937$) allowed the clear differentiation between saffron samples stored up to 4 years and saffron stored for more than 4 years (5–14 years; Ordoudi et al., 2015). Furthermore, the findings pointed out that saffron quality deterioration was mainly due to variations in the content of sugars bound to crocetin, glucose in picrocrocin, free sugars and fatty acids, which are the principal saffron metabolites responsible for its properties (Ordoudi et al., 2015).

10.6 CONCLUSIONS AND FUTURE PERSPECTIVES

Nowadays, culinary herbs and spices have gained special significance in the food authenticity field due to their high value, as well as their increasing worldwide use as flavorings in foods and beverages. This chapter has shown the combination of analytical and chemometric techniques as an effective tool to guarantee the authenticity of seasonings, including geographical origin traceability and discrimination of seasonings according to post-harvest processing. Metabolomics approaches have also been reported as a high-throughput strategy for quality control, to detect adulteration, as well as other authenticity applications such as the assessment of storage time effect on the metabolic profile of plant-derived condiments.

As the metabolomics workflow begins with the sample preparation, metabolite extraction from condiments has become a crucial step to obtain the maximum compositional information. Currently, conventional extraction methods such as hydrodistillation or maceration have been replaced by more environmental-friendly and simpler alternatives. Particularly, automatized and free-solvent HS-SPME has turned into the "golden standard" procedure for further GC-based fingerprinting, whereas UAE (among others) is the extraction method of choice for LC or NMR metabolomics approaches due to its time effectiveness and suitability for multi-sample extraction.

According to the comprehensive overview of recent methodologies for authentication purposes described in this chapter, the analytical tools currently used to resolve issues related to seasoning authenticity include chromatographic methods such as GC, UHPLC and HPLC mainly combined with MS or MS/MS detection, but HRMS, using Orbitrap or QTOF mass spectrometers, is starting to be more used due to its higher selectivity and rapid untargeted analysis in which complete metabolite identification and quantification is not necessarily needed. Besides, ^{1}H NMR spectroscopy has emerged as a promising authentication tool due to its strengths, such as its minimal sample pretreatment and reproducibility.

Chemometric tools including unsupervised techniques such as PCA and HCA, as well as supervised approaches such as PLS-DA, LDA and OPLS-DA, are the multivariate analyses with the most significant presence in investigating condiment authenticity. They also allowed the reliable identification of potential markers responsible for sample discrimination.

The increasing economically motivated adulteration or mislabeling of seasonings shows that there is a continuous need for the development and refinement of current analytical methods, as well as the development of new technologies. Further advances in this field would lead to the validation of selected markers using targeted methodologies, the development of metabolomics approaches to be easily implemented in routine quality control analyses, as well as the exploration of other innovative statistical methods such as DD-SIMCA or biomimetic pattern recognition (BPR), among others.

10.7 ACKNOWLEDGMENT

ARP acknowledges the Grant FPU18/05133 funded by MCIN/AEI/10.13039/501100011033.

10.8 DECLARATION OF INTEREST STATEMENT

The authors declare no conflict of interest.

REFERENCES

Abd Ghafar, S.Z., Mediani, A., Maulidiani, M., Rudiyanto, R., Mohd Ghazali, H., Ramli, N.S. and Abas, F. (2020) 'Complementary NMR- and MS-based metabolomics approaches reveal the correlations of phytochemicals and biological activities in Phyllanthus acidus leaf extracts', *Food Research International*, 136, p. 109312. doi:10.1016/j.foodres.2020.109312.

Aliakbarzadeh, G., Parastar, H. and Sereshti, H. (2016) 'Classification of gas chromatographic fingerprints of saffron using partial least squares discriminant analysis together with different variable selection methods', *Chemometrics and Intelligent Laboratory Systems*, 158, pp. 165–173. doi:10.1016/j.chemolab.2016.09.002.

Aliakbarzadeh, G., Sereshti, H. and Parastar, H. (2016) 'Pattern recognition analysis of chromatographic fingerprints of Crocus sativus L. secondary metabolites towards source identification and quality control', *Analytical and Bioanalytical Chemistry*, 408(12), pp. 3295–3307. doi:10.1007/s00216-016-9400-8.

Arrizabalaga-Larrañaga, A., Campmajó, G., Saurina, J., Núñez, O., Santos, F.J. and Moyano, E. (2021) 'Determination of capsaicinoids and carotenoids for the characterization and geographical origin authentication of paprika by UHPLC—APCI—HRMS', *Lwt*, 139, p. 110533. doi:10.1016/j.lwt.2020.110533.

Ávila, M., Zougagh, M., Escarpa, A. and Ríos, Á. (2009) 'Determination of alkenylbenzenes and related flavour compounds in food samples by on-column preconcentration-capillary liquid chromatography', *Journal of Chromatography A*, 1216(43), pp. 7179–7185. doi:10.1016/j.chroma.2009.08.053.

Awin, T., Mediani, A., Maulidiani, Shaari, K., Faudzi, S.M.M., Sukari, M.A.H., Lajis, N.H. and Abas, F. (2016) 'Phytochemical profiles and biological activities of Curcuma species subjected to different drying methods and solvent systems: NMR-based metabolomics approach', *Industrial Crops and Products*, 94, pp. 342–352. doi:10.1016/j.indcrop.2016.08.020.

Azmir, J., Zaidul, I.S.M., Rahman, M.M., Sharif, K.M., Mohamed, A., Sahena, F., Jahurul, M.H.A., Ghafoor, K., Norulaini, N.A.N. and Omar, A.K.M. (2013) 'Techniques for extraction of bioactive compounds from plant materials: A review', *Journal of Food Engineering*, 117(4), pp. 426–436. doi:10.1016/j.jfoodeng.2013.01.014.

Bakkali, F., Averbeck, S., Averbeck, D. and Idaomar, M. (2008) 'Biological effects of essential oils—A review', *Food and Chemical Toxicology*, 46(2), pp. 446–475. doi:10.1016/j.fct.2007.09.106.

Barata, L.M., Andrade, E.H., Ramos, A.R., de Lemos, O.F., Setzer, W.N., Byler, K.G., Maia, J.G.S. and da Silva, J.K.R. (2021) 'Secondary metabolic profile as a tool for distinction and characterization of cultivars of black pepper (Piper nigrum L.) cultivated in Pará State, Brazil', *International Journal of Molecular Sciences*, 22, p. 890. doi:10.3390/ijms22020890.

Barbosa, S., Campmajó, G., Saurina, J., Puignou, L. and Núñez, O. (2020) 'Determination of phenolic compounds in paprika by ultrahigh performance liquid chromatography—tandem mass spectrometry: application to product designation of origin authentication by chemometrics', *Journal of Agricultural and Food Chemistry*, 68, pp. 591–602. doi:10.1021/acs.jafc.9b06054.

Barbosa, S., Saurina, J., Puignou, L. and Núñez, O. (2020a) 'Classification and authentication of paprika by UHPLC-HRMS fingerprinting and multivariate calibration methods (PCA and PLS-DA)', *Foods*, 9, p. 486. doi:10.3390/foods9040486.

Barbosa, S., Saurina, J., Puignou, L. and Núñez, O. (2020b) 'Targeted UHPLC—HRMS (Orbitrap) polyphenolic and capsaicinoid profiling for the chemometric characterization and classification of paprika with protected designation of origin (PDO) attributes', *Molecules*, 25, p. 1623. doi:10.3390/molecules25071623.

Becerra-Martínez, E., Florentino-Ramos, E., Pérez-Hernández, N., Gerardo Zepeda-Vallejo, L., Villa-Ruano, N., Velázquez-Ponce, M., García-Mendoza, F. and Bañuelos-Hernández, A.E. (2017) '^{1}H NMR-based metabolomic fingerprinting to determine metabolite levels in serrano peppers (Capsicum annum L.) grown in two different regions', *Food Research International*, 102, pp. 163–170. doi:10.1016/j.foodres.2017.10.005.

Biancolillo, A., Maggi, M.A., De Martino, A., Marini, F., Ruggieri, F. and D'Archivio, A.A. (2020) 'Authentication of PDO saffron of L'Aquila (Crocus sativus L.) by HPLC-DAD coupled with a discriminant multi-way approach', *Food Control*, 110, p. 107022. doi:10.1016/j.foodcont.2019.107022.

Black, C., Haughey, S.A., Chevallier, O.P., Galvin-King, P. and Elliott, C.T. (2016) 'A comprehensive strategy to detect the fraudulent adulteration of herbs: The oregano approach', *Food Chemistry*, 210, pp. 551–557. doi:10.1016/j.foodchem.2016.05.004.

Cagliani, L.R., Culeddu, N., Chessa, M. and Consonni, R. (2015) 'NMR investigations for a quality assessment of Italian PDO saffron (Crocus sativus L.)', *Food Control*, 50, pp. 342–348. doi:10.1016/j.foodcont.2014.09.017.

Cetó, X., Sánchez, C., Serrano, N., Díaz-Cruz, J.M. and Núñez, O. (2020) 'Authentication of paprika using HPLC-UV fingerprints', *Lwt*, 124, p. 109153. doi:10.1016/j.lwt.2020.109153.

Cetó, X., Serrano, N., Aragó, M., Gámez, A., Esteban, M., Díaz-Cruz, J.M. and Núñez, O. (2018) 'Determination of HPLC-UV fingerprints of Spanish paprika (Capsicum annuum L.) for its classification by linear discriminant analysis', *Sensors*, 18, p. 4479. doi:10.3390/s18124479.

Chaji, S., Ajal, E.A., Olmo-García, L., Serrano-García, I., Carrasco-Pancorbo, A. and Bajoub, A. (2021) 'Metabolomic approaches applied to food authentication: from data acquisition to biomarkers discovery', in Galanakis, C.M. (ed.) *Food Authentication and Traceability*. Academic Press, pp. 331–378. doi:10.1016/b978-0-12-821104-5.00011-8.

Chemat, F., Rombaut, N., Sicaire, A.G., Meullemiestre, A., Fabiano-Tixier, A.S. and Abert-Vian, M. (2017) 'Ultrasound assisted extraction of food and natural products. Mechanisms, techniques, combinations, protocols and applications. A review', *Ultrasonics Sonochemistry*, 34, pp. 540–560. doi:10.1016/j.ultsonch.2016.06.035.

Christaki, E., Bonos, E., Giannenas, I. and Florou-Paneri, P. (2012) 'Aromatic plants as a source of bioactive compounds', *Agriculture*, 2(3), pp. 228–243. doi:10.3390/agriculture2030228.

Consonni, R., Ordoudi, S., Cagliani, L., Tsiangali, M. and Tsimidou, M. (2016) 'On the traceability of commercial saffron samples using ^{1}H-NMR and FT-IR metabolomics', *Molecules*, 21, p. 286. doi:10.3390/molecules21030286.

Cuadros-Rodríguez, L., Ruiz-Samblás, C., Valverde-Som, L., Pérez-Castaño, E. and González-Casado, A. (2016) 'Chromatographic fingerprinting: An innovative approach for food "identitation" and food authentication—A tutorial', *Analytica Chimica Acta*, 909, pp. 9–23. doi:10.1016/j.aca.2015.12.042.

D'Archivio, A.A., Di Pietro, L., Maggi, M.A. and Rossi, L. (2018) 'Optimization using chemometrics of HS-SPME/GC—MS profiling of saffron aroma and identification of geographical volatile markers', *European Food Research and Technology*, 244(9), pp. 1605–1613. doi:10.1007/s00217-018-3073-9.

D'Archivio, A.A., Giannitto, A., Maggi, M.A. and Ruggieri, F. (2016) 'Geographical classification of Italian saffron (Crocus sativus L.) based on chemical constituents determined by high-performance liquid-chromatography and by using linear discriminant analysis', *Food Chemistry*, 212, pp. 110–116. doi:10.1016/j.foodchem.2016.05.149.

Danlami, J.M., Arsad, A., Zaini, M.A.A. and Sulaiman, H. (2014) 'A comparative study of various oil extraction techniques from plants', *Reviews in Chemical Engineering*, 30(6), pp. 605–626. doi:10.1515/revce-2013–0038.

Deborde, C., Fontaine, J.-X., Jacob, D., Botana, A., Nicaise, V., Richard-Forget, F., Lecomte, S., Decourtil, C., Hamade, K., Mesnard, F., Moing, A. and Molinié, R. (2019) 'Optimizing 1D ^{1}H-NMR profiling of plant samples for high throughput analysis: extract preparation, standardization, automation and spectra processing', *Metabolomics*, 15(3), p. 28. doi:10.1007/s11306-019-1488-3.

Di Bella, G., Potortì, A.G., Ben Tekaya, A., Beltifa, A., Ben Mansour, H., Sajia, E., Bartolomeo, G., Naccari, C., Dugo, G. and Lo Turco, V. (2019) 'Organic contamination of Italian and Tunisian culinary herbs and spices', *Journal of Environmental Science and Health, Part B*, 54(5), pp. 345–356. doi:10.1080/03601234.2019.1571364.

Di Donato, F., D'Archivio, A.A., Maggi, M.A. and Rossi, L. (2021) 'Detection of plant-derived adulterants in saffron (Crocus sativus L.) by HS-SPME/GC-MS profiling of volatiles and chemometrics', *Food Analytical Methods*, 14, pp. 784–796. doi:10.1007/s12161–020–01941-x.

Dona, A.C., Kyriakides, M., Scott, F., Shephard, E.A., Varshavi, D., Veselkov, K. and Everett, J.R. (2016) 'A guide to the identification of metabolites in NMR-based metabonomics/metabolomics experiments', *Computational and Structural Biotechnology Journal*, 14, pp. 135–153. doi:10.1016/j.csbj.2016.02.005.

Dowlatabadi, R., Farshidfar, F., Zare, Z., Pirali, M., Rabiei, M., Khoshayand, M.R. and Vogel, H.J. (2017) 'Detection of adulteration in Iranian saffron samples by ^{1}H NMR spectroscopy and multivariate data analysis techniques', *Metabolomics*, 13, p. 19. doi:10.1007/s11306–016–1155-x.

Ellis, D.I., Muhamadali, H., Haughey, S.A., Elliott, C.T. and Goodacre, R. (2015) 'Point-and-shoot: rapid quantitative detection methods for on-site food fraud analysis—moving out of the laboratory and into the food supply chain', *Analytical Methods*, 7(22), pp. 9401–9414. doi:10.1039/C5AY02048D.

Farag, M.A., El-Kersh, D.M., Rasheed, D.M. and Heiss, A.G. (2017) 'Volatiles distribution in Nigella species (black cumin seeds) and in response to roasting as analyzed via solid-phase microextraction (SPME) coupled to chemometrics', *Industrial Crops and Products*, 108, pp. 564–571. doi:10.1016/j.indcrop.2017.07.011.

Farag, M.A., Fayek, N.M. and Reidah, I.A. (2018) 'Volatile profiling in Rhus coriaria fruit (sumac) from three different geographical origins and upon roasting as analyzed via solid-phase microextraction', *PeerJ*, 2018(7), pp. 1–16. doi:10.7717/peerj.5121.

Farag, M.A., Hegazi, N., Dokhalahy, E. and Khattab, A.R. (2020) 'Chemometrics based GC-MS aroma profiling for revealing freshness, origin and roasting indices in saffron spice and its adulteration', *Food Chemistry*, 331, p. 127358. doi:10.1016/j.foodchem.2020.127358.

Farag, M.A., Labib, R.M., Noleto, C., Porzel, A. and Wessjohann, L.A. (2018) 'NMR approach for the authentication of 10 cinnamon spice accessions analyzed via chemometric tools', *LWT—Food Science and Technology*, 90, pp. 491–498. doi:10.1016/j.lwt.2017.12.069.

Farag, M.A., Mohsen, E. and El-Gendy, A.E.N.G. (2018) 'Sensory metabolites profiling in Myristica fragrans (Nutmeg) organs and in response to roasting as analyzed via chemometric tools', *Lwt*, 97, pp. 684–692. doi:10.1016/j.lwt.2018.08.002.

Feng, X., Kong, W., Wei, J., Zhen, O.Y. and Yang, M. (2014) 'HPLC fingerprint analysis combined with chemometrics for pattern recognition of ginger', *Pharmaceutical Biology*, 52(3), pp. 362–367. doi:10.3109/13880209.2013.837493.

Feng, X., Wang, H., Wang, Z., Huang, P. and Kan, J. (2022) 'Discrimination and characterization of the volatile organic compounds in eight kinds of huajiao with geographical indication of China using electronic nose, HS-GC-IMS and HS-SPME-GC—MS', *Food Chemistry*, 375, p. 131671. doi:10.1016/j.foodchem.2021.131671.

Florentino-Ramos, E., Villa-Ruano, N., Hidalgo-Martínez, D., Ramírez-Meraz, M., Méndez-Aguilar, R., Velásquez-Valle, R., Zepeda-Vallejo, L.G., Pérez-Hernández, N. and Becerra-Martínez, E. (2019) '^{1}H NMR-based fingerprinting of eleven Mexican Capsicum annuum cultivars', *Food Research International*, 121, pp. 12–19. doi:10.1016/j.foodres.2019.03.025.

Freitas, J.V.B., Alves Filho, E.G., Silva, L.M.A., Zocolo, G.J., de Brito, E.S. and Gramosa, N.V. (2018) 'Chemometric analysis of NMR and GC datasets for chemotype characterization of essential oils from different species of Ocimum', *Talanta*, 180, pp. 329–336. doi:10.1016/j.talanta.2017.12.053.

Gad, H.A. and Bouzabata, A. (2017) 'Application of chemometrics in quality control of Turmeric (Curcuma longa) based on Ultra-violet, Fourier transform-infrared and ^{1}H NMR spectroscopy', *Food Chemistry*, 237, pp. 857–864. doi:10.1016/j.foodchem.2017.06.022.
Galvin-King, P., Haughey, S.A. and Elliott, C.T. (2018) 'Herb and spice fraud; the drivers, challenges and detection', *Food Control*, 88, pp. 85–97. doi:10.1016/j.foodcont.2017.12.031.
Gao, B., Qin, F., Ding, T., Chen, Y., Lu, W. and Yu, L. (Lucy) (2014) 'Differentiating organically and conventionally grown oregano using ultraperformance liquid chromatography mass spectrometry (UPLC-MS), headspace gas chromatography with flame ionization detection (headspace-GC-FID), and flow injection mass spectrum (FIMS)', *Journal of Agricultural and Food Chemistry*, 62(32), pp. 8075–8084. doi:10.1021/jf502419y.
Gu, F., Wu, G., Fang, Y. and Zhu, H. (2018) 'Nontargeted metabolomics for phenolic and polyhydroxy compounds profile of pepper (Piper nigrum L.) products based on LC-MS/MS analysis', *Molecules*, 23(8), p. 1985. doi:10.3390/molecules23081985.
Guijarro-Díez, M., Nozal, L., Marina, M.L. and Crego, A.L. (2015) 'Metabolomic fingerprinting of saffron by LC/MS: Novel authenticity markers', *Analytical and Bioanalytical Chemistry*, 407(23), pp. 7197–7213. doi:10.1007/s00216-015-8882-0.
Gul, I., Nasrullah, N., Nissar, U., Saifi, M. and Abdin, M.Z. (2018) 'Development of DNA and GC-MS fingerprints for authentication and quality control of Piper nigrum L. and its adulterant Carica papaya L.', *Food Analytical Methods*, 11, pp. 1209–1222. doi:10.1007/s12161-017-1088-7.
Hegazi, N.M., Khattab, A.R., Frolov, A., Wessjohann, L.A. and Farag, M.A. (2022) 'Authentication of saffron spice accessions from its common substitutes via a multiplex approach of UV/VIS fingerprints and UPLC/MS using molecular networking and chemometrics', *Food Chemistry*, 367, p. 130739. doi:10.1016/j.foodchem.2021.130739.
Horn, B., Esslinger, S., Fauhl-Hassek, C. and Riedl, J. (2021) '^{1}H NMR spectroscopy, one-class classification and outlier diagnosis: A powerful combination for adulteration detection in paprika powder', *Food Control*, 128, p. 108205. doi:10.1016/j.foodcont.2021.108205.
Iha, M.H. and Trucksess, M.W. (2019) 'Management of mycotoxins in spices', *Journal of AOAC International*, 102(6), pp. 1732–1739. doi:10.5740/jaoacint.19–0117.
Jimenez-Carvelo, A.M. and Cuadros-Rodríguez, L. (2021) 'Data mining/machine learning methods in foodomics', *Current Opinion in Food Science*, 37, pp. 76–82. doi:10.1016/j.cofs.2020.09.008.
Jo, S., Song, Y., Jeong, J.H., Hwang, J. and Kim, Y. (2020) 'Geographical discrimination of Allium species (garlic and onion) using ^{1}H NMR spectroscopy with multivariate analysis', *International Journal of Food Properties*, 23(1), pp. 241–254. doi:10.1080/10942912.2020.1722160.
Jung, Y., Lee, J., Kim, H.K., Moon, B.C., Ji, Y., Ryu, D.H. and Hwang, G.S. (2012) 'Metabolite profiling of Curcuma species grown in different regions using ^{1}H NMR spectroscopy and multivariate analysis', *Analyst*, 137(23), pp. 5597–5606. doi:10.1039/c2an35397k.
Kabak, B. and Dobson, A.D.W. (2017) 'Mycotoxins in spices and herbs—An update', *Critical Reviews in Food Science and Nutrition*, 57(1), pp. 18–34. doi:10.1080/10408398.2013.772891.
Kalogiouri, N.P., Aalizadeh, R., Dasenaki, M.E. and Thomaidis, N.S. (2020) 'Application of high resolution mass spectrometric methods coupled with chemometric techniques in olive oil authenticity studies—A review', *Analytica Chimica Acta*, 1134, pp. 150–173. doi:10.1016/j.aca.2020.07.029.
Karabagias, I.K., Koutsoumpou, M., Liakou, V., Kontakos, S. and Kontominas, M.G. (2017) 'Characterization and geographical discrimination of saffron from Greece, Spain, Iran, and Morocco based on volatile and bioactivity markers, using chemometrics', *European Food Research and Technology*, 243, pp. 1577–1591. doi:10.1007/s00217-017-2866-6.
Kim, H.K. and Verpoorte, R. (2010) 'Sample preparation for plant metabolomics', *Phytochemical Analysis*, 21, pp. 4–13. doi:10.1002/pca.1188.
Kucharska-Ambrożej, K. and Karpinska, J. (2020) 'The application of spectroscopic techniques in combination with chemometrics for detection adulteration of some herbs and spices', *Microchemical Journal*, 153, p. 104278. doi:10.1016/j.microc.2019.104278.
Liang, J., Sun, J., Chen, P., Frazier, J., Benefield, V. and Zhang, M. (2021) 'Chemical analysis and classification of black pepper (Piper nigrum L.) based on their country of origin using mass spectrometric methods and chemometrics', *Food Research International*, 140, p. 109877. doi:10.1016/j.foodres.2020.109877.
Liang, T., Wei, F., Lu, Y., Kodani, Y., Nakada, M., Miyakawa, T. and Tanokura, M. (2015) 'Comprehensive NMR analysis of compositional changes of black garlic during thermal processing', *Journal of Agricultural and Food Chemistry*, 63(2), pp. 683–691. doi:10.1021/jf504836d.

Liu, J., Chen, N., Yang, J., Yang, B., Ouyang, Z., Wu, C., Yuan, Y., Wang, W. and Chen, M. (2018) 'An integrated approach combining HPLC, GC/MS, NIRS, and chemometrics for the geographical discrimination and commercial categorization of saffron', *Food Chemistry*, 253, pp. 284–292. doi:10.1016/j.foodchem.2018.01.140.

Li, Y., Zhang, C., Pan, S., Chen, L., Liu, M., Yang, K., Zeng, X. and Tian, J. (2020) 'Analysis of chemical components and biological activities of essential oils from black and white pepper (Piper nigrum L.) in five provinces of southern China', *LWT—Food Science and Technology*, 117, p. 108644. doi:10.1016/j.lwt.2019.108644.

Lopez, P., Van Sisseren, M., De Marco, S., Jekel, A., De Nijs, M. and Mol, H.G.J. (2015) 'A straightforward method to determine flavouring substances in food by GC-MS', *Food Chemistry*, 174, pp. 407–416. doi:10.1016/j.foodchem.2014.11.011.

Luca, S.V., Minceva, M., Gertsch, J. and Skalicka-Woźniak, K. (2021) 'LC-HRMS/MS-based phytochemical profiling of Piper spices: Global association of piperamides with endocannabinoid system modulation', *Food Research International*, 141, p. 110123. doi:10.1016/j.foodres.2021.110123.

Macchia, M., Ceccarini, L., Molfetta, I., Cioni, P.L. and Flamini, G. (2013) 'Studies on saffron (Crocus sativus L.) from Tuscan Maremma (Italy): Effects of geographical origin, cultivation environment and drying method on volatile emission', *International Journal of Food Science and Technology*, 48(11), pp. 2370–2375. doi:10.1111/ijfs.12227.

Mais, E., Alolga, R.N., Wang, S.L., Linus, L.O., Yin, X. and Qi, L.W. (2018) 'A comparative UPLC-Q/TOF-MS-based metabolomics approach for distinguishing Zingiber officinale Roscoe of two geographical origins', *Food Chemistry*, 240, pp. 239–244. doi:10.1016/j.foodchem.2017.07.106.

Maldini, M., Montoro, P., Addis, R., Toniolo, C., Petretto, G.L., Foddai, M., Nicoletti, M. and Pintore, G. (2016) 'A new approach to discriminate Rosmarinus officinalis L. plants with antioxidant activity, based on HPTLC fingerprint and targeted phenolic analysis combined with PCA', *Industrial Crops and Products*, 94, pp. 665–672. doi:10.1016/j.indcrop.2016.09.042.

Mandrone, M., Marincich, L., Chiocchio, I., Petroli, A., Gođevac, D., Maresca, I. and Poli, F. (2021) 'NMR-based metabolomics for frauds detection and quality control of oregano samples', *Food Control*, 127, p. 108141. doi:10.1016/j.foodcont.2021.108141.

Manukyan, A. (2019) 'Secondary metabolites and their antioxidant capacity of Caucasian endemic thyme (Thymus transcaucasicus Ronn.) as affected by environmental stress', *Journal of Applied Research on Medicinal and Aromatic Plants*, 13, p. 100209. doi:10.1016/j.jarmap.2019.100209.

Masi, E., Taiti, C., Heimler, D., Vignolini, P., Romani, A. and Mancuso, S. (2016) 'PTR-TOF-MS and HPLC analysis in the characterization of saffron (Crocus sativus L.) from Italy and Iran', *Food Chemistry*, 192, pp. 75–81. doi:10.1016/j.foodchem.2015.06.090.

Matsushita, T., Zhao, J.J., Igura, N. and Shimoda, M. (2018) 'Authentication of commercial spices based on the similarities between gas chromatographic fingerprints', *Journal of the Science of Food and Agriculture*, 98(8), pp. 2989–3000. doi:10.1002/jsfa.8797.

Maulidiani, Abas, F., Khatib, A., Shitan, M., Shaari, K. and Lajis, N.H. (2013) 'Comparison of Partial Least Squares and Artificial Neural Network for the prediction of antioxidant activity in extract of Pegaga (Centella) varieties from ^{1}H Nuclear Magnetic Resonance spectroscopy', *Food Research International*, 54, pp. 852–860. doi:10.1016/j.foodres.2013.08.029.

Medina, S., Perestrelo, R., Silva, P., Pereira, J.A.M. and Câmara, J.S. (2019) 'Current trends and recent advances on food authenticity technologies and chemometric approaches', *Trends in Food Science & Technology*, 85, pp. 163–176. doi:10.1016/j.tifs.2019.01.017.

Modupalli, N., Naik, M., Sunil, C.K. and Natarajan, V. (2021) 'Emerging non-destructive methods for quality and safety monitoring of spices', *Trends in Food Science and Technology*, 108, pp. 133–147. doi:10.1016/j.tifs.2020.12.021.

Morozzi, P., Zappi, A., Gottardi, F., Locatelli, M. and Melucci, D. (2019) 'A quick and efficient non-targeted screening test for saffron authentication: application of chemometrics to gas-chromatographic data', *Molecules*, 24, p. 2602. doi:10.3390/molecules24142602.

Negi, A., Pare, A. and Meenatchi, R. (2021) 'Emerging techniques for adulterant authentication in spices and spice products', *Food Control*, 127, p. 108113. doi:10.1016/j.foodcont.2021.108113.

Núñez, N., Vidal-Casanella, O., Sentellas, S., Saurina, J. and Núñez, O. (2020) 'Characterization, classification and authentication of turmeric and curry samples by targeted LC-HRMS polyphenolic and curcuminoid profiling and chemometrics', *Molecules*, 25, p. 2942. doi:10.3390/molecules25122942.

Oms-Oliu, G., Odriozola-Serrano, I. and Martín-Belloso, O. (2013) 'Metabolomics for assessing safety and quality of plant-derived food', *Food Research International*, 54(1), pp. 1172–1183. doi:10.1016/j.foodres.2013.04.005.

Ordoudi, S.A., Cagliani, L.R., Lalou, S., Naziri, E., Tsimidou, M.Z. and Consonni, R. (2015) '^{1}H NMR-based metabolomics of saffron reveals markers for its quality deterioration', *Food Research International*, 70, pp. 1–6. doi:10.1016/j.foodres.2015.01.021.

Palmieri, S., Pellegrini, M., Ricci, A., Compagnone, D. and Lo Sterzo, C. (2020) 'Chemical composition and antioxidant activity of thyme, hemp and coriander extracts: A comparison study of maceration, Soxhlet, UAE and RSLDE techniques', *Foods*, 9, p. 1221. doi:10.3390/foods9091221.

Petrakis, E.A., Cagliani, L.R., Polissiou, M.G. and Consonni, R. (2015) 'Evaluation of saffron (Crocus sativus L.) adulteration with plant adulterants by ^{1}H NMR metabolite fingerprinting', *Food Chemistry*, 173, pp. 890–896. doi:10.1016/j.foodchem.2014.10.107.

Pieri, V., Sturm, S., Seger, C., Franz, C. and Stuppner, H. (2012) '^{1}H NMR-based metabolic profiling and target analysis: a combined approach for the quality control of Thymus vulgaris', *Metabolomics*, 8, pp. 335–346. doi:10.1007/s11306-011-0317-0.

Rajabi, M., Haji-Esfandiari, S., Barfi, B. and Ghanbari, H. (2014) 'Ultrasound-assisted temperature-controlled ionic-liquid dispersive liquid-phase microextraction method for simultaneous determination of anethole, estragole, and para-anisaldehyde in different plant extracts and human urine: A comparative study', *Analytical and Bioanalytical Chemistry*, 406(18), pp. 4501–4512. doi:10.1007/s00216–014–7848-y.

Reale, S., Biancolillo, A., Gasparrini, C., Di Martino, L., Di Cecco, V., Manzi, A., Di Santo, M. and D'Archivio, A.A. (2021) 'Geographical discrimination of bell pepper (Capsicum annuum) spices by (HS)-SPME/GC-MS aroma profiling and chemometrics', *Molecules*, 26, p. 6177. doi:10.3390/molecules26206177.

Rivera-Pérez, A., López-Ruiz, R., Romero-González, R. and Garrido Frenich, A. (2020) 'A new strategy based on gas chromatography—high resolution mass spectrometry (GC—HRMS-Q-Orbitrap) for the determination of alkenylbenzenes in pepper and its varieties', *Food Chemistry*, 321, p. 126727. doi:10.1016/j.foodchem.2020.126727.

Rivera-Pérez, A., Romero-González, R. and Garrido Frenich, A. (2021a) 'Application of an innovative metabolomics approach to discriminate geographical origin and processing of black pepper by untargeted UHPLC-Q-Orbitrap-HRMS analysis and mid-level data fusion', *Food Research International*, 150, p. 110722. doi:10.1016/j.foodres.2021.110722.

Rivera-Pérez, A., Romero-González, R. and Garrido Frenich, A. (2021b) 'Determination and occurrence of alkenylbenzenes, pyrrolizidine and tropane alkaloids in spices, herbs, teas, and other plant-derived food products using chromatographic methods: Review from 2010–2020', *Food Reviews International*, pp. 1–27. doi:10.1080/87559129.2021.1929300.

Rivera-Pérez, A., Romero-González, R. and Garrido Frenich, A. (2021c) 'Persistent organic pollutants (PCBs and PCDD/Fs), PAHs, and plasticizers in spices, herbs, and tea—A review of chromatographic methods from the last decade (2010–2020)', *Critical Reviews in Food Science and Nutrition*, pp. 1–21. doi:10.1080/10408398.2021.1883546.

Rivera-Pérez, A., Romero-González, R. and Garrido Frenich, A. (2021d) 'Feasibility of applying untargeted metabolomics with GC-Orbitrap-HRMS and chemometrics for authentication of black pepper (Piper nigrum L.) and identification of geographical and processing markers', *Journal of Agricultural and Food Chemistry*, 69(19), pp. 5547–5558. doi:10.1021/acs.jafc.1c01515.

Rivera-Pérez, A., Romero-González, R. and Garrido Frenich, A. (2022) 'A metabolomics approach based on ^{1}H NMR fingerprinting and chemometrics for quality control and geographical discrimination of black pepper', *Journal of Food Composition and Analysis*, 105, p. 104235. doi:10.1016/j.jfca.2021.104235.

Rocchi, R., Mascini, M., Faberi, A., Sergi, M., Compagnone, D., Di Martino, V., Carradori, S. and Pittia, P. (2019) 'Comparison of IRMS, GC-MS and E-Nose data for the discrimination of saffron samples with different origin, process and age', *Food Control*, 106, p. 106736. doi:10.1016/j.foodcont.2019.106736.

Rubert, J., Lacina, O., Zachariasova, M. and Hajslova, J. (2016) 'Saffron authentication based on liquid chromatography high resolution tandem mass spectrometry and multivariate data analysis', *Food Chemistry*, 204, pp. 201–209. doi:10.1016/j.foodchem.2016.01.003.

Satyal, P., Murray, B.L., McFeeters, R.L. and Setzer, W.N. (2016) 'Essential oil characterization of thymus vulgaris from various geographical locations', *Foods*, 5(4), pp. 1–12. doi:10.3390/foods5040070.

Saviano, G., Paris, D., Melck, D., Fantasma, F., Motta, A. and Iorizzi, M. (2019) 'Metabolite variation in three edible Italian Allium cepa L. by NMR-based metabolomics: a comparative study in fresh and stored bulbs', *Metabolomics*, 15(8). doi:10.1007/s11306-019-1566-6.

Senizza, B., Rocchetti, G., Ghisoni, S., Busconi, M., De Los Mozos Pascual, M., Fernandez, J.A., Lucini, L. and Trevisan, M. (2019) 'Identification of phenolic markers for saffron authenticity and origin: An untargeted metabolomics approach', *Food Research International*, 126, p. 108584. doi:10.1016/j.foodres.2019.108584.

Sereshti, H., Ataolahi, S., Aliakbarzadeh, G., Zarre, S. and Poursorkh, Z. (2018) 'Evaluation of storage time effect on saffron chemical profile using gas chromatography and spectrophotometry techniques coupled with chemometrics', *Journal of Food Science and Technology*, 55(4), pp. 1350–1359. doi:10.1007/s13197-018-3046-9.

Sgorbini, B., Bicchi, C., Cagliero, C., Cordero, C., Liberto, E. and Rubiolo, P. (2015) 'Herbs and spices: Characterization and quantitation of biologically-active markers for routine quality control by multiple headspace solid-phase microextraction combined with separative or non-separative analysis', *Journal of Chromatography A*, 1376, pp. 9–17. doi:10.1016/j.chroma.2014.12.007.

Sobolev, A.P., Carradori, S., Capitani, D., Vista, S., Trella, A., Marini, F. and Mannina, L. (2014) 'Saffron samples of different origin: An NMR study of microwave-assisted extracts', *Foods*, 3, pp. 403–419. doi:10.3390/foods3030403.

Sporkert, F. and Pragst, F. (2000) 'Use of headspace solid-phase microextraction (HS-SPME) in hair analysis for organic compounds', *Forensic Science International*, 107(1–3), pp. 129–148. doi:10.1016/S0379–0738(99)00158-9.

Suzuki, R., Kasuya, Y., Sano, A., Tomita, J., Maruyama, T. and Kitamura, M. (2022) 'Comparison of various commercially available cinnamon barks using NMR metabolomics and the quantification of coumarin by quantitative NMR methods', *Journal of Natural Medicines*, 76, pp. 87–93. doi:10.1007/s11418-021-01554-6.

Tabtabaei, S., D'Archivio, A.A., Maggi, M.A., Brutus, M., Bajracharya, D.H., Konakbayeva, D., Soleimani, A., Brim, H. and Ashktorab, H. (2019) 'Geographical classification of Iranian and Italian saffron sources based on HPLC analysis and UV—Vis spectra of aqueous extracts', *European Food Research and Technology*, 245, pp. 2435–2446. doi:10.1007/s00217-019-03352-7.

US Food and Drug Administration (2018) *Compliance policy guide sec. 525.750 Spices—Definitions*. Available at: www.fda.gov/regulatory-information/search-fda-guidance-documents/cpg-sec-525750-spices-definitions.

Vallverdú-Queralt, A., Regueiro, J., Martínez-Huélamo, M., Rinaldi Alvarenga, J.F., Leal, L.N. and Lamuela-Raventos, R.M. (2014) 'A comprehensive study on the phenolic profile of widely used culinary herbs and spices: Rosemary, thyme, oregano, cinnamon, cumin and bay', *Food Chemistry*, 154, pp. 299–307. doi:10.1016/j.foodchem.2013.12.106.

Vidal-Casanella, O., Núñez, N., Sentellas, S., Núñez, O. and Saurina, J. (2020) 'Characterization of turmeric and curry samples by liquid chromatography with spectroscopic detection based on polyphenolic and curcuminoid contents', *Separations*, 7, p. 23. doi:10.3390/separations7020023.

Villa-Ruano, N., Ramírez-Meraz, M., Méndez-Aguilar, R., Zepeda-Vallejo, L.G., Álvarez-Bravo, A., Pérez-Hernández, N. and Becerra-Martínez, E. (2019) '1 H NMR-based metabolomics profiling of ten new races from Capsicum annuum cv. serrano produced in Mexico', *Food Research International*, 119, pp. 785–792. doi:10.1016/j.foodres.2018.10.061.

Wang, Y., Harrington, P. de B. and Chen, P. (2020) 'Metabolomic profiling and comparison of major cinnamon species using UHPLC—HRMS', *Analytical and Bioanalytical Chemistry*, 412(27), pp. 7669–7681. doi:10.1007/s00216-020-02904-1.

Windarsih, A., Rohman, A. and Swasono, R.T. (2018) 'Application of H-NMR metabolite fingerprinting and chemometrics for the authentication of Curcuma longa adulterated with Curcuma manga', *Journal of Applied Pharmaceutical Science*, 8(6), pp. 75–81. doi:10.7324/JAPS.2018.8610.

Windarsih, A., Rohman, A. and Swasono, R.T. (2019) 'Application of ^{1}H-NMR based metabolite fingerprinting and chemometrics for authentication of Curcuma longa adulterated with C. heyneana', *Journal of Applied Research on Medicinal and Aromatic Plants*, 13, p. 100203. doi:10.1016/j.jarmap.2019.100203.

Yang, Z., Baldermann, S. and Watanabe, N. (2013) 'Recent studies of the volatile compounds in tea', *Food Research International*, 53(2), pp. 585–599. doi:10.1016/j.foodres.2013.02.011.

Yudthavorasit, S., Wongravee, K. and Leepipatpiboon, N. (2014) 'Characteristic fingerprint based on gingerol derivative analysis for discrimination of ginger (Zingiber officinale) according to geographical origin using HPLC-DAD combined with chemometrics', *Food Chemistry*, 158, pp. 101–111. doi:10.1016/j.foodchem.2014.02.086.

Zgoła-Grześkowiak, A. and Grześkowiak, T. (2011) 'Dispersive liquid-liquid microextraction', *TrAC—Trends in Analytical Chemistry*, 30(9), pp. 1382–1399. doi:10.1016/j.trac.2011.04.014.

11 Authenticity and Chemometrics of Legumes

Rubén Agregán, Mirian Pateiro, Paulo E. S. Munekata, Theodoros Varzakas, Rubén Domínguez, Noemí Echegaray, and José M. Lorenzo

CONTENTS

11.1 INTRODUCTION

Food authentication is an area of research including a compendium of techniques that encompass a wide range of disciplines that, combined with product labeling, help increase consumer confidence. Today, this field is on the rise, attracting the attention of the authorities and the media (Bansal, Sharma, and Malhotra 2021), who see in these techniques a way to combat food fraud more efficiently. Especially interesting is the authentication of those plant foodstuff such as legumes, which have become relevant in recent times as a source of protein to replace meat for sustainability and health reasons (Huschek et al. 2018). Thus, different analytical techniques have the ability to detect adulterations and fraud in these foods. Chromatographic techniques are a fast and reliable way of similar compound separation in food matrices, which have multiple analytical challenges due to their complexity (Cseháti et al. 2005). In addition to the high proteins and fiber content (Rebello, Greenway, and Finley 2014), legumes have a large number of metabolites that include numerous bioactive compounds, such as flavonoids, phenolic acids, and tannins (Ha et al. 2014), which can be separated by chromatographic methods and compose a characteristic fingerprint of the plant that helps to authenticate it against other species or varieties. The coupling of liquid and gas chromatography to mass spectrometry (LC-MS and GC-MS, respectively) as a fundamental part of the proteomic and metabolomics approaches has contributed positively in recent years to the classification and discrimination of legumes (Scippa et al. 2010; Farag et al. 2016; Pérez-Martín, Bustamante-Rangel, and Delgado-Zamarreño 2017; Fayek et al. 2021).

The study of the elemental profile of foods is another approach capable of establishing differences between similar organisms. Thus, the mineral content of plants, which derives from the soil, can be used to find their botanical origin, helping to detect fraud in legumes, whose quality is often associated with a specific geographical area of production. Similarly, studying the isotopes present in legume tissues can reveal crucial information for sorting samples. Some methods such as isotope ratio mass spectrometry (IRMS) measure changes in stable isotope ratios produced by factors such as climatic conditions and soil geography (Danezis et al. 2016). Vibrational spectroscopy is another applicable methodology in the evaluation of food authenticity. Instrumental techniques using near- and mid-infrared radiation (NIR and MIR, respectively) have been used to discover fingerprints on plant materials, including legumes (Xu et al. 2015; Dıblan, Kadiroğlu, and Aydemir 2018; Innamorato et al. 2019).

The data provided by all these analytical methods may hide relevant information for the correct identification of the samples and their classification. A multivariate analysis of the data can detect specific markers in a multivariate data set and significantly infer in the result of food authentication studies. Therefore, this type of analytical instrument, also called chemometrics, allows optimal application of analytical methods by extracting and interpreting a great amount of complex data, identifying patterns and developing calibration models (Kamal and Karoui 2015). This

 DOI: 10.1201/9781003266167-11

chapter aims to offer an up-to-date description of the contribution of instrumental analysis and chemometric techniques to legume authentication.

11.2 NEED FOR AUTHENTICATION IN LEGUMES

Lying has been associated with food since ancient times, as with many other areas of daily life, since the lack of truth is something inherent to human being, used as a way of defense or to rise above an established system. Thus, for instance, the adulteration of food is a very widespread fraudulent act, where writings record how in the Roman Empire some practices deviated from the correct standard, such as mixing spoiled food with fresh food, substituting inferior local products for other expensive from foreign regions, or using dyes or masking additives for deceptive purposes (Schieber 2018). These unethical activities have been happening throughout history until today, where despite the controls and laws established in this regard, fraud is still present, affecting all food supply chains, being a recurring problem for the entire food industry that is situated between the scenario of being a victim and being responsible (Robson et al. 2021).

Food fraud can be defined as the marketing of food for profit and—here is the key—intent to deceive. This deception is considered a crime when there is an organized society behind it with the purpose of offending the defenseless consumer (Elliott 2014). Food fraud unfortunately affects all food sectors, affecting a great diversity of foods (Black et al. 2016; Poonia et al. 2017; Alan Reilly 2018; Kowalska, Soon, and Manning 2018; Gimonkar, E. Van Fleet, and Boys 2021). Furthermore, modern food processing, as well as globalization, have not helped to tackle this problem but to aggravate it, propitiating a fraud that goes beyond local borders; in other words, a fraud on an international scale (Montanari and Pisanello 2015).

The boom in plant-based foods as a result of the world's population growth created an opportunity for organizations that engage in food fraud. Legume plants are considered one of the most important crops worldwide due to their high protein content, serving as partial substitutes for animal protein. In fact, they represent the second most important plant group in the food pyramid after cereals (Madesis et al. 2012). One of the most common frauds that affect this group of plants is the premeditated mixing of their seeds with others from commercially more valuable popular varieties, fraudulently increasing the quality of the product. This adulteration is also carried out with seeds of species similar to the initial one in shape and color. An example of this manipulation is frequently seen in the legume commonly called Fava, with protected designation of origin Fava Santorinis (*Lathyrus clymenum*), whose grains are partially replaced by others of the same genus but of different species, such as *L. cicera, L. ochrus* and *L. sativus*. The use of other genera, such as *Vicia* and *Pisum*, has also been reported. Another example of frequent adulteration related to the origin of the product is the substitution of lentil (*Lens culinaris*) seeds by others of the genus *Vicia* (Oliveira, Amorim, and Luísa Azevedo 2017). Examples of recently food adulterations are shown in Table 11.1.

The continuous intention of trying to deceive the system and the consumer by introducing of strange and/or undeclared ingredients, the alteration of some stage during the processing of the product, or simply the use of raw materials that do not correspond to the nature of the product have led the authorities to develop authentication protocols with which to guarantee the reliability on the label. Modern instrumentation, advances in basic sciences, and information and communication technologies provide the necessary means to implement these protocols, allowing a precise and clarifying measurement of the provenance of food (Georgiou and Danezis 2015). The European Union (EU) is particularly demanding in this sense through the creation of legislation that aims to identify foods of a special quality or reputation, granting them a quality seal, including the PDO (protected designation of origin), which links the products with the geographic area where they are made; the PGI (protected geographical indication), which associates the products with a geographic area where at least one step in the production process was carried out; and the TSG (traditional specialities guaranteed), which guarantees a specific traditional elaboration of the product (Danezis et al. 2016). Some examples of legumes with these quality seals are the previously mentioned Fava Santorinis (Greece; Ganopoulos, Madesis et al. 2012), the common bean variety Plake Megalosperma Prespon (Greece; Ganopoulos, Bosmali et al. 2012), or the bean known as Fesols de Santa Pau (Spain), which have the PDO quality recognition. On the other hand, some PGIs can also be distinguished, such as the chickpea (*Cicer arietinum*) Garbanzo de Escacena (Spain), the lentil Lenteja de Tierra de Campos (Spain), or the kidney bean Fagiolo di Sorana (Italy).

11.3 CHEMOMETRICS AS A COMPLEMENTARY DISCIPLINE IN FOOD AUTHENTICATION: DESCRIPTION OF THE MOST USED METHODS

Chemometrics, also called chemoinformatics, is a discipline of chemical science established in the 1970s by the researchers Svante Wold, Bruce L. Kowalski and D.L. Massart, which uses mathematical tools, including statistics, and other methodologies to improve the design or selection of measurement procedures and experiments, as well as to optimize the chemical data collected (Héberger 2008).

Fingerprint-based modern methodologies for the evaluation of food traceability, geographical definition, and the detection of fraud and adulterations generate a large amount of difficult understanding data, which requires complex statistical analysis capable of extracting chemical relevant information. The use of chemometrics thus allows a dimensional reduction of data, providing a clearer prospect that favors its interpretation. For this "data clarification" process, available chemometric tools can be organized in several categories, called exploratory analysis, classification/discriminant analysis, and regression analyses/prediction models (Medina et al. 2019).

11.3.1 Exploratory Analysis

Exploratory analysis, also known as non-supervised approach, comprises the use of algorithms for clustering data without prior information of the group relationship, allowing data being visualized for the establishment of differences and similarities (Medina et al. 2019). Two exploratory methods highlight over the rest in food fingerprinting, principal component analysis (PCA) and hierarchical cluster analysis (HCA).

- *PCA*: This chemometric method allows to reduce the dimensionality of the data sets, increasing the interpretability but minimizing the loss of information. This drastic reduction in the amount of data can be achieved by generating new variables that are linear functions of those in the original data set, the so-called principal components. These new variables would uncorrelated with each other and would successively maximize variance (Jollife and Cadima 2016). Geladi and Linderholm (2020) explain the mathematical model in detail, collecting also relevant information about the method.

- *HCA*: In this exploratory method, the experimental samples are organized in groups, considering their nearness in multidimensional space (Medina et al. 2019). The purpose of applying HCA to the samples of an experiment is to obtain a collection of hierarchically ordered subsets. It is, therefore, an attempt to represent the available information in a proximity measure. In contrast, in a non-hierarchical cluster analysis, the proximity function being reconstructed would be dichotomous, since an object pair could be either within the same set or not (Köhn and Hubert 2015). Two main approaches are responsible for solving the sample clustering in HCA, the so-called agglomerative and divisive approaches, whose result is depicted in a dendrogram (plot displaying the organization of samples and their relationships in tree form; Granato et al. 2018). The work published by Köhn and Hubert (2015) is recommended reading to understand in-depth the descriptive clustering method that HCA intends.

11.3.2 Classification/Discriminant Analysis

The classification of samples and their discrimination, considering all pattern of features, are carried out using supervised methods, including partial least squares–discriminant analysis (PLS-DA), orthogonal projections to latent structures modeling–discriminant analysis (OPLS-DA), linear discriminant analysis (LDA), artificial neural networks (ANNs), support vector machines (SVMs), and soft independent modeling of class analogies (SIMCA). From the mentioned discriminant approaches, PLS-DA and LDA are preferably used in food fingerprinting (Medina et al. 2019).

- *PLS-DA*: The use of this discriminant chemometric method can be focused on both predictive and descriptive modeling, as well as for discriminative variable selection. PLS-DA combines discriminant analysis with data dimension reduction, so its use is specially indicated in the treatment of high-density data sets. Moreover, the non-assumption that the data have to be fitted to a specific distribution make it possible to use the algorithm in varied scientific fields (Lee, Liong, and Jemain 2018). Brereton and Lloyd described this algorithm in a recent article (Brereton and Lloyd 2014). Definitely, PLS-DA offers a simple interpretation of complex data sets through a low-dimensional score plot that illustrates the separation among different groups (Worley, Halouska, and Powers 2013).

- *LDA*: Similar to PLS-DA, LDA focuses on reducing the dimension of data, creating a subspace with a lower information density (Xanthopoulos, Pardalos, and Trafalis 2013). According to this technique, the first step would be to calculate the separability among different classes (between-class variance or different classes). Next, the distance among the mean and the samples of each class (within-class variance or within-class matrix) would be calculated. Finally, it would proceed to construct the lower dimensional space that maximizes the between-class variance and minimizes the within-class variance (Tharwat et al. 2017). There are two types of LDA, the class-dependent and the class-independent methods, both with their respective algorithms, which are described in detail in that article.

11.3.3 Regression Analysis/Prediction Models

This category encompasses different techniques, such as multiple linear regression (MLR) or partial least square–regression (PLS-R). The latter technique is one of the most used in this category. It intends the prediction of a set of dependent variables from a set of independent variables, also known as predictors. The extraction of a set of orthogonal factors, called latent variables, from these predictors allow to achieve such prediction. There are two principal variants of this technique. In one of them, the roles of dependent and independent variables are separated (the most used), while in the other, there is no such distinction, assigning the same roles to both type of variables (Abdi 2010). The use of PLS-R is especially useful when the number of independent variables greatly exceeds the number of dependent variables (Medina et al. 2019).

11.4 INCORPORATION OF CHEMOMETRICS WITH ANALYTICAL TECHNIQUES FOR LEGUME FINGERPRINTING

11.4.1 Omic Technologies

In recent decades, the academy and food science industry has become very interested in developing strategies for the identification of ingredients in complex food matrices. In this area of analytical research, strategies based on omic technologies have provided promising results since the end of the 20th century. Proteomics and metabolomics are suitable technologies for generating molecular fingerprints with which to certify the provenance of foods, production or cultivation processes. The use of these two omic technologies can also be coupled with chemometric analysis to deal with the large amount of information generated and thus achieve a better interpretation of the data (Table 11.2). The principles of these technologies, the techniques used, and examples of how the chemometric discipline is integrated into legume authentication studies are provided below.

11.4.1.1 Proteomics

The proteome is a highly dynamic system in continuous change, being affected by different factors (Ortea, O'Connor, and Maquet 2016). The characterization of the set of proteins that conform a food sample such as a legume, as well as their quantification, may help to detect deviations from the standard quality, establishing a kind of seal or unique pattern for each sample. Thus, proteomics provides the appropriate tools to address significant aspects that determine the food legitimacy, such as the identification of the species and geographical origin (Bianchi and Careri 2021), especially relevant aspects in the legume authentication. There are overall two main approaches to challenge the proteomic analysis: the top-down approach, which focuses on the characterization of intact proteins, that is, without enzymatic digestion to prior spectrometric analysis of the sample, and the bottom-up approach, which involves a previous enzymatic digestion step (Figure 11.1) and which, due to the high infrastructure and experience requirements needed by the top-down approach, is the more widely used (Creydt and Fischer 2018).

The search for characteristic protein fragments in food is a widely used strategy in authentication processes, which allows the detection of changes in the proteome, frequently associated with deceptive practices (Agregán, Echegaray, López-Pedrouso et al. 2021). These fragments are found and characterized using qualitative proteomics, being usually applied through the approaches known as peptide mass fingerprinting (PMF) and peptide fragmentation fingerprinting (PFF; I. Ortea 2021). In both approaches, mass spectrometry plays a fundamental role as the technique that will allow the obtaining of these protein markers (Böhme et al. 2019). In PMF, peptides are simply subjected to mass spectrometry after their separation/isolation by two-dimensional gel electrophoresis (2-DGE) and subsequent enzymatic digestion. In the case of PFF, mass spectrometry is used in tandem (MS/MS) to produce fragment-ion data from the different peptides, thus increasing the reliability in the protein identification. This identification is achieved by using databases, which provide a relationship between the mass spectra obtained and the proteins (Gallardo, Ortea, and Carrera 2013). However, the large amount of data generated often hinders their identification. Chemometrics has been shown to play a pivotal role in qualitative analysis by using proximity measures (Szymańska et al. 2015).

The matrix-assisted laser desorption/ionization–time of flight–mass spectrometry (MALDI-ToF-MS) technique is very useful to quickly generate spectra, profiles and/or fingerprints of foods

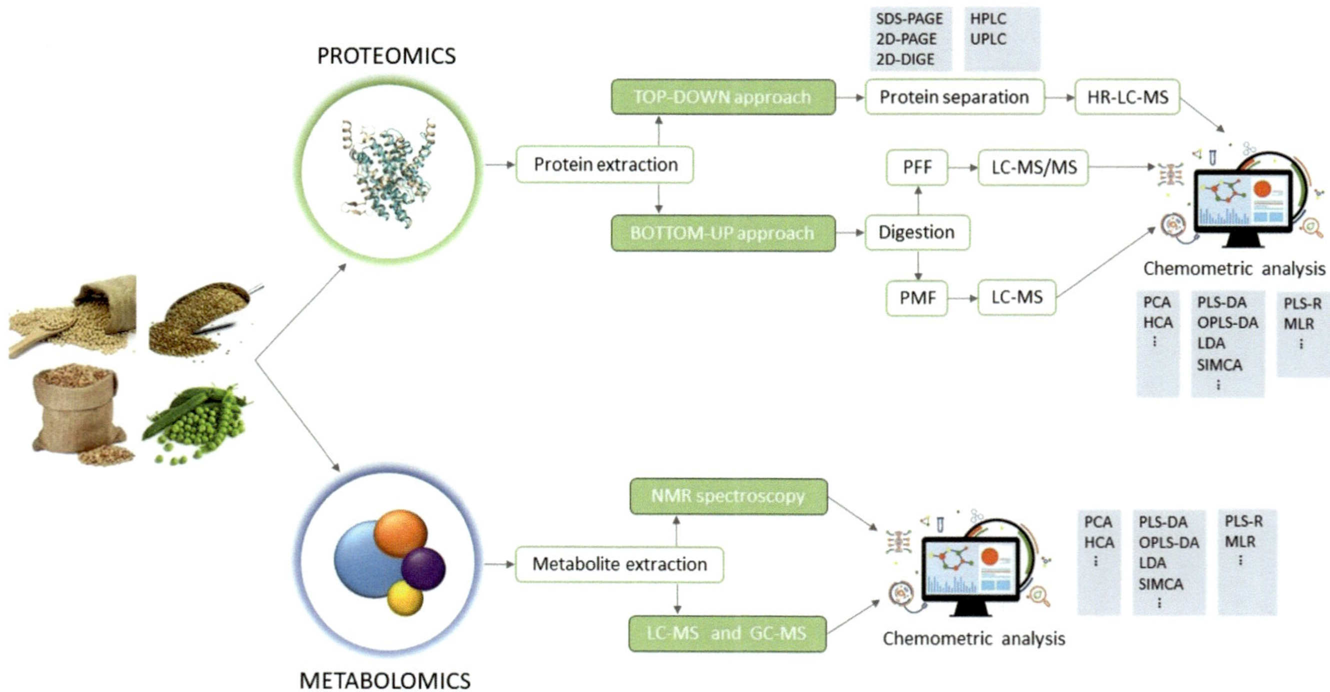

Figure 11.1 Summary workflows based on proteomic and metabolomic technologies for legume authentication. *Abbreviations*: 2D-PAGE, two-dimensional–polyacrylamide gel electrophoresis; 2D-DIGE, two-dimensional–differential gel electrophoresis; HCA, hierarchical cluster analysis; HPLC, high-performance liquid chromatography; HR-LC-MS, high-resolution–liquid chromatography coupled to mass spectrometry; LDA, linear discriminant analysis; LMR, multiple linear regression; NMR, nuclear magnetic resonance; OPLS-DA, orthogonal projections to latent structures modeling–discriminant analysis; PCA, principal component analysis; PFF, peptide fragmentation fingerprinting; PLS-DA, partial least squares–discriminant analysis; PLS-R, partial least squares–regression; PMF, peptide mass fingerprinting; SDS-PAGE, sodium dodecyl sulphate–polyacrylamide gel electrophoresis; SIMCA, soft independent modeling of class analogies; UPLC, ultra-performance liquid chromatography.

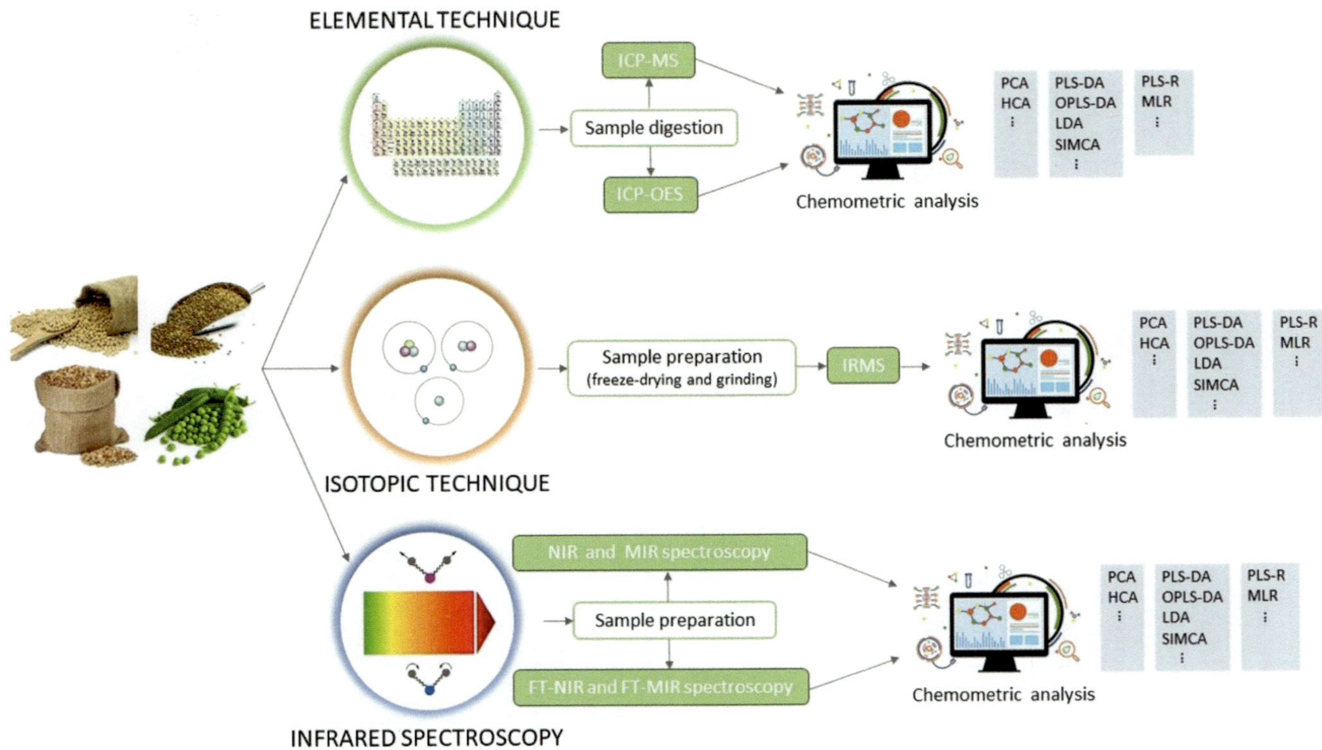

Figure 11.2 Summary workflows based on elemental and isotopic techniques and infrared spectroscopy for legume authentication. *Abbreviations*: FT-MIR, Fourier transform–mid-infrared; FT-NIR, Fourier transform–near-infrared; HCA, hierarchical cluster analysis; ICP-MS, inductively coupled plasma–mass spectroscopy; ICP-OES, inductively coupled plasma–optical emission spectroscopy; IRMS, isotope ratio mass spectrometry; LDA, linear discriminant analysis; MIR, mid-infrared; MLR, multiple linear regression; NIR, near-infrared; OPLS-DA, orthogonal projections to latent structures modeling-discriminant analysis; PCA, principal component analysis; PLS-DA, partial least squares–discriminant analysis; PLS-R, partial least squares–regression; SIMCA, soft independent modeling of class analogies.

(Zambonin 2021). Once these protein markers have been established, the reference sample is compared with the study sample (a strategy known as comparative proteomics). Caprioli et al. (2010) carried out the differentiation of seven different varieties of lentils using this proteomic approach. The MALDI-ToF technique allowed the establishment of the respective protein fingerprints, and a subsequent multivariate method of PCA achieved the detailed ordering of the varieties according to the similarity of the selected markers.

In another strategy widely used in proteomics, samples can be compared based on their level of protein expression. This is known as quantitative proteomics, and it can establish important differences regarding the origin, growing conditions and handling of legumes. Scippa et al. (2010) managed to discriminate between lentil landraces using this approach. Of the 122 protein spots identified in their study by MALDI-ToF-MS and/or nano liquid chromatography coupled to electrospray ionization–linear ion trap tandem mass spectrometry (nLC-ESI-LIT MS/MS), using 2-DGE for their previous resolution, 103 were found with different expression levels. However, a PCA helped to discard 79 of those proteins by having an irrelevant weight in the discrimination of legumes, meaning a considerable reduction in the sample size. On the other hand, the depth of the multivariate analysis detailed the specific protein spots mostly correlated in separating the commercial lentil varieties from those indigenous. Therefore, a comprehensive proteome analysis supported by the chemometric treatment of the data obtained might facilitate the selection of candidate biomarkers for the positive detection of events in plant materials. However, the analytical protocol could be considerably shortened if the study were focused on a specific protein or limited set of proteins previously recognized as of interest. This approach, known as targeted proteomics, would allow a large set of samples to be compared with high precision and reproducibility (Arora and Somasundaram 2019).

11.4.1.2 Metabolomics

As for proteins, the identification and quantification of metabolites in a biological sample provides essential information on the state of the system, which can be useful for the evaluation of possible changes in the reference standard. An analytic strategy based on metabolomics can be approached through two different categories, the known as untargeted and targeted metabolomics. The first approach focuses on the simultaneous detection of many unknown metabolites, while the second does on the identification and quantification of previously selected metabolites (Chen, Zhong, and Zhu 2020). Both approaches, with their strengths and weaknesses, serve as analytic platforms capable of detecting compounds generally below 1500 Da (Castro-Puyana and Herrero 2013), and estimating their quantity in food.

Now, nuclear magnetic resonance (NMR) spectroscopy and mass spectrometry techniques agglutinate most of the analysis in metabolomic research, allowing fraud to be addressed and verifying the authenticity of legumes. Similar to proteomic studies, chemometric tools play a prominent role in simplifying the interpretative process of data, thus easing the task of the analyst. Longobardi et al. (2017) showed the value of multivariate analysis in the discrimination of lentil samples from different provenance. Several of these chemometric methods were assessed for their performance, which included SIMCA, k-nearest neighbor (k-NN), PCA-LDA and PLS-DA. The patterns of compounds produced by a proton nuclear magnetic resonance (^{1}H-NMR) spectroscopy-based approach, the type of NMR used primarily in metabolomic studies (Emwas et al. 2019), were introduced in the aforementioned multivariate techniques, generating models that assisted in understanding the data. PCA-LDA provided the best results, recognizing the origin of 100% of the lentils examined, with a cross-validation prediction ability of 96.7% and external prediction rates of 100% and 95.3% on the test set and by a Monte Carlo cross-validation, respectively. On the other hand, good discriminations were also reported with the k-NN and PLS-DA methods. Farag et al. (2021) similarly used chemometrics for the selection of markers in the authentication process of legume sprouts of the Fabaceae family. A PCA satisfactorily provided the main fingerprints for correct discrimination among samples of the genera *Vicia* and *Lens*. However, OPLS-DA models were additionally constructed to further confirm this selection, contributing with more specific information on the differentiation between the two groups of samples. Both multivariate techniques showed a better classification of legumes when data from the aromatic region were used exclusively, since the variations between sprouts originated mainly from secondary metabolites.

Despite the ability to find metabolic markers in food, the use of NMR is limited to certain shortcomings of the technique, such as low sensitivity, which narrows the range of potentially detectable compounds. On the contrary, mass spectrometry–based metabolomics improves this parameter, also achieving a good score in others, such as operating speed and dynamic range (Agregán, Echegaray, Nawaz et al. 2021). Generally, mass spectrometry is combined with separation techniques to achieve the most possible chemical analysis of biological systems. Thus, binary couplings can be found involving liquid and gas chromatography, and capillary electrophoresis. Llorach et al. (2019) used LC-MS to classify chickpeas, lentils and white beans (*Phaseolus vulgaris*) with PGI quality seal according to their profile of bioactive compounds. Thus, a comparative untargeted metabolomics was applied, supported by an exploratory analysis of the data obtained using the liquid chromatography coupled to electrospray ionization–linear trap quadrupole–orbitrap mass spectrometry (LC–ESI–LTQ–Orbitrap MS) technique. A total of 43 compounds belonging to different classes of phytochemicals, classified into polyphenols, α-galactosides, fatty acyls, prenol lipids, nucleosides, and organic compounds, were identified after applying PCA and HCA to the data. The PCA score plot displayed a great separation among the different legumes but a strong clustering between samples of each legume, especially in lentils and chickpeas. On the other hand, the performance of an HCA with a heatmap plot revealed two main clusters (lentils vs. chickpeas and white beans) with subdivision among chickpeas and beans. The metabolomic analysis found specific biomarkers for each legume class, although there was also a shared pattern. Another study used the same approach on the same legumes, relying only on isoflavone content to establish discriminant patterns between different samples. PCA models helped differentiate among subspecies. Additionally, other multivariate analysis, such as HCA and PLS-DA reported on the different species and subspecies of beans and on the different varieties and sub-varieties of lentils, and even on the different ecotypes within the same variety of lentils (Pérez-Martín, Bustamante-Rangel, and Delgado-Zamarreño 2017). This study positively evaluates the alternative of exploring the isoflavone profile instead of analyzing the set of phytochemicals present in legume tissues in order to find biomarkers that help in the detection of fraud by adulteration.

The potential of LC-MS and GC-MS techniques for discrimination between legumes was addressed in seeds of the genus *Trigonella*. Farag et al. (2016) found that the ultra-performance liquid chromatography coupled to mass spectrometry (UPLC-MS) technique is more efficient to classify these seeds taxonomically than the GC-MS technique. A first evaluation of the data delivered by ultra-performance liquid chromatography coupled to quadrupole time of flight–mass spectrometry (UPLC-qToF-MS) consisted of an HCA of the masses detected, revealing a dendrogram with two clusters that showed a closeness among the species *Trigonella caerulea* and *T. corniculata* and a distance of both with respect to the species *T. foenum-graecum*. A subsequent PCA applied to the extracted intensities suggested a greater contribution of the secondary metabolites C-flavonoids and saponins to the separation of the samples. A similar approach via UPLC-MS coupled with HCA, PCA and OPLS-DA successfully ordered species of the genus *Vicia*, popularly known as vetch. A wide variety of compounds were identified, including flavonoids, procyanidins, prodelphinidins, anthocyanins, stilbenes, dihydrochalcones, phenolic acids, coumarins, alkaloids and fatty acids. Specifically, flavonoids and fatty acids were the most abundant groups in the seeds of these legumes, and their quantitative differences weighted more than other variables for the separation of species (Fayek et al. 2021).

The use of mass spectrometry alone, without being coupled to any separation technique, can also provide metabolomic patterns for the identification of foods. The proton transfer reaction–mass spectrometry (PTR-MS) technique is an example of this approach, quickly providing volatile-based fingerprints, even at very low concentrations (Danezis et al. 2016). However, the separation ability of this technique is not comparable to that of the GC-MS technique, for instance. In addition, the chemical structure of the detected compound often remains unclear owing to incomplete fragmentation of the molecule (Majchrzak et al. 2018). MALDI-ToF-MS can also be used alone, without being coupled to liquid chromatography, providing a rapid analysis, with good sensitivity but with poor reproducibility (Ren et al. 2018).

11.4.2 Elemental Techniques

The study of elemental profiles is an increasingly widespread methodology to assess the authenticity in food, including plants. Although the mineral content of plants is directly affected by factors such as fertilization, harvest, botanical origin, soil type, pollution and production year, the variation produced in mineral content is less than that produced due to the production area and the geographical region, since the elemental patterns are strongly linked to the geology of the cultivation area. Thus, obtaining of elemental fingerprints may be very useful to authenticate an agricultural product such as legumes according to a specific region (Danezis et al. 2016).

Inductively coupled plasma–mass spectroscopy (ICP-MS) and inductively coupled plasma–optical emission spectroscopy (ICP-OES) techniques (Figure 11.2) can be used effectively in the identification and measurement of differences in mineral content of legumes and thus discriminate between varieties of the same species, a very important issue in the detection of fraud and the protection of quality marks.

ICP-MS instruments can detect food elements at very low concentrations, in the order of trace or even ultra-trace, with good precision and sensitivity (Aceto 2016). Nevertheless, the technique produces interferences, usually caused by isobaric ions. For instance, ^{58}Fe may interfere with ^{58}Ni and ^{40}Ar with ^{40}Ca. Other factors involved in the interference production may be the sample preparation, the double-charged ions such as $^{166}Er^2$ during the $^{83}Kr^+$ determination, the well-known matrix effect of analytical processes, the space-charge effect, and the ionization interference, among others (Deng et al. 2017). The ability of ICP-MS to find elemental patterns in legumes was tested in a type of fava bean indigenous of the Greek island of Santorini (Fava Santorinis). Once the data corresponding to the trace elements (Th, Sc, B, Al, Cr, Mn, Fe, Co, Ni, Cu, Zn, Se, Cf, Ba and Tl) and rare earth elements (REEs; Y, La, Ce, Pr, Nd, Sm, Eu, Gd, Dy, Ho, Er and Yb) were delivered by the ICP-MS, the chemometric tools of orthogonal projection analysis (OPA), PLS-DA, Mahalanobis distance (MD) and k-NN were applied for their mathematical treatment. Of all techniques, the OPA showed the highest precision for the correct geographic identification of the fava both using the data from the analysis of the REEs alone and in conjunction with the trace elements. The latter approach provided the best ordering of legumes, with 100% accuracy (Drivelos et al. 2014). Zhao, Wang, and Yang (2020) similarly used the ICP-MS technique for the classification of peanut (*Arachis hypogaea*) kernels according to the production area. The k-NN, LDA and SVM methods were applied to the data delivered by the technique, effectively matching the peanut kernel with the cultivation province. SVM showed to be especially suitable for discrimination among peanuts from different grown cities.

The ICP-OES technique is less sensitive than ICP-MS due to spectral interferences caused by ionic emission lines and molecular emission bands, which prevent the correct identification and quantification of elements. However, it is often an attractive option owing to its multi-element detection ability and the low cost of the implementation and instrument maintenance (Deng et al. 2017). Foschi, D'Archivio, and Rossi (2020) used this technique for the geographic ordering and authentication of Italian lentil landraces of the nearby areas of Castelluccio di Norcia, Colfiorito and Santo Stefano di Sessanio, through the detection and quantification of 15 elements (Al, Ba, B, Cu, Mn, Mo, Ni, Sr, Zn, Ca, Fe, Na, P, K and Mg), which provided adequate discrimination of the varieties studied. The PCA exploratory strategy and the LDA and PLS-DA discriminants helped to distinguish the provenance of the samples, reaching an accuracy of 100% and 92.6%, respectively, in the prediction. Furthermore, a class modeling performed by SIMCA further strengthened the authentication of the samples, reaching a result very close to that of the exploratory and discriminant models. The four SIMCA models carried out displayed a sensitivity of 100% both in calibration and prediction, while specificity ranged between 91.8% and 100%, being able to authenticate with high efficiency the landraces with high quality seal of PGI Santo Stefano di Sessanio y Castelluccio di Norcia. Di Donato et al. (2022) carried out a similar study for the discrimination of high-value Italian varieties of chickpea according to the area employed for their cultivation. Up to 60 samples were characterized with respect to the content of ten minerals (Ca, K, P, Mg, Mo, Cu, Fe, Mn, Zn and Sr) via ICP-OES. The application of the LDA discriminant method and the SIMCA class modeling method were capable of correctly classifying the chickpea samples despite the limited number of elements analyzed, the proximity of the production areas, and the severity of the validation (55%:45% training:test). In light of these recent results in obtaining patterns based on chemical elements in two widely consumed legumes, such as lentils and chickpeas, it is deduced the ICP-OES approach coupled with the chemometric analysis of the data might be an excellent strategy to find out the traceability of legumes and authenticate varieties.

11.4.3 Isotopic Techniques

The identification of plant foods can be determined from their composition in stable isotopes, since their variation is closely linked to factors such as origin and climatic conditions. Thus, for example, hydrogen and oxygen isotopic data for organic matter in plants are linked to hydrogen and oxygen isotopic data of the water in the region of culture, and nitrogen and carbon isotopes are related to climate and agricultural practices (Drivelos and Georgiou 2012). It is important to highlight that the specific isotope fingerprints found in food might be used for the verification of crops from organic agriculture (Caio, Chalk, and Magalhães 2015; Caio and Chalk 2017), activity currently booming, aimed at promoting a better use of the natural resources in the soil, maximizing fertility and biologic activity, while protecting consumer health by not using synthetic fertilizers and pesticides and environment by minimizing the use of non-renewable resources (Shirzadi Babakan, Karimi, and Sultana 2021). Hence, the authentication of legumes grown under these conditions takes on special relevance in order to protect the consumer and a sector still in consolidation. It has been noted that $\delta^{18}O$ analysis (isotopic composition is usually expressed as delta values) of plant-derived nitrate could provide valuable information in the authentication process of these vegetable foods. In this line, the compound specific isotope analysis of nitrate was shown to remove limitations of bulk tissue $\delta^{18}N$ analysis when trying to classify plant materials according to the use or not of synthetic fertilizers during cultivation (Laursen et al. 2016).

Isotopic fingerprinting can be carried out using the IRMS (Figure 11.2), thermal ionization mass spectrometry (TIMS) and multi-collector–inductively coupled plasma–mass spectroscopy (MCICP-MS) techniques (Georgiou and Danezis 2015), which can also be used in combination with other techniques, such as NMR or gas chromatography, thus improving the determination of authentication parameters (Danezis et al. 2016). As for the previous technologies discussed in this chapter, the data delivered by these techniques requires chemometrics, which once again plays a leading role in the treatment and interpretation of the results. Evaluating the parameters of a single isotope rarely provides reliable information. However, several stable isotopes may contain remarkable information for authentication purposes as long as the data provided by the analytical technique is assessed by multivariate methods (Laursen et al. 2016). This approach has been efficiently used in legume authentication studies, when discriminating by agriculture or geographical origin. This is the case of a study carried out by using IRMS for the prediction of the provenance of some lentils according to the stable isotope ratios of $\delta^{13}C$, $\delta^{15}N$, $\delta^{2}H$, $\delta^{18}O$, and $\delta^{34}S$ found in the freeze-dried plant, as well as in the proteins and starch, which showed disparate values depending on the origin of samples, with the exception of $\delta^{15}N$. A PCA of the data delivered by the technique grouped the lentils in two differentiated clusters. Nevertheless, the existence of overlapping areas forced the application of discriminant chemometric techniques, such as PLS-DA and k-NN algorithm, for a more accurate prediction of the production place (F. Longobardi et al. 2015). Considering these results, it is shown that the IRMS technique might have a high potential in solving problems related to the production area of plant foodstuff. This is possible thanks to the great capacity of the technique in key aspects related to the origin of two or more samples, such as the ability to discriminate between chemically identical samples from different areas, the supply of information associated with the provenance of a sample through database comparison, and the determination of a probable origin by geospatial mapping of predicted data (Benson and Jones 2021).

11.4.4 Infrared (IR) Spectroscopy

IR spectroscopy is another widely used compound detection method for the authentication of plant materials (Bisutti et al. 2019; Yan et al. 2019; Nardecchia et al. 2020; Shawky, Abu El-Khair, and Selim 2020). Almost all organic compounds have associated a unique and characteristic IR spectrum, which can be used to certify their presence in different media by comparing the test spectra (unknown compounds) with the spectra of the reference compounds (Thompson 2018). This type of spectroscopy is based on the measurement of the molecule vibrations in the IR region of the electromagnetic spectrum, located between the visible (Vis) region and that of microwave radiations, where each vibratory frequency is exclusive of a functional group or molecular structure, indirectly indicating its presence in the sample. Therefore, the sum of all the measurements corresponding to the different functional groups forms an exclusive molecular fingerprint. Depending on the distance regarding the extremes of the electromagnetic region, two types of IR spectroscopy are distinguished, the NIR and MIR spectroscopy (Figure 11.2), along the wavelengths 750 and 2500 nm and between 2500 and 25,000 nm, respectively (Cozzolino 2012).

NIR spectroscopy is an easy-to-use technology that requires minimal or no sample preparation. In contrast, the reliability of the results can be affected by hydrogen bonding and sample temperature. Moreover, the dominant peaks occurring at the wavelengths 1440 and 1930 nm, corresponding to water, make it difficult to couple to chemometric analysis. Although the main disadvantage of this technology is its low sensitivity to minor constituents, being 0.1% for most of the compounds (Woodcock et al. 2008). Even so, the NIR supported by chemometric analysis may provide a correct authentication of plant foods. Xu et al. (2015) studied the feasibility of a rapid and non-destructive method for the non-targeted detection of adulterants in kudzu (*Pueraria lobata*) starch, a legume native to East Asia, using NIR spectroscopy along with the one-class partial least squares (OCPLS) class modeling technique. Taking second-order derivative (D2) spectra enhanced the specificity of the model, detecting adulterations even at a level of 0.5% or more for talcum powder adulterant and 4% or more for the cheaper starches from sweet potato, potato, maize and cassava, used to reduce the starch content of kudzu.

Fourier transform infrared (FT-IR) spectroscopy represents an improvement of IR technology for the detection of compounds, developed precisely to overcome the limitations inherent to dispersive instruments, enhancing parameters that include operating speed, sensitivity, operational simplicity and sample preparation (Dutta 2017). Dıblan, Kadiroğlu, and Aydemir (2018) characterized the legumes soybean (*Glycine max*), white bean, chickpea, red lentil and pinto bean in this way, applying Fourier transform mid-infrared (FT-MIR) spectroscopy to the extracts obtained from the use of different solvents. This improved IR spectroscopy provided a spectrum of the main functional groups present in each legume and a PCA model constructed with two main components successfully classified each sample. The results displayed the suitability of this type of spectroscopic analysis for the fast and reliable characterization of legumes, detecting diverse functional groups, such as carbonyl, hydroxyl, amide, and amine. Innamorato et al. (2019) went one step further, comparing FT-MIR and Fourier transform-near-infrared (FT-NIR) spectroscopies during a study aimed at discriminating lentils according to their origin. The application of PCA, LDA, wavelet packet transform (WPT), and PLS-DA to the instrumental data spectrum revealed that the information provided by the MIR spectral range was higher than that of the NIR to predict the specific provenance of the legumes. Nevertheless, good predictive models were also obtained through the WPT-LDA approach with the FT-NIR spectroscopy data. On the other hand, the authors did not report any synergies when using the combined data from both types of spectroscopy.

11.5 CONCLUSION

The combination of analytical methods, such as those exposed throughout the chapter, along with chemometrics has shown to be an approach able to process huge and complex data sets such as those from the analysis of legume matrices. The use of multivariate techniques in the authentication processes of these vegetables has greatly helped researchers to distinguish patterns or biomarkers for the detection of fraud related to possible adulterations or mislabeling due to intentional errors regarding the origin of the plants. Thus, the most recent applications of k-NN, SIMCA, PLS-DA and LDA have proven to be reliable techniques in this matter within food analysis, and although their use is constantly growing, it is still a challenge for the analyst to choose the right method that better solves the analytical problem. The future of food authentication and quality assurance will largely depends on the use of these types of tools for the processing and interpretation of data obtained through analytical techniques.

Table 11.1 Some Examples of Adulterations Found in the Market of Legumes

Legume	Adulteration	Reference
Chickpea (*Cicer arietinum*) and red gram (*Cajanus cajan*)	Addition of the lower-quality and cheaper legume *Lathyrus sativus*	(Paradkar, Singhal, and Kulkarni 2003)
Chickpea	Addition of flours from the lower-quality and cheaper legumes *L. sativus* and pea (*Pisum sativum*) in the production of chickpea flour	(Dattatreya, Nanjegowda, and Viswanath 2011)
Lathyrus clymenum (Fava)	Mixture with the lower-quality and cheaper legumes *Vicia fava* (common Fava), peas, *Lathyrus cicera*, *L. ochrus*, and *L. sativus*	(Ganopoulos, Madesis et al. 2012)
Alfalfa (*Medicago sativa*) and red clover (*Trifolium pratense*)	Accidental mixture with the lower-quality legume yellow trefoil (*Medicago lupulina*)	(Ganopoulos, Madesis, and Tsaftaris 2012)
Alfalfa	Addition of the lower-quality and cheaper legumes white sweet clover (*Melilotus albus*) and yellow sweet clover (*M. officinalis*)	(Luo et al. 2014)
Split red lentil (*Lens culinaris*)	Addition of the lower-quality and cheaper legume common vetch (*Vicia sativa*)	(Thavarajah et al. 2012)
Peanut (*Arachis hypogaea*)	Blend of peanut oil with cheaper oils (e.g., soybean, sunflower, canola and palm oils)	(Zhu, Wang, and Chen 2017)
	Blend of peanut oil with the cheaper rapeseed oil	(Tian et al. 2019)
Black gram (*Vigna mungo*)	Addition of cheaper wheat flour in the production of black gram *papad* flour	(Amane and Ananthanarayan 2019)
Soybean (*Glycine max*)	Addition of melamine	(Dorne et al. 2013)

Table 11.2 Coupling of Chemometrics to Analytical Techniques in Legume Authentication Studies

Legume	Target	Instrumental Technique	Chemometric Technique	Reference
Lentil (*Lens culinaris*)	Characterization of different legume varieties using protein fingerprints	MALDI-ToF-MS	PCA	(Caprioli et al. 2010)
Lentil	Discrimination between legume landraces based on most abundant proteins	2-DGE, MALDI-ToF MS, and nLC-ESI-LIT MS/MS	PCA	(Scippa et al. 2010)
Lentil	Contribution of the metabolite fingerprint in the prediction of the origin of the legume	^{1}H-NMR	SIMCA, k-NN, PCA-LDA, and PLS-DA	(Longobardi et al. 2017)
Chickpea (*Cicer arietinum*), fenugreek (*Trigonella foenum-graecum*), fava (*Vicia faba*), and lentil	Classification of legume sprouts according the identified metabolites	^{1}H-NMR	PCA and OPLS-DA	(Farag et al. 2021)
Chickpea, lentil, and white bean (*Phaseolus vulgaris*)	Identification of the most discriminating metabolites to establish differences between legumes	LC–ESI–LTQ–Orbitrap MS	PCA and HCA	(Llorach et al. 2019)
Chickpea, lentil, and bean	Classification of different varieties according to their isoflavone contents	LC-ESI MS	PCA, HCA, and PLS-DA	(Pérez-Martín, Bustamante-Rangel, and Delgado-Zamarreño 2017)
Fenugreek (*Trigonella foenum-graecum*), blue fenugreek (*T. caerulea*), and *cultivated fenugreek* (*T. corniculata*)	Establishment of differences between legumes species based on the metabolite profile	UPLC-MS and GC-MS	PCA, HCA and OPLS-DA	(Farag et al. 2016)
Vetch (*Vicia*)	Establishment of differences between legumes species based on the metabolite profile	UHPLC-ESI-qToF MS	PCA, HCA, and OPLS-DA	(Fayek et al. 2021)
Fava Santorinis (*Lathyrus clymenum*) and other yellow split peas	Discrimination among PDO legumes and other similar using REEs and trace elements	ICP-MS	OPA, MD, k-NN, and PLS-DA	(Drivelos et al. 2014)
Peanut (*Arachis hypogaea*)	Legume classification regarding the place of origin through multi-element fingerprinting	ICP-MS	PCA, k-NN, LDA, and SVM	(Zhao, Wang, and Yang 2020)

(Continued)

Table 11.2 (Continued)

Legume	Target	Instrumental Technique	Chemometric Technique	Reference
Lentil	Legume classification regarding the place of origin according to multi-elemental composition	ICP-OES	PCA, LDA, PLS-DA, SIMCA	(Foschi, D'Archivio, and Rossi 2020)
Chickpea	Legume classification regarding the place of origin using multi-elemental analysis	ICP-OES	LDA and SIMCA	(Di Donato et al. 2022)
Lentil	Prediction of the origin of the legume according to stable isotope ratios	IRMS	PCA, PLS-DA, k-NN	(Longobardi et al. 2015)
Kudzu (*Pueraria lobata*)	Detection of adulterants in legume starch	NIR spectroscopy	OCPLS	(Xu et al. 2015)
Soybean (*Glycine max*), white bean, chickpea, red lentil, and pinto bean	Characterization and discrimination of legumes according to their functional groups	FT-MIR spectroscopy	PCA	(Dıblan, Kadiroğlu, and Aydemir 2018)
Lentil	Legume discrimination according to geographical origin	FT-NIR and FT-MIR.	PCA, LDA, WPT, and PLS-DA	(Innamorato et al. 2019)

Abbreviations: ^{1}H-NMR, proton nuclear magnetic resonance; 2-DGE, two-dimensional gel electrophoresis; GC-MS, gas chromatography coupled to mass spectrometry; HCA, hierarchical cluster analysis; k-NN, k-nearest neighbor; LC–ESI–LTQ–Orbitrap MS, liquid chromatography coupled to electrospray ionization–linear trap quadrupole–Orbitrap mass spectrometry; LC-ESI MS, liquid chromatography coupled to electrospray ionization mass spectrometry; LDA, linear discriminant analysis; MALDI-ToF MS, matrix-assisted laser desorption–ionization time of flight–mass spectrometry; MD, Mahalanobis distance; nLC-ESI-LIT MS/MS, nano liquid chromatography coupled to electrospray ionization–linear ion trap tandem mass spectrometry; OCPLS, one-class partial least squares; OPA, orthogonal projection analysis; OPLS-DA, orthogonal projections to latent structures modeling–discriminant analysis; PCA, principal component analysis; PLS-DA, partial least squares–discriminant analysis; SIMCA, soft independent modeling of class analogies; SVM, support vector machines; UPLC-MS, ultra-performance liquid chromatography coupled to mass spectrometry; UHPLC-ESI-qToF MS, ultra-high performance liquid chromatography coupled to electrospray ionization–quadrupole time of flight mass spectrometry; WPT, wavelet packet transform.

REFERENCES

Abdi, Hervé. 2010. "Partial Least Squares Regression and Projection on Latent Structure Regression (PLS Regression)." *Wiley Interdisciplinary Reviews: Computational Statistics* 2 (1): 97–106. https://doi.org/10.1002/wics.51.

Aceto, Maurizio. 2016. "The Use of ICP-MS in Food Traceability." In *Advances in Food Traceability Techniques and Technologies: Improving Quality Throughout the Food Chain*, edited by Montserrat Espiñeira and Francisco J. Santaclara, 137–164. Amsterdam: Elsevier Inc. https://doi.org/10.1016/B978-0-08-100310-7.00008-9.

Agregán, Rubén, Noemí Echegaray, Asad Nawaz, Christophe Hano, Gholamreza Gohari, Mirian Pateiro, and José M. Lorenzo. 2021. "Foodomic-Based Approach for the Control and Quality Improvement of Dairy Products." *Metabolites* 11 (12). https://doi.org/10.3390/metabo11120818.

Agregán, Rubén, Noemí Echegaray, María López-Pedrouso, Rana Muhammad Aadil, Christophe Hano, Daniel Franco, and José M. Lorenzo. 2021. "Proteomic Advances in Cereal and Vegetable Crops." *Molecules* 26 (16): 4924. https://doi.org/10.3390/molecules26164924.

Alan Reilly. 2018. "Overview of Food Fraud in the Fisheries Sector." *FAO Fisheries and Aquaculture Circular*, 1165.

Amane, Dhanashree, and Laxmi Ananthanarayan. 2019. "Application of Two-Dimensional Gel Electrophoresis Technique for Protein Profiling of Indian Black Gram Varieties and Detection of Adulteration in Black Gram-Based Food Products Using Comparative Proteomics." *Food Chemistry: X* 3 (September). https://doi.org/10.1016/j.fochx.2019.100051.

Arora, Anjali, and Kumaravel Somasundaram. 2019. "Targeted Proteomics Comes to the Benchside and the Bedside: Is It Ready for Us?" *BioEssays* 41 (2). https://doi.org/10.1002/bies.201800042.

Benson, Sarah, and Kylie Jones. 2021. "Isotope Ratio Mass Spectrometry." In *Analytical Techniques in Forensic Science*, edited by Rosalind Wolstenholme, Susan Jickells, and Shari Forbes, 267–293. Hoboken: John Wiley & Sons. https://doi.org/10.1002/9781119373421.ch11.

Bansal, Sangita, Kanika Sharma, and Era V. Malhotra. 2021. "Food Authentication: Basics and Detection Methods". In *Handbook of Research on Food Processing and Preservation Technologies*, Vol. 5, edited by Monika Sharma, Megh R. Goyal, and Preeti Birwal, 193–218. New York: Apple Academic Press.

Bianchi, Federica, and Maria Careri. 2021. "Mass Spectrometry in Food Authenticity and Traceability." In *Food Authentication and Traceability*, edited by Charis M. Galanakis, 101–130. Amsterdam: Elsevier. https://doi.org/10.1016/B978-0-12-821104-5.00008-8.

Bisutti, Vittoria, Roberta Merlanti, Lorenzo Serva, Lorena Lucatello, Massimo Mirisola, Stefania Balzan, Sandro Tenti, et al. 2019. "Multivariate and Machine Learning Approaches for Honey Botanical Origin Authentication Using near Infrared Spectroscopy." *Journal of Near Infrared Spectroscopy* 27 (1): 65–74. https://doi.org/10.1177/0967033518824765.

Black, Connor, Simon A. Haughey, Olivier P. Chevallier, Pamela Galvin-King, and Christopher T. Elliott. 2016. "A Comprehensive Strategy to Detect the Fraudulent Adulteration of Herbs: The Oregano Approach." *Food Chemistry* 210 (November): 551–557. https://doi.org/10.1016/j.foodchem.2016.05.004.

Böhme, Karola, Pilar Calo-Mata, Jorge Barros-Velázquez, and Ignacio Ortea. 2019. "Recent Applications of Omics-Based Technologies to Main Topics in Food Authentication." *TrAC—Trends in Analytical Chemistry* 110 (January): 221–232. https://doi.org/10.1016/j.trac.2018.11.005.

Brereton, Richard G., and Gavin R. Lloyd. 2014. "Partial Least Squares Discriminant Analysis: Taking the Magic Away." *Journal of Chemometrics* 28 (4): 213–225. https://doi.org/10.1002/cem.2609.

Caio, Teves Inácio, and Phillip Michael Chalk. 2017. "Principles and Limitations of Stable Isotopes in Differentiating Organic and Conventional Foodstuffs: 2. Animal Products." *Critical Reviews in Food Science and Nutrition* 57 (1): 181–196. https://doi.org/10.1080/10408398.2014.887056.

Caio, Teves Inácio, Phillip Michael Chalk, and Alberto M.T. Magalhães. 2015. "Principles and Limitations of Stable Isotopes in Differentiating Organic and Conventional Foodstuffs: 1. Plant Products." *Critical Reviews in Food Science and Nutrition* 55 (9): 1206–1218. https://doi.org/10.1080/10408398.2012.689380.

Caprioli, Giovanni, Gloria Cristalli, Eugenio Ragazzi, Laura Molin, Massimo Ricciutelli, Gianni Sagratini, Roberta Seraglia, Yanting Zuo, and Sauro Vittori. 2010. "A Preliminary Matrix-Assisted Laser Desorption/Ionization Time-of-Flight Approach for the Characterization of Italian Lentil Varieties." *Rapid Communications in Mass Spectrometry* 24 (19): 2843–2248. https://doi.org/10.1002/rcm.4711.

Castro-Puyana, María, and Miguel Herrero. 2013. "Metabolomics Approaches Based on Mass Spectrometry for Food Safety, Quality and Traceability." *TrAC—Trends in Analytical Chemistry* 52: 74–87. https://doi.org/10.1016/j.trac.2013.05.016.

Chen, Li, Fanyi Zhong, and Jiangjiang Zhu. 2020. "Bridging Targeted and Untargeted Mass Spectrometry-Based Metabolomics via Hybrid Approaches." *Metabolites* 10 (9): 1–19. https://doi.org/10.3390/metabo10090348.

Cozzolino, Daniel. 2012. "Recent Trends on the Use of Infrared Spectroscopy to Trace and Authenticate Natural and Agricultural Food Products." *Applied Spectroscopy Reviews* 47 (7): 518–530. https://doi.org/10.1080/05704928.2012.667858.

Creydt, Marina, and Markus Fischer. 2018. "Omics Approaches for Food Authentication." *Electrophoresis* 39 (13): 1569–1581. https://doi.org/10.1002/elps.201800004.

Cseháti, Tibor, Esther Forgács, Zdenek Deyl, and Ivan Miksik. 2005. "Chromatography in Authenticity and Traceability Tests of Vegetable Oils and Dairy Products: A Review." *Biomedical Chromatography* 19 (3): 183–190. https://doi.org/10.1002/bmc.486.

Danezis, Georgios P., Aristidis S. Tsagkaris, Federica Camin, Vladimir Brusic, and Constantinos A. Georgiou. 2016. "Food Authentication: Techniques, Trends & Emerging Approaches." *TrAC—Trends in Analytical Chemistry* 85 (December): 123–132. https://doi.org/10.1016/j.trac.2016.02.026.

Dattatreya, Akshatha M., Divyashree K. Nanjegowda, and Prema Viswanath. 2011. "Microscopic Detection of Adulteration of Bengal Gram (*Cicer arietinum*) Flour with Other Legume Flour Based on the Seed Testa Macrosclereids." *Journal of Food Science and Technology* 48 (1): 114–119. https://doi.org/10.1007/s13197-010-0168-0.

Deng, Yujia, Xi Wu, Yunfei Tian, Zhirong Zou, Xiandeng Hou, and Xiaoming Jiang. 2017. "Sharing One ICP Source for Simultaneous Elemental Analysis by ICP-MS/OES: Some Unique Instrumental Capabilities." *Microchemical Journal* 132 (May): 401–405. https://doi.org/10.1016/j.microc.2017.02.024.

Dıblan, Sevgin, Pınar Kadiroğlu, and Levent Yurdaer Aydemir. 2018. "FT-IR Spectroscopy Characterization and Chemometrics Evaluation of Legumes Extracted with Different Solvents." *Food and Health* 4 (2): 80–88. https://doi.org/10.3153/fh18008.

Di Donato, Francesca, Francesco Squeo, Alessandra Biancolillo, Leucio Rossi, and Angelo Antonio D'Archivio. 2022. "Characterization of High Value Italian Chickpeas (*Cicer arietinum* L.) by Means of ICP-OES Multi-Elemental Analysis Coupled with Chemometrics." *Food Control* 131 (January). https://doi.org/10.1016/j.foodcont.2021.108451.

Dorne, Jean Lou, Daniel R. Doerge, Marc Vandenbroeck, Johanna Fink-Gremmels, Wim Mennes, Helle K. Knutsen, Francesco Vernazza, Laurence Castle, Lutz Edler, and Diane Benford. 2013. "Recent Advances in the Risk Assessment of Melamine and Cyanuric Acid in Animal Feed." *Toxicology and Applied Pharmacology* 270 (3): 218–229. https://doi.org/10.1016/j.taap.2012.01.012.

Drivelos, Spiros A., and Constantinos A. Georgiou. 2012. "Multi-Element and Multi-Isotope-Ratio Analysis to Determine the Geographical Origin of Foods in the European Union." *TrAC—Trends in Analytical Chemistry* 40 (November): 38–51. https://doi.org/10.1016/j.trac.2012.08.003.

Drivelos, Spiros A., Kevin Higgins, John H. Kalivas, Serkos A. Haroutounian, and Constantinos A. Georgiou. 2014. "Data Fusion for Food Authentication. Combining Rare Earth Elements and Trace Metals to Discriminate 'Fava Santorinis' from Other Yellow Split Peas Using Chemometric Tools." *Food Chemistry* 165 (December): 316–322. https://doi.org/10.1016/j.foodchem.2014.03.083.

Dutta, Aastha. 2017. "Fourier Transform Infrared Spectroscopy." In *Spectroscopic Methods for Nanomaterials Characterization*, edited by Sabu Thomas, Raju Thomas, Ajesh K. Zachariah, and Raghvendra Kumar Mishra, 73–93. Amsterdam: Elsevier. https://doi.org/10.1016/B978-0-323-46140-5.00004-2.

Elliott, Chris. 2014. *Elliott Review into the Integrity and Assurance of Food Supply Networks-Final Report: A National Food Crime Prevention Framework*. London: Department for Environment, Food & Rural Affairs Food Standards Agency.

Emwas, Abdul Hamid, Raja Roy, Ryan T. McKay, Leonardo Tenori, Edoardo Saccenti, G. A. Nagana Gowda, Daniel Raftery, et al. 2019. "NMR Spectroscopy for Metabolomics Research." *Metabolites* 9 (7): 123. https://doi.org/10.3390/metabo9070123.

Farag, Mohamed A., Dalia M. Rasheed, Matthias Kropf, and Andreas G. Heiss. 2016. "Metabolite Profiling in *Trigonella* Seeds via UPLC-MS and GC-MS Analyzed Using Multivariate Data Analyses." *Analytical and Bioanalytical Chemistry* 408 (28): 8065–8078. https://doi.org/10.1007/s00216-016-9910-4.

Farag, Mohamed A., Mohamed G. Sharaf El-Din, Mohamed A. Selim, Asmaa I. Owis, Sameh F. Abouzid, Andrea Porzel, Ludger A. Wessjohann, and Asmaa Otify. 2021. "Nuclear Magnetic Resonance Metabolomics Approach for the Analysis of Major Legume Sprouts Coupled to Chemometrics." *Molecules* 26 (3). https://doi.org/10.3390/molecules26030761.

Fayek, Nesrin M., Reham Hassan Mekky, Clarice Noleto Dias, Matthias Kropf, Andreas G. Heiss, Ludger A. Wessjohann, and Mohamed A. Farag. 2021. "UPLC-MS Metabolome-Based Seed Classification of 16 Vicia Species: A Prospect for Phyto-Equivalency and Chemotaxonomy of Different Accessions." *Journal of Agricultural and Food Chemistry* 69 (17): 5252–5266. https://doi.org/10.1021/acs.jafc.0c06054.

Foschi, Martina, Angelo Antonio D'Archivio, and Leucio Rossi. 2020. "Geographical Discrimination and Authentication of Lentils (*Lens culinaris* Medik.) by ICP-OES Elemental Analysis and Chemometrics." *Food Control* 118 (December). https://doi.org/10.1016/j.foodcont.2020.107438.

Gallardo, José Manuel, Ignacio Ortea, and Mónica Carrera. 2013. "Proteomics and Its Applications for Food Authentication and Food-Technology Research." *TrAC—Trends in Analytical Chemistry* 52: 135–141. https://doi.org/10.1016/j.trac.2013.05.019.

Ganopoulos, Ioannis, Irene Bosmali, Panagiotis Madesis, and Athanasios Tsaftaris. 2012. "Microsatellite Genotyping with HRM (High Resolution Melting) Analysis for Identification of the PGI Common Bean Variety Plake Megalosperma Prespon." *European Food Research and Technology* 234 (3): 501–508. https://doi.org/10.1007/s00217-011-1653-z.

Ganopoulos, Ioannis, Panagiotis Madesis, and Athanasios Tsaftaris. 2012. "Universal ITS2 Barcoding DNA Region Coupled with High-Resolution Melting (HRM) Analysis for Seed Authentication and Adulteration Testing in Leguminous Forage and Pasture Species." *Plant Molecular Biology Reporter* 30 (6): 1322–1328. https://doi.org/10.1007/s11105-012-0453-3.

Ganopoulos, Ioannis, Panagiotis Madesis, Nikos Darzentas, Anagnostis Argiriou, and Athanasios Tsaftaris. 2012. "Barcode High Resolution Melting (Bar-HRM) Analysis for Detection and Quantification of PDO 'Fava Santorinis' (*Lathyrus clymenum*) Adulterants." *Food Chemistry* 133 (2): 505–512. https://doi.org/10.1016/j.foodchem.2012.01.015.

Geladi, Paul, and Johan Linderholm. 2020. "Principal Component Analysis." In *Comprehensive Chemometrics. Chemical and Biochemical Data Analysis*, edited by Steven Brown, Romà Tauler, and Beata Walczak, Second, 2:17–37. Amsterdam: Elsevier. https://doi.org/10.1016/B978-0-12-409547-2.14892-9.

Georgiou, Constantinos A., and Georgios P. Danezis. 2015. "Elemental and Isotopic Mass Spectrometry." In *Comprehensive Analytical Chemistry*, edited by Yolanda Picó, 68:131–243. Amsterdam: Elsevier. https://doi.org/10.1016/B978-0-444-63340-8.00003-0.

Gimonkar, Shweta, Erin E. Van Fleet, and Kathryn A. Boys. 2021. "Dairy Product Fraud." In *Food Fraud*, edited by Rosalee S. Hellberg, Karen Everstine, and Steven A. Sklare, 249–279. Amsterdam: Elsevier. https://doi.org/10.1016/B978-0-12-817242-1.00014-2.

Granato, Daniel, Jânio S. Santos, Graziela B. Escher, Bruno L. Ferreira, and Rubén M. Maggio. 2018. "Use of Principal Component Analysis (PCA) and Hierarchical Cluster Analysis (HCA) for Multivariate Association between Bioactive Compounds and Functional Properties in Foods: A Critical Perspective." *Trends in Food Science and Technology* 72 (February): 83–90. https://doi.org/10.1016/j.tifs.2017.12.006.

Ha, Tae Joung, Byong Won Lee, Ki Hun Park, Seong Hun Jeong, Hyun Tae Kim, Jong Min Ko, In Youl Baek, and Jin Hwan Lee. 2014. "Rapid Characterisation and Comparison of Saponin Profiles in the Seeds of Korean Leguminous Species Using Ultra Performance Liquid Chromatography with Photodiode Array Detector and Electrospray Ionisation/Mass Spectrometry (UPLC-PDA-ESI/MS) Analysis." *Food Chemistry* 146: 270–277. https://doi.org/10.1016/j.foodchem.2013.09.051.

Héberger, Károly. 2008. "Chemoinformatics—Multivariate Mathematical—Statistical Methods for Data Evaluation." In *Medical Applications of Mass Spectrometry*, edited by Károly Vékey, András Telekes, and Akos Vertes, 141–169. Amsterdam: Elsevier. https://doi.org/10.1016/B978-044451980-1.50009-4.

Huschek, Gerd, Josephine Bönick, Dietrich Merkel, Doreen Huschek, and Harshadrai Rawel. 2018. "Authentication of Leguminous-Based Products by Targeted Biomarkers Using High Resolution Time of Flight Mass Spectrometry." *LWT—Food Science and Technology* 90 (April): 164–171. https://doi.org/10.1016/j.lwt.2017.12.034.

Innamorato, Valentina, Francesco Longobardi, Vincenzo Lippolis, Marina Cortese, Antonio F. Logrieco, Lucia Catucci, Angela Agostiano, and Annalisa De Girolamo. 2019. "Tracing the Geographical Origin of Lentils (*Lens culinaris* Medik.) by Infrared Spectroscopy and Chemometrics." *Food Analytical Methods* 12 (3): 773–779. https://doi.org/10.1007/s12161-018-1406-8.

Jollife, Ian T., and Jorge Cadima. 2016. "Principal Component Analysis: A Review and Recent Developments." *Philosophical Transactions of the Royal Society A: Mathematical, Physical and Engineering Sciences* 374 (2065). https://doi.org/10.1098/rsta.2015.0202.

Kamal, Mohammad, and Romdhane Karoui. 2015. "Analytical Methods Coupled with Chemometric Tools for Determining the Authenticity and Detecting the Adulteration of Dairy Products: A Review." *Trends in Food Science and Technology* 46 (1): 27–48. https://doi.org/10.1016/j.tifs.2015.07.007.

Köhn, Hans-Friedrich, and Lawrence J. Hubert. 2015. "Hierarchical Cluster Analysis." In *Wiley StatsRef: Statistics Reference Online*, 1–13. Hoboken: John Wiley & Sons, Ltd. https://doi.org/10.1002/9781118445112.stat02449.pub2.

Kowalska, Aleksandra, Jan Mei Soon, and Louise Manning. 2018. "A Study on Adulteration in Cereals and Bakery Products from Poland Including a Review of Definitions." *Food Control* 92 (October): 348–356. https://doi.org/10.1016/j.foodcont.2018.05.007.

Laursen, Kristian Holst, Luana Bontempo, Federica Camin, and Andreas Rossmann. 2016. "Advances in Isotopic Analysis for Food Authenticity Testing." In *Advances in Food Authenticity Testing*, edited by Gerard Downey, 227–252. Amsterdam: Elsevier. https://doi.org/10.1016/B978-0-08-100220-9.00009-6.

Lee, Loong Chuen, Choong Yeun Liong, and Abdul Aziz Jemain. 2018. "Partial Least Squares-Discriminant Analysis (PLS-DA) for Classification of High-Dimensional (HD) Data: A Review of Contemporary Practice Strategies and Knowledge Gaps." *Analyst* 143 (15): 3526–3539. https://doi.org/10.1039/c8an00599k.

Llorach, Rafael, Claudia Favari, David Alonso, Mar Garcia-Aloy, Cristina Andres-Lacueva, and Mireia Urpi-Sarda. 2019. "Comparative Metabolite Fingerprinting of Legumes Using LC-MS-Based Untargeted Metabolomics." *Food Research International* 126 (December). https://doi.org/10.1016/j.foodres.2019.108666.

Longobardi, Francesco, Grazia Casiello, Marina Cortese, Matteo Perini, Federica Camin, Lucia Catucci, and A. Agostiano. 2015. "Discrimination of Geographical Origin of Lentils (*Lens culinaris* Medik.) Using Isotope Ratio Mass Spectrometry Combined with Chemometrics." *Food Chemistry* 188 (May): 343–349. https://doi.org/10.1016/j.foodchem.2015.05.020.

Longobardi, Francesco, Valentina Innamorato, Annalisa Di Gioia, Andrea Ventrella, Vincenzo Lippolis, Antonio F. Logrieco, Lucia Catucci, and Angela Agostiano. 2017. "Geographical Origin Discrimination of Lentils (*Lens culinaris* Medik.) Using ^{1}H NMR Fingerprinting and Multivariate Statistical Analyses." *Food Chemistry* 237 (December): 443–448. https://doi.org/10.1016/j.foodchem.2017.05.159.

Luo, D., W.X. Liu, Y.R. Wang, J.Y. Zhang, and Z.P. Liu. 2014. "Development of a Rapid One-Step PCR Protocol to Distinguish between Alfalfa (*Medicago sativa*) and Sweet Clover (*Melilotus* spp.) Seeds." *Seed Science and Technology* 42 (2): 237–246. https://doi.org/10.15258/sst.2014.42.2.12.

Madesis, Panagiotis, Ioannis Ganopoulos, Argiriou Anagnostis, and Athanasios Tsaftaris. 2012. "The Application of Bar-HRM (Barcode DNA-High Resolution Melting) Analysis for Authenticity Testing and Quantitative Detection of Bean Crops (Leguminosae) without Prior DNA Purification." *Food Control* 25 (2): 576–582. https://doi.org/10.1016/j.foodcont.2011.11.034.

Majchrzak, Tomasz, Wojciech Wojnowski, Martyna Lubinska-Szczygeł, Anna Różańska, Jacek Namieśnik, and Tomasz Dymerski. 2018. "PTR-MS and GC-MS as Complementary Techniques for Analysis of Volatiles: A Tutorial Review." *Analytica Chimica Acta* 1035 (December): 1–13. https://doi.org/10.1016/j.aca.2018.06.056.

Medina, Sonia, Rosa Perestrelo, Pedro Silva, Jorge A.M. Pereira, and José S. Câmara. 2019. "Current Trends and Recent Advances on Food Authenticity Technologies and Chemometric Approaches." *Trends in Food Science and Technology* 85 (March): 163–176. https://doi.org/10.1016/j.tifs.2019.01.017.

Montanari, Francesco, and Daniele Pisanello. 2015. "Responsibilities for Food Information to Consumers: Much Ado about Nothing?" *European Food and Feed Law Review* 10 (2): 107–113.

Nardecchia, Alessandro, Regina Presutto, Remo Bucci, Federico Marini, and Alessandra Biancolillo. 2020. "Authentication of the Geographical Origin of 'Vallerano' Chestnut by Near Infrared Spectroscopy Coupled with Chemometrics." *Food Analytical Methods* 13 (9): 1782–1790. https://doi.org/10.1007/s12161-020-01791-7.

Oliveira, Manuela, Maria Isabel Amorim, and & Luísa Azevedo. 2017. "Authentication of Plant Food Products: Under the Magnification of Botany Forensics." *Nova Acta Científica Compostelana (Bioloxía)* 24: 45–62.

Ortea, I. 2021. "Omics and Food Authentication." In *Foodomics: Omic Strategies and Applications in Food Science*, edited by Jorge Barros-Velázquez, 204–221. Cambridge: RSC Publishing.

Ortea, Ignacio, Gavin O'Connor, and Alain Maquet. 2016. "Review on Proteomics for Food Authentication." *Journal of Proteomics* 147 (September): 212–225. https://doi.org/10.1016/j.jprot.2016.06.033.

Paradkar, Manish M., Rekha S. Singhal, and Pushpa R. Kulkarni. 2003. "Detection of *Lathyrus sativus* in Processed Chickpea- and Red Gram-Based Products by Thin Layer Chromatography." *Journal of the Science of Food and Agriculture* 83 (7): 727–730. https://doi.org/10.1002/jsfa.1361.

Pérez-Martín, Lara, Myriam Bustamante-Rangel, and M. Milagros Delgado-Zamarreño. 2017. "Classification of Lentils, Chickpeas and Beans Based on Their Isoflavone Content." *Food Analytical Methods* 10 (5): 1191–1201. https://doi.org/10.1007/s12161-016-0675-3.

Poonia, Amrita, Alok Jha, Rajan Sharma, Harikesh Bahadur Singh, Ashwini Kumar Rai, and Nitya Sharma. 2017. "Detection of Adulteration in Milk: A Review." *International Journal of Dairy Technology* 70 (1): 23–42. https://doi.org/10.1111/1471-0307.12274.

Rebello, Candida. J., Frank Lyons Greenway, and John Weldon Finley. 2014. "A Review of the Nutritional Value of Legumes and Their Effects on Obesity and Its Related Co-Morbidities." *Obesity Reviews* 15 (5): 392–407. https://doi.org/10.1111/obr.12144.

Ren, Jun Ling, Ai Hua Zhang, Ling Kong, and Xi Jun Wang. 2018. "Advances in Mass Spectrometry-Based Metabolomics for Investigation of Metabolites." *RSC Advances* 8 (40): 22335–22350. https://doi.org/10.1039/c8ra01574k.

Robson, Kelsey, Moira Dean, Simon Haughey, and Christopher Elliott. 2021. "A Comprehensive Review of Food Fraud Terminologies and Food Fraud Mitigation Guides." *Food Control* 120 (February). https://doi.org/10.1016/j.foodcont.2020.107516.

Schieber, Andreas. 2018. "Introduction to Food Authentication." In *Modern Techniques for Food Authentication*, edited by Da-Wen Sun, Second, 1–21. Amsterdam: Elsevier. https://doi.org/10.1016/B978-0-12-814264-6.00001-3.

Scippa, Gabriella Stefania, Mariapina Rocco, Manuela Ialicicco, Dalila Trupiano, Vincenzo Viscosi, Michela Di Michele, Simona Arena, Donato Chiatante, and Andrea Scaloni. 2010. "The Proteome of Lentil (*Lens culinaris* Medik.) Seeds: Discriminating between Landraces." *Electrophoresis* 31 (3): 497–506. https://doi.org/10.1002/elps.200900459.

Shawky, Eman, Rasha A. Abu El-Khair, and Dina A. Selim. 2020. "NIR Spectroscopy-Multivariate Analysis for Rapid Authentication, Detection and Quantification of Common Plant Adulterants in Saffron (*Crocus sativus* L.) Stigmas." *LWT* 122 (March). https://doi.org/10.1016/j.lwt.2020.109032.

Shirzadi Babakan, Ali, Firoozeh Karimi, and Selima Sultana. 2021. "Land Suitability Evaluation for Organic Agriculture: A Case Study in the US." In *Exploring and Optimizing Agricultural Landscapes*, edited by Lothar Mueller, Viktor G. Sychev, Nikolai M. Dronin, and Frank Eulenstein, 329–347. Cham: Springer. https://doi.org/10.1007/978-3-030-67448-9_14.

Szymańska, Ewa, Jan Gerretzen, Jasper Engel, Brigitte Geurts, Lionel Blanchet, and Lutgarde M.C. Buydens. 2015. "Chemometrics and Qualitative Analysis Have a Vibrant Relationship." *TrAC—Trends in Analytical Chemistry* 69 (June): 34–51. https://doi.org/10.1016/j.trac.2015.02.015.

Tharwat, Alaa, Tarek Gaber, Abdelhameed Ibrahim, and Aboul Ella Hassanien. 2017. "Linear Discriminant Analysis: A Detailed Tutorial." *AI Communications* 30 (2): 169–190. https://doi.org/10.3233/AIC-170729.

Thavarajah, Pushparajah, Dil Thavarajah, G. A.S. Premakumara, and Albert Vandenberg. 2012. "Detection of Common Vetch (*Vicia sativa* L.) in Lentil (*Lens culinaris* L.) Using Unique Chemical Fingerprint Markers." *Food Chemistry* 135 (4): 2203–2206. https://doi.org/10.1016/j.foodchem.2012.06.124.

Thompson, James M. 2018. *Infrared Spectroscopy*. Singapore: Pan Stanford Publishing.

Tian, Lili, Yuanyuan Zeng, Xiuqian Zheng, Yahuang Chiu, and Tristan Liu. 2019. "Detection of Peanut Oil Adulteration Mixed with Rapeseed Oil Using Gas Chromatography and Gas Chromatography—Ion Mobility Spectrometry." *Food Analytical Methods* 12 (10): 2282–2292. https://doi.org/10.1007/s12161-019-01571-y.

Woodcock, Tony, Colette C. Fagan, Colm P. O'Donnell, and Gerard Downey. 2008. "Application of Near and Mid-Infrared Spectroscopy to Determine Cheese Quality and Authenticity." *Food and Bioprocess Technology* 1 (2): 117–129. https://doi.org/10.1007/s11947-007-0033-y.

Worley, Bradley, Steven Halouska, and Robert Powers. 2013. "Utilities for Quantifying Separation in PCA/PLS-DA Scores Plots." *Analytical Biochemistry* 433 (2): 102–104. https://doi.org/10.1016/j.ab.2012.10.011.

Xanthopoulos, Petros, Panos M. Pardalos, and Theodore B. Trafalis. 2013. "Linear Discriminant Analysis." In *Robust Data Mining*, 27–33. New York: Springer. https://doi.org/10.1007/978-1-4419-9878-1_4.

Xu, Lu, Wei Shi, Chen Bo Cai, Wei Zhong, and Kang Tu. 2015. "Rapid and Nondestructive Detection of Multiple Adulterants in Kudzu Starch by Near Infrared (NIR) Spectroscopy and Chemometrics." *LWT—Food Science and Technology* 61 (2): 590–595. https://doi.org/10.1016/j.lwt.2014.12.002.

Yan, Hui, Yi Chao Xu, Heinz W. Siesler, Bang Xing Han, and Guo Zheng Zhang. 2019. "Hand-Held Near-Infrared Spectroscopy for Authentication of Fengdous and Quantitative Analysis of Mulberry Fruits." *Frontiers in Plant Science* 10 (November). https://doi.org/10.3389/fpls.2019.01548.

Zambonin, Carlo. 2021. "MALDI-TOF Mass Spectrometry Applications for Food Fraud Detection." *Applied Sciences* 11 (8). https://doi.org/10.3390/app11083374.

Zhao, Haiyan, Feng Wang, and Qingli Yang. 2020. "Origin Traceability of Peanut Kernels Based on Multi-Element Fingerprinting Combined with Multivariate Data Analysis." *Journal of the Science of Food and Agriculture* 100 (10): 4040–4048. https://doi.org/10.1002/jsfa.10449.

Zhu, Wenran, Xin Wang, and Lihua Chen. 2017. "Rapid Detection of Peanut Oil Adulteration Using Low-Field Nuclear Magnetic Resonance and Chemometrics." *Food Chemistry* 216 (February): 268–274. https://doi.org/10.1016/j.foodchem.2016.08.051.

12 Authenticity and Chemometrics of Mushrooms

Sofia Agriopoulou, Hesham el Enshashy, and Theodoros Varzakas

CONTENTS

12.1 INTRODUCTION

For many centuries there was no specific definition of the mushroom or consensus on its uses, but given its value and its products, there should be a clear and comprehensive definition. In this regard, experts have described two definitions. The first is a descriptive-functional definition that refers to a creation without leaves, without buds, without flowers that nevertheless forms fruit and is used as food, as a tonic and as medicine, while the whole creation is valuable. Another descriptive definition defines a mushroom as "macrophage with a distinctive fruiting body". It can grow above the ground (epigeous) or grow entirely in the ground (hypogenous) and be large enough to be seen with the naked eye and collected by hand. Mushrooms can generally be grouped into four categories: (1) edible mushrooms (e.g., *Agaricus bisporus*); (2) medicinal mushrooms (e.g., *Ganoderma lucidum* [*Lingzhi*]); (3) poisonous mushrooms (e.g., *Amanita phalloides*); and (4) other mushrooms, which include many mushrooms whose properties are not well defined (El Sheikha and Hu, 2018).

Mushroom cultivation can be considered as an effective means of extracting resources from agricultural solid recyclable waste. The most important benefit associated with their cultivation is, in fact, that they can convert agricultural and other organic waste into nutrients. This procedure of biotransformation is part of an environmentally sound protection strategy and contributes to the production of protein-rich foods that, if used properly, can contribute to the global diet and provide marketable products in poor areas (Giusti et al., 2021a).

Mushrooms have been considered a delicacy for centuries due to their distinctive texture, aroma and excellent taste, while relevant research has proven unique benefits for nutrition and health. The popularity of mushrooms has grown steadily in recent decades, leading to a 30-fold increase in world production of edible mushrooms, with *Lentinus edodes* (shiitake) and *Pleurotus ostreatus* (oyster) representing more than 40% of the corresponding total supply (Tagkouli et al., 2021). Many *Pleurotus* species are currently produced worldwide (e.g., *P. ostreatus, P. pulmonarius, P. eryngii, P. djamor and P. citrinopileatus*) originating from lignocellulosic residues as substrates.

Of particular interest is their cultivation in by-products of olive mills and wineries (i.e., materials that are difficult to manage due to their high content of organic compounds, especially lipids and polyphenols), which is in line with the concepts of biodegradation and utilization of waste (Koutrotsios et al.,

 DOI: 10.1201/9781003266167-12

2018). At least 58 lignocellulosic substrates from agro-industrial waste have been used to grow *Pleurotus* spp. with promising results from mixtures of cottonseed husk, wheat straw, sawdust, sugar cane and ground limestone that have achieved a biological yield of 99.8%–104.3% (Tarrillo et al., 2021).

The characteristic odor of mushrooms is the combined effect of different groups of compounds, such as alcohols, aldehydes, ketones, acids, hydrocarbons, esters and heterocyclic, aromatic and sulfur compounds (Sun et al., 2020; Aisala et al., 2019; Tietel and Masaphy, 2018) while their relative abundance may vary between species, strains, growth stage and part of the mushroom under consideration (e.g., pileus and stipe; Feng et al., 2021). One of the most expensive foods in the world is the mushroom-truffle (*Tuber* spp.). With a unique and intense aroma. Their aroma is influenced by the species, the maturity, the genotype, the microbial communities that inhabit the fruiting bodies, climate and soil composition (Šiškovič et al., 2021). Fresh truffles are extremely fragile and have a shelf life of 7–10 days, which is also the time when they show their best sensory properties (Campo et al., 2017).

12.2 NUTRITIONAL BENEFITS AND RISKS OF EATING MUSHROOMS

There are 14,000 known species of mushrooms, of which 3000 are safe for human consumption. Mushrooms are a rich source of nutrients, especially proteins, dietary fiber, carbohydrates, minerals (sodium, potassium, calcium, magnesium and phosphorus), alkaloids, small-chain fatty acids, enzymes, leukocytes and enzymes. Mushrooms contain 20%–45% protein (dry weight), are low in lipids and contain a significant amount of essential amino acids such as valine, leucine, threonine, methionine, lysine, phenylalanine and tryptophan. The quality of mushrooms in protein is better than the quality of most proteins of plant origin (Fernandes et al., 2021; USDA, 2018). The digestibility of mushroom proteins is good and ranks lower than that of animal proteins but far above other foods such as vegetables, fruits and milk (El Sheikha and Hu, 2018). The composition of amino acids and proteins found in mushrooms is very similar to that of animal proteins, so they are much more beneficial than vegetable proteins. Among the edible species, about 200 are wild and 35 are cultivated commercially (Meenu and Xu, 2019).

Polysaccharides β-glucans are important components of the cell walls of mushrooms. In recent years, their health effects have been studied in cellular and experimental models, to add scientifically based knowledge to their long-term beneficial effects and to clarify their mechanisms (Cerletti et al., 2021).

The nutritional benefits of mushrooms vary from species to species. For example, the mushroom *Macrohyporia cocos*, which is an edible and medicinal species, has been shown by phytochemical studies to contain terpenes and polysaccharides, which have beneficial biological properties, such as a prebiotic effect, through the modulation of the composition of intestinal microbes (Khan et al., 2018), anti-hyperlipidemic actions (Miao et al., 2016), anticancer actions (Lee et al., 2018) hepatoprotective actions (Wu et al., 2018) and the effects of differentiation of adipocytes and osteoblasts (Lee et al., 2018). Table 12.1 shows the bioactive compounds of various mushrooms.

Consumption of mushrooms is not without risks. Mushroom poisoning is actually an important form of disease caused by toxins and can affect the gastrointestinal, neurological, renal and hepatic systems (Govorushko et al., 2019; White et al., 2019). In the presence of highly poisonous species, the

Table 12.1 Bioactive Compounds of Different Mushrooms

Family/Genus/Species of Mushroom	Bioactive Compounds	Reference
Pleurotus	Ergosterol, vitamin D_2	(Bekiaris et al., 2020)
Pleurotus spp. (oyster)	Chitin, β-glucans, water-soluble vitamins, ergosterol (provitamin D2), phenolic acids, ergothionein, lovastatin	(Tagkouli et al., 2021)
Bachu	Glycoproteins, terpenoids, polyphenols	(Ma et al., 2020)
Macrohyporia cocos	Terpenes, polysaccharides	(Wang et al., 2019a)
Ganodermataceae	Triterpenes, polysaccharides, phenols, steroids	(Li et al., 2020)
Poria cocos	Triterpenes, polysaccharides	(Zhu et al., 2019)

liver undergoes irreparable damage that can be fatal. In addition, some edible species can become toxic if collected, transported, stored and cooked (Gawlikowski et al., 2015; Nieminen et al., 2020).

While there are known species, such as *Amanita phalloides*, that cause life-threatening poisoning, in fact there is also accumulated evidence of poisoning associated with species that have been considered edible and traditionally consumed (Nieminen et al., 2020).

Heavy metals can enter the food chain through the consumption of edible mushrooms. Therefore, it is important to assess the levels of heavy metals and report potential contamination that may be dangerous to human health. Edible mushrooms are a valuable component of our daily diet; however, due to the significant increase in the levels of trace elements in the environment in recent decades, mushrooms can accumulate high levels of heavy metals.

The results of the study of Huang et al. (2015), which investigated 285 samples, nine species (*Lentinus edodes, Auricularia auricula, Pleurotus ostreatus, Tremella fuciformis, Flammulina velutipes, Agrocybe chaxinggu, Armillaria mellea, Agaricus bisporus* and *Pholiota nameko*) showed that the edible mushrooms collected from Beijing markets all contained cadmium (Cd), arsenic (As), mercury (Hg), lead (Pb), iron (Fe) and zinc (Zn). In addition, the estimated weekly trace element intake was compared with the temporarily tolerated weekly intake as established by the World Health Organization (WHO) and the UN Food and Agricultural Organization (FAO).

The results of the study showed that the estimated weekly intake of Cd, As, Hg, Pb, Fe and Zn from the consumption of the examined edible mushrooms was lower than the temporary tolerated weekly intake (Huang et al., 2015).

In a recent review, Zhang et al. (2019), provided an up-to-date review of the findings on arsenic and various species of arsenic in Chinese mushrooms, collecting data from published sources in English or Chinese related to different environments and health hazards. This natural environmental pollutant, the chronic exposure of which can cause various types of cancer, has been found in several mushroom families (Zhang et al., 2020).

12.3 USES AND APPLICATIONS OF MUSHROOMS

Although taxonomists have identified mushrooms as fungi, mushrooms can be classified in the group of vegetables based on their use (Feeney et al., 2014). Edible mushrooms are tasty and nutritious macro-fungi with fruiting bodies and mycelia, both wild and cultivated, belonging to the classes of Basidiomycetes and Ascomycetes (Cerletti et al., 2021).

Mushrooms offer excellent meat alternatives for vegetarian meals and can be found on the market as fresh, dried or canned (Giusti et al., 2021b). Tea (sleep friendly), pies, jellies, candies and various drinks are some of the products produced from mushrooms (Wang et al., 2019a). Figure 12.1 depicts the most notable edible mushrooms in the world (Feeney et al., 2014).

There are about 2000 known medicinal mushrooms with a variety of health effects (El Sheikha and Hu, 2018; Figure 12.2). From a medical point of view, mushrooms are valuable foods for health, as they have bioactive compounds of high medicinal value such as lectins, polysaccharides, phenols, polyphenols, terpenoids, ergosterols and volatile organic compounds. The positive health effects of these biomolecules include many forms such as antimicrobial, hepatoprotective, antioxidant, antidiabetic, anti-hypercholesterolemia, anticancer and immunomodulatory (Wasser, 2014).

Medicinal mushrooms have a long history of use in conventional oriental therapies, especially in China, Korea and Japan. A medicinal mushroom that deserves attention is *Lingzhi*. Lingzhi is the Chinese name given to the *Ganoderma* mushroom family. *Lingzhi* is called the "herb of God" and is a traditional Chinese fungus that has been widely used to enhance human health and longevity in China and other East Asian countries. Various food and health products made from different tissues (mycelia, spores and fruiting body) of *Lingzhi* with important properties have gained enormous commercial value (Wang et al., 2019b).

12.4 ECONOMIC IMPORTANCE

Statistics from the FAO show that China is the largest producer of edible mushrooms, followed by Italy, the United States, the Netherlands, and Poland (FAO, 2016). Today, there is a growing awareness of health among populations around the world and a growing trend for vegan foods and natural foods in the diet. In addition, mushrooms have multiple health benefits and are a relevant component of these diets. Thus, the global mushroom market is projected to show significant growth.

The value of the mushroom growing market reached \$16.7 billion in 2020 and is projected to increase at a compound annual growth rate (CAGR) of 4.0% and reach \$20.4 billion in 2025. Button-type mushrooms make up 40% of the world's mushroom production, with *Agaricus bisporus*

Figure 12.1 The most notable edible mushrooms in the world. (a) *Agaricus bisporus*; (b) *Agrocybe aegreta*; (c) *Auricularia auricular*; (d) *Flammulina velutipes*; (e) *Hericium erinaceus*; (f) *Hypsizygus marmoreus*; (g) *Lentinula edodes*; (h) *Oudemansiella raphanipes*; (i) *Pleurotus eryngii*; (j) *Pleurotus giganteus*; (k) *Pleurotus ostreatus*; and (l) *Tremella fuciformis*.

Source: Feeney et al. (2014).

Figure 12.2 A selected collection of well-known species of medicinal mushrooms. (a) *Calvatia gigantean*; (b) *Cordyceps jiangxiensis*; (c) *Cordyceps sobolifera*; (d) *Coriolus versicolor*; (e) *Ganoderma densizonatum*; (f) *Ganoderma hainanense*; (g) *Ganoderma lucidum*; (h) *Ganoderma sinense*; (i) *Lycoperdon wrightii*; (j) *Ophiocordyceps sinensis*; (k) *Scleroderma citrinum*; (l) *Trametes sanguinea*.

Source: El Sheikha and Hu (2018).

(white button mushrooms) being one of the most popular cultivated mushroom species worldwide (Siwulski et al., 2020).

12.5 PROBLEMS

Traditional methods of recognizing mushrooms based on morphological characteristics depend mainly on professional staff. In some cases, even taxonomists find it difficult to correctly identify the genus of mushrooms (Wang et al., 2019b). Recognition by mycologists is based on the observation of phenotypic characters, macroscopic structures of the fruiting body or the microscopic structure of spores (Hibbett et al., 2016).

However, many poisonous mushrooms look like edible ones, and all genera containing poisonous mushrooms also include many non-poisonous and edible mushrooms (Bever et al., 2020). In addition, phenotypic characters are strongly influenced by environmental conditions, hybridization process, encryption and convergent evolution (Raja et al., 2017a; Schoch et al., 2014). Therefore, the phenotypic approach may not always work well and is particularly difficult for processed products (dried, canned or composite matrices of multiple species) or clinical specimens (Raja et al., 2017b).

Button-type mushrooms are relatively cheap mushrooms in terms of cultivation, with low calorie content, and they contain important vitamins and minerals (niacin, riboflavin, phosphorus, copper, zinc, magnesium). In addition, button mushrooms contain bioactive compounds, which exert anti-mutagenic, chemoprotective and hypocholesterolemic action. *Agaricus bisporus* mushrooms are best known for their quality flavors. However, a negative aspect of this mushroom is that it is very vulnerable, resulting in a short shelf life (Muszyńska et al., 2017; Mushroom Cultivation Market, 2020).

12.6 TYPES OF MUSHROOM ADULTERATION

12.6.1 Adulteration Related to the Geographical Origin of Mushrooms

Geographical indication is used to determine the origin of a product, when its characteristics or quality are a result of geographical origin (Cassago et al., 2021). As food products that are widely consumed, the traceability of the origin of mushrooms and their products is an important step in ensuring their quality and safety (Qi et al., 2018a). In terms of quality, nutrients and medicinal effects, they obviously vary depending on the geographical origin (Fernandes et al., 2016; Tesanovic et al., 2017; Yao et al., 2018a).

The exact authenticity of the geographical origin of mushrooms from incorrect marking or falsification is becoming increasingly vital for both consumers and producers to improve the issue of food safety and quality assurance (Chung et al., 2018).

12.6.2 Fraud Related to Mushroom Species

Different species show different drug effects. More than 2000 species of *Lingzhi* mushrooms have been recorded but only red *Lingzhi* (*Ganoderma lucidum*) and black *Lingzhi* (*G. sinensis*) show the most amazing health-boosting results. In other words, the economic value of *G. lucidum* and *G. sinensis* is significant, but the supply chain of their commodities is complex.

Due to the similarity between the species, adulterated materials are often fraudulently added to the market (Wang et al., 2019b). Often these additions are of lower quality or are additions of different species (e.g., of Chinese species to European mushroom species; Casale et al., 2016).

12.6.3 Fraud Related Mushroom Production (Cultivated and Wild)

Today, due to the high similarity of wild and cultivated edible mushrooms, it is difficult to distinguish them based on their morphological characteristics (Yao et al., 2020). Wild cultivated *G. lucidum*, commonly grown in high mountains and forests, is rare in nature and is therefore very beloved and expensive for its medicinal value. Due to high demand, it is currently very difficult to find high quality *G. lucidum* and grow it naturally.

Due to their different growing conditions, wild and cultivated *G. lucidum* may contain different levels of effective chemicals that affect their quality and medicinal effect. Since many *G. lucidum* products are available in various compositions, such as capsules and powder, it is difficult to determine the product of wild growth by its natural appearance, aroma or taste (Zhu and Tan, 2016).

Of all the wild mushrooms in the world, there are few that are as valuable and sought after as *Boletus edulis* Bull. and allied/related species ((*Boletus edulis* Bull and allied/related species (BEAS)), commonly called porcini mushrooms. *Porcini* represent almost all the wild mushrooms available on the Italian market. A significant portion of porcini is dried, however, most of the porcini sold on the Italian market or exported from Italy are imported from Eastern Europe and China where they are collected, dried on site and then subjected to first selection.

Unfortunately, among the imported *porcini* attributed to BEAS the presence of different, less valuable mushrooms is evident, some of which are not edible or not traded according to certain national laws (Casale et al., 2016).

12.6.4 Analytical Methods for the Rapid Detection of Adulteration in Mushrooms

Traditional analytical techniques/assays used to measure bioactive compounds in mushrooms can be laborious, time consuming and costly. Process analytical technology (PAT) provides valuable data on chemical processes to be used for monitoring and optimization purposes (Cullen et al., 2014).

In PAT, the combination of appropriate measuring devices with multifactorial statistical analysis (chemometrics) creates tools that can quickly, accurately and usually non-destructively assess the quality, quantity and certain functional properties of various organic compounds (Roussel et al., 2014)

To date, distinction of the geographical origin of edible or medicinal mushrooms (e.g., *Lentinula edodes, Tricholoma matsutake, Boletus edulis, Auricularia auricular, Wolfiporia extensa, Ganoderma lucidum, Flammulina velutipes*) has been made by measurement of chemical elements in the fruiting body of mushrooms by spectroscopic, spectrometric artificial olfactory and/or chromatographic techniques in combination with various chemometric approaches (Chung et al., 2018).

Table 12.2 presents recent studies (2014–2021) related to authenticity and quality control in mushrooms and then describes for each study its success and how it was achieved in combination with chemometrics.

Table 12.2 Analytical Methods for Assessing the Authenticity of Mushrooms

A/A	Family/Genus/ Species of Mushroom	Scope	Method of Analysis	Chemometric Methodology	Reference
1.	*Ganoderma lucidum*	Investigation of *Ganoderma* acids from different varieties and origins of *Ganoderma* species	HPLC-UV/ ESI-MS/MS	PLS	(Zhang et al., 2019)
2.	*Ganoderma lucidum*	Identification and quality control of *Ganoderma lucidum*	HPLC-PAD, HPLC-MS^n	CA, PCA, SA	(Wu et al., 2017)
3.	*Ganoderma* (*Lingzhi*)	Validation of the purity of commercial products	FTIR	*	(Choong et al., 2014b)
4.	*Ganoderma lucidum*	Distinctions between wild and cultivated *Ganoderma lucidum*	FTIR	PCDA, PLS-DA	(Zhu and Tan, 2016)
5.	Four species of *Ganoderma*	Determination of fatty acids	GC-MS	PCA, HCA	(Tokul-Olmez et al., 2018)
6.	Ganodermataceae	Species distinction	ATR-FTIR, UV-Vis	t-SNE, PLS-DA, RF	(Li et al., 2020a)
7.	*Lignosus*	Geographical origin	FTIR	*	(Choong et al., 2014a)
8.	*Pleurotus ostreatus, P. eryngii*	Determination of volatile compounds	HS-SPME, GC-MS	*	(Tagkouli et al., 2021)
9.	*Pleurotus ostreatus*	Evaluation of individual species	FT-MIR, FT-NIR, FT-Raman	PCA, PLSR	(Baeva et al., 2020)
10.	*Pleurotus ostreatus, P. eryngii, P.s nebrodensis*	Accurate determination/ prediction of their content in glucans and ergosterol	ATR-FTIR	PCA, PLSR	(Bekiaris et al., 2020)

(*Continued*)

Table 12.2 (Continued)

A/A	Family/Genus/ Species of Mushroom	Scope	Method of Analysis	Chemometric Methodology	Reference
11.	*Boletus griseus, Leccinum rugosiceps, B. umbriniporus, B. edulis*	Traceability of 247 Boletaceae mushrooms	UV-Vis, FTIR	PLS-DA, GS-SVM	(Yao et al., 2018a)
12.	*Boletus speciosus, Leccinum rugosiceps, B. umbriniporus, B. tomentipes, B. edulis*	Traceability of 230 Boletaceae mushrooms	FT-IR and ICP-AES	SVM	(Yao et al., 2018b)
13.	Boletaceae mushrooms	Geographical traceability of Boletaceae mushrooms	FT-IR, UV and ICP- AES	SVM	(Yao et al., 2019)
14.	*Boletus edulis*	Geographical traceability	ICP-AES, UV-Vis and FT-MIR	PLS-DA, GS-SVM	(Qi et al., 2018a)
15.	*Boletus edulis* and *B. tomentipes*	Geographical origin	FT-MIR	PCA, HCA, PLS-DA	(Qi et al., 2017)
16.	*Boletus tomentipes*	*Boletus tomentipes* data collections from different regions	ICP-AES	PCA	(Wang et al., 2015)
17.	*Lingzhi*	Classification of *Lingzhi* species	ATR-FTIR	RF, SVM, PLS-DA	(Wang et al., 2019b)
18.	*Bachu*	Classification of *Bachu* mushroom	FT-IR	PLS-GS-SVM, PLS-ELM	(Ma et al., 2020)
19.	192 samples from six different species (*Agaricus bisporus, Pleurotus* × hybrid, *Lentinula edodes, Hypsizygus tessellatus, Pleurotus eryngii, Hericium erinaceus*)	Quantification of macronutrients and classification	FT-NIRS	PLSR, iPLSR	(Mikola et al., 2020)
20.	*Agaricus bisporus*	Geographical distinction of the mushroom *Agaricus bisporus*	IRMS	PLS-DA	(Chung et al., 2018)
21.	Dried slices *Lentinus edodes*	Geographical origin	IRMS	OPLS-DA	(Chung et al., 2019)
22.	1500 samples from 22 species of mushrooms	Geographical origin	FAAS, ICP-AES	CA	(Brzezicha-Cirocka et al., 2019)
23.	*Poria cocos* (Schw.) Wolf	Analysis of bioactive components	HPLC-PAD, HPLC-MS	SA, CA, PCA	(Wu et al., 2016)
24.	*Poria cocos* (Schw.) Wolf	Geographical origin and quality	UHPLC-QTOF-MS/MS	PCA, OPLS-DA, HCA	(Zhu et al., 2018)
25.	*Poria cocos*	Prediction of polysaccharide content and antioxidant activity of *Poria cocos*	NIR	GA-PLSR	(Yi et al., 2020)

A/A	Family/Genus/ Species of Mushroom	Scope	Method of Analysis	Chemometric Methodology	Reference
26.	*Macrohyporia cocos*	Geographical origin	FTIR and LC	PLS-DA	(Wang et al., 2019a)
27.	Dried mushrooms (*porcini*)	Detection of fraudulent additions (*Tylopilus* spp., *Boletus violaceofuscus*)	NIR	PLS-DM, UNEQ, SIMCA	(Casale et al., 2016)
28.	*Amanita*	Species differentiation	LC-HRMS	codification DNA	(Jansson et al., 2018)
29.	*Macrohyporia cocos*	Geographical origin of wild and cultivated mushrooms	FTIR, LC	PLS-DA, RF	(Wang et al., 2021)
30.	17 samples of wild mushrooms	Determination of phenolic substances	HPLC/MS	*	(Yahia et al., 2017)
31.	*Tuber aestivum, T. brumale, T. magnatum, T. melanosporum*	Differentiation between species and geographical origin of fresh and lyophilized truffles	HS-SPME/GC-MS	*	(Šiškovič et al., 2021)
32.	*Poria cocos, Polyporus umbellatus*	Geographical origin	HPLC	HCA, PCA, LDA	(Liu et al., 2019)
33.	75 samples of mushroom-truffle	Geographical origin	FT-NIR	PCA, LDA, lin. SVM, quad. SVM, SSD, RF, k-NN	(Segelke et al., 2020)
34.	*Porcini* mushrooms	Differentiation between species and geographical origin	UV, FTIR	PLS-DA, k-NN, SVM, RF	(Qi et al., 2018b)
35.	*Porcini* mushrooms	Differentiation between species and overall prediction of polyphenols	FT-MIR	PLS-DA	(Li et al., 2020b)
36.	*Thelephora ganbajun*	Geographical origin	ICP-MS	PCA, LDA, HCA	(Chen et al., 2021)

* *Not determined.*
Abbreviations: ATR, analysis of variance attenuated total reflectance; CA, components analysis; FAAS, flame atomic absorption spectroscopy; FT-MIR, Fourier transform mid-infrared spectroscopy; FT-NIR, Fourier transform near-infrared spectroscopy; FT-Raman, Fourier transform Raman spectroscopy; FTIR, Fourier transform infrared spectroscopy; GA-PLSR, genetic algorithm partial least squares regression; GC-MS, gas chromatography–mass spectroscopy; GS-SVM, grid-search support vector machine; HCA, hierarchical cluster analysis; HPLC-MS^n, high-performance liquid chromatography–mass spectrometry; HPLC-PDA, high-performance liquid chromatography–photodiode array detection; HPLC-UV/ESI-MS/MS, high-performance liquid chromatography–ultraviolet electrospray ionization–tandem mass spectrometry; HPLC, high-performance liquid chromatography; HS-SPME, headspace–solid phase microextraction; ICP-AES, inductively coupled plasma–atomic emission spectroscopy; ICP-MS, inductively coupled plasma–mass spectrometry; iPLSR, interval partial least squares regressions; IRMS, isotope ratio mass spectrometry; LC, liquid chromatography; LC-HR-MS, liquid chromatography–high resolution mass spectrometry; LDA, linear discriminant analysis; (lin. SVM), linear support vector machine; NIR, near-infrared spectroscopy; OPLS-DA, orthogonal partial least squares discriminant analysis; PCA, principal component analysis; PCDA, principal component discriminant analysis; PLS-DA, partial least squares–discriminant analysis; PLS-DM, partial least squares–density modeling; PLS-ELM, partial least squares extreme learning machine; PLS-GS-SVM, partial least squares grid–search support vector machine; PLSR, partial least squares regression; quad. SVM, quadratic support vector machine; RF, random forest; SA, similarity analysis; SPME-GC/MS, headspace–solid phase microextraction–gas chromatography; SSD, subspace discriminant; SVM, support vector machines; t-SNE, t-distributed stochastic neighbor embedding; UNEQ, unequal dispersed classes; UPLC-QTOF-MS, ultra-performance liquid chromatography–quadrupole time of flight–mass spectrometry; UV, ultraviolet; UV-Vis, ultraviolet–visible.

12.6.5 Chemometrics

Multivariate chemometric techniques have been found to be very useful in locating the geographical origin of mushrooms (Brzezicha-Cirocka et al., 2019). Relevant approaches include PCA, FA, CA and LDA techniques in combination with various analytical techniques such as ICP-AES, CV-AAS and ICP-MS DRC (Jarzyńska et al., 2011; Jarzyńska et al., 2012; Falandysz et al., 2017); GC-MS (Tagkouli et al., 2021); α- and γ-spectroscopy (Cocchi et al., 2017; Szymańska et al., 2018); NIR (Casale et al., 2016); FT-NIR (Fu et al., 2017); and LC-QTOF-MS (Chan et al., 2017).

12.6.6 Liquid Chromatography (LC)

Chromatography is the main method for determining the various components of herbs and mushrooms (Choong et al., 2014a). The aim of the study by Wu et al. (2016) was to introduce a new method based on similarity analysis (SA), cluster analysis (CA) and principal component analysis (PCA) to determine the quality of different *Poria cocos* (Schw.) Wolf samples obtained from Yunnan, Hubei, Guizhou, Fujian, Henan, Guangxi, Anhui and Sichuan regions in China (Wu et al., 2016).

The study of *Ganoderma* acids (GAs) from different varieties and origins of *Ganoderma* species was performed by Zhang et al. (2019) to reveal the correlations between their chemical fingerprints and their replication activities, using high-performance ultraviolet liquid chromatography with electrospray ionization coupled to sequential mass spectrometry, high-ultraviolet liquid chromatography HPLC-UV/ESI-MS/MS) in combination with PLS chemometric analysis (Zhang et al., 2019).

The cytotoxicity assay was performed on three cancer cell lines (SGC-7901, HT-29 and Hep G2). Chemometric analysis was applied to correlate their chemical compositions and corresponding bioactivities. Sixteen peaks (representing 70% of the total peaks) were identified as GAs and their content ranged from 0.368 to 10.8 µg/g in various species of *Ganoderma*. The results of the study showed that the correlations between chemical fingerprints and the antiproliferative activities of different species of *Ganoderma* give a remarkable picture of the true bioactive components of chemical markers for quality evaluation of *Ganoderma* mushrooms and provide a good guide for studying chemical spectrum–bioactivity relationship.

In a study by Yahia et al. (2017) phenolic compounds were identified by liquid chromatography–mass spectrometry in 17 species of wild mushrooms in central Mexico, and their antioxidant activity and bioactive compounds were determined (Yahia et al., 2017). All species analyzed were found to have antioxidant activity in vitro, and a wide range of phenolic and organic compounds were identified.

The study by Wu et al. (2017) established a fingerprint evaluation system based on SA, CA and PCA chemometric techniques for the identification and quality control of *G. lucidum*. Fifteen samples from nine Chinese provinces were analyzed by high-performance liquid chromatography with a photodiode array detector (photodiode array detection [HPLC-PAD] and high-performance liquid chromatography–tandem mass spectrometry [HPLC-MS^n]). Forty-seven compounds were detected by HPLC, of which 42 compounds were temporarily identified by comparing retention times and mass spectrometry data with those of the reference compounds and reviewing the literature.

Ganoderic acid B, 3,7,15-trihydroxy-11,23-dioxolanost-8,16-dien-26-vicic acid, lucidic acid A, ganoderic acid G and 3,7-oxo-12-DM acetylganoderic acid were considered to be the marker compounds for distinguishing samples of different quality according to CA and PCA techniques (Wu et al., 2017).

12.6.7 Gas Chromatography (GC)

The research of Tagkouli et al. (2021) dealt with the effect of genetic (species, strains) and environmental factors (substrate) on the volatile profiles of eight mushroom strains *Pleurotus eryngii* and *P. ostreatus* were grown on wheat straw or substrates enriched with winery or olive mill products by the solid phase extraction microextraction technique with gas chromatography coupled to mass spectrometry (headspace–solid phase microextraction–gas chromatography–mass spectrometry [HS-SPME-GC-MS]).

Selected samples were additionally baked. More than 50 compounds were identified in fresh mushroom samples, with *P. ostreatus* showing higher concentrations but lower numbers of volatile compounds compared to *P. eryngii*. Baking resulted in the partial elimination of volatile compounds and the formation of pyrazines, Strecker aldehydes and sulfur compounds. The analysis of the basic components in the data obtained succeeded in distinguishing between raw and cooked mushrooms, as well as between *Pleurotus* species and strains, but not between the different culture media. Ketones, alcohols and toluene were mainly responsible for distinguishing between *P. ostreatus* strains, while aldehydes and fatty acid methyl esters contributed more to the separation of *P. eryngii* strains (Tagkouli et al., 2021).

12.6.8 DNA Methodologies for Determining the Authenticity of Mushrooms

DNA analysis that complements the phenotypic observations is the method of choice for the specific identification of mushroom species thanks to its speed and relatively low cost (Raja et al., 2017a; Badotti et al., 2017). Current methods rely almost entirely on PCR-amplified rRNA gene assays, particularly in the ITS domain (Hibbett et al., 2016), which has been selected as a universal DNA marker for fungi due to the large variation between of species (Schoch et al., 2012). The ITS region (~600–800 bp) contains two non-coding regions (ITS-1 and ITS-2) that are separated by the highly conserved 5.8S rRNA subunit. In general, this index follows the sequence and is compared with public DNA reference libraries, which are imperative tools for classification assignments (Creedy et al., 2020).

High-performance liquid chromatography–mass spectrometry in combination with multifactorial statistical analysis and DNA coding was used for the spatial identification of five species of mushrooms of the genus *Amanita* from four European countries and the geographical origin of the samples (Jansson et al., 2018).

12.6.9 Near-Infrared Spectroscopy with Fourier Transform (Fourier Transform Near-Infrared Spectroscopy [FT-NIR])

Fourier transform near-infrared spectroscopy (FT-NIRS) was used to quantify the macro-components (dry matter, protein, carbohydrates) of different cultivated mushroom species. PLSR prediction models were generated using 192 mushroom samples. In addition, six species of mushrooms (*Agaricus bisporus, Lentinula edodes, Hericium erinaceus, Pleurotus* × hybrid, *Hypsizygus tessellatus, L. edodes* and *H. erinaceus*) were successfully distinguished by the LDA method based solely on NIR spectra (Mikola et al., 2020).

The recognition of *Tylopilus* spp. in mixing with BEAS is very difficult, not only for workers working in the manual selection and packaging of mushrooms, but also for mycologists. The presence of *Tylopilus* spp. and other foreign species (i.e., edible species but not belonging to the BEAS) in dried *porcini* mushrooms represent a serious problem of adulteration. Verification of the authenticity of dried *porcini* mushrooms using near-infrared NIRS spectroscopy, in combination with chemometric techniques, partial least squares–density modeling (PLS-DM), unequal dispersed classes (UNEQ) and soft independent modeling of class analogy (SIMCA), was conducted through the study of Casale et al. (2016). For this purpose, 80 samples of dried mushrooms including BEAS, *Tylopilus* spp. and *Boletus violaceofuscus* were analyzed by NIRS spectroscopy. NIR spectra were used to develop reliable and efficient categorization models using the PLS-DM method and the two most commonly used categorization model techniques, UNEQ and SIMCA. The results showed that NIR spectroscopy in combination with the chemometric technique of categorization models can be proposed as an effective analytical strategy for checking the authenticity of dried porcini mushrooms (Casale et al., 2016).

The antioxidant capacity and polysaccharide content of *Poria cocos* were evaluated by NIRS in combination with chemometric methods by Yi et al. (2020). Various methods for pre-processing the spectra and selecting the variables have been carefully optimized to improve the accuracy and robustness of the models. Advanced regression models of least squares with genetic algorithms (genetic algorithm–partial least squares regression [GA-PLSR]; R^2 >0.94) achieved very satisfactory results, indicating their applicability in the evaluation of quality and control of *P. cocos* (Yi et al., 2020).

12.6.10 Fourier Transform Mid-Infrared Spectroscopy (MIR)

Li et al. (2020a) differentiated species in Ganodermataceae mushrooms accoding to a new methodology.

Li et al. (2020b) found a quick and easy way to distinguish species and predict polyphenol content using Fourier transform (FT-MIR) medium infrared spectra derived from different morphological parts of *porcini* mushrooms. The raw media spectra of *porcini* mushrooms of the four species (*Boletus edulis, Leccinum rugosiceps, B. omentipes, B. umbriniporus*) are shown in Figure 12.3.

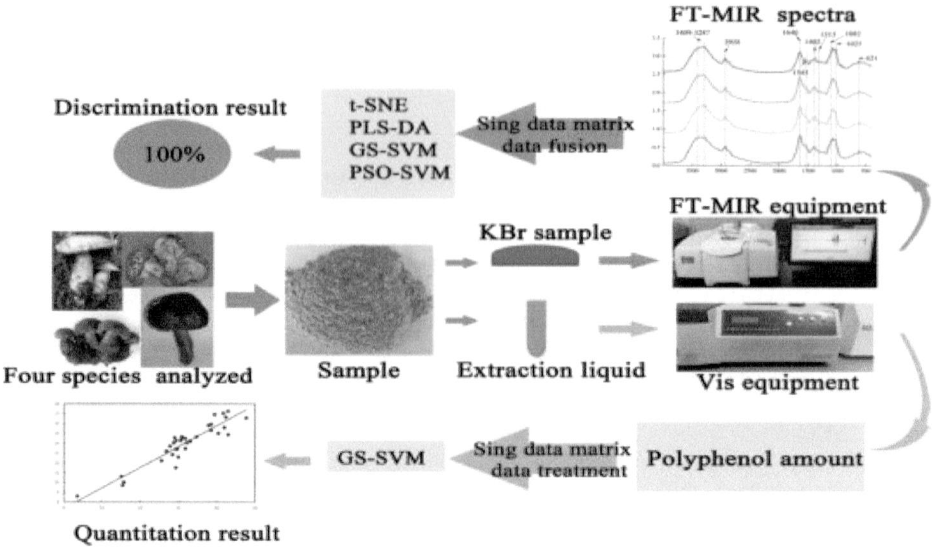

Figure 12.3 The raw spectra of pileus and stipes from four types of *porcini* mushrooms. *Source*: Li et al. (2020).

In addition to using low-level data merging, three types of variable attributes were selected simultaneously to complete the mid-level data merging, which showed that the partial least squares–discriminant analysis (PLS-DA) model was based on a variable characteristic selected from the maximum Q^2 (cum). Grid-search support vector machine (GS-SVM) models and particle swarm optimization support vector machines (PSO-SVM) based on low data integration had 100% accuracy of distinction allowing each mushroom to be classified into its true species. In addition, this study measured the polyphenol content, soluble in alcohol in four *porcini* mushrooms, contributed to the tendency of polyphenols to accumulate in various parts of the mushroom, and predicted the polyphenol content for four *porcini* mushrooms.

Boletus edulis and *B. tomentipes* are two well-known species of mushrooms that are widely consumed in Yunnan province in China due to their high nutritional and medicinal properties. Fourier transform mid-infrared spectroscopy (FT-MIR) can determine the unique fingerprint of a sample and further analyze its quality when combined with appropriate chemistry. The work of Qi et al. (2017) described the development of a methodology for the identification and differentiation of *B. edulis* and *B. tomentipes* mushrooms from different geographical locations based on FT-MIR spectroscopy and chemometric determinations.

PCA analysis, hierarchical cluster analysis (HCA) and PLS-DA analysis enabled the identification and differentiation of mushroom samples according to their unique metabolic spectral fingerprints. The range of 1800–400 cm^{-1}, which showed significant characteristics of the mushroom samples, was selected for the next analysis. Unsupervised PCA and HCA showed that mushroom samples from different geographical locations could be effectively identified. In addition, the supervised PLS-DA method was used to predict unknown mushroom samples successfully based on the calibration model developed. In conclusion, the FT-MIR technique combined with appropriate chemometric determinations can be used as an effective and rapid strategy for the quality control of both *Boletus edulis* and *B. tomentipes* mushrooms according to their geographical locations (Qi et al., 2017).

The data fusion strategy can combine chemical information from different sensor systems and therefore provide a deep and comprehensive perspective on understanding complex data variables. Using this strategy, Qi et al. (2018a) have combined three sensor techniques—inductively coupled plasma–atomic emission spectroscopy (ICP-AES), ultraviolet-visible spectroscopy (UV-Vis) and Fourier transform mid-infrared spectroscopy (FT-MIR)—with chemometric data to determine the traceability of the origin of *Boletus edulis* mushrooms. The results of this study showed that the combination of multiple sensory systems can lead to two goals: first, in the improved traceability of the origin of *B. edulis* mushrooms; and second, in their quality control (Qi et al., 2018a).

Boletaceae mushrooms are edible wild mushrooms that have high nutritional value, delicious taste and great economic value and are distributed in Yunnan province in China (Yao et al., 2019). This work combined the use of infrared UV spectroscopy with FTIR for the traceability of 247 samples of Boletaceae mushrooms in combination with chemometrics, providing a high synergistic result for the traceability of Boletaceae mushrooms.

Compared to a simple spectroscopy technique, the data fusion strategy can obviously improve the classification performance of PLS-DA models and grid-search–support vector machine (GS-SVM) models for both species' traceability as well as for geographical origin. In addition, the PLS-DA and GS-SVM models provided 100% accuracy for item traceability and had reliable rating parameters. For traceability of geographical origin, the prediction accuracy via the PLS-DA model was only 64.63%, but the GS-SVM model based on data merging was 100%.

FT-IR spectroscopy and inductively coupled plasma–atomic emission spectroscopy (ICP-AES) were used to distinguish five species of Boletaceae fungi using the support vector machine (Yao et al., 2019). The accuracy of the correct classification of the samples into calibration kits and test kits was 100% and 98.68%, respectively. The results showed that the merging of mid-level data of FT-IR and ICP-AES could provide a higher synergistic result for the differentiation of different species of Boletaceae mushrooms, which could benefit for further authentication and quality control of edible mushrooms.

FT-IR spectroscopy, ultraviolet (UV) spectroscopy and ICP-AES spectrometry were used for traceability of mushroom samples from eight different geographical origins in combination with GS-SVM multifactorial statistical analysis (Yao et al., 2019). In particular, the prediction accuracy in the SVM model from the data merging of the three instruments was 99.04%, which was higher than the simple technique and the data merging of two spectroscopy techniques. This result showed that the data merging strategy in conjunction with the SVM can provide a highly synergistic result for the traceability of the geographical origin of Boletaceae mushrooms.

Distinguishing between wild and cultivated mushrooms is important for evaluating their quality. The study by Yao et al. (2020) was designed to investigate the merging of UV and FTIR data in combination with the SVM or random forest model, to distinguish between the wild and cultivated mushrooms *Phlebopus portentosus* and *Wolfiporia cocos*, two popular edible mushrooms in China and other East Asian countries. For low- and medium-level data fusion classification models, the validation set accuracy was 90.91% and 98.70%, respectively.

However, parameter c was too large, which means the models were unreliable. For high-level data merging, the results showed that the validation set accuracy was 100% and the sensitivity and specificity were significantly increased compared to the simple technique. Therefore, merging high-level data in conjunction with random forest was an effective method for distinguishing wild and cultivated mushrooms *Wolfiporia cocos* and *Phlebopus portentosus* and could be applied to protect the steady growth of the edible mushroom market (Yao et al., 2020).

The active chemical components of a limited number of samples (only three samples) of the mushroom *Lignosus* spp. using FTIR and two-dimensional Fourier transform infrared spectroscopy (2DIR) were measured to determine their geographical origin (Choong et al., 2014a). The 2DIR optical and multicolor spectra provided structural information of the chemical components analyzed and demonstrated a robust and useful approach to identifying different geographical origins, although the samples used were raw materials from neighboring areas.

Similarly, Choong et al. (2014b) engaged in the commercial verification of *Ganoderma* (*lingzhi*) mushroom products with FTIR and two-dimensional Fourier transform correlation spectroscopy (Choong et al., 2014b). This study compared three types of commercial *Ganoderma* products and demonstrated the potential of developing a verification tool to validate the purity of commercial medicinal products and mushroom products.

Effective identification of Bachu mushrooms using FT-IR spectroscopy in combination with partial least squares–grid-search–support vector machine (PLS-GS-SVM) was performed by Ma et al. (2020). FT-IR spectral data of mushroom pile and strain were obtained separately for different types and amino acid contents to improve mushroom differences, which were combined with statistical modeling algorithms for fast and efficient *Bachu* mushroom identification.

After measuring the pileus and stipes of 35 samples of *Bachu* mushrooms, 47 samples of coral fungi and 39 samples of *Lentinus edodes* separately, two classification groups were created: Bachu-coral fungi and Bachu–*L. edodes*. The results showed that the classification accuracy of the PLS-GS-SVM model was 98.78% for the Bachu-coral fungus pileus data and 98.78% for the stipe. For the Bachu–*L. edodes* group, the classification accuracy for the pileus was 98.78% and 100% for the stipe. For the partial least squares–extreme learning machine (PLS-ELM) model, it was 94.87% for the

Bachu-coral fungus pileus data and 77.81% for the stipe. For the Bachu–*L. edodes* group, the classification accuracy for the pileus is 99.45% and 99.18% for the stipe.

12.7 ATTENUATED TOTAL REFLECTANCE-FOURIER TRANSFORM INFRARED SPECTROSCOPY (ATR-FTIR)

The study by Bekiaris et al. (2020) was performed using Fourier transform infrared spectroscopy with attenuated total reflectance (ATR-FTIR) with the aim of obtaining a chemical image of mushrooms from *Pleurotus ostreatus, P. eryngii* and *P. nebrodensis* strains cultured in by-products of the oil industry and the development of chemometric tools for the accurate determination/prediction of their content in glucans and ergosterol (Bekiaris et al., 2020)

Applying optimized chemometric models that predict such contents with good predictive power ($R^2 = 0.80–0.84$) and accuracy, ATR-FTIR spectroscopy successfully characterized mushroom samples, detecting differences related to species or substrates were used, dividing them into groups by PCA analysis. The study findings showed that ATR-FTIR spectroscopy could be used as a potential technology analysis tool in the mushroom industry to characterize mushrooms and evaluate their content in bioactive compounds.

Wang et al. (2019b) collected 120 fresh *Lingzhi* fruiting bodies, which were analyzed by ATR-FTIR in combination with RF, SVM and PLS-DA as classification models to distinguish five different *Lingzhi* specimens. This study focused on comparing the ability of the three classification models to provide a reliable and fast method of analysis for the classification of species, as well as to provide qualitative estimates of *Lingzhi*. So compared to the RF model, the results of the SVM and PLS-DA methods were more satisfactory, and their accuracy for the test set was for both methods 100%.

Li et al. (2020a) developed a method of distinguishing Ganodermataceae species. For this purpose, fingerprints were obtained from ten species of Ganodermataceae mushrooms with ATR-FTIR and UV-Vis spectroscopy. Different chemometric modes of discrimination were applied, such as t-distributed stochastic neighbor embedding (t-SNE), PLS-DA and RF, and data fusion strategies were combined with these chemometric modes to achieve better distinction (Li et al., 2020a).

Both PLS-DA and RF were the best models in distinguishing Ganodermataceae when medium-level data merging was used. Both developed approaches could be described as a "simple, green and accurate method" of mushroom sorting. In particular, they were able to distinguish species among marketing mushroom varieties and to detect adulterations of Ganodermataceae mushrooms with other inferior mushrooms.

12.8 IRMS (ISOTOPE RATIO MASS SPECTROMETRY)

The article by Chung et al. (2019) aimed to measure the constant isotope ratios ($\delta^{13}C$, $\delta^{15}N$, $\delta^{18}O$, $\delta^{34}S$) with an isotope ratio mass spectrometer in order to establish a discrete method for the authenticity of the geographical origin of the dried slices of the mushroom *Lentinus edodes* (shiitake) from China and Korea in combination with orthogonal partial least squares (OPLS) (Chung et al., 2019). The method of discrimination using OPLS-DA showed excellent predictability (Q^2cum = 0.888) for the distinction of geographical origin of dried shiitake slices.

The same constant isotope ratios were used by Chung et al. (2018) aiming at verifying the geographical traceability of the mushroom *Agaricus bisporus* in combination with chemometric approaches (Chung et al., 2018). Ten varieties from 15 mushroom farms in six regions of Korea were measured with a fixed isotope ratio mass spectrometer. The values of $\delta^{13}C$, $\delta^{15}N$, $\delta^{18}O$, and $\delta^{34}S$ were significantly different between the mushrooms collected from the six cultivation areas ($P < 0.05$). Multidimensional drawings including $\delta^{15}N$ and $\delta^{34}S$ showed a clear geographical distinction between the Saedo and Saehan varieties produced in the Buyeo, Nonsan, Eumseong, Boryeong and Gyeongju areas. The PLS-DA models showed good discrimination for the Saedo variety (Rx2 = 0.798 and $Q^2 = 0.563$) and Saehan (Rx2 = 0.819 and $Q^2 = 0.894$).

12.9 RAMAN SPECTROSCOPY

In the study of Baeva et al. (2020), FT Raman, FT-MIR and FT-NIR spectroscopy were used in combination with PCA and PLSR chemistry techniques to characterize the fruiting bodies (basidiocarps) of the cultivated mushroom *Pleurotus ostreatus* and yielded the characteristic bands of protein and glucan. The sampling was performed using PCA while the PLSR models created were used to predict the protein and total glucan content based on the reported spectroscopic data.

In conclusion, the authors state that the combination of vibration spectroscopy with chemistry is a powerful tool for the evaluation of edible and medicinal mushrooms and that this approach can be proposed as an alternative to common methods of analysis.

12.10 ICP-AES (INDUCTIVELY COUPLED PLASMA–ATOMIC EMISSION SPECTROSCOPY)

The study by Wang et al. (2015) was performed using inductively coupled plasma–atomic emission spectroscopy (ICP-AES) with the aim of developing a protocol for the determination of the mineral content of *Boletus tomentipes* from different parts of China. Applying the PCA chemometric model, ICP-AES spectroscopy was able to explain 77% of the total variance (Wang et al., 2015). Triterpenoid acids were the most important chemical components in the samples and the results of the study showed that for the quality control of *Poria cocos*, the HPLC fingerprint and the HPLC-MS^n evaluation system based on SA, CA and PCA is a simple, fast and efficient method.

12.11 CONCLUSIONS

The demand of consumers for high-quality mushrooms free from any type of fraud has increased the research about their rapid detection and adulteration. The high similarities between the species result in the appearance on the market of adulterated mushrooms that are usually of lower quality or are additions of Chinese species to European species. Moreover, some types of mushrooms are very expensive, and the financial gain that can result from their adulteration is great. Adulteration of mushrooms related to the geographical origin, to mushroom production (cultivated and wild) and to mushroom species. In this chapter, recent, novel and rapid analytical techniques for traceability and adulteration of mushrooms and their quality control, in combination with chemometric approaches, have been presented.

REFERENCES

Aisala, H., Sola, J., Hopia, A., Linderborg, K.M. and Sandell, M. 2019. Odor-contributing volatile compounds of wild edible Nordic mushrooms analyzed with HS—SPME—GC—MS and HS—SPME—GC—O/FID. Food Chemistry, 283, 566–578.

Badotti, F., de Oliveira, F.S., Garcia, C.F., Vaz, A.B.M., Fonseca, P.L.C., Nahum, L.A., Oliveira, G. and Góes-Neto, A. 2017. Effectiveness of ITS and sub-regions as DNA barcode markers for the identification of Basidiomycota (Fungi). BMC Microbiology, 17, 42.

Baeva, E., Bleha, R., Sedliaková, M., Sushytskyi, L., Švec, I., Čopíková, J., Jablonsky, I., Klouček, P. and Synytsya, A. 2020. Evaluation of the cultivated mushroom *Pleurotus ostreatus* Basidiocarps using vibration spectroscopy and chemometrics. Applied Sciences, 10, 8156.

Bekiaris, G., Tagkouli, D., Koutrotsios, G., Kalogeropoulos, N. and Zervakis, G.I. 2020. *Pleurotus* mushrooms content in glucans and ergosterol assessed by ATR-FTIR spectroscopy and multivariate analysis. Foods, 9, 535.

Bever, C.S., Adams, C.A., Hnasko, R.M., Cheng, L.W. and Stanker, L.H. 2020. Lateral flow immunoassay (LFIA) for the detection of lethal amatoxins from mushrooms. PLOS One, 15, e0231781.

Brzezicha-Cirocka, J., Grembecka, M., Grochowska, I., Falandysz, J. and Szefer, P. 2019. Elemental composition of selected species of mushrooms based on a chemometric evaluation. Ecotoxicology and Environmental Safety, 173, 353–365.

Campo, E., Marco, P., Oria, R., Blanco, D. and Venturini, M. E. 2017. What is the best method for preserving the genuine black truffle (*Tuber melanosporum*) aroma? An olfactometric and sensory approach. LWT, 80, 84–91.

Casale, M., Bagnasco, L., Zotti, M., Di Piazza, S., Sitta, N. and Oliveri, P.A. 2016. NIR spectroscopy-based efficient approach to detect fraudulent additions within mixtures of dried porcini mushrooms. Talanta, 160, 729–734.

Cassago, A.L.L., Artêncio, M.M., Giraldi, J.D.M.E. and Da Costa, F.B. 2021. Metabolomics as a marketing tool for geographical indication products: a literature review. European Food Research and Technology, 1–17.

Cerletti, C., Esposito, S. and Iacoviello, L. 2021. Edible mushrooms and Beta-Glucans: Impact on human health. Nutrients, 13, 2195.

Chan, K.-M., Gar-LeeYue, G., Li, P., Chun-Wai Wong, E., Kin-Ming Lee, J., Kennelly, E.J. and Bik-San Lau, C. 2017. Screening and analysis of potential anti-tumor components from the stipe of *Ganoderma* sinense using high-performance liquid chromatography/time-of-flight mass spectrometry with multivariate statistical tool. Journal of Chromatography A, 1487, 162–167.

Chen, L.P., Zhu, H.Y., Li, Y.F., Zhang, Y., Zhang, W., Yang, L.C., Yin, H., Dong, C.Y. and Wang, Y. 2021. Combining multielement analysis and chemometrics to trace the geographical origin of *Thelephora ganbajun*. Journal of Food Composition and Analysis, 96, 103699.

Choong, Y.K., Sun, S.Q., Zhou, Q., Lan, J., Lee, H.L. and Chen, X.D. 2014b. Verification of Ganoderma (lingzhi) commercial products by Fourier transform infrared spectroscopy and two-dimensional IR correlation spectroscopy. Journal of Molecular Structure, 1069, 60–72.

Choong, Y.K., Xu, C.H., Lan, J., Chen, X.D. and Jamal, J.A. 2014a. Identification of geographical origin of Lignosus samples using Fourier transform infrared and two-dimensional infrared correlation spectroscopy. Journal of Molecular Structure, 1069, 188–195.

Chung, I.M., Han, J.G., Kong, W.S., Kim, J.K., An, M.J., Lee, J.H., An, Y.J., Jung, M.Y. and Kim, S.H. 2018. Regional discrimination of *Agaricus bisporus* mushroom using the natural stable isotope ratios. Food Chemistry, 264, 92–100.

Chung, I.M., Kim, J.K., Han, J.G., Kong, W.S., Kim, S.Y., Yang, Y.J., An, Y.J., Kwon, C., Chi, H.Y., Jung, M.Y. and Kim, S.H. 2019. Potential geo-discriminative tools to trace the origins of the dried slices of shiitake (*Lentinula edodes*) using stable isotope ratios and OPLS-DA. Food Chemistry, 295, 505–513.

Cocchi, L., Kluza, K., Zalewska, T., Apanel, A. and Falandysz, J. 2017. Radioactive caesium (134Cs and 137Cs) in mushrooms of the genus *Boletus* from the Reggio Emilia in Italy and Pomerania in Poland. Isotopes in Environmental and Health Studies, 53, 620–627.

Creedy, T.J., Norman, H., Tang, C.Q., Qing Chin, K., Andujar, C., Arribas, P., O'Connor, R.S., Carvell, C., Notton, D.G. and Vogler, A.P. 2020. A validated workflow for rapid taxonomic assignment and monitoring of a national fauna of bees (Apiformes) using high throughput DNA barcoding. Molecular Ecology Resources, 20, 40–53.

Cullen, P.J., O'Donnell, C.P. and Fagan, C.C. 2014. Benefits and challenges of adopting PAT for the food industry. In Process analytical technology for the food industry; O'Donnell, C.P., Fagan, C., Cullen, P.J., Eds.; Springer: New York, NY, pp. 1–5.

El Sheikha, A.F. and Hu, D.M. 2018. How to trace the geographic origin of mushrooms? Trends in Food Science and Technology, 78, 292–303.

Falandysz, J., Chudzińska, M., Barałkiewicz, D., Drewnowska, M. and Hanć, A. 2017. Toxic elements and bio-metals in *Cantharellus* mushrooms from Poland and China. Environmental Science and Pollution Research, 24, 11472–11482.

FAO. 2016. Top 10 country production of mushrooms and truffles 2016. www.fao.org/faostat/en/#rankings/countries_by_commodity (accessed 18.7.2021).

Feeney, M. J., Miller, A. M. and Roupas, P. 2014. Mushrooms—biologically distinct and nutritionally unique: Exploring a "Third Food Kingdom". Nutrition Today, 49, 301–307.

Feng, T., Yang, M., Ma, B., Zhao, Y., Zhuang, H., Zhang, J. and Chen, D. 2021. Volatile profiles of two genotype *Agaricus bisporus* species at different growth stages. Food Research International, 140, 1–9.

Fernandes, Â., Petrović, J., Stojković, D., Barros, L., Glamočlija, J., Soković, M., Martins, A. and Ferreira, I.C. 2016. *Polyporus squamosus* (Huds.) Fr from different origins: Chemical characterization, screening of the bioactive properties and specific antimicrobial effects against *Pseudomonas aeruginosa*. LWT-Food Science and Technology, 69, 91–97.

Fernandes, T., Garrine, C., Ferrão, J., Bell, V. and Varzakas, T. 2021. Mushroom nutrition as preventative healthcare in Sub-Saharan Africa. Applied Science, 11, 4221.

Fu, H., Yin, Q., Xu, L., Wang, W., Chen, F. and Yang, T. 2017. A comprehensive quality evaluation method by FT-NIR spectroscopy and chemometric: fine classification and untargeted authentication against multiple frauds for Chinese *Ganoderma lucidum*. Spectrochimica Acta Part A: Molecular Biomolecular Spectroscopy, 182, 17–25.

Gawlikowski, T., Romek, M. and Satora, L. 2015. Edible mushroom-related poisoning: A study on circumstances of mushroom collection, transport, and storage. Human and Experimental Toxicology, 34, 718–724.

Giusti, A., Ricci, E., Gasperetti, L., Galgani, M., Polidori, L., Verdigi, F., Narducci, R. and Armani, A. 2021a. Molecular identification of mushroom species in Italy: An ongoing project aimed at reinforcing the control measures of an increasingly appreciated sustainable food. Sustainability, 13, 238.

Giusti, A., Ricci, E., Gasperetti, L., Galgani, M., Polidori, L., Verdigi, F., Narducci, R. and Armani, A. 2021b. Building of an Internal Transcribed Spacer (ITS) gene dataset to support the Italian health service in mushroom Identification. Foods, 10, 1193.

Govorushko, S., Rezaee, R., Dumanov, J. and Tsatsakis, A. 2019. Poisoning associated with the use of mushrooms: A review of the global pattern and main characteristics. Food Chemical Toxicology, 128, 267–279

Hibbett, D., Abarenkov, K., Kõljalg, U., Öpik, M., Chai, B., Cole, J., Wang, Q., Crous, P., Robert, V., Helgason, T., et al. 2016. Sequence-based classification and identification of Fungi. Mycologia, 108, 1049–1068.

Huang, Q., Jia, Y., Wan, Y., Li, H. and Jiang, R. 2015. Market survey and risk assessment for trace metals in edible fungi and the substrate role in accumulation of heavy metals. Journal of Food Science, 80, 1612–1618.

Jansson, D., Wolterink, A., Bergwerff, L., Hough, P., Geukens, K. and Åstot, C. 2018. Source attribution profiling of five species of Amanita mushrooms from four European countries by high resolution liquid chromatography-mass spectrometry combined with multivariate statistical analysis and DNA-barcoding. Talanta, 186, 636–644.

Jarzyńska, G. and Falandysz, J. 2011. The determination of mercury in mushrooms by CV- AAS and ICP-AES techniques. Journal of Environmental Science and Health Part A, 46, 569–573.

Jarzyńska, G. and Falandysz, J. 2012. Trace elements profile of Slate Bolete (*Leccinum duriusculum*) mushroom and associated upper soil horizon. Journal of Geochemical Exploration, 121, 69–75.

Khan, I., Huang, G., Li, X., Leong, W., Xia, W. and Hsiao, W.L.W. 2018. Mushroom polysaccharides from *Ganoderma lucidum* and *Poria cocos* reveal prebiotic functions. Journal of Functional Foods, 41, 191–201.

Koutrotsios, G., Kalogeropoulos, N., Kaliora, A.C. and Zervakis, G.I. 2018. Toward an increased functionality in oyster (*Pleurotus*) mushrooms produced on grape marc or olive mill wastes serving as sources of bioactive compounds. Journal of Agriculture and Food Chemistry, 66, 5971–5983.

Lee, S., Choi, E., Yang, S., Ryoo, R., Moon, E., Kim, S. and Kim, K.H. 2018. Bioactive compounds from sclerotia extract of *Poria cocos* that control adipocyte and osteoblast differentiation. Bioorganic Chemistry 81, 27–34.

Lee, S., Lee, S., Roh, H., Song, S., Ryoo, R., Pang, C., Baek, K. and Kim, K. 2018. Cytotoxic constituents from the sclerotia of *Poria cocos* against human lung adenocarcinoma cells by inducing mitochondrial apoptosis. Cells, 7, 116.

Liu, J., Zhou, J., Zhang, Q.Q., Zhu, M.H., Hua, M.L. and Xu, Y.H. 2019. Monosaccharide analysis and fingerprinting identification of polysaccharides from *Poria cocos* and *Polyporus umbellatus* by HPLC combined with chemometrics methods. Chinese Herbal Medicines, 11, 406–411.

Li, X.P., Li, J., Li, T., Liu, H. and Wang, Y. 2020b. Species discrimination and total polyphenol prediction of porcini mushrooms by Fourier transform mid-infrared (FT-MIR) spectrometry combined with multivariate statistical analysis. Food science and nutrition, 8, 754–766.

Li, X.P., Li, J., Liu, H. and Wang, Y.Z. 2020a. A new analytical method for discrimination of species in Ganodermataceae mushrooms. International Journal of Food Properties, 23, 227–240.

Ma, L., Gao, R., Han, H., Chen, C., Yan, Z., Zhao, J., Lv, X., Chen, C. and Xie, L. 2020. Efficient identification of Bachu mushroom by flourier transform infrared (FT-IR) spectroscopy coupled with PLS-GS-SVM. Optik, 224, 165712.

Meenu, M. and Xu, B. 2019. Application of vibrational spectroscopy for classification, authentication and quality analysis of mushroom: A concise review. Food Chemistry, 289, 545–557.

Miao, H., Zhao, Y., Vaziri, N.D., Tang, D., Chen, H., Chen, H., Khazaeli, M., Tarbiat-Boldaji, M., Hatami, L. and Zhao, Y. 2016. Lipidomics biomarkers of diet-induced hyperlipidemia and its treatment with *Poria cocos*. Journal of Agriculture and Food Chemistry, 64, 969–979.

Mikola, E., Geösel, A., Stefanovits-Bányai, É. and Fodor, M. 2020. Quantitative determination of macro components and classification of some cultivated mushrooms using near-infrared spectroscopy. Journal of Food Processing and Preservation, 44, 14540.

Mushroom Cultivation Market. 2020. Mushroom Cultivation Market by Type (Button Mushroom, Oyster Mushroom, Shiitake Mushroom, Other Types), By Phase, By Region (North America, Europe, Asia Pacific, South America, Rest of the World)—Global Forecast to 2025. Report, pp. 1–141. www.researchandmarkets.com/reports/5018601 (accessed 7.7.2021).

Muszyńska, B., Piotrowska, J., Krakowska, A., Gruba, A., Kała, K., Sułkowska-Ziaja, K., Kryczyk, A. and Opoka, W. 2017. Study of physiologically active components in different parts of fruiting bodies of varieties of *Agaricus bisporus* (white mushroom). European Food Research and Technology, 243, 2135–2145.

Nieminen, P. and Mustonen, A.M. 2020. Toxic potential of traditionally consumed mushroom species-a controversial continuum with many unanswered questions. Toxins, 12, 639.

Qi, L., Li, J., Liu, H., Li, T. and Wang, Y. 2018b. An additional data fusion strategy for the discrimination of porcini mushrooms from different species and origins in combination with four mathematical algorithms. Food and Function, 9, 5903–5911.

Qi, L., Liu, H., Li, J., Li, T. and Wang, Y. 2018a. Feature fusion of ICP-AES, UV-Vis and FT-MIR for origin traceability of *Boletus edulis* mushrooms in combination with chemometrics. Sensors, 18, 241.

Qi, L.M., Zhang, J., Liu, H.G., Li, T. and Wang, Y.Z. 2017. Fourier transform mid-infrared spectroscopy and chemometrics to identify and discriminate *Boletus edulis* and *Boletus tomentipes* mushrooms. International Journal of Food Properties, 20, S56-S68.

Raja, H.A., Baker, T.R., Little, J.G. and Oberlies, N.H. 2017b. DNA barcoding for identification of consumer-relevant mushrooms: A partial solution for product certification? Food Chemistry 214, 383–392.

Raja, H.A., Miller, A.N., Pearce, C.J. and Oberlies, N.H. 2017a. Fungal identification using molecular tools: A primer for the natural products research community. Journal of Nature Products, 80, 756–770.

Roussel, S., Preys, S., Chauchard, F. and Lallemand, J. 2014. Multivariate data analysis (chemometrics). In process analytical technology for the food Industry; O'Donnell, C.P., Fagan, C., Cullen, P.J., Eds.; Springer: New York, NY, pp. 7–59.

Schoch, C.L., Seifert, K.A., Huhndorf, S., Robert, V., Spouge, J.L., Levesque, C.A. and Chen, W. 2012. Fungal barcoding consortium. Nuclear ribosomal internal transcribed spacer (ITS) region as a universal DNA barcode marker for Fungi. Proceedings of the National Academy of Sciences of the United States of America, 109, 6241–6246.

Schoch, C.L., Seifert, K.A., Huhndorf, S., Robert, V., Spouge, J.L., Levesque, C.A. and Chen, W. 2014. Finding needles in haystacks: Linking scientific names, reference specimens and molecular data for Fungi. Database, 109, 6241–6246.

Segelke, T., Schelm, S., Ahlers, C. and Fischer, M. 2020. Food authentication: Truffle (*Tuber* spp.) species differentiation by FT-NIR and chemometrics. Foods, 9, 922.

Šiškovič, N., Strojnik, L., Grebenc, T., Vidrih, R. and Ogrinc, N. 2021. Differentiation between species and regional origin of fresh and freeze-dried truffles according to their volatile profiles. *Food Control,* 123, 107698.

Siwulski, M., Budka, A., Rzymski, P., Gąsecka, M., Kalač, P., Budzyńska, S., Magdziak, Z., Niedzielski, P., Mleczek, P. and Mleczek, M. 2020. Worldwide basket survey of multielemental composition of white button mushroom *Agaricus bisporus*. Chemosphere, 239, 124718.

Sun, L., Zhang, Z., Xin, G., Sun, B., Bao, X., Wei, Y., Zhao, X. and Xu, H. 2020. Advances in umami taste and aroma of edible mushrooms. Trends in Food Science and Technology, 96, 176–187.

Szymańska, K., Falandysz, J., Skwarzec, B. and Strumińska-Parulska, D. 2018. 210Po and 210Pb in forest mushrooms of genus *Leccinum* and topsoil from northern Poland and its contribution to the radiation dose. Chemosphere, 213, 133–140.

Tagkouli, D., Bekiaris, G., Pantazi, S., Anastasopoulou, M.E., Koutrotsios, G., Mallouchos, A., Zervakis, G.I. and Kalogeropoulos, N. 2021. Volatile profiling of *Pleurotus eryngii* and *Pleurotus ostreatus* mushrooms cultivated on agricultural and agro-industrial by-products. Foods, 10, 1287

Tarrillo, E., Garcia, L.M. and Bustamante, D.E. 2021. Composting of the waste of the heart of palm agroindustry for the cultivation of edible mushrooms. Cogent Food and Agriculture, 7, 1905976.

Tešanović, K., Pejin, B., Šibul, F., Matavulj, M., Rašeta, M., Janjušević, L. and Karaman, M. 2017. A comparative overview of antioxidative properties and phenolic profiles of different fungal origins: fruiting bodies and submerged cultures of *Coprinus comatus* and *Coprinellus truncorum*. Journal of food Science and Technology, 54, 430–438.

Tietel, Z. and Masaphy, S. 2018. Aroma-volatile profile of black morel (*Morchella importuna*) grown in Israel. Journal of the Science of Food and Agriculture, 98, 346–353.

Tokul-Olmez, O., Kaplaner, E., Ozturk, M., Ullah, Z. and Duru, M.E. 2018. Fatty acid profile of four *Ganoderma* species collected from various host trees with chemometric approach. Biochemical Systematics and Ecology, 78, 91–97.

USDA. 2018. Full report (all nutrients): 11260, mushrooms, white, raw a. https://ndb. nal.usda.gov/ndb/foods/show/11260?fgcd=&manu=&format=Full&count=& max=25&offset=2600&sort=ndb&order=asc&qlookup=&ds=&qt=&qp=&qa= &qn=&q=&ing= (Accessed 18.7.2021).

Wang, Q.Q., Huang, H.Y. and Wang, Y.Z. 2019a. Geographical authentication of *Macrohyporia cocos* by a data fusion method combining ultra-fast liquid chromatography and Fourier transform infrared spectroscopy. Molecules, 24, 1320.

Wang, Q., Wang, Y.Z. and Wang, Y. 2021. Comparison of geographical traceability of wild and cultivated *Macrohyporia Cocos* with different data fusion approaches. Journal of Analytical Methods in Chemistry. https://doi.org/10.1155/2021/581899.

Wang, X.M., Zhang, J., Li, T., Li, J.Q., Wang, Y.Z. and Liu, H.G. 2015. ICP-AES Determination of mineral content in *Boletus tomentipes* collected from different sites of China. Spectroscopy and Spectral Analysis 35, 1398–1403.

Wang, Y.Y., Li, J.Q., Liu, H.G. and Wang, Y.Z. 2019b. Attenuated total reflection-Fourier transform infrared spectroscopy (ATR-FTIR) combined with chemometrics methods for the classification of *Lingzhi* species. Molecules, 24, 2210.

Wasser, S.P. 2014. Medicinal mushroom science: current perspectives, advances, evidences, and challenges. Biomedical Journal, 37, 345–356.

White, J., Weinstein, S.A., De Haro, L., Bédry, R., Schaper, A., Rumack, B.H. and Zilker, T. 2019. Mushroom poisoning: A proposed new clinical classification. Toxicon, 157, 53–65.

Wu, K., Fan, J., Huang, X., Wu, X. and Guo, C. 2018. Hepatoprotective effects exerted by *Poria cocos* polysaccharides against acetaminophen-induced liver injury in mice. International Journal of Biological Macromolecules, 114, 137–142.

Wu, L.F., Wang, K.F., Mao, X., Liang, W.Y., Chen, W.J., Li, S., Qi, Q., Cui, Y.P. and Zhang, L.Z. 2016. Screening and analysis of the potential bioactive components of *Poria cocos* (Schw.) wolf by HPLC and HPLC-MSn with the aid of chemometrics. Molecules, 21, 227.

Wu, L., Liang, W., Chen, W., Li, S., Cui, Y., Qi, Q. and Zhang, L. 2017. Screening and analysis of the marker components in *Ganoderma lucidum* by HPLC and HPLC-MS^n with the aid of chemometrics. Molecules, 22, 584.

Yahia, E.M., Gutiérrez-Orozco, F. and Moreno-Pérez, M.A. 2017. Identification of phenolic compounds by liquid chromatography-mass spectrometry in seventeen species of wild mushrooms in Central Mexico and determination of their antioxidant activity and bioactive compounds. Food Chemistry, 226, 14–22.

Yao, S., Li, J., Li, T., Duan, Z. and Wang, Y. 2019. Geographical traceability of Boletaceae mushrooms using data fusion of FT-IR, UV, and ICP-AES combined with SVM. International Journal of Food Properties, 22, 414–426.

Yao, S., Li, T., Liu, H., Li, J. and Wang, Y. 2018a. Traceability of Boletaceae mushrooms using data fusion of UV—visible and FTIR combined with chemometrics methods. Journal of the Science of Food and Agriculture, 98, 2215–2222.

Yao, S., Li, J., Li, T., Liu, H. and Wang, Y. 2018b. Discrimination of Boletaceae mushrooms based on data fusion of FT-IR and ICP—AES combined with SVM. International Journal of Food Properties, 21, 255–266.

Yao, S., Li, J.Q., Duan, Z.L., Li, T. and Wang, Y.Z. 2020. Fusion of Ultraviolet and infrared spectra using support vector machine and random forest models for the discrimination of wild and cultivated mushrooms. Analytical Letters, 53, 1019–1033.

Yi, Y., Hua, H., Sun, X., Guan, Y. and Chen, C. 2020. Rapid determination of polysaccharides and antioxidant activity of *Poria cocos* using near-infrared spectroscopy combined with chemometrics. Spectrochimica Acta Part A: Molecular and Biomolecular Spectroscopy, 240, 118623.

Zhang, C., Fu, D., Chen, G. and Guo, M. 2019. Comparative and chemometric analysis of correlations between the chemical fingerprints and anti-proliferative activities of ganoderic acids from three *Ganoderma* species. Phytochemical Analysis, 30, 474–480.

Zhang, J., Barałkiewicz, D., Wang, Y., Falandysz, J. and Cai, C. 2020. Arsenic and arsenic speciation in mushrooms from China: a review. Chemosphere, 246, 125685.

Zhu, L.-X., Xu, J., Wang, R.-J., Li, H.-X., Tan, Y.-Z., Chen, H.-B., Dong, X.-P. and Zhao, Z.-Z. 2018. Correlation between quality and geographical origins of *Poria cocos* revealed by qualitative fingerprint profiling and quantitative determination of triterpenoid acids. Molecules 23, 2200.

Zhu, L.X., Xu, J., Wu, Y., Su, L.F., Lam, K.Y.C., Qi, E.R., Dong, X.P., Chen, H.B., Liu, Y.D. and Zhao, Z.Z. 2019. Comparative quality of the forms of decoction pieces evaluated by multidimensional chemical analysis and chemometrics: *Poria cocos*, a pilot study. Journal of Food and Drug Analysis, 27, 766–777.

Zhu, Y. and Tan, T.L. 2016. Penalized discriminant analysis for the detection of wild-grown and cultivated *Ganoderma lucidum* using Fourier transform infrared spectroscopy. Spectrochimica Acta Part A: Molecular and Biomolecular Spectroscopy, 159, 68–77.

13 Authenticity and Chemometrics of Nuts

Fayyaz Salih Hussain, Najma Memon, Nilgün Yenil,
Fadim Yemiş, and Tahir Mehmood

CONTENTS

13.1 INTRODUCTION

Nuts are dry, fruit-based seeds that contain nutrient-dense complex food matrices that are generally rich in bioactive compounds, unsaturated fatty acids, and have a hard outer wall when the seed becomes mature. The consumption and production of nuts vary from place to place depending on geographical variations. Nuts play a very important role in a healthy lifestyle by providing essential nutrients, as well as adding great taste. Nuts are widely used as ingredients for many food products including chocolate, bakery, and confectionary products, where these nuts are consumed as whole nuts, raw or with salt in chocolate or as a roasted form. Consumption of plant foods is generally included in a healthy diet that can prevent many chronic diseases due to the presence of fiber and nutrients. The benefits of the dietary constituents of nuts have gained more importance beyond just the previously described micronutrients and macronutrients. Nowadays, the discovery and identification of new classes of dietary constituents can help to develop new or modified diets that can be useful in the treatment and prevention of chronic diseases. Importantly, nuts are generally good for health due to their inverse effects associated with diseases related to some type of cancers, cardiovascular disease (CVD) and diabetes. A fat-modified diet can also show considerable effects on cardiovascular risk, other components in the diet such as plant protein, soya protein and dietary fiber can give more protective effects than lowering lipid effects in the diet. Dietary constituents that can produce good effects, and will promote the development of diets that are even more beneficial for the treatment and prevention of cardiovascular and other chronic diseases (Kris-Etherton et al., 1999). Nuts are a major part of the recommended dietary guidelines in Canada, the United States, and Europe. Nuts contain minerals, vitamins, monounsaturated, polyunsaturated fatty acids, and fat-soluble bioactive including phospholipids, sphingolipids, TAG, PUFA, MUFA, monoacylglycerols, diacylglycerols, esters, sterol, tocopherols, phytosterols, chlorophylls, phytosterols, squalene,

 DOI: 10.1201/9781003266167-13

terpenoids, carotenoids, tocotrienols, flavonoids, phytosterols and fiber, which are reasonably attributed to the benefit of peanuts and other nuts. Other macronutrients include protein, fat, and carbohydrates. Micronutrients include vitamins, minerals, and phytochemicals such as hydroxycinnamic acid, hydroxybenzoic acid, flavonoids like flavanols, flavones, flavanones, isoflavones, flavanols, anthocyanins, catechins, lignans, anthocyanidins, naphthoquinones and stilbenes, hydrolysable tannins such as gallotannins, ellagitannins, proanthocyanidins, condensed tannins, phenolic aldehydes, ellagic acid, phytoestrogens, phytates, terpenes and coumestan alkaloids (Chen and Blumberg, 2008; Alasalvar and Bolling, 2015; Rose and Oliveira, 1997; Maguire et al., 2004; Phillips et al., 2005; Fang et al., 2005; Shahidi and Miraliakbari, 2006; Alasalvar et al., 2011; Bolling et al., 2011; Shahidi and nutraceuticals, 2004). Carotenoids are pigments with characteristic colors from yellow to red. They are present in algae, plants, fungi, fish, birds, and the cuticles of insects or crustaceans. Based on their functions, carotenoids can be divided into two major groups: carotenes such as lycopene, α-carotene and β-carotene; and xanthophylls such as zeaxanthin and lutein. Carotenoids cannot be synthesized inside the human or animal body; they can only be taken up via the diet. Nutritionists and food scientists show great interest in carotenoids, because of bioactive compounds present in carotenoids have antioxidant and anti-inflammatory effects. Antioxidant properties can protect humans from cancers, low immune response, arthritis, certain cardiovascular diseases, and premature aging. The major sources of carotenoids are fresh vegetables, fruits, and juices; in particular, α-carotene and β-carotene are present in yellow-orange vegetables. For example, orange fruits such as mandarins provide α-cryptoxanthin. Lutein is present in dark vegetables and in egg yolk (Langi et al., 2018; Kiokias et al., 2016; Rao and Rao, 2007; Saini et al., 2015; Fernández-García et al., 2012).

Nut derivatives in their crushed forms are used as an ingredient in many food industries. Due to their small amount of harvest, and high demand, grain powders such as chickpeas, lentils, and legumes are often mixed into these food additives. Although the amount of product is increased, but the quality of the product affected by adding these kinds of grain powders, which have physical similarities. Therefore, raw or processed food products need to be analyzed for reliable process control, quality assurance, identification of flavor and non-flavor compounds, and to provide accurate authentication and traceability confirmations.

13.1.1 Nuts and Their Geographical Distribution

Nuts are very good source of fiber, protein, antioxidants, fats, and other nutrients. Figure 13.1 shows various types of most consumed nuts. A few nut types are described in upcoming subsections.

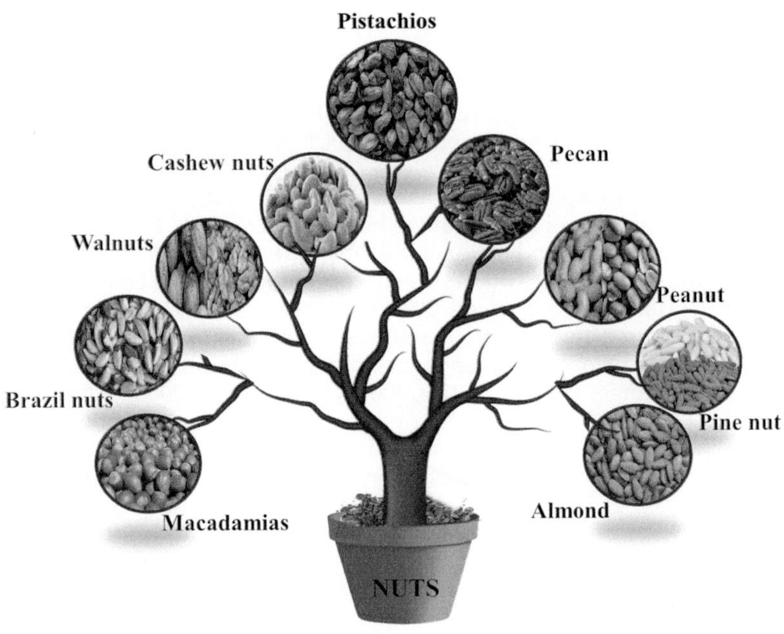

Figure 13.1 The most commonly used types of nuts.

13.1.1.1 Peanuts

Groundnuts, which are also known as peanuts (*Arachis hypogea*), are edible seeds of legumes. Technically, peanuts are known as peas from the family Fabaceae (legumes or beans). Peanuts are generally known for being rich in oil content, protein, and fiber. They are widely used in the production of roasted peanuts, sauce, oil, butter, paste, snack products, milk, beverages, cheese, confections, desserts and in soups. They are consumed all around the world in many cuisines, especially in traditional foods. They are considered a complete dietary source for expeditioners in remote areas like Antarctica, while trekking and in space. Peanuts are grown in warm climates. They are harvested, washed, and dried under the sun, and are de-shelled in order to eat. China produces more than 45% of the world's peanuts followed by India 16% and the United States 5%. There are four different types of peanuts, i.e., Spanish, Valencia, Runner and Virginia (Dheir et al., 2020; Arya et al., 2016; Guimón et al., 2012). Peanuts contain 50% type of peanuts) w/w fat content, and that is the reason peanuts were previously considered nuts that contain unhealthy fat. Recently, however, peanuts have gained importance as an energy-dense functional food containing flavonoids, vitamin E, magnesium, and copper. Peanuts contain high amounts of mono- and polyunsaturated fatty acids and a low amount of saturated fatty acids, antioxidants, complex carbohydrates, fiber and considerable amounts of phenolics (Francisco et al., 2008). Different clinical studies show peanuts reduce triglycerides, low-density lipoproteins and total cholesterol (Kris-Etherton et al., 1999). Chemical profiling of peanuts shows they contain 50% w/w monounsaturated fatty acids, 33% w/w paraformaldehyde and 14% w/w saturated fatty acids (Feldman, 1999). A peanuts diet is beneficial especially in terms of cholesterol; compared with an olive oil diet, the presence of high monounsaturated fatty acids can lower the body cholesterol up to 11%, reduce triglycerides low-density lipoprotein cholesterol (LDL) up to 14%, thus it is beneficial against coronary heart disease (Pelkman et al., 2004). Peanuts contain 20 amino acids and are high in arginine (Arya et al., 2016). As compared to beans, peanuts contain more protein; which reaches 50% after oil extraction (Zhao et al., 2011). Peanut proteins are excellent for their water retention capacity, high solubility, emulsification, and stability and foaming capacity (Wu et al., 2009).

13.1.1.2 Almonds

Almonds (*Prunus amygdalis* var. *dulcis*) belong to the family Rosaceae, are one of the oldest and most popular nuts consumed and cultivated nuts worldwide. The annual production of almond nuts is approximately 1.7 million metric tons. Almonds are consumed in the form of raw nuts or in snacks; they are sold with or without peels. Almonds are generally eaten roasted, powdered in milk (almond milk), and used in preparations of different foods such as confectionery and bakery items. Almonds are generally cultivated in countries having subtropical and temperate climates with hot dry summers and mild winters. The largest producer of almonds is the United States, with an estimated production of one metric ton during 2020–2021. The state of California holds a 77% share of nut production worldwide. Apart from the United States, almonds are cultivated in Spain, Australia, Morocco, Iran and Italy (Kodad et al., 2008; Yada et al., 2013; Esfahlan and Jamei, 2012; de Souza and Koblitz, 2022). Chemical profiling shows that almonds contain lipids (>49%), protein (21.2%), carbohydrate (21.6%), fiber (12.5%) and total sugar (4.4%). Almonds Lipids are mostly mono- and polyunsaturated fatty acids, and they do not participate in the formation of cholesterol in the human body (Yada et al., 2013). Generally, almond oil contains palmitic, linoleic, stearic and oleic acids (Čolić et al., 2017b). The fatty acid composition may vary depending on climate, variety, genotype, environmental condition, harvest time and stage of maturity (Matthäus and Özcan, 2020; Banjanin et al., 2021; Özcan and Technology, 2022). Almonds also contain different micro- and macronutrients including minerals, tocopherols, polyphenols, different bioactive components, phytosterols and vitamin E. The protein composition of almonds ranges from 16 to 23 g/100 g (Ahrens et al., 2005). Albumin in almonds constitutes about 88%–91% of the total protein present; lysine, threonine and methionine are present in abundance. The share of essential amino acids in the protein of almond is about 30%. A diet containing almonds reduces the risk of colon cancer in rats, and in humans, it reduces LDL cholesterol and increases HDL cholesterol. Almond extracts have the capacity for free radical scavenging, perhaps due to the presence of phenolic compounds including flavonoids (Siriwardhana and Shahidi, 2002; Moure et al., 2007; PINELO et al., Wijeratne et al., 2006; Sfahlan et al., 2009). Almond hulls are a rich source of ursolic, oleanolic, betulinic, phenolic acid and flavanols, glycoside, vanillic acid, catechin, glucoside, protocatechuic acid, galactoside, naringenin, glucoside, *p*-hydroxybenzoic acid, 3β-O-methyl quercetin, rhamnoglucoside and rhamnoglucoside (Esfahlan et al., 2010; Sang et al., 2002).

13.1.1.3 Pine Nuts

Pine nuts (*Pinus koraiensis*) are derived from pine trees, which are believed to have existed in northern Asia since about 180 million years ago, during the Triassic period of the Mesozoic era (Awan and Pettenella, 2017; Lanner, 1981; Millar and Pinus, 1998; Keeley, 2012). The most common pine nut species consumed by humans are stone pine (*Pinus pinea*), siberian pine (*Pinus sibirica*) and chilgoza pine (*Pinus gerardiana*). Twentynine species of pine nuts are listed as edible nuts by the UN Food and Agricultural Organization (FAO). China, Russia, Pakistan and Korea are among the larger exporting countries of pine nuts. Pine nuts are widely used in different cuisines, in their original form and as an extracted oils (Conifers, 1998; Destaillats et al., 2010). Pine nuts are a healthy source of nutrients, and their daily consumption can lower the risk of myocardial infarction and coronary artery disease. The chemical composition of pine nuts varies due to geographical and climatic conditions. Pine nuts contain 44.9% fatty acids, 31.9% protein and considerable amounts of vitamin B_1, magnesium, potassium and phosphorus (Sagrero-Nieves and Agriculture, 1992; Nergiz and Dönmez, 2004). Pine nuts are very rich in Δ5-UPIFA (Δ5-unsaturated polymethylene-interrupted fatty acids) including phenolic, sciadonic and oxalic acids (Wolff and Bayard, 1995; Destaillats et al., 2010). The oil yield from pine nuts ranges from 45 to 65 g/100 g of nuts; about 99.4% of extracted oil is based on nonpolar lipids that are mainly triacylglycerol (58 different TAGs), with the remaining 0.6% depending on polar lipids. Typically, 50% of these fatty acids are polyunsaturated fatty acids with a major contribution of linoleic acid; 40% are monounsaturated fatty acids with dominantly oleic acid; and the remaining 10% are typically saturated fatty acids (Ryan et al., 2006). Pine nuts also contain α-tocopherols, γ-tocopherols, δ-tocopherols, squalene and phytosterols. Pine nuts help in lowering plasma cholesterol and low-density lipid levels by inhibiting cholesterol absorption in the intestine with the help of phytosterols. These compounds are also beneficial in antiatherogenic, antioxidants and as anti-inflammatories. Low amounts of phenolic compounds are present in pine nuts, including gallic acid, lignans, secoisolariciresinol and matairesinol (de Jong et al., 2003; Gylling et al., 2014; Ferguson et al., 2016).

13.1.1.4 Walnuts

Walnuts are species of the genus *Juglans* belonging to the family Juglandaceae. Walnuts were considered beneficial for health and medical purposes from ancient times dating back to 7000 bc. Walnuts are not only consumed as food but are also used in shelter, shade and wood. The husks of walnuts are used to produce different dyes (McDaniel and Jadeja, Little, 1978; Elias, 1980; Petrides, 1998). China and the United States are the largest producers of walnuts in the world, contributing more than 75% of the total world production. It is considered the sixth most valuable tree crop and fruit in the United States (Blessington et al., 2012; Pollack and Perez, 2008). Walnuts are a rich source of nutrition contain vitamins (vitamin B_6, vitamin C, tocopherols, folate and niacin), protein, minerals, fat and a rich source of phenolic compounds (tocopherols and phytosterols) with higher content of α-linolenic acid. Consumption of walnuts reduces low-density lipoprotein and plasma cholesterol (Hepburn et al., 1986; Tapsell et al., 2004; Davis et al., 2007). The synergistic effects of anti-inflammatory and antioxidant components present in walnuts could reduce the risk of different age-related diseases. Walnuts contain high amounts of antioxidants including folate, ellagic acid, melatonin, proanthocyanidins juglone, gamma tocopherol, n-3 α-linolenic acid, selenium and omega-3 fatty acid. In addition, walnuts contain 141 mg/g protein, 70.5 mg/g fiber, 11% of the daily value of magnesium and 10% of phosphorus (Chauhan and Chauhan, 2020; Anderson et al., 2001; Fukuda et al., 2003; Halvorsen et al., 2006; Reiter et al., 2005; Chauhan and Chauhan, 2012; Ren et al., 2007; Erdinest et al., 2012; Reifen et al., 2015). The antioxidant efficacy of walnuts is found to be the best among many seeds due to the presence of a large amount of phenolic content (Mishra et al., 2010). Walnuts contain 13 g of polyunsaturated fatty acid out of total 18 g of fat per 28 g of walnut. The polyunsaturated fat composition of 13 g contains about 2.5 g of α-linolenic acid, which is the precursor of docosahexaenoic and eicosapentaenoic acids, which are best known for their anti-inflammatory properties by regulating inflammatory cytokines, COX-2 and NO production (Chauhan and Chauhan, 2012; Ren et al., 2007; Chauhan and Chauhan, 2020). Different parts of the walnuts are used to cure gastrointestinal problems, toothache, ulcers, fevers, snake bites and kidney ailments (Hutchens, 1992; Nolan and Robbins, 2000).

13.2 NUT QUALITY AND SAFETY

The quality parameters of nuts are generally used to determine its shelf life, including sweetness, crispness, rancidity, bitterness, oiliness, firmness and collectively the flavor's intensity. The quality and shelf life of nuts is affected during preharvest and postharvest (Piscopo et al.,

2010; KOAZE et al., 2002; Kazantzis et al., 2003; Walton et al., 2009). Shelf life is the time period during which any medicine, food or drink remains fit for consumption or for sale (Irtwange et al., 2009). Preharvest factors include different ecological conditions during production orchard practices, while postharvest factors are all practices performed after the removal of nuts including drying, packaging, storage and transportation (Ajith et al., 2015; Walton and Wallace, 2004; Srichamnong and Srzednicki, 2015). These all require great care in order to avoid any chemical or physical damage or alteration in nuts quality. Storage conditions and pretreatment are one of the most important factors that can affect the quality of nuts. Another is that several oxidative reactions could take place due to improper storage conditions (Gama et al., 2018).

Nuts are susceptible to different types of microbial contamination including bacteria, fungi, and different viruses. Microbes like Escherichia *coli*, *Salmonella* spp. and *Aspergillus flavus*, as well as mildew and mold, can cause severe food safety risk (Uesugi and Harris, 2006; Blessington et al., 2012). Other than microbes, fungi and viruses, insects also cause damage to tree nuts, including *Cydia pomonella* (Codling moth), *Plodia interpunctella* (Indianmeal moth) *and Amyelois transitella* (orange worm; Mitcham et al., 2004). Fungal contamination related to aflatoxins, belonging to mycotoxins in seeds, can cause hepatotoxic and carcinogenic activity in human (Arrus et al., 2005; De Mello and Scussel, 2009; Pacheco et al., 2007). The most common fungal species isolated from Brazil nuts are *Aspergillus nomius, A. flavus, A. niger, Penicillium glabrum, Fusarium oxysporum, A. parasiticus, P. citrinum*, and *A. tamarii* (Reis et al., 2012; Baquião, 2015). The eggs of orange worms are even hidden inside dried nuts for a long time period and can be found in dried nuts or any product containing almonds. Nutrient contents and moisture facilitate the growth of microbes and fungus in nuts; this can be minimized by drying, which can inhibit the growth of microbes, but it is not much effective on spores, so the process of disinfection is required in order to ensure the safety of nuts or nut-based products in order to extend shelf life (Chen et al., 2021; Yu et al., 2018a). Traditionally, there are two broad categories available for disinfestation or disinfection: chemical and thermal. Fumigation is commonly used as a chemical treatment for tree nuts. The common pesticides or disinfectants used are aluminum phosphide, magnesium phosphide, sulfuryl fluoride and methyl bromide, but nowadays there are growing concerns about pesticide or chemical residue remaining on food as a potent human carcinogen. Thermal treatment is also another option. Heating with a high temperatures or hot water bleaching can kill the *Salmonella*, up to a 4-log reduction. Oil roasting at high temperature can also reduce *Salmonella* levels up to 4-log (Cuervo et al., 2016; Pan et al., 2019). Microwave processes are also effective in food processing, including pasteurization, tempering of food, or pre-cooking of bacon; these processes also have many benefits including reduction in processing time, economically viable, retain maximum nutritional quality, energy efficiency and so on. The possible chemical treatment options available for nuts are acid solutions or sprays (pecans, almonds, sesame seeds, pine nuts); chlorine such as bleach or sodium hypochlorite (hazelnuts, coconut, walnuts, pecans); chlorine dioxide gas (sesame seeds, almonds); cinnamon oil (almonds); ethanol sprays (almonds); hydrogen peroxide spray (almonds and walnuts); methyl bromide, ozone (pistachios); propylene oxide (cashews, pecans, almonds, macadamias); and peracetic acid (pecans, coconut, almonds, walnuts, hazelnuts; Beuchat et al., 2013; Ha et al., 2015; Olaimat et al., 2017; Pao et al., 2006; Salazar et al., 2018; Beuchat et al., 2012; Blessington et al., 2013; Walter et al., 2009; Golden et al., 2019; Wihodo et al., 2005; Tsai et al., 2017; Schade et al., 1977; Akbas et al., 2006; Frelka and Harris, 2015; Pearson et al., 2018; Weller et al., 2013). Some reported thermal treatment options for nuts are controlled atmosphere and heat (almonds); hot air or dry roasting (peanuts, almond, sesame seeds and pecans); hot air and 70% ethanol sprays (pecans, walnuts, almonds and pistachios); hot water (pecans and almonds); chlorine and hot water (pecans); indirect, such as water or silicon oil bath (almond, peanut butter, tree nut butters, hazelnut and cashew); infrared (almonds); distilled water and gas catalytic infrared (almond); dry roasting and gas catalytic infrared (almond); electrolyzed water and gas catalytic infrared (almonds); superheated steam and gas catalytic infrared (almonds); infrared and near infrared (pine nuts and almonds); lactic acid and near infrared (almonds, pine nuts); microwave (peanut butter); moist air convection (pistachios and almonds); presoak and moist air convection (pistachio); oil roasting (pecans, walnuts, almonds, peanuts); radio frequency (almonds); and ethanol spray and radio frequency (almonds, walnuts, pecans, pistachios; Beuchat et al., 2012; Brar and Danyluk, 2019; Poirier et al., 2014; Beuchat et al., 2013; Yang et al., 2010; Zhang et al., 2017; Cuervo et al., 2016; Abd et al., 2012; Kharel et al., 2018; Keller et al., 2012; Bingol et al., 2011; Brandl et al., 2008; Bari et al., 2010; Venkitasamy et al., 2017; Song et al., 2019; Jeong et al., 2009; Meyer and Vaughn, 1969; Sanders and Calhoun, 2014; Gao et al., 2011; Li et al., 2014; Salazar et al., 2018). Non-thermal treatment includes high-intensity 405 nm light (almonds); high-pressure processing (peanut, sesame seeds, almonds and butter); intense 200–1100 nm pulsed light (sesame seeds); irradiation via electron beam (peanut butter, pecans, and almonds); irradiation with gamma rays (peanut butter, almonds, sesame seeds, walnuts, pistachios); irradiation with x-rays (walnuts and almonds); and non-thermal plasma (pistachios and almonds; D'SOUZA et al., 2012; D'Souza et al., 2014; Goodridge et al., 2006; Grasso et al., 2010; Wuytack et al., 2003; Beuchat et al., 2013; Hvizdzak et al., 2010; Osaili and Al-Nabulsi, 2016; Hertwig et al., 2017). Some new technologies have emerged nowadays for the disinfection of nuts such as ozone, electrolyzed oxide water and nanoparticles. Photocatalytic nanoparticles with or without a combination of different materials can generate reactive oxygen species against several bacterial, fungal, and viral species. These nanoparticles include silver, copper, zinc and titanium dioxide nanoparticles; they have great potential and are considered candidates due to their bactericidal properties.

Although most food frauds may seem innocent to the user, some are known to bring many health risks. Nuts are very suitable to expose for fraud because of likely change with old and expired stocks. At this point, the chemical composition ("profiling" or "fingerprint") of foodstuffs comes to the fore, and the recording of multivariate parameters effective in these markings is carried out with supervised and unsupervised analysis methods (Medina et al., 2019c; Esteki et al., 2021b). Food fingerprints depend on many variations caused by differences in metabolite levels, geographical origin of foods, storage conditions or adulteration trials. Accordingly, different chemometric approaches have been described in the literature to optimize experimental design, and it has been reported that useful insights have been obtained from large and complex data sets in food validation challenges, but the information is very limited (Granato et al., 2017; Medina et al., 2019a).

Food authenticity is the legal recognition of prominent characteristics of an original food based on its uniqueness, safety, quality, and traceability guaranteed by a regulatory certification framework. This certification is usually a rigorous and bureaucratic process that involves government, the scientific community, and producers and consumers of food.

13.3 EXPERIMENTAL WORKFLOW FOR NUT AUTHENTICATION PURPOSES

The food authentication process is very important to verify and authenticate the originality of food. Nuts have complex matrices including non-volatile, semi-volatile and volatile compounds. Nuts are among the more complex food matrices, and the nut authentication process (Figure 13.2) requires a carefully selected integrated workflow of well-known sampling processes techniques,

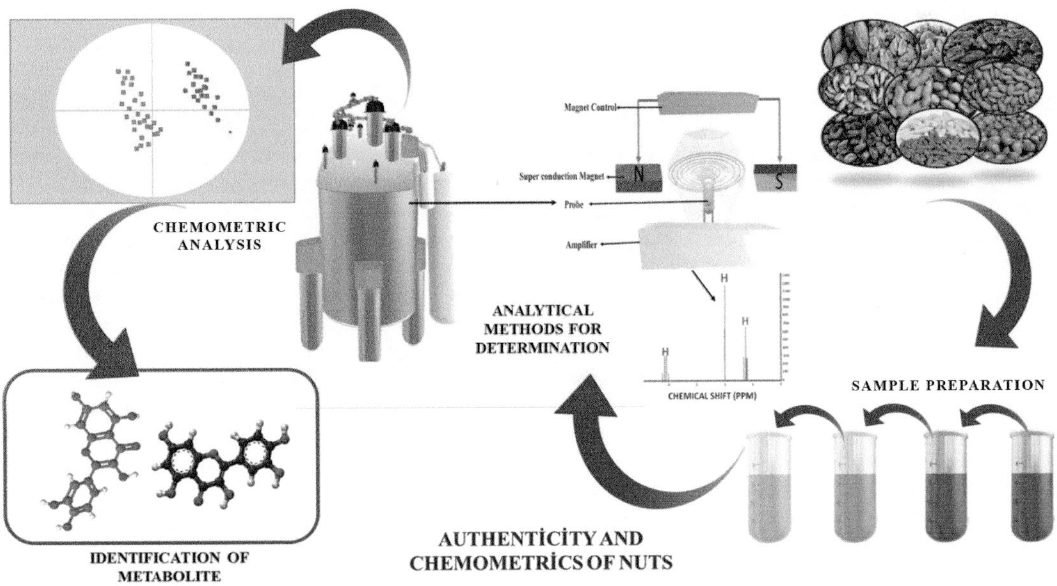

Figure 13.2 Workflow chart for chemical analysis for authenticity protocols of nuts.

best suitable analytical methods, unbiased and reliable chemometric analysis, and clear metabolite identification (Medina et al., 2019c; Granato et al., 2018b; Ríos-Reina et al., 2018).

13.4 SAMPLE EXTRACTION TECHNIQUES

As earlier discussed, foods are complex structures and contain trace amounts of some components. For this reason, appropriate initial preparations must be carefully planned in the experimental workflow process for analysis. The sample preparation step for food fingerprint analysis is highly important, and the applied extraction techniques at this stage directly affect the quality of analytical results by increasing the selectivity and sensitivity of the methods.

Extraction is the process of isolating the relevant analytes from a given matrix and their precleaning and concentrating and making them ready for instrumentation for analytical separation and/or detection. In extraction, the extractant first meets with the sample matrix. Subsequently, the analytes are transported between the two phases. In conventional extraction processes, the analytes are generally removed from the matrix by partitioning the relevant analytes into suitable liquid solvents or solid sorbents. There are numerous extraction methods reported in the literature. Extraction types can be subdivided into four main groups: liquid-liquid extraction (LLE; Daniel et al., 1989; de Albuquerque Mendes et al., 2019b); solid phase extraction (SPE); stir bar sorptive extraction (SBSE); and microwave-assisted extraction (MAE). The use of the LLE technique needs usually long extraction times and has different applications, including liquid-liquid microextraction (LLME); disperse liquid-liquid microextraction (DLLME; Arroyo-Manzanares et al., 2013); ionic liquids (IL); and pressurized liquid extraction (PLE). Although the SPE technique is very similar to LLE, it is eco-friendly. It has a wide range of applications and is quite good in the recovery efficiency of the supernatant.

SPE techniques have different application procedures such as micro–solid phase extraction (μSPE); dispersive solid phase extraction (dSPE); solid phase microextraction (SPME; Reyes-Garces et al., 2017); and microextraction by packed syringe. The stir bar sorptive extraction (SBSE technique needs little solvent volume and short extraction times (SPME), magnetic solid phase extraction (MESP) and SBSE are different miniaturized formats for SPE and are commercially available. Its various applications, including stir cake sorptive extraction (SCSE); rotating-disc extraction (RDSE); sorbent, quick, easy, cheap, effective, rugged and safe extraction (QuEChERS; Alcántara-Durán et al., 2019; Delgado-Zamarreño et al., 2016); and micro quick, easy, cheap, effective, rugged and safe (μ-QuEChERS), are listed in the literature. The MAE technique has good extraction efficiency, short extraction times, and is inexpensive and environmentally friendly compared to traditional extraction techniques. Microextraction techniques (ME) such as SPME, MEPS and μ-QuEChERS take the lead among these extraction applications for the determination of food fingerprints because they have several advantages like reduced labor time and costs, as well as low sample and solvent volume requirements. This also is a sign of being environmentally friendly. Therefore, their microlevel usage in solvents and samples contributes to green chemistry studies while protecting human health (Gómez-Ríos and Pawliszyn, 2014).

13.4.1 Liquid-Liquid Extractions

Many foodstuffs such as nuts and their derivatives, sunflower seeds, sesame and even coffee beans are very rich foods in terms of free fatty acids that act as pro-oxidants in oil (Orm et al., 2020). This means that high fatty acid content related to the foodstuff is actually very sensitive to oxidation. The liquid-liquid extraction (LLE) technique, traditionally a simple technique and having low recoveries of supernatant, is generally used for the deacidification of nut-type foods with high oil content; that is, for the separation of free fatty acids, as well as some contaminants such as phthalates. However, for this extraction, which usually has to be worked with more than one solvent, a lot of solvent and time are needed to recover the supernatant that lies between the immiscible phases such as the organic and aqueous phases (Campmajó et al., 2019b; Pereira et al., 2020). These disadvantages of the method paved the way for the development of more advanced LLE methods. In a study controlled with the LLE method, it was shown that there is a balance between the efficiency of free fatty acid extraction using solvents containing 82% to 90% by weight ethanol and neutral oil losses (Orm et al., 2020).

The liquid-liquid microextraction (LLME) is an improved version of the LLE method, and requires much less solvent compared to the main method; it is cost-effective and more environmentally friendly. Although it is possible to work with lower solvent consumptions, the extraction time is long and the enrichment factors in the recovery of supernatants are limited. It is noted that there are many different applications of LLME, such as single-drop microextraction (SDME), dispersive liquid-liquid microextraction (DLLME), solvent-bar microextraction (SBME), hollow fiber liquid-phase microextraction (HF-LPME), and electro-membrane extraction (EME; Hansen and Pedersen-Bjergaard, 2019).

Dispersive liquid-liquid microextraction (DLLME) as a new mode of LLME was developed by Mohammad Rezaee et al. in 2006 (Rezaee et al., 2006). DLLME is a simple and effective method and involves isolating analytes from foodstuffs by two-stage extraction. First, the analytes are extracted from the solid matrix with the appropriate solvent or solvent mixtures. Then, it is aimed to obtain high-yield extracts by dispersing this extract in a lower density liquid. The main purpose of the dispersion has already been to increase the contact surface between the extract and the sample to obtain a high-yield supernatant. It has been reported that the determination of total aflatoxins (AFs) in commercial peanut samples was successfully determined by DLLME, which is simple, inexpensive, accurate and far from interference effects. Also, no limitations were noted for the extraction solvent to be high density (Amoli-Diva et al., 2015). An appropriate extraction solvent for DLLME should have a good emulsification efficiency in the aqueous sample, high affinity for compounds of interest, low solubility in water, low density and low vapor pressure to prevent loss during agitation. Since toxic organic solvents such as chlorobenzene, chloroform and carbon tetrachloride are used in the DLLME method, this situation creates many risks in terms of the environment and human health. For this reason, ionic liquids (IL) have been developed as a green chemistry alternative and are used in DLLME as conventional solutions. ILs have many advantages such as negligible vapor pressure, good solubility, non-flammability, high thermal stability, and adjustable viscosities. Also, they have a wide temperature range in the liquid phase, making them suitable for use in the DLLME method. Apart from using the DLLME method alone in the literature, many modified implementations of the original DLLME are also encountered. These can be listed as a solidification of floating organic drop (DLLME-SFO), ultrasound-assisted DLLME (UA-DLLME), vortex-assisted DLLME (VA-DLLME), alcohol-assisted DLLME (AA-DLLME), solvent and terminated DLLME (ST-DLLME), and they are used in conjunction with different chromatographic, spectroscopic or hyphenated techniques to analyze water and food samples (Medina et al., 2019c). Sequential extraction applications including both DLLME and SPE have been developed for the extraction of total AFs in pistachios and followed by a surfactant-enhanced spectrofluorimetric detection method. Improvements to this method showed that the extraction procedure is simple, has a low solvent consumption, unique enrichment in a short time, and good reproducibility and repeatability for low-cost and high-accuracy detection of AFs (Amoli-Diva et al., 2015). There is a wide range of solvent choices for PLE, which has both faster extraction processes and a lower solvent consumption for the sample preparation of solids (Delgado-Zamarreño et al., 2004). Generally, solvents used in PLE are ethanol, ethyl acetate, D-limonene and water. Among them, water is the most environmentally friendly solvent. While using water as a solvent PLE, extraction techniques can be applied in different ways, such as pressurized hot-water extraction (PHWE), hot water extraction (HWE), subcritical water extraction (SWE) and high-temperature water extraction (HTWE; Alvarez-Rivera et al., 2020).

13.4.2 Solid Phase Extraction (SPE)

The solid phase extraction (SPE) technique, which is one of the commonly used sample preparation methods to capture analytes and separate them from the sample matrix and which can be used in several application fields of sample preparation, was developed as an alternative application to LLE in the 1970s (Otles, 2016). It is still an emerging sample preparation field in modern analytical methods and is receiving more attention due to their many advantages over traditional methods. Because isolation, concentration, purification, and cleaning are the main approaches in the application of this method, homogenization, extraction and removal attempts, which are necessary and important applications for the analysis of food compounds, can be carried out easily. SPE works on the principle of passing the liquid sample through a sorbent. During the cleaning step, many analytes from the food matrix remain on the sorbent. This advantage makes SPE a powerful technique that offers researchers around the world different extraction procedures and optimization pathways. High recovery level, reduction in analysis time by decreasing the procedure steps and easy automation, adaptability with chromatographic analysis and a decrease in the organic solvent amount are the main advantages of this method. SPE methods can be investigated in four main application categories: reverse phase SPE, normal phase SPE, ion exchange SPE and mixed-mode SPE. The reverse phase SPE works in the least selective retention mechanism. It is very useful for extracting analytes that are very diverse in structure within the same sample. Conversely, it is not convenient for the separation of structurally similar molecules. Reverse phase contains a non-polar stationary phase with a polar or moderately polar mobile phase and is usually used for aqueous samples where the main interaction force is Van Der Waals forces. Normal phase SPE contains a polar stationary phase and retention resulting from the interaction between polar functional groups of the analyte and polar groups on the sorbent surface. Ion exchange SPE can be used for compounds that are charged in a solution. In mixed-mode SPE, two different functional groups are used on the same sorbent. Isolation, which takes place in both the reverse phase and cation exchange, is a common example of this type of SPE (Otles, 2016). Alumina, magnesium, silicate or graphitized carbon can be used as a sorbent in this method. But the used most suitable sorbent type is silicium in terms of modification and stability. It has disadvantages such as large sample solution volume, elution solvent network and time consumption. These kinds of defects decrease the accuracy of the method. Therefore, methods such as micro solid phase extraction (μ-SPE), dispersive solid phase extraction (dSPE), magnetic solid phase extraction (MSPE), immune affinity solid phase extraction (IA-SPE) and molecularly imprinted solid phase extraction (MI-SPE), which are much more advanced compared to the basic SPE technique, have gained popularity in food analysis. It is known that with the high selectivity and sensitivity properties of these improved SPE techniques, they have much less consumption of both sorbent and solvent, much shorter extraction time, and it is much easier to remove interfering compounds and taking clearly identifiable signals in chromatographic analysis while comparing to the initial earlier SPE method.

13.4.3 Solid Phase Microextraction (SPME)

In traditional extraction techniques, the main objective is to achieve high sensitivity and to ensure the use of simple calibration methods validated by the recovery of the analytes. But following the behavior of many analytes in complex matrices is difficult with conventional extraction methods. Our knowledge of microextraction in analytical purity sample preparation dates to the 1990s. Today, the microextraction methodology is highly developed and advanced and serves with numerous applications in different forms for different matrices.

Microextraction can be defined as an extraction method in which the liquid or solid extractant has a significantly smaller volume (typically <100 μL or <100 mg) compared to the sample (>1 mL). Solid phase microextraction (SPME) is a versatile sample preparation technique that has been shown to be an easy and effective method of analysis suitable, but not exhaustive, for a wide variety of matrices. Applications of SPME range from food to all samples of the environment and even biomedical and pharmaceutical analysis. SPME technology has appeared in various geometries and configurations such as fiber, thin film, in-tube and dispersed particles. The fiber and thin film geometries of the SPME techniques are mainly developed and designed for various applications in specific research areas. Depending on these geometries and configurations, the existence of fiber SPME (FSPME), which is the first developed application of SPME, thin film SPME (TFME), dispersed magnetic nanoparticles based SPME (d-SPME) and stir bar sorptive extraction (SBSE) formats were mentioned by Reyes Garces in 2018 (Boyacı et al., 2018). Although the main disadvantage of the SPME technique is its limited commercially available stationary phases (fiber materials), there are many important advantages to its application in food analysis, as follows:

- Its selectivity based on differences in K and/or H, D;
- Its good quantitative sensitivity in GC analysis.
- It supports green technology with its solvent-free usage feature for GC analysis.
- Its flexibility in configuration and automation sufficiency.
- Its small size and convenience of sample integrations and preparations.
- Simple and soft extraction approach.
- Free concentration determinations.
- Permission of binding and monitoring of reaction.
- Unconventional calibration and optimization.

Much scientific literature reports that there are numerous new materials developed for SPME fiber coatings. Some of these are graphene aerogel-mesoporous carbon (GA-MC) composites, pyrrole, titanium oxide composites, amine (NH_2)-functionalized material, carbon nanotubes (CNTs), graphitic carbon nitride (g-CN) and boron nitride nanotubes/sheets, molecularly imprinted polymers (MIPs), metal-organic frameworks (MOFs), zeolitic imidazole frameworks (ZIFs), sol-gel coatings, aptamer-functionalized SPME fibers, ILs fiber coatings and biocompatible SPME coatings (Perestrelo et al., 2019; Boyacı et al., 2018). Early SPME applications consisted of a fused fiber silica rod supporting a polymeric coating suitable for direct dipping (DI), headspace (HS) and membrane protection extraction. Later, in-tube SPME, a variant of traditional SPME and introduced in 1997, was advanced with some alterations as a wire-in-tube with a stainless-steel wire put into the extraction capillary (Roy et al., 2019). The significant reduction of capillary volume, increased the concentration of target analytes, and the suitability of automation, especially for LC instrumentation, increase the extraction yield improved with this modification. In-tube SPME is mostly used in polymer application areas, metal nanoparticle coatings and phosphopeptide extractions. In addition to in-tube SPME, recent developments in advanced techniques such as in-needle SPME, in-tip SPME or vacuum-assisted headspace SPME (Vac-HS-SPME) have been reported. In-tip SPME, a solid sorbent made of silica and/or a methacrylate monolith, is packaged in a pipette tip, and it is possible to process many samples simultaneously. It is reported that the recovery rates of in-tip SPME are better than fiber SPME and in-tube SPME (Medina et al., 2019c). In addition, Vac-HS-SPME, a popular and a suitable sub-technique of SPME, allows for the characterization of volatile organic compounds in various food sample content analyses (Kalogiouri et al., 2021).

In-needle SPME technique encompasses more advanced methods such as solid phase dynamic extraction (SPDE), microextraction by packaged syringe (or sorbent; MEPS), and fiber-packed needle microextraction (FNME). There are many advantages, such as large coating volume, concentration capacity, lower extraction time, and better reproducibility and mechanical stability in these extraction applications. Introduced in 2004, MEPS is a miniaturization of the traditional SPE technique using a significantly much lower sample volume and less extraction solvent. Recent applications of MEPS include opiate extraction from whole blood, the extraction of the psychoactive substance dichloropane from saliva, drug extraction from human plasma by using polyacrylonitrile/graphene oxide nanofibers, and the extraction of estrogen from human urine by using molecularly imprinted polymer. MEPS advantages are mainly silica-based C18, C8 and C2 (reverse phase), silica (normal phase), restricted access material (RAM), carbon, polystyrene-divinylbenzene copolymer (PS-DVB), molecularly imprinted polymers (MIPs), mixed mode (C8/SCX), cation exchange (SCX) type commercially available packable sorbents, all of which are repeatedly used and easily processed of their enrichment factors. The widespread use of MEPS in food analysis is due to these valuable advantages. In a conclusion, it is noted that MEPS is a frequently used technique for detecting food frauds. Therefore, custom MEPS sorbents with combined MIPs have come to the fore in food quality analysis in recent years.

13.4.4 Thin Film SPME (TFME)

Thin film SPME (TFME) was developed in 2001 with the sensitivity of SPME as another variant of traditional SPME, utilizing a larger volume of extracting phase and with a larger surface area. The aim of TFME is that the effective surface area of the extraction phase that interacts with the sample is enlarged. In this way, TFME achieves the best sensitivity, providing fast extraction kinetics and the extraction efficiency increases without surrendering the extraction time. Thin film SPME was used for the determination of estrogens in biological samples and the extraction of specific organic volatiles such as benzene, pyridine, 1-pentanol and octane from environmental samples (Kalogiouri et al., 2021). TFME has two main applications for instrumental compatibility: thermally stable TFME and solvent-stable TFME. Thermally stable TFME conforms to GC and aims to improve the sensitivity of SPME for GC-based applications. Similarly, solvent-stable TFME (referred to as TFME blade) is to improve the sensitivity of SPME for LC-based applications. TFME is a high-yield, sensitive and potent sample preparation method and can be utilized in various research areas, such as doping analysis and clinical analyses, for instance, where fast turnaround times are particularly desired, as well as in environmental and food applications.

13.4.5 Stir Bar Sorptive Extraction (SBSE)

Stir bar sorptive extraction (SBSE) is another extraction practice of SPME very similar to TFME. It involves an equilibrium-based non-exhaustive extraction. TFME utilizes a larger volume of extraction phase to magnify the sensitivity of the extraction. But SBSE primarily focuses on increasing the sensitivity of extraction by use of a thicker extraction phase; therefore, it takes a longer equilibration time than a typical TFME. Equilibrium-based studies or in cases where the best efficiency from a given application is required encounters long extraction times. It has been reported that SBSE is a kind of extraction technique and needs a long extraction time like 24 h. SBSE consists of a magnetic stir bar or stainless-steel wire that is coated with polydimethylsiloxane (PDMS), placed in the core of a glass jacket (usually 10–20 mm in length) and an outer layer of extractive phase covering the glass jacket. During the extraction time this stir bar is kept in an aqueous medium, and in the final step the stir bar is taken from the aqueous medium to get the analytes that are desorbed thermally or with solvent. Additionally, this configuration allows self-stirring of the sample by the extractive bar in cases where extraction from a given collected sample takes place on a stirring plate (Boyacı et al., 2018). It is not recognized for food analysis due to the number of commercially available coatings and the difficulty of achieving full automation. Fibers such as PDMS, polyacrylate (PA) and ethylene glycol–modified silicone (EG-silicon) used in SPME can also be used as extractive material in SBSE application. The SBSE method, which helps for large-volume samples to be extracted with on-site applications, also can improve extraction kinetics. It is noted that in SBSE applications, the use of thin coatings may destroy the coatings. On the other hand, the employment of thick coatings for extraction causes long extraction times, which are not desired in many cases. Therefore, significant effort has been devoted to improving extraction phase stability on the stirring bar. Many improvements have been made in SBSE application to increase the selectivity and/or efficiency of the extraction process and overcome all troublesome by using MIPs, monolithics, ethylene glycol modified silicon (EG-silicone), magnetic ionic liquid (MIL), nanosilica polydimethylsiloxane, polyurethane (PU) and polyacrylate (PA). For example, to enhance the coating stability, first the wire surface with a stable layer of polydopamine is modified via an in situ polymerization reaction; then a highly stable surface cross-linked layer of poly(EGDMA-AA) is synthesized on the bar, with a coating thickness of 4 µm, as has been reported in a previous report. Additionally, some improvements have also been made to overcome the long extraction times associated with SBSE methods. For example, the use of magnetic nanoparticles (MNPs) with a stir bar without being permanently fixed on its surface is one of them. Another approach to improve long extraction times in SBSE studies is the use of monolithic extraction phase coated stir bars. High extraction efficiency and relatively shorter extraction times are achieved by using monolithic, which provides high permeability and large surface area, as well as fine-tuning the extraction phase chemistry based on the analytes of interest.

There are different forms of modified applications of SBSE. They are named stir cake sorptive extraction (SCSE) and rotating-disc sorbent extraction (RDSE), which may increase the exchange-efficiency mass, and present higher recovery rates and better repeatability and reproducibility. The data on the use of SBSE in the sensorial analysis provided useful information about the flavor fingerprints and profiles of food, especially in wine samples. New adsorptive microextraction (AµE), bar adsorptive microextraction (BAµE), multi-spheres adsorptive microextraction (MSAµE) and micro quick, easy, cheap, effective, rugged and safe (µ-QuEChERS) are the new and advanced processes of SBSE. While AµE uses nanostructured sorbents (e.g., activated carbons, polymers) that operate under the floating sampling technology, the spherical substrates are fixed to the surface of the device through thermal treatment in MSAµE. These techniques have a major advantage in their flexibility to adapt the most suitable sorbent for each target analyte or class of compounds while compared to SBSE application (quick, easy, cheap, effective, rugged, and safe (QuEChERS; Medina et al., 2019c).

13.4.6 Quick, Easy, Cheap, Effective, Rugged, and Safe Extraction (QuEChERS)

The QuEChERS extraction method—which is an abbreviation of the words quick, easy, cheap, effective, rugged and safe—was introduced for the first time by Anastassiades and Lehotay in 2003 as a selective and sensitive pesticide residue detection method in environment and agri-foods and especially for the extraction of veterinary drugs from animal tissue (Medina et al., 2019c). This method, which focuses on multi-extra pesticide analysis in matrices, increases its popularity today for extracting many extracts from a variety of foods. The QuEChERS method, which does not require the use of external energy, is one of the remarkable methods that are low cost. It has a short-term process and a moderate number of samples that are made of 10–20 parts extracted in 30–40 min by a single analyst that can be analyzed at the same time. It has an impact on green chemistry studies, as it only requires the use of relatively small amounts of low-toxicity and non-halogenated solvents and reagents and simple laboratory equipment. The QuEChERS technique is a two-step application combined with extraction based on partitioning of water via salting out by using the combination of anhydrous $MgSO_4$ together with NaCl in 4:1 ratio and dispersive solid phase extraction (dSPE) that is a minimal and very fast clean-up step with adsorbents in order to remove matrix interferences (Fernandes et al., 2014). Therefore, extraction, separation and purification can be performed together to create a sample preparation with minimal loss of analyte in intermediate step processes such as filtration, evaporation and dehydration. Acetonitrile (ACN) is a very efficient solvent to isolate a broad range of pesticides and has been used as an organic solvent for extraction as well as acetone (ACE) and ethyl acetate (ETAC; Kokosa and Pharmacy, 2021).

The QuEChERS method has many advantages, including higher recoveries (>85%), and is possible for a wide polarity and volatility range of pesticides, including notoriously difficult analytes. Very accurate and precise results are possible due to the use of an internal standard (ISDT) that is required to correct for water content differences and volume fluctuations from commodity to commodity. It is possible to obtain about 10–20 samples simultaneously with a high yield in very short periods of about 30–40 minutes. It proves to be an environmentally friendly application, as chlorinated solvents are not used at all, and other solvents are less often used. The method does not require technical skills and training and very little laboratory equipment is enough, and it is quite robust as extract cleaning is done to remove organic acids, and can even be performed in a small mobile laboratory with a single worker if necessary (Maštovská and Lehotay, 2003).

QuEChERS applications for food matrices primarily work toward the detection of food additives. It has been reported that the QuEChERS extraction procedure was used for the detection of additives such as ractopamine in pork or ferulic acid metabolites in milk samples. The (μ-QuEChERS) technique, introduced to the quality control of baby foods, has been developed as a cost-effective and miniaturized, quite a new method producing minimal waste disposal compared to the conventional extraction procedure (Maštovská and Lehotay, 2003; Medina et al., 2019c). There is quite valid information to taking the determination of PAHs and other organic compounds in atmospheric particulate matter by using QuEChERS type extraction in the literature (Albinet et al., 2013). In other work, there is an isolation of some steroids from the herbal medicinal products via QuEChERS extraction combined with gas chromatography instruments (Schmidt, 2015).

Mycotoxins are a concern for food safety. Identification of mycotoxins, especially in pistachios, is very difficult due to the low concentrations and heterogeneous distribution of mycotoxins and on the other hand, pistachio is a complex food matrix due to its large number of components and high lipid content (53%). In a work related to mycotoxins identification, it was reported that mycotoxins can be easily extracted from the solid matrix to the liquid phase by QuEChERS extraction, and it is possible to separate a wide variety of analytes with high yields in a very short time. There are valuable evidence has been shown by Mateus and coworkers that the QuEChERS method that followed ultra-high performance liquid chromatography combined with time-of-flight mass spectrometry (UHPLC–ToF–MS) for analysis and was followed by Z-Sep d-SPE clean-up, provided the best analytical performance for aflatoxins (AFB1, AFB2, AFG1 and AFG2), ochratoxin A (OTA), zearalenone (ZEA), toxin T2 (T2) and toxin HT-2 (HT2) in pistachios. There are literature data showing that the QuEChERS method used in analyte extraction has been successfully applied in HPLC-DAD analyses of tocopherol and free phytosterols in almond, cashew nut, hazelnut, peanut, tiger nut, sunflower seeds and pistachio samples (Delgado-Zamarreño et al., 2016). In another work, 14 mycotoxins were identified by using QuEChERS and DLLME consecutive extraction steps combined with UHPLC-MS/MS instrumentation analysis (Arroyo-Manzanares et al., 2013).

13.4.7 Microwave-Assisted Extraction

The extraction used in the recovery of many different analytes is usually carried out using a conventional method such as Soxhlet extraction (Ballard et al., 2010). This tradition, which requires a long time and a large amount of solvent and increased risk of degradation of thermo-labile constituents, has recently been replaced by more modern and environmentally friendly methods such as enzyme assisted extraction (EAE), ultrasound assisted extraction (UAE) and microwave assisted extraction (MAE). Microwave extraction gains its power from combining microwave beams with conventional solvent extraction. Among these techniques, MAE has attracted attention over the past 15–20 years as a very effective and green extraction technique due to its many advantages such as reduced solvent, time and energy consumption and high recovery of analytes (Pinto et al., 2021; Gallo et al., 2010). On the other hand, when compared to the traditional Soxhlet techniques, MAE supplies higher extraction yields in a shorter time regarding total phenolic and tocopherol contents of some nut derivatives (Özbek et al., 2018). The MAE system contains a flask, a condensing tube, a microwave oven and the magnetic stirrer. During the extraction time the solvent and sample are contained in sealed extraction flasks under controlled temperature and pressure conditions. The closed flasks authorize the solvent temperature to rise well above its boiling point, which shortens extraction time and subsequently increases extraction efficiency (Ballard et al.,

2010). There are different factors that can affect the extraction efficiency of MAE such as microwave power, type and composition of solvent, extraction time, the particle size of sample, solvent-to-solid ratio, soaking time and extraction cycles (Valdés et al., 2015). Among these solvent type plays an important role in MAE. Generally, solvent selection in MAE should be based on microwave absorption properties and solubility of the target compound. In MAE, water and ethanol, which are completely environmentally friendly and polar solvents and have a high dispersion factor, are commonly used as invaluable solvents (Pinto et al., 2010). It is known that water is a perfect organic solvent because it offers a higher dispersion factor for MAE application and is supported in green chemistry studies wherever it is used (Medina et al., 2019c).

MAE extractions combined with HPLC-DAD-ESI-MS/MS equipment were introduced by Valdes and coworkers in 2015 for the isolation of some phenolic compounds with excellent yields from almond skin by-products. They declared that the best solvent in this extraction was ethanol for the determination of seven different almond cultivars, and optimization of HPLC-DAD-ESI-MS/MS detection procedure was excellent to have a high yield of the chemical composition of them. Green and effective aqueous enzymatic process assisted microwave extraction (AEME) to prepare macadamia nut oil have been noted. It is reported that higher amounts of more odoriferous oxygenated terpenes in the macadamia nuts can be extracted by AEME while compared to Soxhlet extraction. Zinc, copper and iron mineralization from pistachio nuts by microwave-assisted extraction (MAE) prior to its determination using flame atomic absorption spectrometry (FAAS) was proved by Khajeh in 2012 (Pinto et al., 2010). It is said that the MAE method is successful in detecting fingerprints in many areas of food analysis and even detecting the disinfection of by-products in meat samples (Medina et al., 2019a).

13.5 ANALYTICAL METHODOLOGIES

Fingerprint detection is very important in food science today. The population is increasing every day; however, the production capacity of products is decreased. Thus, more products are needed for people to survive. As a result, increased demand for food leads to food fraud. This situation inevitably leads to the emergence of food fraud. Because enough food needs to be supplied to people, this is getting harder and harder. For this reason, food fraud is encountered in almost all food products. Accordingly, of course, scientific investigations are gaining momentum and new methods are being developed every day for the detection of fraud. Many detection methods can be listed, including important technologies for food fingerprinting, such as chromatographic, hyphenated techniques and spectroscopic, spectrometric and electrochemical approaches. It is known that these methods are successful in detecting food fraud, and much reported research that we have has proved this achievement. Although spectroscopic and spectrometric-based studies attract more attention today, other techniques including chromatographic and hyphenated techniques also have made an undeniable contribution to the detection of food fraud (Taylan et al., 2020).

13.5.1 Chromatographic and Hyphenated Techniques

Chromatographic separation is based on the concentration balance of the respective components between the two immiscible phases. The stationary phase is immobilized in a column or fixed on a support, while the mobile phase is forced to pass over the stationary phase. Thus, compounds are separated by directing differential migrations (Pinto-Martinez et al., 2018).

The chromatographic methods can be classified based on the nature of the mobile phase (gas or liquid chromatography), the nature of the stationary phase (liquid, solid or bonded liquid), the mechanism of separation (adsorption, ion exchange, exclusion, partition, modified partition), and the technique used (column or planar; Rodrigues, 2015). Hyphenated techniques are combinations of spectroscopic techniques (e.g., NMR, FTIR and MASS) with chromatographic techniques (e.g., CE, GC or LC) or the coupling of separation techniques with spectroscopic detection techniques. Analytes detected by profiling must be clearly identified and quantified by using GC and LC detection of chromatography before the spectrometric analysis by such as UV, fluorescence, NMR, or MS. It is possible to determine that an analyte is present (at a known detection level) or not with certainty with the correct and working properly analytical equipment. If there is no analyte in the mixture no signal such as color change, titration loop, or chromatographic peaks are observed (Duxbury, 2005). Therefore, the chemists can easily report that there are none of the analyte presents in any mixture based on the detection and validation that gives experimental procedure, process, instrumental application, analysis approach and strategy (Peris-Vicente et al., 2015). Fingerprint analysis that helps the detection of authentications and food characterizations can be performed using both chromatographic and hyphenated techniques. Fingerprinting techniques can be classified according to the type of fingerprint that can be obtained including mass spectrometry (MS) fingerprinting, chromatographic fingerprinting, electrophoretic fingerprinting and spectroscopic fingerprinting (Esteki et al., 2021b). The specific signals of all components in any mixture can be analyzed both qualitatively and quantitatively with special detection systems selected in accordance with the chromatographic application. There are different chromatographic separation analyses such as capillary electrophoresis (CE), gas chromatography (GC) and liquid chromatography (LC) for the separation of different components in the mixtures. Chromatographic fingerprint analyses are usually applied through the food chain and are accomplished by a combination with chemometric methods in order to control food authentication. It is recognized that this combined application between chromatography and chemometry is a powerful tool for food fraud detection and encourages all research to establish the use of this methodology.

Capillary electrophoresis (CE) works in an applied electric field. It is an electrophoretic technique that separates charged analytes according to their ability to move in a conductive medium in response to an electric field. Many features such as high resolution, speed and low mobile phase volume requirement, and very affordable cost are among the advantages of CE. In contrast, it should be noted that CE is not suitable for trace compound analysis as it has low reproducibility and sensitivity (Medina 2019c). It is known that the CE is very successful with high resolution in a single step to separate the proteins. The technique is particularly well suited for polar and ionic compounds in complex polar matrices (Beretov et al., 2014). Different combined applications of CE such as diode array sensing (CE-DAD) and CE-MS models are encountered in the detection of some drinking water contaminants and the separation of metabolites in fruit juice or wine, and even in detecting contaminants in meat products (Medina 2019c). The CE-MS method provides automatic and fast separation with high resolution. It requires only small sample volumes, is very robust and uses an inexpensive capillary. It is also compatible with most buffers and analytes. It allows sequence analysis via MS/MS with platform-independent sample separation facilitating the independent entry of different sequencing platforms for peptide sequencing of CE-MS–defined biomarkers from highly complex mixtures. Unfortunately, CE-MS is not suitable for the analysis of high MW proteins (>20 kDa) as they tend to precipitate in acidic buffers often used for CE-MS analysis. Only a small sample volume can be loaded into the capillary, leading to lower selectivity compared to LC (Beretov et al., 2014). Capillary electrophoresis–mass spectrometry (CE-MS) is workable for charged analytes such as small molecular weight nucleic acids, amino acids, carboxylic acids, sugar phosphates, charged lipids or peptides.

GC chromatography, in which the mobile phase is gaseous, and the stationary phase is solid or a liquid immobilized on a solid inert support that is widely used for volatile, semi-volatile or thermally stable analytes or labor-intensive chemical derivatization techniques for separation, has many advantages such as high-resolution separation property and extensive utilization by the food industry for analyzing target compounds, volatile compounds and metabolomics (targeted and untargeted). It is one of the most popular separation methods in food science for being in foodstuff profiling, authenticity, quality assessment and safety compared to LC. In GC, the sample mixture is evaporated in a heated chamber (without decomposition) and the compounds are separated according to their selective interaction with the column material (stationary phase). The compounds pass through the column under high pressure using a mobile phase, an inert gas such as He and N_2. It is noted that GC has been identified in foodstuff profiling, authenticity, quality assessment and safety. Additionally, 2D GC or GC×GC, characterized by the combination of two columns with different polarities in tandem by a modulator interface, is known as two-dimensional gas chromatography and has been used as a multidimensional analytical technique for food analysis for over 20 years. Coping with the co-eluting problem, GC×GC provides greater sensitivity, resolution, selectivity and rapid analysis than the traditional 1D GC method. 1D GC and/or 2D GC can be hyphenated to different detectors such as flame ionization detector (FID), quadrupole mass spectrometry (qMS) and time-of-flight mass spectrometry (TOF-MS). Among them, MS-based GC analysis has become the most robust technology to assess the authenticity and safety of a great diversity of food products because mass spectrometry analysis is easier to automate with GC usage for bioprocess monitoring and control (Cabrera-López et al., 2019). For example, in the mass spectrometer, for vitamins, a parent ion can be observed and several ion fragments in a specific pattern but adding ions of the other compound can be exaggerated signals to the ion pattern of the mass spectrum. To solve these kinds of confusing, exaggerated signals a very high degree of specificity of the analysis can be obtained by using a combined GC-MS method that has more capabilities for food analysis. Furthermore, there are several advantages in the biological sample analysis specifically of GC/MS compared to LC/MS. That is, in the literature, the overwhelming superiority of GC in amino acid oxidation, DNA methylation and drug testing studies is mentioned (Vivekanandan-Giri et al., 2011; Qi et al., 2014; Lynch, 2017). On the other hand, it is reported that both GC and LC are very convenient chromatographic separations for pesticide determination that are usually extracted via fast approaches such as QuEChERS extractions (Artavia et al., 2021). Thus, the more specific an analytical method that is preferred for the analysis is, the higher the confidence level for the analytes because many researchers aim for a higher specificity degree in the analysis while using a single technique or instrument.

LC is another separation technique in which a liquid mobile phase is forced to pass through the solid stationary phase. To identify and quantify bioactive compounds such as fat-soluble and water-soluble vitamins and in the determination of carotenoid and polyphenol profiles for the characterization of some foods of interest due to possible biological activity, LC separation techniques are widely used. LC has many advantages such as automated sample preparation (e.g., filtration, dilution and even derivatization) and transfers to chromatographs with suitable flow injection analysis systems as interfaces. Therefore, it has a powerful effect in the food safety research field.

LC can be classified according to the principle of retention: (1) absorption, (2) reverse phase liquid-liquid partition and (3) ion exchange, ion pairing, gel filtration or affinity chromatography. LC can also be classified by the operating pressure: for example (1) fast protein liquid chromatography (FPLC), (2) high-performance liquid chromatography (HPLC) and (3) ultra-high-performance liquid chromatography (UHPLC).

FPLC is used largely for laboratory-scale protein purification and operates at relatively high flow rates (up to 20 mL/min) and low pressures (<50 bar). It uses glass or plastic columns packed with small particles (5–40 µm).

HPLC is a stable technique used to analyze a wide variety of molecules such as carbohydrates, organic acids, alcohols and amino acids. It is an evolved technique of LC and uses stainless-steel columns packed with smaller particles (3–5 µm) at high operating pressures (up to 400 bar). The HPLC column is packed with a stationary phase composed of irregularly or spherically shaped particles, a porous monolithic layer or a porous membrane (Rodríguez-Duran et al., 2017). HPLC can be examined in two different subclass applications based on the polarity of the mobile and stationary phases. In the first one, defined as normal phase liquid chromatography, the stationary phase is more polar than the mobile phase (e.g., toluene as the mobile phase, silica as the stationary phase). The second one is the opposite, termed reverse phase liquid chromatography (e.g., water-methanol mixtures as the mobile phase and C18 octadecyl silyl as the stationary phase; Prado and Weber, 2016).

In recent years, UHPLC, a miniature form of HPLC that uses smaller columns and reduced solvent volume, has been developed to allow faster separation of target analytes. UHPLC requires much less analysis time and solvent compared to LC and is similarly high resolution. UHPLC has packed columns with sub-2 µm particles, and its operating system works at higher pressures (1000 bar). UHPLC has attracted a lot of attention because of its effective and successful results in the profiling of metabolomic foods in many food authenticities studies. It is possible to further develop the potential of UHPLC and use the UHPLC system in conjunction with a mass spectrometer for this that gives excellent results. There may be the possibility of simultaneous presence of multiple overlapping target analytes. In this case, liquid chromatography can be used by sizing to solve the problem of overlapping of analytes. For this purpose, multidimensional liquid-liquid chromatography 2D (LC×LC) was developed and reported to be suitable for the analysis of complex food samples (Medina et al., 2019c).

Today, liquid chromatography/mass spectrometry (LC/MS), the direct coupling of of chromatographic separation of analytes dissolved in a liquid phase with mass spectrometric detection/identification, becomes necessary in the chemical analysis of food contents. Analytes are separated by hydrophobic, polar or ionic interactions. Obtained components from LC are given directly to MS via an interface that removes the solvent and ionizes the analytes to produce ions like a positive or a negative charge. These separated ions are detected on their mass-to-charge (m/z) ratios by the mass spectrometer according to the working principle of mass spectroscopy. It has fast scanning speeds with a high degree of multiplexing and many compounds can be measured in a single analytical run (Pitt, 2009). Atmospheric pressure ionization (API), atmospheric pressure chemical ionization (APCI) and electrospray ionization (ESI) are existed to have the comprehensive application of LC/MS as sophisticated and robust type interfaces. Thus, utilization of these kinds of versatile interfaces has enhanced the application field of LC/MS analysis. It is known that LC equipped with MS detector is a very remarkable application in monitoring food contamination such as naturally occurring toxins (mycotoxins, marine biotoxins, cyanogenic glycosides and toxins from poisonous fungi) and environmental pollutants, as well as advanced clinical biochemistry research. Numerous different LC-MS platforms have been used, including LC-MS, LC-MS/MS, LC-TOF, LC-QTOF and LC-Orbitrap, especially for toxicology tests (Lynch, 2017). Furthermore, reverse phase LC/MS provides a rapid screening procedure for hydrolysis products such as phosphonic acids and thiodiglycol without troublesome isolation and derivatization requirements compared to GC/MS. It can be applied to a wide range of biological molecules with highly sensitive and accurate tests, and it is stated that biomarkers of poisoning by chemical warfare agents can be easily analyzed using more advanced LC instruments such as LC/MS/MS (McHugh et al., 2011; Timperley et al., 2005).

In many other works, it is widely reported that LC-MS is also a common application in the detection of pesticide residues in foods and is even used to analyze more complex molecules, such as peptides from allergy proteins. It can be added that it is very successful in detecting food frauds (Artavia et al., 2021). For research of biopharmaceutical drug development, drug metabolism and toxicology studies, quantification of drugs in biological fluids (plasma, urine, tissue, etc.), pharmacokinetic studies, bioavailability studies, doping control, quantification of biogenic amines and therapeutic drug monitoring, LC/MS is a well-suited separation technique. Hyphenated techniques combined with ICP-MS, such as LC, can be used to perform multi-element speciation (Artavia et al., 2021).

13.5.2 Spectroscopic Approaches

Food fraud such as dilution, substitution and misleading labeling and origin information for greater financial gain is a major problem in all countries. Food adulteration by relabeling old or expired nuts, replacing them with cheaper ones or using other foods as additives causes serious problems for human health. There is a continual increase in food fraud, in terms of both the materials used for adulteration and the food that is adulterated. This situation makes it difficult to ensure food safety and quality. Moreover, different growing and climatic conditions characteristically affect the chemical composition (fingerprint) of nuts (Valdés et al., 2018; Esteki et al., 2021a; Esteki et al., 2019). Therefore, many analytical methods need to be developed to check the origin and authenticity of foods, to determine their chemical composition and prevent adulteration (Esteki et al., 2019). But, each of the analytical methods has some advantages and disadvantages, and the most appropriate method should be chosen to distinguish original food products from non-original ones (Valletta et al., 2021). To detect a wide variety of adulterants, as discussed earlier, sophisticated analytical instruments such as chromatographic or mass spectrometry is often used. For instance, by performing the analysis of fatty acids, cis-trans isomers, sterene, sterol in GC with a capillary column, it is possible to determine the adulteration and frauds in vegetable and seed oils. Although the fatty acid analysis with GC gives characteristic results for vegetable oils, the fatty acid structure of hazelnut, almond and pumpkin seed oils are similar both to each other and to olive oil. For this reason, effective results cannot be obtained in determining the fatty acid profile of these nuts by GC. This situation is overcome by measuring the amount of triglyceride (triacylglycerol [TAG]) components that the oils contained in different proportions by the high-pressure liquid chromatography (HPLC) method. HPLC analysis based on TAG components gives more characteristic results than GC in adulteration and fraud in vegetable seed oils. In addition, the fact that α and γ tocopherol fractions in oils differ in terms of amount in some studies performed with HPLC indicates a more reliable detection of adulteration in olive oil and hazelnut oil samples. In a study, total α and γ tocopherol levels in hazelnuts were found to be quite higher than in natural olive oil. The analyses performed with these devices are sensitive and accurate, but they are very expensive and not eco-friendly because they require the use of solvents. There is an urgent need to develop easy-to-use, fast, eco-friendly and inexpensive analytical methods to keep up with the rapid change in food cheating with fast results. In these respects, methods based on spectroscopic techniques provide these advantages needed to ensure food safety. The examination of herbal products and foods in terms of adulteration, fingerprints, authenticity and quality is carried out by combining the results of spectroscopic analysis and chemometric approaches (López et al., 2014b; Kucharska-Ambrożej and Karpinska, 2020; Danezis et al., 2016b). Figure 13.3 elaborates the process of fingerprint identification.

In spectroscopic methods, absorbed or emitted energy is formed as a result of the interaction of electromagnetic radiation and the components in the food. Much information about

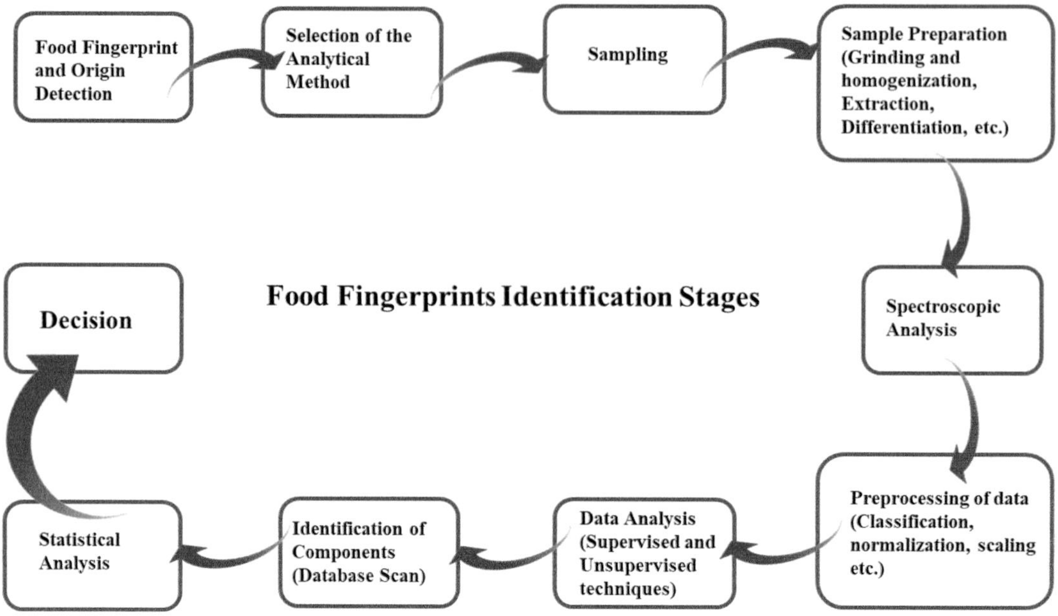

Figure 13.3 Food fingerprint identification stages.

the composition and properties of the food are obtained by the identification of wavelength or frequency in the energy spectrums. Nuclear magnetic resonance (NMR), electron spin resonance (ESR), x-ray spectroscopy, electron spectroscopy, fluorescence spectroscopy and ultraviolet-visible spectroscopy (UV-Vis), as well as vibrational spectroscopy methods such as infrared (IR), near-infrared (NIR), mid-infrared (MIR), Fourier transform infrared (FTIR), Raman and hyperspectral imaging (HSI), are the spectroscopic techniques used for detection of food adulteration, fingerprints, authenticity and quality (Medina et al., 2019b; Oliveira et al., 2019; Márquez et al., 2016).

When investigating the chemical profile of fatty acids, chromatography, Fourier-transform infrared spectroscopy and mass spectrometry instruments are more preferred. Apart from these methods, NMR spectroscopy is also a very suitable technique for the evaluation of fatty acids, since it does not require toxic organic solvents or long preparation time, and samples can be analyzed directly. NMR devices can give very fast results in purity and quantitative analysis, geographical origin, chemical profile and adulteration detection (Oliveira et al., 2019; Ravaglia et al., 2019; Sudhakar et al., 2021). In addition, when NMR results were evaluated together with chemometric methods, the differences and similarities of the samples can be observed more easily (Ravaglia et al., 2019).

In a study to detect the adulteration carried out by mixing different oils into Brazil nut oil, fatty acid profiles of Brazil nut, chia, flaxseed, sesame and soybean oils were investigated with ^{1}H-NMR combined with chemometric analysis (principal component analysis [PCA]) and vegetable-derived fatty acids and adulteration were successfully detected in commercial samples (Ravaglia et al., 2019).

While NMR spectroscopy is based on the electromagnetic radiation absorbed by the nucleus during its nuclear spin under a magnetic field, electron spin resonance (ESR) spectroscopy is based on the absorption of microwave light by an unpaired electron held in a strong magnetic field (Oliveira et al., 2019). Free radicals, are formed by decomposition with reasons such as storage conditions and duration, heat treatment and expiration date and contain unpaired electrons, can be easily detected by ESR in nuts and their oils (Cämmerer and Kroh, 2009; Jiang et al., 2020).

X-ray techniques are non-invasive methods used for internal quality assessment. Recently these techniques have been used for internal quality research of agricultural products such as pistachios, wheat kernels, pineapple, olives and tomatoes. X-rays have a low wavelength (0.01–10 nm) and high-energy electromagnetic radiation (120 eV to 120 keV) and give results in a very short time (Buthelezi et al., 2019). With x-ray fluorescence (XRF), elemental analysis is carried out for the characterization of some plant foods, and trace elements are also detected. Although it is an expensive technique, this analysis is promising for food validation studies and origin determination (Rajapaksha et al., 2017). Hirano et al. used x-ray–computed tomography (X-ray–CT) and ^{1}H-NMR-computed tomography (^{1}H-NMR-CT) imaging techniques to detect internal mold in peanuts. They reported that NMR spectroscopy can perfectly predict the internal mold in peanuts. However, the high installation cost of the instrument limits its use (Hirano et al., 1998).

Electron spectroscopy is a widely used method for surface characterization in chemical analyses and is also known as X-ray photoelectron spectroscopy (XPS). The sample surface to be analyzed is irradiated with photons in the X-ray energy range. The irradiated atoms emit electrons (photoelectrons) and give information about the chemical composition of the surface depending on the emitted electrons. Its biggest advantage is that it has the flexibility to work with a wide variety of samples (Ratner and Castner, 2009). Surface analysis is of great importance in food powders such as baby food and milk powder. The fat, protein, lactose and casein content of these products can be examined by surface analysis. Parameters such as oil oxidation, food powder solubility, allergic effects and color stability are important for food quality and safety (Chen et al., 2022; Faldt et al., 1993). For example, Zhao et al. noted that the surface composition of soybean protein powders influences particle-to-particle interactions as well as particle-to-water and particle-to-air interactions in food processes (Zhao et al., 2015).

Fluorescence spectroscopy is a spectroscopic technique used to detect the amount of compound from the intensity of the emission spectrum (at the emission wavelength) that occurs as a result of the excitation of a compound at a fixed excitation wavelength.

UV-Vis spectroscopy is a widely applied technique in food quality control. UV-Vis spectroscopy is suitable for correlating optical parameters with geographic regions or food varieties. Although this method has some limitations for characterization and confirmation, it gives good results when combined with NIR or MIR spectroscopy (Puertas and Vázquez, 2019).

Menevseoglu et al., in their study to detect adulteration of peanut and dried green pea in pistachio, added peanut or dried green pea in varying amounts of 5%–40% to pistachio samples and analyzed them with a portable FT-IR and UV-Vis instruments. Before analysis, chlorophyll and fatty acids from prepared pistachio, peanut, dried green pea, pistachio + peanut, and pistachio + dried green pea samples were extracted with a hexane:ethanol:methanol solution mixture. FT-IR spectra and UV-Vis spectra were recorded in the range of 4000–700 cm^{-1} and 200–800 nm, respectively. Spectra data were analyzed via soft independent modeling of class analogy (SIMCA), which had distinctive results in low levels of adulteration (5%). It has been determined by the partial least squares regression (PLSR) method that FT-IR has an excellent performance in estimating the amount of adulteration in pistachio (Menevseoglu et al., 2021).

Fourier transform infrared (FTIR) and Raman spectroscopy have an important role in elucidating the structure and functional groups of compounds found in foods. Also, these techniques help to investigate the oils of the nuts and to detect if they have been mixed with other low-value oils. Both techniques provide easy, fast, accurate, non-destructive, eco-friendly and economic analyses without causing damage to the samples (Valasi and Pappas, 2021; Rohman and Che Man, 2009).

Scattering occurs when the sample molecules excited by exposure to an intense light beam such as a laser do not absorb the light. The wavelength of the resulting scattering is mostly the same as the wavelength of the incident light, and only a small part of it shifts to different wavelengths. The shift of light to different wavelengths is Raman scattering, also known as inelastic scattering. Raman spectroscopy, which is one of the vibrational spectroscopic methods and investigates Raman scattering spectroscopically, is a method that gives information about the vibrational energy levels of molecules interacting with light (Xu et al., 2020). Table 13.1 shows the application of various spectroscopic techniques in determining the adulteration of nuts. IR spectral data has been widely documented as compared to other techniques (e.g., NMR and ICP).

13.5.3 Mass Spectrometry Technologies

Mass spectrometry (MS) is an analytical method and gives sensitive, repeatable, accurate and specific results in a food fraud determination. For this reason, there is a growing interest in MS methodologies in food authentication and traceability. MS provides both qualitative and quantitative analysis by ionizing and classifying food components according to their mass/charge ratio. The ion source present in the mass spectrometer ensures the ionization of the molecules. The ions are classified and separated by electric and magnetic fields in the mass analyzer according to their mass and charge (Valletta et al., 2021). The MS technique can be used in tandem with various techniques, and these techniques provide a great advantage in determining the quality and

Table 13.1 Applications of Spectroscopic Techniques in Detection of Adulteration or Authenticity of Nuts Using a Chemometric Approach

Nuts	Purpose of Analysis	Food Fraud	Fingerprint Component	Spectroscopic Technique	Predictors	Chemometric Approach	Reference
Almond	Investigating the potential of near-infrared spectroscopy (NIRS) as a portable instrument that can detect the lowest presence of bitter almonds mixed into commercial sweet almond batches	Addition of bitter almond to sweet almond	Amygdalin determination	Near-infrared spectroscopy (NIRS)	Spectral data from 950 to 1650 nm	PLSDA	(Torres et al., 2021)
	Determination of adulterants in almond powder adulterated with apricot and peanut powders by FT-IR and FT-NIR spectroscopic techniques	Addition of apricot kernel and peanut powder to almond powder	Non-targeted component	Fourier transform infrared (FT-IR) spectroscopy and Fourier transform near-infrared (FT-NIR) spectroscopy	Spectral data from 600 to 4000 cm^{-1} and 4000 to 10000 cm^{-1}	OCPLS and DD SIMCA	(Faqeerzada et al., 2020)
	Determination of peanut adulteration in almond via analysis of their multi-element fingerprints by inductively coupled plasma–optical emission spectrometry (ICP-OES)	Addition of peanut to almond powder	Multielemental analysis with B, Na, Mg, K, Ca, Fe, Cu, Cu, Zn and Sr	Inductively coupled plasma–optical emission spectrometry (ICP-OES)	–	PCA, PCALDA, PLS, LSSVM	(Esteki et al., 2017b)
Pistachio	Investigation of the performance of NIR spectroscopy combined with chemometric analysis as an alternative method to detect adulteration of green pea and spinach in pistachio	Addition of green pea and spinach to pistachio	Non-target component	Near-infrared spectroscopy (NIRS)	Spectral data from 908 to 1695 nm	PCA, PLSR	(Genis et al., 2021)

Nuts	Purpose of Analysis	Food Fraud	Fingerprint Component	Spectroscopic Technique	Predictors	Chemometric Approach	Reference
	Determination of peanut and green pea as adulterant in pistachio using Fourier transform infrared (FTIR) spectroscopy and ultraviolet-visible spectroscopy (UV-Vis)	Addition of peanut and green pea to pistachio	Non-target component	Fourier transform infrared (FT-IR) spectroscopy and ultraviolet-visible spectroscopy (UV-Vis)	Spectra data from 4000 to 700 cm^{-1} for FT-IR and spectra data from 200 to 800 nm for UV-Vis	SIMCA, PLSR	(Menevseoglu et al., 2021)
	Determination of green pea and peanut adulteration in pistachio by using portable FT-MIR and FT-NIR spectroscopy	Addition of peanut and green pea to pistachio	Non-target component	Fourier transform near-infrared (FT-NIR) spectroscopy and Fourier transform mid-infrared (FT-MIR)	Spectra data from 4000 to 650 cm^{-1} for FT-MIR and spectra data from 3920 to 7400 cm^{-1} for FT-NIR	PLSR, SIMCA	(Aykas and Menevseoglu, 2021)
	Determination of green pea adulteration in roasted, ground pistachio products by using polymerase chain reaction–capillary electrophoresis (PCR-CE) via DNA-based test	Addition of green pea to pistachio	DNA analysis	Polymerase chain reaction–capillary electrophoresis (PCR-CE)	–	–	(Sen et al., 2020)
	Detection of green pea and spinach adulteration in *Pistacia vera* by using laser-induced breakdown spectroscopy (LIBS)	Addition of green pea and spinach to pistachio	Fe, Mg, Ca, Na, K and P determination	Laser-induced breakdown spectroscopy (LIBS)	Spectra data from 186 to 900 nm	PCA, PLS	(Sezer et al., 2019)
	Investigation and detection of green pea adulteration in pistachio nut granules with FT-Raman hyperspectral imaging	Addition of green pea granules to pistachio nut granules	Fatty acid composition	FT-Raman hyperspectral imaging	Spectra data from 3700 to 200 cm^{-1}	PCA, PLSR	(Eksi-Kocak et al., 2016)

(Continued)

Table 13.1 (Continued)

Nuts	Purpose of Analysis	Food Fraud	Fingerprint Component	Spectroscopic Technique	Predictors	Chemometric Approach	Reference
Hazelnut	Analysis of hazelnut paste adulterated with almond paste and chickpea flour	Addition of almond paste and chickpea flour to hazelnut paste	Non-target component	NIR and FT-Raman spectroscopy	Spectra data from 3650 to 12000 cm^{-1} for NIR and spectra data from 290 to 3200 cm^{-1} for FT-IR	SIMCA	(Márquez et al., 2016)
	Analysis of hazelnut paste adulterated with almond paste and chickpea flour	Addition of almond paste and chickpea flour to hazelnut paste	Non-target component	FTIR	Spectra data from 4000 to 600 cm^{-1}	ROC, PCC, SIMCA	(López et al., 2014a)
	Analysis of hazelnut paste adulterated with almond paste and chickpea flour	Addition of almond paste and chickpea flour to hazelnut paste	Non-target component	NIR	Spectra data from 3650 to 12000 cm^{-1}	SIMCA	(López et al., 2014b)
Brazil nut	Botanical origin determination of Brazil nut, chia, linseed, sesame (toasted and raw) and soybean oils and analysis of Brazil nut oil adulterated with soybean oil by nuclear magnetic resonance (^1H-NMR)	Addition of soybean oil to Brazil nut oil	Fatty acid composition	^1H-NMR	Spectral data from 6.5 to 0.5 ppm	PCA	(Ravaglia et al., 2019)
Walnut	Detection of sunflower oil and rapeseed oil in walnut oil	Addition of sunflower oil and rapeseed oil to walnut oil	–	FTIR	Spectral data from 3050 to 700 cm^{-1}	PCR, PLSR	(El Mouftari et al., 2022)
	Origin determination of walnuts by inductively coupled plasma–mass spectrometry	Mislabeling of walnut origin	Multi-elemental analyses of Al, Ba, Co, Cu, Fe, Mo, Ni, and Sr etc.	HR-ICP-MS	–	PCA, SVM	(Segelke et al., 2020)

Nuts	Purpose of Analysis	Food Fraud	Fingerprint Component	Spectroscopic Technique	Predictors	Chemometric Approach	Reference
	Detection of soybean oil in walnut oil	Addition of soybean oil to walnut oil	Fatty acid composition	FTIR and fluorescence spectroscopy	Spectra data from 4000 to 650 cm^{-1} for FTIR and emission spectra data from 370 to 700 nm for fluorescence spectroscopy	PCA, SIMCA and PLS	(Li et al., 2015)
	Detection of sunflower oil in walnut oil	Addition of sunflower oil to walnut oil	Non-target component	Fluorescence spectroscopy	Fluorescence spectra data from 250 to 700 nm	PLSR	(Ge et al., 2014)
Cashew nut	Detection of allergenic nuts such as peanut, Brazil nut, macadamia nut and pecan nut in cashew nut using a portable near-infrared spectrophotometer	Addition of peanut, Brazil nut, macadamia nut and pecan nut to cashew nut	Non-target component	NIR	Spectra data from 908 to 1676 nm	GA, PLSDA, SIMCA	(Miaw et al., 2022)
Pumpkin	Preventing fraud and mislabeling of Austrian pumpkin seed oil	Mislabeling of pumpkin seed oil	Elemental analysis of Y, La, Ce, Pr, Nd, Sm, Eu, Gd, Tb, Dy, Ho, Er, Tm, Yb, Lu, Th, Ca, Cu, Zn, Rb, Sr, Ba, Pb and U	ICP-MS	–	PLS, SIMCA, SVM	(Zettl et al., 2017)
Sunflower	Monitoring and tracking the oxidation process of hazelnuts, cashews, almonds, macadamia nuts, sunflower seeds, watermelon seeds, pine seeds and peanuts oil by Fourier transform infrared spectroscopy	Oil oxidation formation	Peroxide, p-anisidine and carbonyl value analysis	FTIR	Spectra data from 400 to 4000 cm^{-1}	PLSR	(Zhou et al., 2021)

(Continued)

Table 13.1 (Continued)

Nuts	Purpose of Analysis	Food Fraud	Fingerprint Component	Spectroscopic Technique	Predictors	Chemometric Approach	Reference
	Spectroscopic characterization of different varieties of sunflower seeds	–	Oleic acid	Near-infrared diffuse reflectance spectroscopy (NIRDRS)	Spectral data from 1100 to 2200 nm	PCA, CA, LDA, PLSDA	(Monferrere et al., 2012)
Peanut	Determination of TAG and fatty acids in peanut oil	–	Triacylglycerols	APCI-MS	Retention time of spectral data from 11.72 to 28.48 min	PCA	(Dong et al., 2015)
	Determination of adulteration and gluten, fatty acid, fiber and omega-3 in different nuts and cereals	Addition of pine nuts, other nuts (almond) and cereals (sesame and flaxseed) to peanuts	Gluten, fatty acid, fiber and omega-3	NIR	Spectral data from 896 to 1686 nm	PCA, PLSDA	(Ghosh et al., 2016)
Pine nut	Determination of pine nut chemical composition by ICP-OES	–	Determination of Ca, Cu, Fe, K, Mg, Mn, P and Zn	ICP-OES	–	PCA, HCA	(Silva et al., 2022)
	Detection of different labeled origins	Misleading labeling of origin as Spanish pine nuts rather than Chinese pine nuts	Fat, fatty acid, water and protein	Near-infrared hyperspectral imaging (NIR-HSI)	Spectra data from 940 to 1625 nm	PCA, MCR, SIMCA	(Ríos-Reina et al., 2021)
	Pine nut species recognition by using NIR spectroscopy and image analysis	Differentiating Mediterranean stone pine nut and Siberian pine nut species	Moisture, fat, protein, fatty acid	NIR	Spectra data from 1100–2300 nm	PLSDA,	(Moscetti et al., 2021)
Coconut	Detection of the use of palm oil as an additive in coconut oil with FTIR spectroscopy	Addition of palm oil to coconut	Fatty acid composition	FTIR	Spectra data from 4000 to 650 cm^{-1}	PLS	(Rohman and Che Man, 2009)
	Analysis of lard adulteration in virgin coconut oil	Addition of lard to coconut	Aroma and chemical component such	FTIR	Spectra data from 4000 to 650 cm^{-1}	PLS	(Tengku Mansor et al., 2011)

authenticity of foods with complex chemical components. MS-based approaches are a powerful technique for the determination of allergens (Zhang et al., 2019), fatty acids (Ciotti et al., 2020; Kesen et al., 2018), volatile oils (Kesen et al., 2018), aflatoxin (Giang and Thien, 2020) and pesticides (Barci et al., 2020; Hidalgo-Ruiz et al., 2021) in nuts. A combination of the MS technique with high-resolution chromatographic techniques, especially GC and LC, is used in food authentication (Danezis et al., 2016b).

Chromatographic techniques are combined with different detector types. One of them is MS. Due to the superiority of mass spectroscopy in the qualitative and quantitative determination of each component in food in detail, using it in combination with chromatographic and other separation techniques allows the development of even more superior analytical techniques. Especially, GC-MS and LC-MS have excellent sensitivity, reproducibility and specificity in food analysis compared to others and are the most popular techniques. In the integration of gas chromatography and mass spectrometry (GC-MS), GC is used to separate volatile and thermally stable compounds such as esters, fatty acids, alcohols, aldehydes, and the terpenes from a complex mixture, while MS allows the determination of each component with high selectivity and sensitivity (Khan and Rahman, 2021). LC-MS also enables the analysis of proteins, peptides and carbohydrates in foods with high selectivity and sensitivity (Suman et al., 2021; Valletta et al., 2021). In the literature, various LC-MS approaches have been used in the determination of non-targeted and/or targeted metabolomics to determine the geographical and botanical origin of different nuts (Ghisoni et al., 2020; Gil Solsona et al., 2018; Klockmann et al., 2016; Klockmann et al., 2017). In addition, double MS instruments such as LC-MS/MS and GC-MS/MS are gradually replacing older instruments for food analyses (Danezis et al., 2016b; Ansari et al., 2012; Roselló-Soto et al., 2019; Aydin and Ulvi, 2019).

Gas chromatography (GC) combined with FT-IR and MS spectroscopy can be used in the analysis of foods. Obtained chromatographic spectra can be searched and compared with existing database spectra of reference compounds in the library. However, when the database of compounds is not available in the library or standard compounds are not available for analyses, combining the GC technique with vibrational techniques (GC-FTIR-MS) helps the identification of unknown peaks. Several possible compound structures are obtained from the ions and mass of an unknown compound, which is dissociated in MS. The FTIR has molecular-specific vibrations and is particularly effective in comparing and confirming functional groups or isomeric structures. Thus, with the obtained FTIR spectra, the structures of the compounds can be further elucidated. More reliable results can be obtained when complementary information about the m/z ratios and possible functional groups of the compounds in FTIR-MS is combined with the retention index of them in GC (Wahl et al., 1994; Nolvachai et al., 2021; Buchbauer et al., 1997; Zavahir et al., 2018).

Mass spectrometers have different ionization sources that provide ion formation after sample introduction. MS involves ionization techniques such as electrospray ionization (ESI) (Oliveira-Alves et al., 2020), electron ionization (EI; Fischer et al., 2017), easy ambient sonic-spray ionization (EASI; Funasaki et al., 2012), atmospheric pressure chemical ionization (APCI; Pérez-Fernández et al., 2017), atmospheric pressure photoionization (APPI; Gómez-Ariza et al., 2006), desorption electrospray ionization (DESI; Hou et al., 2022) and matrix-assisted laser desorption/ionization (MALDI; Jayasinghe and Caruso, 2013) have been used in various analysis of nuts. The selection of the ionization source is quite important. For example, ESI for ionization of highly polar compounds, such as amino acids, peptides, vitamins and phenolic compounds at atmospheric pressure; APPI and APCI for low- and medium-polarity food components such as carotenoids, steroids, flavonoids and flavors; MALDI for creating ions single-charged from large molecules such as proteins, carbohydrates and organic acids; and DESI and laser-assisted ionization sources are used to ionize the samples directly at atmospheric pressure (Cao et al., 2020). MS techniques involving these ion sources can be used in combination with chromatographic or non-chromatographic techniques for food analyses. Non-chromatographic MS techniques have also been used to detect food authenticity and adulteration (De Ceglie et al., 2014; Calvano et al., 2012).

One of the food fingerprints that varies from food to food is the isotope fingerprint and is determined by isotope ratio mass spectrometer (IRMS) devices. A food's origin, authenticity and product labeling can be perfectly verified using isotope fingerprints (Bianchi et al., 2018). This verification process can be followed using isotopes of carbon, nitrogen, sulfur, oxygen and hydrogen called stable light isotopes (bio elements). Strontium, lead and neodymium, which are in the heavy isotopes group, are also isotopes used in food authentication. The analysis of isotope ratios is performed by methods such as isotope ratio mass spectrometry (IRMS), multi-collector–inductively coupled plasma–mass spectrometry (MC-ICP-MS) and thermal ionization mass spectrometry

(TIMS). IRMS connected with the elemental analyzer, equilibration instruments, pyrolizer, GC or HPLC are used for the determination of light isotope ratios; MC-ICP-MS and TIMS for the measurement of heavy isotopes; and NMR (nuclear magnetic resonance) with deuterium probe for the determination of $^{2}H/^{1}H$ ratio (Danezis et al., 2016b).

13.5.4 Electronic Sensors

Consumers prefer not only safe and nutritious foods but also quality products that appeal to their senses. For this reason, sensory tests are an important parameter in the food industry. Sensory quality parameters can be controlled in two ways: traditional sensory testing by well-trained panelists and analysis using instrumented methods (Danezis et al., 2016b; Danezis et al., 2016a).

The appearance, aroma, taste and texture of food are important parameters that determine food quality. In the traditional method, it is important that the panelists are trained and there is enough reiteration. In addition, the statistical analyses of the observation results should be made in detail (Danezis et al., 2016a).

In the evaluations of the panelists, there are differences in sensory tests due to various conditions and individual differences. To obtain reliable results, panelists must have a certain level of sensory perception and motivation and be trained in the use of standard and systematic sensory methods. Therefore, this traditional method is time-consuming and expensive. Instrumental tests using perception sensors (electronic sensors) instead of panelists have attracted a lot of attention in recent years because in the analyses made with this device, the measurement procedure is quite simple, fast and inexpensive (Danezis et al., 2016b).

Electronic sensors are instrumental devices that cross-detect with multiple sensors and work by imitating human senses (biomimetic). Gas chromatography olfactometry (GCO), electronic tongue (e-tongue), electronic nose (e-nose) and electronic eye (e-eye) are electronic sensors. For example, the e-nose detects volatile compounds in a food sample with the help of a semi-selective gas sensor that mimics the sense of smell. When the volatile aromatic components in the headspace of the food are adsorbed on the sensor surface, the signal converted to a digital value is recorded and the analysis is calculated according to the statistical models (Danezis et al., 2016a; Danezis et al., 2016b; Wadood et al., 2020). Similarly, the e-tongue works by imitating the sense of taste via a liquid sensor, and the e-eye works by imitating the sense of sight via computer vision systems and colorimetric techniques. When the results obtained with electronic sensors are combined with chemometric analysis, they can be used to detect food originality, origin and additives in various foods such as wine, beer, honey, tea, olive oil and cheese (Danezis et al., 2016a; Danezis et al., 2016b; Medina et al., 2019b).

To detect the adulteration of virgin coconut oil, Marina et al. used an e-nose based on surface acoustic wave sensor technology. By adding deodorized palm kernel olein to virgin coconut oil, they were able to detect pure and adulterated coconut oil via e-nose along with principal component analysis (PCA; Marina et al., 2010).

13.6 THE ROLE OF CHEMOMETRIC APPROACHES IN THE CHARACTERIZATION OF FOOD FINGERPRINTS

Various food profiling techniques are used for food authenticity and profiling detection. Genomic, proteomic, metabolomic and isotopolomic techniques are among the food profiling techniques. In the genomic approach, genetic profiles are detected using nuclear, mitochondrial or plastidic DNA. In the proteomics approach, the characterization of food and food products is provided by analyzing proteins and peptides. In the metabolomic approach, deep information such as metabolites about the general composition of food products is provided, while in the isotopolomic approach, isotope ratios and rare earth elements are analyzed (Ghisoni et al., 2020; Portaluri et al., 2021; von Wuthenau et al., 2022). Briefly, as a result of the analysis of food products with food profiling techniques, various databases create to reveal data such as geographical origin, production type and diversity and to easily determine many quality parameters. Combining the obtained databases with chemometric approaches makes food fingerprint analyses even easier.

It is quite complex to analyze food components from raw spectra obtained by spectroscopic methods, especially vibration spectroscopy, and chemometric data analysis methods are applied to make spectral data meaningful. There are three main goals in chemometric analysis: data description, qualitative analysis (e.g., discrimination, classification and clustering) and quantitative analysis (e.g., regression and prediction; see Figure 13.4). Spectral data contains a lot of physical and chemical information and are different for distinct sample types (Xu et al., 2020). According to the purpose of the study, qualitative or quantitative analysis can be made by applying various

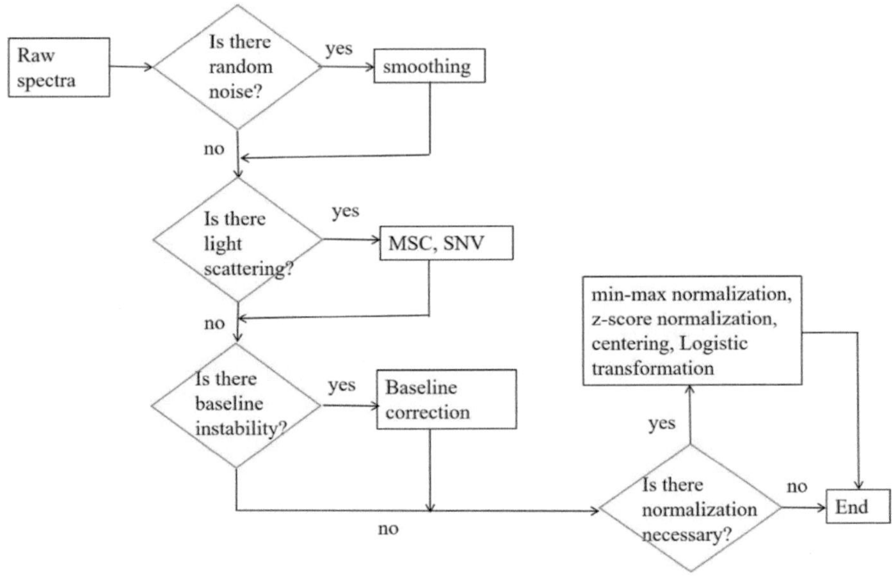

Figure 13.4 Selection of preprocessing techniques for a spectral data set.

Source: Reproduced with permission from Xu et al. (2020).

chemometric models and some preprocessing techniques to the data (Teixeira and Sousa, 2019). The qualitative analysis contains a multivariate classification model. The multivariate classification method, also known as the pattern recognition method, includes various chemometric methods that provide classification and separate of samples (Xu et al., 2020). With chemometric analysis, the relationships between complex data are revealed and converted into analytical parameters that can be used mathematically (Teixeira and Sousa, 2019). The multivariate classification method is grouped into two as supervised and unsupervised pattern recognition. The main difference that separates these two recognition systems is whether there is additional prior knowledge. A pattern recognition system consists of three steps: data set and preprocessing, data representation and final decision (Xu et al., 2020).

In summary, chemometric analyses provide a great convenience in the analysis of their originality, geographical origin and whether they have been adulterated and spoiled or not in a wide variety of food samples by using the spectral information of the components in the food. These tricks are determined by the correlation of spectral fingerprinting of food components and chemometric analysis such as principal component analysis (PCA), classification techniques (e.g., partial least squares–discriminant analysis [PLS-DA]) and regression models (e.g., partial least squares regression [PLSR]/principal component regression [PCR]; Oliveira et al., 2019).

13.6.1 Exploratory Analysis

The exploratory analysis that searches for possible data sets regardless of the nature of the analyzed samples is used to discover unidentified compounds in unlabeled data (Xu et al., 2020; Teixeira and Sousa, 2019). The exploratory analysis approach known as non-supervised pattern recognition reveals connections between samples, between variables, and between samples and variables. By investigating the relationship between samples, it is revealed whether there is a consistent distribution or trend in the sample distribution with previous information. For example, in the detection of authentication, samples that are distributed differently and separated from the main group should be removed (Callao and Ruisánchez, 2018).

In the relationship between variables, it is important which variables provide complementary, similar and more information. The link between samples and variables reveals which variables are important and which are not. Thus, the number of experiments is reduced, and the database is simplified. The most popular exploratory analysis useful for discovering hidden structures in unlabeled data is PCA (Medina et al., 2019b; Callao and Ruisánchez, 2018).

PCA is a data processing method that aids in classification and reduces the size of the data set, so it is a data compression and visualization method. When a score value is calculated for the spectral data of each sample, the samples examined in the graphical representation of the scores are clustered close to each other if they have similar characteristics. In examples of different qualities, the graphs obtained are distantly related. With PCA analysis, while the information about the sample available in the data set is preserved, the number of interrelated variables in large numbers is successfully reduced. Thus, the newly formed variable set shows linearity with the spectrum itself and can represent the information in the spectrum with less data (Xu et al., 2020; Medina et al., 2019b). Several analytical techniques in combination with PCA have been described in the literature to detect food fraud. For example, fatty acid profiles of 15 almond cultivars grown in the same ecological conditions in Turkey were analyzed by the GC-FID technique, and the usability of the obtained results in species verification of almond cultivars was determined by PCA analysis (Yildirim et al., 2016). In another study on different almond cultivars grown in Serbia, phenolic and fatty acid components of almonds were determined by UHPLC-DAD-MS/MS and GC-FID techniques, respectively, and characterization studies of almonds in terms of phenolic components and fatty acids were performed with PCA (Čolić et al., 2017a).

Another exploratory analysis technique is applying a clustering algorithm (CA) to classify different samples, and in cluster analysis, samples or variables are related to others based on their similarity. In short, CA analysis is the process of dividing all the data in the sample data matrix into several groups based on their similarity or difference and groups are identified based on their similarity. CA shows only the relationships between closely related samples rather than relationships among all samples. It also does not show the relationship between samples and variables. However, the CA uses all the information in the data and completes the PCA (Callao and Ruisánchez, 2018). In the study of Kafaoğlu et al., 17 trace metals were investigated using inductively coupled plasma-atomic emission spectrometry (ICP-OES) and inductively coupled plasma-mass spectrometry (ICP-MS) in hazelnuts (raw and dry roasted), almonds (raw and dry roasted), sunflower seeds (black and white), peanuts (raw and dry roasted), cashew nuts, Brazil nuts, walnuts, chickpeas (raw and dry roasted), pumpkin seeds (raw and dry roasted) and pistachio nuts; nuts were classified by PCA, CA and linear discriminant analysis (LDA) analyses (Kafaoğlu et al., 2014).

Hierarchical cluster analysis (HCA) makes a similar classification and establishes a proximity relationship between clusters. HCA analysis, which is a discriminative method, determines the boundaries between samples in different clusters, but does not determine the statistical reliability limits for each cluster. By performing HCA analysis, the spectra obtained from the samples are evaluated according to the similarity or difference relationship. Results are shown in a dendrogram and interpreted according to the clusters and subsets in which the samples are scattered (Medina et al., 2019b; Xu et al., 2020). In one study, phenolic compound, fatty acid, phospholipid and sugar contents in cashew nut extracts were determined by negative mode coupled to a hybrid mass spectrometer with quadrupole and time-of-flight analyzers (ESI (–)-Q/TOFMS) combined with PCA and HCA for fingerprinting of cashew nuts (de Albuquerque Mendes et al., 2019a).

13.6.2 Class Modeling and Discriminant Analysis

Supervised methods create a classification model based on a predefined sample class to classify and distinguish unknown samples based on experimental data. Supervised pattern recognition techniques have been used in food analysis to evaluate the results of spectroscopic, chromatographic and sensor applications for fingerprint analysis, profiling, quality control, authentication and adulteration detection (Berrueta et al., 2007; Medina et al., 2019b; Mishra et al., 2021). These pattern recognition techniques, which are mostly applied for qualitative purposes, can be listed linear discriminant analysis (LDA), partial least squares–discriminant analysis (PLSDA), orthogonal projections to latent structures modeling discriminant analysis (OPLSDA), k-nearest neighbors (kNN), artificial neural networks (ANN), support vector machines (SVM), least squares–support vector machine (LSSVM), soft independent modeling of class analogies (SIMCA) and unequal dispersed classes (UNEQ; Medina et al., 2019b; Teixeira and Sousa, 2019; Taylan et al., 2021).

LDA is one of the most common supervised pattern recognition methods in food analysis and has an important role in the detection of geographical origin, botanical origin, cultivation system and adulteration. In LDA, which is used to determine the linear definition functions, in order to optimize the class separation, the classes must follow the multivariate normal distribution and separate linearly. In this method, while the rate of variance between classes is maximized, the rate of variance within the class is minimized (Medina et al., 2019b; Kamal et al., 2021; Berrueta et al., 2007; Granato et al., 2018a).

Since the number of variables doesn't exceed the number of objects in the LDA technique, it causes a limitation. Overcoming this problem is possible by using methods such as PLS. PLS modeling is a multivariate method. To optimize the discrimination between different sample groups in PLS, the information in X and Y are evaluated together and the optimal number of variables is estimated by using cross-validation or external test sets and maximizing the relationship of the two variables. This method is suitable for data sets with few objects. In PLSDA, while the X matrix consists of the original data, a Y matrix is created with 1 and 0. The closer the element of a particular column in Y is to 1 and the elements of other columns to 0, the closer that object is considered a member of a particular class, and a model is developed for each class (Kamal et al., 2021; Berrueta et al., 2007; Chevallier et al., 2006).

OPLSDA uses multivariate data to distinguish between two or more classes. It is superior to the PLSDA method because a single component is used as an estimator for the class, while the other components explain the variation orthogonal to the initial estimation component (Kamal et al., 2021; Westerhuis et al., 2010).

In the kNN method, using the Euclidean distance, the distance between unknown objects and each object in the training set is determined for class assignment of them. In the class assignments, closest distance and the group with the majority are preferred. The mathematical simplicity of the method provides the advantage. When the number of samples is large, the computational slowdown is its disadvantage (Berrueta et al., 2007; Karthikeya et al., 2020).

ANN is a non-linear method designed by imitating the pattern recognition and information storage processes performed by the brain and nervous system. ANN consists of many artificial neurons capable of data processing and computation (Yu et al., 2018b; Elmasry et al., 2012).

SVM is a supervised learning technique used for both classification and regression. It is a method that combines both classes by classifying the training set and unknown samples (Berrueta et al., 2007).

One of the most used classes modeling techniques is SIMCA. In SIMCA, each category is modeled independently with the use of PCA, and these categories are explained by a different number of principal components, the number of which is determined by cross-validation (Berrueta et al., 2007). One of the advantages of SIMCA is that it determines whether an object can be classified into one, several or no classes. If the SIMCA analysis results in the object belonging to more than one class, it indicates that the data used do not distinguish between classes or there may be an error in the measurement. The disadvantage of SIMCA is that it requires preliminary analysis by PCA (Granato et al., 2018a).

UNEQ, a supervised pattern recognition method used when the number of variables is low, is a classroom modeling technique based on the assumption of multivariate normally distributed groups. According to the similarity of objects in a certain category, different classification functions are developed for each of the classes. An outlier test is used to decide whether the unknown object belongs to a particular class or not. If the value of the generalized distance from the center of the class exceeds the critical value, the object is an outlier and not part of the class (Berrueta et al., 2007; Derde and Massart, 1986). In literature, in order to reveal the origin of walnuts from different regions of Iran, fatty acid components were investigated by GC-FID method and geographical fingerprints of walnut species were determined by multivariate data analysis PCA, PCA-LDA and SIMCA (Esteki et al., 2017a). In another study by Taylan et al., Raman spectroscopy was combined with PCA, HCA, PLSR and GILS chemometric methods for the detection of adulteration of green peas in pistachio nut and thus creating a new methodology for the rapid, inexpensive and reliable detection of green pea fraud in pistachio nut (Taylan et al., 2021).

In a study of Campmajó et al., non-targeted chromatographic fingerprints of natural, fried or toasted nuts (almonds, cashew nuts, hazelnuts, macadamia nuts, peanuts, pine nuts, pistachios, pumpkin seeds, sunflower seeds and walnuts) were classified by HPLC-UV combined with PCA and PLSDA (Campmajó et al., 2019a).

13.6.3 Regression Analyses/Prediction Models

Multivariate calibration methods allow the determination of macro and micro components in mixtures. Since there is no need for selective measurements in such quantitative analyses, the necessity of sample preparation is reduced. Thus, the analytical performance of the instruments in the analyses and the reliability of obtained the data can be increased. Multilinear regression calibration (MLRC), principal component regression (PCR), partial least squares regression (PLSR), classification and regression tree (CART), genetic inverse least squares (GILS), artificial neural network (ANN) and support vector machines (SVM) are among the chemometric calibration methods.

Multivariate calibration methods have many advantages over univariate calibrations. In the multi-component analysis of the sample, the measurement of each component can be made one by one with the single-variable calibration method, while all the components with the multi-variable calibration method can be analyzed simultaneously. Less time spent on the multivariate calibration method is a great advantage for research. In the multivariate calibration method, estimates can be finalized by repeating the measurements and calculating the average. This results in a decrease in the standard deviation. In multivariate calibration methods, errors can be detected and unknown interferences in the analyzed sample can be eliminated. In univariate calibration methods, the presence of interfering species causes an incorrect determination of the analyte concentration. To eliminate the effect of interfering species, the analyte must be physically separated from the matrix or selective measurements must be used. By applying the multivariate calibration method, by choosing more variables, the chance of obtaining the desired level of calibration graphic curve increases and the effect of nonlinear types caused by interferences can be reduced. MLRC, which is the most basic form of linear regression, is a calibration method in which a single dependent variable is expressed from multiple independent variables.

PCR is a two-stage and multivariate method that first decomposes the spectral value by principal component analysis (PCA) and then uses selected principal components as predictors by MLR (Elmasry et al., 2012; Liu et al., 2015).

PLSR is a multivariate calibration model based on a linear algorithm. It is used to estimate the dependent variables from the set of independent variables. It is a very reliable method when there is a high correlation between spectral variables (Elmasry et al., 2012; Liu et al., 2015).

CART is a non-parametric statistical technique in which classification and regression trees are generated. In CART analysis, decision trees can be created for both classification and regression purposes. Calculated dependent variable classes form the classification tree. However, regression trees do not have predefined classes (Choubin et al., 2018; Bonifazi et al., 2021). GILS is an application of optimization and search methods based on the principles of natural evolution and selection. This model creates multivariate calibration models with a reduced data set (Üner et al., 2011).

13.7 CONCLUSION

The demand to feed the globe's growing population has put a strain on food production and food systems around the world. Food fraud occurs not only in terms of quantity and spoiled food but also in terms of adulteration. Food adulteration and food authentication are the two most difficult concerns facing food supply systems today. Nuts are an important part of the diet, and their popularity is growing as new health benefits are discovered. Triglycerides make up the majority of nuts, and conjugated structures provide health benefits. Other small components, including vitamins, sterols and metal ions, are believed to be effective in the authentication of nuts.

In recent years, a lot of work has been done to authenticate the origin and originality of nuts. Complex determination techniques like chromatography coupled with mass spectrometry or other spectroscopic techniques have been reported in the literature. No matter how advanced the analytical techniques, sample preparation techniques have also advanced to clean the very complex nut samples. Spectroscopic techniques as stand-alone techniques are also reported, but the data generated from chromatographic techniques has wider acceptance in food as well as analytical science. Nevertheless, how to clean the sample is, chromatographic techniques coupled with mass spectroscopy generates a lot of data, but it is not conclusive to decide on the quality or originality of the nuts. The current era has witnessed the use of various chemometrics approaches like LDA, which is one of the most common supervised pattern recognition methods in food analysis and has an important role in the detection of geographical origin, botanical origin, cultivation system and adulteration. Likewise, many other approaches are reported depending upon the type of data on nuts and the decisive targets on nuts. For example, CART is a non-parametric statistical technique in which classification and regression trees are generated. In CART analysis, decision trees can be created for both classification and regression purposes.

This chapter indicates that nut authenticity is a real issue that has been effectively addressed by employing improved (greener) sample preparation processes and chromatographic techniques followed by appropriate chemometric techniques to determine components. This method, however, is still laboratory-based and time-consuming. Simpler techniques, such as electronic sensors combined with chemometrics, could be developed to assist cargo officials in detecting adulteration at the terminal. Researchers may create a sensor array device based on minimal sample preparation for detecting chosen components in nut samples and integrate chemometrics into

Android-based applications. This authenticity detection unit could be used to conduct on-site testing of nuts.

REFERENCES

Abd, S. J., Mccarthy, K. L. & Harris, L. J. J. 2012. Impact of storage time and temperature on thermal inactivation of Salmonella Enteritidis PT 30 on oil-roasted almonds. *Journal of Food Science, 77,* M42–M47.

Ahrens, S., Venkatachalam, M., Mistry, A. M., Lapsley, K. & Sathe, S. K. 2005. Almond (Prunus dulcis L.) protein quality. *Plant Foods for Human Nutrition,* 60, 123–128.

Ajith, S., Pramod, S., Prabha Kumari, C., Potty, V. J. 2015. Effect of storage temperatures and humidity on proximate composition, peroxide value and iodine value of raw cashew nuts. *Journal of Food Science and Technology,* 52, 4631–4636.

Akbas, M. Y., Ozdemir, M. J. I. 2006. Effectiveness of ozone for inactivation of Escherichia coli and Bacillus cereus in pistachios. *Journal of Food Science and Technology,* 41, 513–519.

Alasalvar, C. & Bolling, B. W. 2015. Review of nut phytochemicals, fat-soluble bioactives, antioxidant components and health effects. *British Journal of Nutrition,* 113, S68–S78.

Alasalvar, C., Pelvan, E. J. E. 2011. Fat-soluble bioactives in nuts. *European Journal of Lipid Science and Technology,* 113, 943–949.

Albinet, A., Tomaz, S. & Lestremau, F. 2013. A really quick easy cheap effective rugged and safe (QuEChERS) extraction procedure for the analysis of particle-bound PAHs in ambient air and emission samples. *Science of the Total Environment,* 450, 31–38.

Alcántara-Durán, J., Moreno-González, D., García-Reyes, J. F. & Molina-Díaz, A. 2019. Use of a modified QuEChERS method for the determination of mycotoxin residues in edible nuts by nano flow liquid chromatography high resolution mass spectrometry. *Food Chemistry,* 279, 144–149.

Alvarez-Rivera, G., Bueno, M., Ballesteros-Vivas, D., Mendiola, J. A. & Ibanez, E. 2020. Pressurized liquid extraction. *Liquid-phase extraction.* Elsevier.

Amoli-Diva, M., Taherimaslak, Z., Allahyari, M., Pourghazi, K. & Manafi, M. H. 2015. Application of dispersive liquid—liquid microextraction coupled with vortex-assisted hydrophobic magnetic nanoparticles based solid-phase extraction for determination of aflatoxin M1 in milk samples by sensitive micelle enhanced spectrofluorimetry. *Talanta,* 134, 98–104.

Anderson, K. J., Teuber, S. S., Gobeille, A., Cremin, P., Waterhouse, A. L. & Steinberg, F. M. 2001. Walnut polyphenolics inhibit in vitro human plasma and LDL oxidation. *The Journal of Nutrition,* 131, 2837–2842.

Ansari, P., Stoppacher, N. & Baumgartner, S. 2012. Marker peptide selection for the determination of hazelnut by LC—MS/MS and occurrence in other nuts. *Analytical and Bioanalytical Chemistry,* 402, 2607–2615.

Arroyo-Manzanares, N., Huertas-Pérez, J. F., Gámiz-Gracia, L. & García-Campaña, A. M. 2013. A new approach in sample treatment combined with UHPLC-MS/MS for the determination of multiclass mycotoxins in edible nuts and seeds. *Talanta,* 115, 61–67.

Arrus, K., Blank, G., Abramson, D., Clear, R. & Holley, R. J. 2005. Aflatoxin production by Aspergillus flavus in Brazil nuts. *Journal of Stored Products Research,* 41, 513–527.

Artavia, G., Cortés-Herrera, C. & Granados-Chinchilla, F. 2021. Selected instrumental techniques applied in food and feed: Quality, safety and adulteration analysis. *Foods,* 10, 1081.

Arya, S. S., Salve, A. R., Chauhan, S. J. 2016. Peanuts as functional food: a review. *Journal of Food Science and Technology,* 53, 31–41.

Awan, H. U. M. & Pettenella, D. 2017. Pine nuts: a review of recent sanitary conditions and market development. *Forests,* 8, 367.

Aydin, S. & Ulvi, M. 2019. Residue levels of pesticides in nuts and risk assessment for consumers. *Quality Assurance and Safety of Crops & Foods,* 11, 539–548.

Aykas, D. P. & Menevseoglu, A. 2021. A rapid method to detect green pea and peanut adulteration in pistachio by using portable FT-MIR and FT-NIR spectroscopy combined with chemometrics. *Food Control,* 121, 107670.

Ballard, T. S., Mallikarjunan, P., Zhou, K. & O'Keefe, S. 2010. Microwave-assisted extraction of phenolic antioxidant compounds from peanut skins. *Food Chemistry,* 120, 1185–1192.

Banjanin, T., Nikolic, D., Uslu, N., Gökmen, F., Özcan, M. M., Milatovic, D., Zec, G., Boškov, Đ., Dursun, N. J. 2021. Physicochemical properties, fatty acids, phenolic compounds, and mineral contents of 12 Serbia regional and commercial almond cultivars. *Journal of Food Processing,* 45, e15015.

Baquião, A. C. 2015. *Fungos e micotoxinas em castanhas do Brasil, da colheita ao armanezamento.* Universidade de São Paulo.

Barci, P. E., Alves, L. D. S., Avellar, Á. A., Cendon, L. R., Dos Santos, P. J., Stringhini, F. M., Prestes, O. D. & Zanella, R. 2020. Modified QuEChERS method for multiresidue determination of pesticides in pecan nuts by liquid chromatography tandem mass spectrometry. *Food Analytical Methods*, 13, 793–801.

Bari, L., Nei, D., Sotome, I., Nishina, I. Y., Hayakawa, F., Isobe, S., Kawamoto, S. 2010. Effectiveness of superheated steam and gas catalytic infrared heat treatments to inactivate Salmonella on raw almonds. *Journal of Food Borne pathogen and Disease*, 7, 845–850.

Beretov, J., Wasinger, V. C., Graham, P. H., Millar, E. K., Kearsley, J. H. & Li, Y. 2014. Proteomics for breast cancer urine biomarkers. *Advances in Clinical Chemistry*, 63, 123–167.

Berrueta, L. A., Alonso-Salces, R. M. & Héberger, K. 2007. Supervised pattern recognition in food analysis. *Journal of Chromatography A*, 1158, 196–214.

Beuchat, L. R., Mann, D. A. & Alali, W. Q. J. 2012. Evaluation of sanitizers for inactivating Salmonella on in-shell pecans and pecan nutmeats. *Journal of Food Protection*, 75, 1930–1938.

Beuchat, L. R., Mann, D. A. & Alali, W. Q. J. 2013. Efficacy of sanitizers in reducing Salmonella on pecan nutmeats during cracking and shelling. *Journal of Food Protection*, 76, 770–778.

Bianchi, F., Giannetto, M. & Careri, M. 2018. Analytical systems and metrological traceability of measurement data in food control assessment. *TrAC Trends in Analytical Chemistry*, 107, 142–150.

Bingol, G., Yang, J., Brandl, M. T., Pan, Z., Wang, H. & Mchugh, T. H. J. 2011. Infrared pasteurization of raw almonds. *Journal of Food Engineering*, 104, 387–393.

Blessington, T., Mitcham, E. J. & Harris, L. J. J. 2012. Survival of Salmonella enterica, Escherichia coli O157: H7, and Listeria monocytogenes on inoculated walnut kernels during storage. *Journal of Food Protection*, 75, 245–254.

Blessington, T., Theofel, C. G., Mitcham, E. J. & Harris, L. J. 2013. Survival of foodborne pathogens on inshell walnuts. *International Journal of Food Microbiology*, 166, 341–348.

Bojko, B., Cudjoe, E., Gómez-Ríos, G. A., Gorynski, K., Jiang, R., Reyes-Garcés, N., Risticevic, S., Silva, É. A., Togunde, O. & Vuckovic, D. 2012. SPME—Quo vadis? *Analytica Chimica Acta*, 750, 132–151.

Bolling, B. W., Chen, C.-Y. O., Mckay, D. L. & Blumberg, J. B. 2011. Tree nut phytochemicals: composition, antioxidant capacity, bioactivity, impact factors. A systematic review of almonds, Brazils, cashews, hazelnuts, macadamias, pecans, pine nuts, pistachios and walnuts. *Nutrition Research Reviews*, 24, 244–275.

Bonifazi, G., Capobianco, G., Gasbarrone, R. & Serranti, S. 2021. Contaminant detection in pistachio nuts by different classification methods applied to short-wave infrared hyperspectral images. *Food Control*, 130, 108202.

Boyacı, E., Bojko, B., Reyes-Garcés, N., Poole, J. J., Gómez-Ríos, G. A., Teixeira, A., Nicol, B. & Pawliszyn, J. 2018. High-throughput analysis using non-depletive SPME: challenges and applications to the determination of free and total concentrations in small sample volumes. *Scientific Reports*, 8, 1–10.

Brandl, M. T., Pan, Z., Huynh, S., Zhu, Y. & Mchugh, T. H. J. 2008. Reduction of Salmonella Enteritidis population sizes on almond kernels with infrared heat. *Journal of Food Protection*, 71, 897–902.

Brar, P. K. & Danyluk, M. 2019. Validation of Enterococcus faecium as a surrogate for Salmonella under different processing conditions for peanuts and pecans. *Food Microbiology*, 80, 9–17.

Buchbauer, G., Jirovetz, L. & Nikiforov, A. 1997. Use of GC-FID, GC-FTIR-MS, and Olfactory Characterisation in the Analysis of Essential Oils and Plant Extracts. *Plant Volatile Analysis*. Springer.

Buthelezi, N. M. D., Tesfay, S. Z., Ncama, K. & Magwaza, L. S. 2019. Destructive and non-destructive techniques used for quality evaluation of nuts: A review. *Scientia Horticulturae*, 247, 138–146.

Cabrera-López, J.-J., Velasco-Medina, J. 2019. Structured approach and impedance spectroscopy microsystem for fractional-order electrical characterization of vegetable tissues. *IEEE Transactions on Instrumentation and Measurement*, 69, 469–478.

Callao, M. P. & Ruisánchez, I. 2018. An overview of multivariate qualitative methods for food fraud detection. *Food Control*, 86, 283–293.

Calvano, C. D., De Ceglie, C., D'Accolti, L. & Zambonin, C. G. 2012. MALDI-TOF mass spectrometry detection of extra-virgin olive oil adulteration with hazelnut oil by analysis of phospholipids using an ionic liquid as matrix and extraction solvent. *Food Chemistry*, 134, 1192–1198.

Cämmerer, B. & Kroh, L. W. 2009. Shelf life of linseeds and peanuts in relation to roasting. *LWT-Food Science and Technology*, 42, 545–549.

Campmajó, G., Navarro, G. J., Núñez, N., Puignou, L., Saurina, J. & Núñez, O. 2019a. Non-targeted HPLC-UV fingerprinting as chemical descriptors for the classification and authentication of nuts by multivariate chemometric methods. *Sensors*, 19, 1388.

Campmajó, G., Núñez, N., Núñez, & Intechopen: London, U. 2019b. The role of liquid chromatography-mass spectrometry in food integrity and authenticity. *Journal of Future Perceptions and Applications*, 3–20.

Cao, G., Li, K., Guo, J., Lu, M., Hong, Y. & Cai, Z. 2020. Mass spectrometry for analysis of changes during food storage and processing. *Journal of Agricultural and Food Chemistry*, 68, 6956–6966.

Chauhan, A. & Chauhan, V. 2012. Potential beneficial effects of a diet with walnuts in aging and Alzheimer's disease. *Brain Aging and Therapeutic Interventions*. Springer.

Chauhan, A. & Chauhan, V. 2020. Beneficial effects of walnuts on cognition and brain health. *Nutrients*, 12, 550.

Chen, C.-Y. O. & Blumberg, J. B. 2008. Phytochemical composition of nuts. 17.

Chen, C., Khir, R., Shen, Y., Wu, X., Zhang, R., Cao, X., Niederholzer, F. & Pan, Z. J. L. 2021. Energy consumption and product quality of off-ground harvested almonds under hot air column drying. *Journal of Agricultural and Food Chemistry*, 138, 110768.

Chen, C. & Pan, Z. 2022. Processing of Tree Nuts. *Postharvest Technology-Recent Advances, New Perspectives and Applications*. IntechOpen.

Chen, Q., Yin, Q., Xie, Q., Evivie, S. E., Zhao, X., Li, B. & Huo, G. 2022. Elucidating the physicochemical properties and surface composition of goat milk-based infant formula powders. *Food Chemistry*, 377, 131936.

Chevallier, S., Bertrand, D., Kohler, A. & Courcoux, P. 2006. Application of PLS-DA in multivariate image analysis. *Journal of Chemometrics: A Journal of the Chemometrics Society*, 20, 221–229.

Choubin, B., Zehtabian, G., Azareh, A., Rafiei-Sardooi, E., Sajedi-Hosseini, F. & Kişi, Ö. 2018. Precipitation forecasting using classification and regression trees (CART) model: a comparative study of different approaches. *Environmental Earth Sciences*, 77, 1–13.

Ciotti, M., Ciccozzi, M., Terrinoni, A., Jiang, W.-C., Wang, C.-B. & Bernardini, S. 2020. The COVID-19 pandemic. *Critical Reviews in Clinical Laboratory Sciences*, 57, 365–388.

Čolić, S. D., Akšić, M. M. F., Lazarević, K. B., Zec, G. N., Gašić, U. M., Zagorac, D. Č. D. & Natić, M. M. 2017a. Fatty acid and phenolic profiles of almond grown in Serbia. *Food Chemistry*, 234, 455–463.

Čolić, S. D., Akšić, M. M. F., Lazarević, K. B., Zec, G. N., Gašić, U. M., Zagorac, D. Č. D. & Natić, M. M. J. F. C. 2017b. Fatty acid and phenolic profiles of almond grown in Serbia. *Food Chemistry*, 234, 455–463.

Lurz, P., Garson, P., and Ogilvie, J. J. 1998. Conifer species mixtures, cone crops and red squirrel conservation. *Seeds, Fruits, and Cones, Forestry: An International Journal of Forest*. 71(1), 67–71.

Cuervo, M. P., Lucia, L. M. & Castillo, A. J. 2016. Efficacy of traditional almond decontamination treatments and electron beam irradiation against heat-resistant Salmonella strains. *Journal of Food Protection*, 79, 369–375.

D'Souza, T., Karwe, M. & Schaffner, D. W. J. 2012. Effect of high hydrostatic pressure and pressure cycling on a pathogenic Salmonella enterica serovar cocktail inoculated into creamy peanut butter. *Journal of food protection*, 75, 169–173.

D'Souza, T., Karwe, M. & Schaffner, D. W. J. 2014. Effect of high hydrostatic pressure on Salmonella inoculated into creamy peanut butter with modified composition. *Journal of food protection*, 77, 1664–1668.

Danezis, G. P., Tsagkaris, A. S., Brusic, V. & Georgiou, C. A. 2016a. Food authentication: state of the art and prospects. *Current Opinion in Food Science*, 10, 22–31.

Danezis, G. P., Tsagkaris, A. S., Camin, F., Brusic, V. & Georgiou, C. A. 2016b. Food authentication: Techniques, trends & emerging approaches. *TrAC Trends in Analytical Chemistry*, 85, 123–132.

Daniel, E. M., Krupnick, A. S., Heur, Y.-H., Blinzler, J. A., Nims, R. W., Stoner, G. D. 1989. Extraction, stability, and quantitation of ellagic acid in various fruits and nuts. *Journal of Food Composition and Analysis*, 2, 338–349.

Davis, L., Stonehouse, W., Loots, D. T., Mukuddem-Petersen, J., Van Der Westhuizen, F. H., Hanekom, S. M. & Jerling, J. C. J. 2007. The effects of high walnut and cashew nut diets on the antioxidant status of subjects with metabolic syndrome. *European Journal of Nutrition*, 46, 155–164.

De Albuquerque Mendes, M. K., Dos Santos Oliveira, C. B., Veras, M. D. A., Araujo, B. Q., Dantas, C., Chaves, M. H., Júnior, C. A. L. & Vieira, E. C. 2019a. Application of multivariate optimization for the selective extraction of phenolic compounds in cashew nuts (Anacardium occidentale L.). *Talanta*, 205, 120100.

De Albuquerque Mendes, M. K., Dos Santos Oliveira, C. B., Veras, M. D. A., Araujo, B. Q., Dantas, C., Chaves, M. H., Júnior, C. A. L. & Vieira, E. C. 2019b. Application of multivariate optimization for the selective extraction of phenolic compounds in cashew nuts (Anacardium occidentale L.). *Talanta*, 205, 120100.

De Ceglie, C., Calvano, C. D. & Zambonin, C. G. 2014. Determination of hidden hazelnut oil proteins in extra virgin olive oil by cold acetone precipitation followed by in-solution tryptic digestion and MALDI-TOF-MS analysis. *Journal of Agricultural and Food Chemistry*, 62, 9401–9409.
De Jong, A., Plat, J. & Mensink, R. P. J. 2003. Metabolic effects of plant sterols and stanols. *The Journal of Nutritional Biochemistry*, 14, 362–369.
De Mello, F. R. & Scussel, V. M. J. 2009. Development of physical and optical methods for in-shell Brazil nuts sorting and aflatoxin reduction. *Journal of Agricultural Science*, 1, 3.
De Souza, T. S. & Koblitz, M. G. B. U. 2022. Almonds—cultivation, production, composition, allergenicity, benefits of consumption, and potential applications. *Agricultural Research Update*, 33.
Delgado-Zamarreño, M., Bustamante-Rangel, M., Sánchez-Pérez, A. & Carabias-Martínez, R. J. J. O. C. A. 2004. Pressurized liquid extraction prior to liquid chromatography with electrochemical detection for the analysis of vitamin E isomers in seeds and nuts. *Journal of Chromatography A*, 1056, 249–252.
Delgado-Zamarreño, M. M., Fernández-Prieto, C., Bustamante-Rangel, M. & Pérez-Martín, L. 2016. Determination of tocopherols and sitosterols in seeds and nuts by QuEChERS-liquid chromatography. *Food Chemistry*, 192, 825–830.
Derde, M.-P. & Massart, D. 1986. UNEQ: a disjoint modelling technique for pattern recognition based on normal distribution. *Analytica Chimica Acta*, 184, 33–51.
Destaillats, F., Cruz-Hernandez, C., Giuffrida, F., Dionisi, F. J. 2010. Identification of the botanical origin of pine nuts found in food products by gas– liquid chromatography analysis of fatty acid profile. *Journal of Agricultural and Food Chemistry*, 58, 2082–2087.
Dheir, I. M., Abu Mettleq, A. S. & Elsharif, A. A. 2020. Nuts types classification using deep learning.
Dong, X., Zhong, J., Wei, F., Lv, X., Wu, L., Lei, Y., Liao, B., Quek, S. & Chen, H. 2015. Triacylglycerol composition profiling and comparison of high-oleic and normal peanut oils. *Journal of the American Oil Chemists' Society*, 92, 233–242.
Duxbury, D. J. F. T. 2005. Flavor analysis integral to product development. 59(2), 60–62.
Eksi-Kocak, H., Mentes-Yilmaz, O. & Boyaci, I. H. 2016. Detection of green pea adulteration in pistachio nut granules by using Raman hyperspectral imaging. *European Food Research and Technology*, 242, 271–277.
El Mouftari, M., Essafi, I., Khalidi, A., Kzaiber, F., Ali, G. A., Mahjoubi, F. & Oussama, A. 2022. Applications of FTIR and chemometrics methods in authenticity analysis of walnut oil. *Emergent Materials*, 1–8.
Elias, T. S. 1980. The complete trees of North America; field guide and natural history.
Elmasry, G., Kamruzzaman, M., Sun, D.-W. & Allen, P. 2012. Principles and applications of hyperspectral imaging in quality evaluation of agro-food products: a review. *Critical Reviews in Food Science and Nutrition*, 52, 999–1023.
Erdinest, N., Shmueli, O., Grossman, Y., Ovadia, H., Solomon, A. 2012. Anti-inflammatory effects of alpha linolenic acid on human corneal epithelial cells. *Investigative Ophthalmology & Visual Science*, 53, 4396–4406.
Esfahlan, A. J., Jamei, R. & Esfahlan, R. J. J. F. C. 2010. The importance of almond (Prunus amygdalis L.) and its by-products. 120, 349–360.
Esfahlan, A. J. & Jamei, R. J. 2012. Properties of biological activity of ten wild almond (Prunus amygdalis L.) species. *Turkish Journal of Biology*, 36, 201–209.
Esteki, M., Farajmand, B., Amanifar, S., Barkhordari, R., Ahadiyan, Z., Dashtaki, E., Mohammadlou, M. & Vander Heyden, Y. 2017a. Classification and authentication of Iranian walnuts according to their geographical origin based on gas chromatographic fatty acid fingerprint analysis using pattern recognition methods. *Chemometrics and Intelligent Laboratory Systems*, 171, 251–258.
Esteki, M., Heydari, E., Simal-Gandara, J., Shahsavari, Z. & Mohammadlou, M. 2021a. Discrimination of pistachio cultivars based on multi-elemental fingerprinting by pattern recognition methods. *Food Control*, 124, 107889.
Esteki, M., Heydari, E., Simal-Gandara, J., Shahsavari, Z. & Mohammadlou, M. J. F. C. 2021b. Discrimination of pistachio cultivars based on multi-elemental fingerprinting by pattern recognition methods. 124, 107889.
Esteki, M., Shahsavari, Z. & Simal-Gandara, J. 2019. Food identification by high performance liquid chromatography fingerprinting and mathematical processing. *Food research international*, 122, 303–317.
Esteki, M., Vander Heyden, Y., Farajmand, B. & Kolahderazi, Y. 2017b. Qualitative and quantitative analysis of peanut adulteration in almond powder samples using multi-elemental fingerprinting combined with multivariate data analysis methods. *Food Control*, 82, 31–41.

Faldt, P., Bergenstahl, B. & Carlsson, G. 1993. The surface coverage of fat on food powders analyzed by ESCA (electron spectroscopy for chemical analysis). *Food Structure*, 12, 10.

Fang, F., Ho, C. T., Sang, S. & Rosen, R. T. J. 2005. Determination of sphingolipids in nuts and seeds by a single quadrupole liquid chromatography—mass spectrometry method. *Journal of Food Lipids*, 12, 327–343.

Faqeerzada, M. A., Lohumi, S., Joshi, R., Kim, M. S., Baek, I. & Cho, B.-K. 2020. Non-Targeted detection of adulterants in almond powder using spectroscopic techniques combined with chemometrics. *Foods*, 9, 876.

Feldman, E. B. 1999. *Assorted monounsaturated fatty acids promote healthy hearts*. Oxford University Press. *The American journal of clinical nutrition*.

Ferguson, J. J., Stojanovski, E., Macdonald-Wicks, L. & Garg, M. L. 2016. Fat type in phytosterol products influence their cholesterol-lowering potential: A systematic review and meta-analysis of RCTs. *Progress in lipid research*, 64, 16–29.

Fernandes, V. C., Domingues, V. F., Mateus, N. & Delerue-Matos, C. J. 2014. Comparison of disposable pipette extraction and dispersive solid-phase extraction in the QuEChERS method for analysis of pesticides in strawberries. *Journal of Chromatographic Science*, 52, 1339–1345.

Fernández-García, E., Carvajal-Lérida, I., Jarén-Galán, M., Garrido-Fernández, J., Pérez-A, A. & Hornero-Méndez, D. 2012. Carotenoids bioavailability from foods: From plant pigments to efficient biological activities. *Food Research International*, 46, 438–450.

Fischer, M., Wohlfahrt, S., Varga, J., Matuschek, G., Saraji-Bozorgzad, M. R., Walte, A., Denner, T. & Zimmermann, R. 2017. Evolution of volatile flavor compounds during roasting of nut seeds by thermogravimetry coupled to fast-cycling optical heating gas chromatography-mass spectrometry with electron and photoionization. *Food Analytical Methods*, 10, 49–62.

Francisco, M. L. D., Resurreccion, A. 2008. Functional components in peanuts. *Crit Rev Food Sci Nutrition*, 48, 715–746.

Frelka, J. C. & Harris, L. 2015. Evaluation of microbial loads and the effects of antimicrobial sprays in postharvest handling of California walnuts. *Food microbiology*, 48, 133–142.

Fukuda, T., Ito, H. & Yoshida, T. J. P. 2003. Antioxidative polyphenols from walnuts (Juglans regia L.). 63, 795–801.

Funasaki, M., Oliveira, R. S., Zanotto, S. P., Carioca, C. R., Simas, R. C., Eberlin, M. N. & Alberici, R. M. 2012. Brazil nut oil: quality control via triacylglycerol profiles provided by easy ambient sonic-spray ionization mass spectrometry. *Journal of Agricultural and Food Chemistry*, 60, 11263–11267.

Gallo, M., Ferracane, R., Graziani, G., Ritieni, A. & Fogliano, V. 2010. Microwave assisted extraction of phenolic compounds from four different spices. *Molecules*, 15, 6365–6374.

Gama, T., Wallace, H. M., Trueman, S. J. & Hosseini-Bai, S. J. S. H. 2018. Quality and shelf life of tree nuts: A review. 242, 116–126.

Gao, M., Tang, J., Villa-Rojas, R., Wang, Y. & Wang, S. 2011. Pasteurization process development for controlling Salmonella in in-shell almonds using radio frequency energy. *Journal of Food Engineering*, 104, 299–306.

Ge, F., Chen, C., Liu, D. & Zhao, S. 2014. Rapid quantitative determination of walnut oil adulteration with sunflower oil using fluorescence spectroscopy. *Food Analytical Methods*, 7, 146–150.

Genis, H. E., Durna, S. & Boyaci, I. H. 2021. Determination of green pea and spinach adulteration in pistachio nuts using NIR spectroscopy. *LWT*, 136, 110008.

Ghisoni, S., Lucini, L., Rocchetti, G., Chiodelli, G., Farinelli, D., Tombesi, S. & Trevisan, M. 2020. Untargeted metabolomics with multivariate analysis to discriminate hazelnut (Corylus avellana L.) cultivars and their geographical origin. *Journal of the Science of Food and Agriculture*, 100, 500–508.

Ghosh, S., Mishra, P., Mohamad, S. N. H., De Santos, R. M., Iglesias, B. D. & Elorza, P. B. 2016. Discrimination of peanuts from bulk cereals and nuts by near infrared reflectance spectroscopy. *Biosystems Engineering*, 151, 178–186.

Giang, L. T. & Thien, T. L. T. 2020. Determination of aflatoxin B1, B2, G1, G2 in cashew nut by UHPLC-HRMS. *Vietnam Journal of Chemistry*, 58, 540–547.

Gil Solsona, R., Boix, C., Ibáñez, M. & Sancho, J. V. 2018. The classification of almonds (Prunus dulcis) by country and variety using UHPLC-HRMS-based untargeted metabolomics. *Food Additives & Contaminants: Part A*, 35, 395–403.

Golden, C. E., Berrang, M. E., Kerr, W. L., Harrison, M. A. E. 2019. Slow-release chlorine dioxide gas treatment as a means to reduce Salmonella contamination on spices. *Innovative Food Science & Emerging Technologies*, 52, 256–261.

Gómez-Ariza, J., Arias-Borrego, A., García-Barrera, T. & Beltran, R. 2006. Comparative study of electrospray and photospray ionization sources coupled to quadrupole time-of-flight mass spectrometer for olive oil authentication. *Talanta*, 70, 859–869.

Gómez-Ríos, G. A. & Pawliszyn, 2014. Solid phase microextraction (SPME)-transmission mode (TM) pushes down detection limits in direct analysis in real time (DART). *Chemical Communications*, 50, 12937–12940.

Goodridge, L., Willford, J. & Kalchayanand, N. 2006. Destruction of Salmonella Enteriditis inoculated onto raw almonds by high hydrostatic pressure. *Food Research International*, 39, 408–412.

Granato, D., Nunes, D. S., Barba, F. 2017. An integrated strategy between food chemistry, biology, nutrition, pharmacology, and statistics in the development of functional foods: A proposal. *Trends in Food Science & Technology*, 62, 13–22.

Granato, D., Putnik, P., Kovačević, D. B., Santos, J. S., Calado, V., Rocha, R. S., Cruz, A. G. D., Jarvis, B., Rodionova, O. Y. & Pomerantsev, A. 2018a. Trends in chemometrics: Food authentication, microbiology, and effects of processing. *Comprehensive Reviews in Food Science and Food Safety*, 17, 663–677.

Granato, D., Putnik, P., Kovačević, D. B., Santos, J. S., Calado, V., Rocha, R. S., Cruz, A. G. D., Jarvis, B., Rodionova, O. Y., Pomerantsev, A. 2018b. Trends in chemometrics: Food authentication, microbiology, and effects of processing. *Comprehensive Reviews in Food Science and Food Safety*, 17, 663–677.

Grasso, E. M., Somerville, J. A., Balasubramaniam, V. & Lee, K. 2010. Minimal Effects of High-Pressure Treatment on Salmonella enterica Serovar Typhimurium Inoculated into Peanut Butter and Peanut Products. *Journal of food science*, 75, E522–E526.

Guimón, J., Guimón, P. J. T. F. & Change, S. 2012. How ready-to-use therapeutic food shapes a new technological regime to treat child malnutrition. *Technological Forecasting and Social Change*, 79, 1319–1327.

Gylling, H., Plat, J., Turley, S., Ginsberg, H. N., Ellegård, L., Jessup, W., Jones, P. J., Lütjohann, D., Maerz, W. & Masana, L. J. A. 2014. Plant sterols and plant stanols in the management of dyslipidaemia and prevention of cardiovascular disease. *Atherosclerosis*, 232, 346–360.

Ha, J.-W., Kang, D.-H. J. A. & Microbiology, E. 2015. Combining lactic acid spray with near-infrared radiation heating to inactivate salmonella enterica serovar enteritidis on almond and pine nut kernels. 81, 4517–4524.

Halvorsen, B. L., Carlsen, M. H., Phillips, K. M., Bøhn, S. K., Holte, K., Jacobs Jr, D. R. & Blomhoff, R. 2006. Content of redox-active compounds (ie, antioxidants) in foods consumed in the United States. *The American Journal of Clinical Nutrition*, 84, 95–135.

Hansen, F. A. & Pedersen-Bjergaard, S. 2019. Emerging extraction strategies in analytical chemistry. *Journal of Analytical Chemistry*, 92, 2–15.

Hepburn, F. N., Exler, J. & Weihrauch, J. L. 1986. Provisional tables on the content of omega-3 fatty acids and other fat components of selected foods. *Journal of the American Dietetic Association.*

Hertwig, C., Leslie, A., Meneses, N., Reineke, K., Rauh, C., Schlüter, O. 2017. Inactivation of Salmonella Enteritidis PT30 on the surface of unpeeled almonds by cold plasma. *Innovative Food Science & Emerging Technologies*, 44, 242–248.

Hidalgo-Ruiz, J. L., Romero-González, R., Vidal, J. L. M. & Frenich, A. G. 2021. Monitoring of polar pesticides and contaminants in edible oils and nuts by liquid chromatography-tandem mass spectrometry. *Food Chemistry*, 343, 128495.

Hirano, S., Okawara, N. & Narazaki, S. 1998. Near infra red detection of internally moldy nuts. *Bioscience, Biotechnology, and Biochemistry*, 62, 102–107.

Hou, J., Zhang, Z., Zhang, L., Wu, W., Huang, Y., Jia, Z., Zhou, L., Gao, L., Long, H. & Lei, M. 2022. Spatial lipidomics of eight edible nuts by desorption electrospray ionization with ion mobility mass spectrometry imaging. *Food Chemistry*, 371, 130893.

Hutchens, A. R. 1992. *A handbook of native American herbs: The pocket guide to 125 medicinal plants and their uses.* Shambhala Publications.

Hvizdzak, A., Beamer, S., Jaczynski, J. & Matak, K. 2010. Use of electron beam radiation for the reduction of Salmonella enterica serovars Typhimurium and Tennessee in peanut butter. *Journal of food protection*, 73, 353–357.

Irtwange, S., Oshodi, A., Engineering & Technology 2009. Shelf-life of roasted cashew nuts as affected by relative humidity, thickness of polythene packaging material and duration of storage. *Research Journal of Applied Sciences, Engineering and Technology*, 1, 149–153.

Jayasinghe, S. B. & Caruso, J. A. 2013. Preliminary investigation of Cu-containing proteins in seeds of Brazil nuts by ICPMS and MALDI-MS methods. *International Journal of Mass Spectrometry*, 356, 33–36.

Jeong, S., Marks, B. P. & Orta-Ramirez, A. 2009. Thermal inactivation kinetics for Salmonella Enteritidis PT30 on almonds subjected to moist-air convection heating. *Journal of food protection*, 72, 1602–1609.

Jiang, S., Xie, Y., Li, M., Guo, Y., Cheng, Y., Qian, H. & Yao, W. 2020. Evaluation on the oxidative stability of edible oil by electron spin resonance spectroscopy. *Food Chemistry*, 309, 125714.

Kafaoğlu, B., Fisher, A., Hill, S. & Kara, D. 2014. Chemometric evaluation of trace metal concentrations in some nuts and seeds. *Food Additives & Contaminants: Part A*, 31, 1529–1538.

Kalogiouri, N. P., Manousi, N., Rosenberg, E., Zachariadis, G. A., Paraskevopoulou, A. & Samanidou, V. 2021. Exploring the volatile metabolome of conventional and organic walnut oils by solid-phase microextraction and analysis by GC-MS combined with chemometrics. *Food Chemistry*, 363, 130331.

Kamal, G. M., Uddin, J., Tahir, M. S., Khalid, M., Ahmad, S. & Hussain, A. I. 2021. Nuclear magnetic resonance spectroscopy in food analysis. *Techniques to measure food safety and quality*. Springer.

Karthikeya, H., Sudarshan, K. & Shetty, D. S. 2020. Prediction of Agricultural Crops using KNN Algorithm. *International Journal of Innovative Science and Research Technology*, 5, 1422–1424.

Kazantzis, I., Nanos, G. D., Stavroulakis, G. G. 2003. Effect of harvest time and storage conditions on almond kernel oil and sugar composition. *Journal of the Science of Food and Agriculture*, 83, 354–359.

Keeley, J. E. J. A. O. F. S. 2012. Ecology and evolution of pine life histories. 69, 445–453.

Keller, S. E., Grasso, E. M., Halik, L. A., Fleischman, G. J., Chirtel, S. J. & Grove, S. F. 2012. Effect of growth on the thermal resistance and survival of Salmonella Tennessee and Oranienburg in peanut butter, measured by a new thin-layer thermal death time device. *Journal of food protection*, 75, 1125–1130.

Kesen, S., Amanpour, A. & Selli, S. 2018. Comparative evaluation of the fatty acids and aroma compounds in selected Iranian nut oils. *European Journal of Lipid Science and Technology*, 120, 1800152.

Khan, M. S. & Rahman, M. S. 2021. Introduction on techniques to measure food safety and quality. *Techniques to measure food safety and quality*. Springer.

Kharel, K., Yemmireddy, V. K., Graham, C. J., Prinyawiwatkul, W. & Adhikari, A. J. L. 2018. Hot water treatment as a kill-step to inactivate Escherichia coli O157: H7, Salmonella enterica, Listeria monocytogenes and Enterococcus faecium on in-shell pecans. 97, 555–560.

Kiokias, S., Proestos, C., Varzakas, T. F. S. 2016. A review of the structure, biosynthesis, absorption of carotenoids-analysis and properties of their common natural extracts. *Current Research in Nutrition and Food Science*, 4, 25–37.

Klockmann, S., Reiner, E., Bachmann, R., Hackl, T. & Fischer, M. 2016. Food fingerprinting: metabolomic approaches for geographical origin discrimination of hazelnuts (Corylus avellana) by UPLC-QTOF-MS. *Journal of Agricultural and Food Chemistry*, 64, 9253–9262.

Klockmann, S., Reiner, E., Cain, N. & Fischer, M. 2017. Food targeting: Geographical origin determination of hazelnuts (Corylus avellana) by LC-QqQ-MS/MS-based targeted metabolomics application. *Journal of Agricultural and Food Chemistry*, 65, 1456–1465.

Koaze, H., Ndaka, D. S., Karanja, P. N., Ishibashi, K.-I., Baba, N. 2002. Changes in quality of dried macadamia nuts during a peak harvest season in Kenya. *Food Science and Technology Research*, 8, 32–35.

Kodad, O., Socias I Company, R. 2008. Variability of oil content and of major fatty acid composition in almond (Prunus amygdalis Batsch) and its relationship with kernel quality. *Journal of agricultural and food chemistry*, 56, 4096–4101.

Kokosa, 2021. A guide to recent trends in green applications of liquid phase microextraction for bioanalytical sample preparations. *Sustainable Chemistry and Pharmacy*, 22, 100478.

Kris-Etherton, P. M., Yu-Poth, S., Sabaté, J., Ratcliffe, H. E., Zhao, G. & Etherton, T. D. 1999. Nuts and their bioactive constituents: effects on serum lipids and other factors that affect disease risk. *The American Journal of Clinical Nutrition*, 70, 504s–511s.

Kucharska-Ambrożej, K. & Karpinska, J. 2020. The application of spectroscopic techniques in combination with chemometrics for detection adulteration of some herbs and spices. *Microchemical Journal*, 153, 104278.

Langi, P., Kiokias, S., Varzakas, T. & Proestos, C. 2018. Carotenoids: From plants to food and feed industries. *Microbial Carotenoids*, 57–71.

Lanner, R. M. 1981. *The piñon pine: A natural and cultural history*. University of Nevada Press.

Li, B., Wang, H., Zhao, Q., Ouyang, J. & Wu, Y. 2015. Rapid detection of authenticity and adulteration of walnut oil by FTIR and fluorescence spectroscopy: A comparative study. *Food Chemistry*, 181, 25–30.

LI, C., Huang, L. & Chen, J. J. F. C. 2014. Comparative study of thermal inactivation kinetics of Salmonella spp. in peanut butter and peanut butter spread. 45, 143–149.

Little, E. L. 1978. *Important forest trees of the United States*. Department of Agriculture, Forest Service.

Liu, D., Zeng, X.-A. & Sun, D.-W. 2015. Recent developments and applications of hyperspectral imaging for quality evaluation of agricultural products: a review. *Critical Reviews in Food Science and Nutrition*, 55, 1744–1757.

López, M. I., Colomer, N., Ruisánchez, I. & Callao, M. P. 2014a. Validation of multivariate screening methodology. Case study: Detection of food fraud. *Analytica Chimica Acta*, 827, 28–33.

López, M. I., Trullols, E., Callao, M. P. & Ruisánchez, I. 2014b. Multivariate screening in food adulteration: untargeted versus targeted modelling. *Food Chemistry*, 147, 177–181.
Lynch, K. 2017. Toxicology: liquid chromatography mass spectrometry. *Mass spectrometry for the clinical laboratory*. Elsevier.
Ma, F. Y., Wei, Z. F., Zhang, M., Shuai, X. X., Du, L. Q. 2022. Optimization of aqueous enzymatic microwave assisted extraction of macadamia oil and evaluation of its chemical composition, physicochemical properties, and antioxidant activities. *European Journal of Lipid Science and Technology*, 2100079.
Maguire, L., O'Sullivan, S., Galvin, K., O'Connor, T., O'Brien, 2004. Fatty acid profile, tocopherol, squalene and phytosterol content of walnuts, almonds, peanuts, hazelnuts and the macadamia nut. *International Journal of Food Sciences and Nutrition*, 55, 171–178.
Marina, A., Che Man, Y. & Amin, I. 2010. Use of the SAW sensor electronic nose for detecting the adulteration of virgin coconut oil with RBD palm kernel olein. *Journal of the American Oil Chemists' Society*, 87, 263–270.
Márquez, C., López, M. I., Ruisánchez, I. & Callao, M. P. 2016. FT-Raman and NIR spectroscopy data fusion strategy for multivariate qualitative analysis of food fraud. *Talanta*, 161, 80–86.
Maštovská, K. & Lehotay, S. 2003. Practical approaches to fast gas chromatography—mass spectrometry. *Journal of Chromatography A*, 1000, 153–180.
Matthäus, B. & Özcan, M. M. 2020. Quantification of sterol contents in almond (Prunus amygdalis L.) oils. *Iranian Journal of Chemistry and Chemical Engineering*, 39.
Mcdaniel, C. & Jadeja, R. Pecan and walnut food safety concerns: A mini review. *Advances in Nutrition and Food Science*, 2021(03).
Mchugh, D., Cameron, C. A., Abdenur, J. E., Abdulrahman, M., Adair, O., Al Nuaimi, S. A., Åhlman, H., Allen, J. J., Antonozzi, I. & Archer, S. 2011. Clinical validation of cutoff target ranges in newborn screening of metabolic disorders by tandem mass spectrometry: a worldwide collaborative project. *Genetics in Medicine volume*, 13, 230–254.
Medina, S., Pereira, J. A., Silva, P., Perestrelo, R. & Câmara, J. S. 2019a. Food fingerprints—A valuable tool to monitor food authenticity and safety. *Food Chemistry*, 278, 144–162.
Medina, S., Perestrelo, R., Silva, P., Pereira, J. A. & Câmara, J. S. 2019b. Current trends and recent advances on food authenticity technologies and chemometric approaches. *Trends in Food Science & Technology*, 85, 163–176.
Medina, S., Perestrelo, R., Silva, P., Pereira, J. A., Câmara, J. S. Current trends and recent advances on food authenticity technologies and chemometric approaches. *Trends in Food Science & Technology*, 85, 163–176.
Menevseoglu, A., Aykas, D. P. & Adal, E. 2021. Non-targeted approach to detect green pea and peanut adulteration in pistachio by using portable FT-IR, and UV—Vis spectroscopy. *Journal of Food Measurement and Characterization*, 15, 1075–1082.
Meyer, M. T. & Vaughn, R. H. 1969. Incidence of Escherichia coli in black walnut meats. *Applied Microbiology*, 18, 925–931.
Miaw, C. S. W., Martins, M. L. C., Sena, M. M. & De Souza, S. V. C. 2022. Screening Method for the Detection of Other Allergenic Nuts in Cashew Nuts Using Chemometrics and a Portable Near-Infrared Spectrophotometer. *Food Analytical Methods*, 1–11.
Millar, C. J. E. & Pinus, B. O. 1998. Early evolution of pines. 69–91.
Mishra, G., Panda, B. K., Ramirez, W. A., Jung, H., Singh, C. B., Lee, S. H. & Lee, I. 2021. Research advancements in optical imaging and spectroscopic techniques for nondestructive detection of mold infection and mycotoxins in cereal grains and nuts. *Comprehensive Reviews in Food Science and Food Safety*, 20, 4612–4651.
Mishra, N., Dubey, A., Mishra, R., Barik, N. 2010. Study on antioxidant activity of common dry fruits. *Food and Chemical Toxicology*, 48, 3316–3320.
Mitcham, E., Veltman, R., Feng, X., De Castro, E. D., Johnson, J., Simpson, T., Biasi, W., Wang, S., Tang, 2004. Application of radio frequency treatments to control insects in in-shell walnuts. *Postharvest Biology and Technology*, 33, 93–100.
Monferrere, G. L., Azcarate, S. M., Cantarelli, M. Á., Funes, I. G. & Camiña, J. M. 2012. Chemometric characterization of sunflower seeds. *Journal of Food Science*, 77, C1018–C1022.
Moscetti, R., Berhe, D. H., Agrimi, M., Haff, R. P., Liang, P., Ferri, S., Monarca, D. & Massantini, R. 2021. Pine nut species recognition using NIR spectroscopy and image analysis. *Journal of Food Engineering*, 292, 110357.
Moure, A., Pazos, M., Medina, I., Domínguez, H. & Parajó, 2007. Antioxidant activity of extracts produced by solvent extraction of almond shells acid hydrolysates. *Food Chemistry*, 101, 193–201.

Nergiz, C. & Dönmez, I. 2004. Chemical composition and nutritive value of Pinus pinea L. seeds. *Journal of Food Chemistry*, 86, 365–368.
Nolan, J. M. & Robbins, M. C. J. H. O. 2000. Ethnobotany and ethnicity in the Ozarks: A reply to jones. 59, 140–142.
Nolvachai, Y., Salzmann, S., Zavahir, J. S., Doetzer, R., Steiner, S., Kulsing, C. & Marriott, P. J. 2021. Structure elucidation using gas chromatography– infrared spectroscopy/mass spectrometry supported by quantum chemical IR spectrum simulations. *Analytical Chemistry*, 93, 15508–15516.
Olaimat, A. N., Al-Holy, M. A., Abu-Ghoush, M. H., Osaili, T. M., Al-Nabulsi, A. A. & Rasco, B. A. 2017. Inhibition of Shigella sonnei and Shigella flexneri in hummus using citric acid and garlic extract. *Journal of Food Science*, 82, 1908–1915.
Oliveira-Alves, S. C., Pereira, R. S., Pereira, A. B., Ferreira, A., Mecha, E., Silva, A. B., Serra, A. T. & Bronze, M. R. 2020. Identification of functional compounds in baru (Dipteryx alata Vog.) nuts: Nutritional value, volatile and phenolic composition, antioxidant activity and antiproliferative effect. *Food Research International*, 131, 109026.
Oliveira, M. M., Cruz-Tirado, J. & Barbin, D. F. 2019. Nontargeted analytical methods as a powerful tool for the authentication of spices and herbs: A review. *Comprehensive Reviews in Food Science and Food Safety*, 18, 670–689.
Orm, R. B., Citeau, M., Comitis, A., Savoire, R., Harscoat-Schiavo, C., Subra-Paternault, P., Carré, P., Leao, J. D., Joffre, & Lipids 2020. Walnut oil deacidification by liquid—liquid extraction with ethanol in a single-and multistage crossflow process. *OCL Oilseeds and fats crops*, 27, 35.
Osaili, T. M. & Al-Nabulsi, A. J. F. C. 2016. Inactivation of stressed Escherichia coli O157: H7 in tahini (sesame seeds paste) by gamma irradiation. 69, 221–226.
Otles, S. 2016. *Handbook of food analysis instruments*. CRC Press.
Özbek, H. N., Fockink, D. H., Yanık, D. K., Göğüş, F. & Łukasik, R. M. 2018. The green biorefinery concept for the valorisation of pistachio shell by high-pressure CO2/H2O system. *Journal of Cleaner Production*, 196, 842–851.
Özcan, M. M. 2022. A review on some properties of almond: impact of processing, fatty acids, polyphenols, nutrients, bioactive properties, and health aspects. *Journal of Food Science and Technology*, 1–12.
Pacheco, A. M., Scussel, V. M. J. J. O. A. & Chemistry, F. 2007. Selenium and aflatoxin levels in raw Brazil nuts from the Amazon basin. 55, 11087–11092.
Pan, Z., Zhang, R. & Zicari, S. 2019. *Integrated processing technologies for food and agricultural by-products*. Academic Press.
Pao, S., Kalantari, A. & Huang, G. 2006. Utilizing acidic sprays for eliminating Salmonella enterica on raw almonds. *Journal of food science*, 71, M14-M19.
Pearson, E. H., Jones, J. & Waite-Cusic, J. G. 2018. Evaluation of Peroxyacetic Acid for Reducing Low Levels of Salmonella on Laboratory-Inoculated and Naturally Contaminated In-Shell Hazelnuts. *Journal of Food Protection*, 81, 254–260.
Pelkman, C. L., Fishell, V. K., Maddox, D. H., Pearson, T. A., Mauger, D. T. & Kris-Etherton, P. M. 2004. Effects of moderate-fat (from monounsaturated fat) and low-fat weight-loss diets on the serum lipid profile in overweight and obese men and women. *The American Journal of Clinical Nutrition*, 79, 204–212.
Pereira, T., Barroso, S., Mendes, S., Amaral, R. A., Dias, J. R., Baptista, T., Saraiva, J. A., Alves, N. M. & Gil, M. M. 2020. Optimization of phycobiliprotein pigments extraction from red algae Gracilaria gracilis for substitution of synthetic food colorants. *Journal of Food Chemistry*, 321, 126688.
Perestrelo, R., Silva, C. L., Silva, P., Medina, S., Pereira, R. & Camara, J. S. 2019. Untargeted fingerprinting of cider volatiles from different geographical regions by HS-SPME/GC-MS. *Microchemical Journal*, 148, 643–651.
Pérez-Fernández, V., Spagnoli, M., Rocco, A., Aturki, Z., Sciubba, F., De Salvador, F. R., Engel, P., Curini, R. & Gentili, A. 2017. Non-aqueous reversed-phase liquid-chromatography of tocopherols and tocotrienols and their mass spectrometric quantification in pecan nuts. *Journal of Food Composition and Analysis*, 64, 171–180.
Peris-Vicente, J., Esteve-Romero, J. & Carda-Broch, S. 2015. Validation of analytical methods based on chromatographic techniques: An overview. *Analytical Separation Science*, 1757–1808.
Petrides, G. A. 1998. *A field guide to western trees: western United States and Canada*. Houghton Mifflin Harcourt.
Phillips, K. M., Ruggio, D. M., Ashraf-Khorassani, M. 2005. Phytosterol composition of nuts and seeds commonly consumed in the United States. 53, 9436–9445.
Pinelo, M., Rubilar, M., Sineiro, J. & Núñez, M. J. P. A. Extraction of antioxidant phenolics from almond hulls. *Journal of Agricultural and Food Chemistry.*, 267–273.

Pinto-Martinez, A. K., Rodriguez-Durán, J., Serrano-Martin, X., Hernandez-Rodriguez, V., Benaim, G. J. A. A. & Chemotherapy 2018. Mechanism of action of miltefosine on Leishmania donovani involves the impairment of acidocalcisome function and the activation of the sphingosine-dependent plasma membrane Ca2+ channel. *Antibacterial Agent and Chromatography*, 62, e01614–e01617.

Pinto, J., Carapito, Â., Amaro, F., Lima, A. R., Carvalho-Maia, C., Martins, M. C., Jerónimo, C., Henrique, R., Bastos, M. D. L. & Guedes De Pinho, P. 2021. Discovery of volatile biomarkers for bladder cancer detection and staging through urine metabolomics. *Metabolites*, 11, 199.

Pinto, M. I., Sontag, G., Bernardino, R. & Noronha, J. J. M. J. 2010. Pesticides in water and the performance of the liquid-phase microextraction based techniques. A review. 96, 225–237.

Piscopo, A., Romeo, F., Petrovicova, B. & Poiana, M. J. S. H. 2010. Effect of the harvest time on kernel quality of several almond varieties (Prunus dulcis (Mill.) DA Webb). 125, 41–46.

Pitt, 2009. Principles and applications of liquid chromatography-mass spectrometry in clinical biochemistry. *The Clinical Biochemist Reviews*, 30, 19.

Poirier, D., Sanders, T. & Davis, 2014. Salmonella surrogate reduction using industrial peanut dry roasting parameters. *Peanut Science*, 41, 72–84.

Pollack, S. & Perez, A. J. W., DC: US Department of Agriculture-Economic Research Service 2008. Fruit and tree nuts situation and outlook yearbook 2008.

Portaluri, V., Thomas, F., Jamin, E., Lorandel, B., Silvestre, V., Akoka, S. & Remaud, G. S. 2021. Vanillin isotopic intramolecular 13C profile through polarization transfer NMR pulse sequence and statistical modelling. *Food Control*, 130, 108345.

Prado, R. & Weber, C. C. 2016. Applications of ionic liquids. *Application, purification, and recovery of ionic liquids*. Elsevier.

Puertas, G. & Vázquez, M. 2019. Fraud detection in hen housing system declared on the eggs' label: An accuracy method based on UV-VIS-NIR spectroscopy and chemometrics. *Food chemistry*, 288, 8–14.

Qi, B.-L., Liu, P., Wang, Q.-Y., Cai, W.-J., Yuan, B.-F. & Feng, Y.-Q. 2014. Derivatization for liquid chromatography-mass spectrometry. *TrAC Trends in Analytical Chemistry*, 59, 121–132.

Rajapaksha, D., Waduge, V., Padilla-Alvarez, R., Kalpage, M., Rathnayake, R., A, A., Frew, R., Abeysinghe, S., Abrahim, A. & Amarakoon, T. 2017. XRF to support food traceability studies: Classification of Sri Lankan tea based on their region of origin. *X-Ray Spectrometry*, 46, 220–224.

Rao, A. V. & Rao, L. G. J. P. R. 2007. Carotenoids and human health. 55, 207–216.

Ratner, B. D. & Castner, D. G. 2009. Electron spectroscopy for chemical analysis. *Surface Analysis: The Principal Techniques*, 2, 374–381.

Ravaglia, L. M., Pizzotti, A. B. C. & Alcantara, G. B. 2019. NMR-based and chemometric approaches applicable to adulteration studies for assessment of the botanical origin of edible oils. *Journal of Food Science and Technology*, 56, 507–511.

Reifen, R., Karlinsky, A., Stark, A. H., Berkovich, Z. & Nyska, A. 2015. α-Linolenic acid (ALA) is an anti-inflammatory agent in inflammatory bowel disease. *The Journal of Nutritional Biochemistry*, 26, 1632–1640.

Reis, T., Oliveira, T., Baquião, A., Gonçalves, S., Zorzete, P. & Corrêa, B. 2012. Mycobiota and mycotoxins in Brazil nut samples from different states of the Brazilian Amazon region. *International Journal of Food Microbiology*, 159, 61–68.

Reiter, R. J., Manchester, L. & Tan, D.-X. 2005. Melatonin in walnuts: influence on levels of melatonin and total antioxidant capacity of blood. *Nutrition*, 21, 920–924.

Ren, J., Chung, S. H. 2007. Anti-inflammatory effect of α-linolenic acid and its mode of action through the inhibition of nitric oxide production and inducible nitric oxide synthase gene expression via NF-κB and mitogen-activated protein kinase pathways. *Journal of Agricultural and Food Chemistry*, 55, 5073–5080.

Reyes-Garces, N., Gionfriddo, E., Gómez-Ríos, G. A., Alam, M. N., Boyaci, E., Bojko, B., Singh, V., Grandy, J. & Pawliszyn, 2017. Advances in solid phase microextraction and perspective on future directions. *Analytical Chemistry*, 90, 302–360.

Rezaee, M., Assadi, Y., Hosseini, M.-R. M., Aghaee, E., Ahmadi, F. & Berijani, S. 2006. Determination of organic compounds in water using dispersive liquid—liquid microextraction. *Journal of Chromatography A* ,1116, 1–9.

Ríos-Reina, R., Callejón, R. & Amigo, J. 2021. Feasibility of a rapid and non-destructive methodology for the study and discrimination of pine nuts using near-infrared hyperspectral analysis and chemometrics. *Food Control*, 108365.

Ríos-Reina, R., García-González, D. L., Callejón, R. M. & Amigo, J. M. 2018. NIR spectroscopy and chemometrics for the typification of Spanish wine vinegars with a protected designation of origin. *Food Control*, 89, 108–116.

Rodrigues, A. 2015. *Simulated moving bed technology: principles, design and process applications*. Butterworth-Heinemann.

Rodríguez-Duran, L., Torres-Mancera, M., Trujillo-Roldán, M., Valdez-Cruz, N., Favela-Torres, E. & Saucedo-Castañeda, G. 2017. Standard instruments for bioprocess analysis and control. *Current Developments in Biotechnology and Bioengineering*. Elsevier.

Rohman, A. & Che Man, Y. 2009. Monitoring of virgin coconut oil (VCO) adulteration with palm oil using Fourier transform infrared spectroscopy. *Journal of Food Lipids*, 16, 618–628.

Rose, D. & Oliveira, V. 1997. Nutrient intakes of individuals from food-insufficient households in the United States. *American Journal of Public Health*, 87, 1956–1961.

Roselló-Soto, E., Martí-Quijal, F. J., Cilla, A., Munekata, P. E., Lorenzo, J. M., Remize, F. & Barba, F. J. 2019. Influence of temperature, solvent and pH on the selective extraction of phenolic compounds from tiger nuts by-products: Triple-TOF-LC-MS-MS characterization. *Molecules*, 24, 797.

Roy, S., Hansen, A. & Sinha, S. 2019. Effective rheology of two-phase flow in a capillary fiber bundle model. *Frontiers in Physics*, 7, 92.

Ryan, E., Galvin, K., O'Connor, T., Maguire, A., O'Brien, N. 2006. Fatty acid profile, tocopherol, squalene and phytosterol content of Brazil, pecan, pine, pistachio and cashew nuts. *International Journal of Food Sciences and Nutrition*, 57, 219–228.

Sagrero-Nieves, L. 1992. Fatty acid composition of Mexican pine nut (Pinus embroiders) oil from three seed coat phenotypes. *Journal of the Science of Food and Agriculture*, 59, 413–414.

Saini, R. K., Nile, S. H. & Park, S. W. 2015. Carotenoids from fruits and vegetables: Chemistry, analysis, occurrence, bioavailability and biological activities. *Food Research International*, 76, 735–750.

Salazar, F., Garcia, S., Lagunas-Solar, M., Pan, Z. & Cullor, J. 2018. Effect of a heat-spray and heat-double spray process using radiofrequency technology and ethanol on inoculated nuts. *Journal of Food Engineering*, 227, 51–57.

Sanders, T. & Calhoun, R. J. P. S. 2014. Effect of oil and dry roasting of peanuts at various temperatures and times on survival of Salmonella and Enterococcus faecium. 41, 65–71.

Sang, S., Cheng, X., Fu, H.-Y., Shieh, D.-E., Bai, N., Lapsley, K., Stark, R. E., Rosen, R. T. & Ho, C. 2002. New type sesquiterpene lactone from almond hulls (Prunus amygdalis Batsch). *Tetrahedron Letters*, 43, 2547–2549.

Schade, J., King Jr, A. J. A. & Microbiology, E. 1977. Methyl bromide as a microbicidal fumigant for tree nuts. 33, 1184–1191.

Schmidt, M. L. 2015. *QuEChERS (quick, easy, cheap, effective, rugged, and safe) extraction-gas chromatography for the analysis of drugs*. Seton Hall University.

Segelke, T., Von Wuthenau, K., Kuschnereit, A., Müller, M.-S. & Fischer, M. 2020. Origin determination of walnuts (Juglans regia L.) on a worldwide and regional level by inductively coupled plasma mass spectrometry and chemometrics. *Foods*, 9, 1708.

Sen, F., Uncu, A. O., Uncu, A. T. & Erdeger, S. N. 2020. The trnL (UAA)-trnF (GAA) intergenic spacer is a robust marker of green pea (Pisum sativum L.) adulteration in economically valuable pistachio nuts (Pistacia vera L.). *Journal of the Science of Food and Agriculture*, 100, 3056–3061.

Sezer, B., Apaydin, H., Bilge, G. & Boyaci, I. H. 2019. Detection of Pistacia vera adulteration by using laser induced breakdown spectroscopy. *Journal of the Science of Food and Agriculture*, 99, 2236–2242.

Sfahlan, A. J., Mahmoodzadeh, A., Hasanzadeh, A., Heidari, R. & Jamei, R. 2009. Antioxidants and antiradicals in almond hull and shell (Amygdalus communis L.) as a function of genotype. *Food Chemistry*, 115, 529–533.

Shahidi, F. & Miraliakbari, H. 2006. Tree nut oils and byproducts: Compositional characteristics and nutraceutical applications. *Nutraceutical and specialty lipids and their co-products*. CRC Press.

Shahidi, F. J. P. I. F. & Nutraceuticals 2004. Antioxidant properties of food phenolics. https://cir.nii.ac.jp/crid/1572543025166784768

Silva, E. F. R., Da Silva Santos, B. R., Minho, L. A. C., Brandão, G. C., De Jesus Silva, M., Silva, M. V. L., Dos Santos, W. N. L. & Dos Santos, A. M. P. 2022. Characterization of the chemical composition (mineral, lead and centesimal) in pine nut (Araucaria angustifolia (Bertol.) Kuntze) using exploratory data analysis. *Food Chemistry*, 369, 130672.

Siriwardhana, S. S. & Shahidi, F. 2002. Antiradical activity of extracts of almond and its by-products. *Journal of the American Oil Chemists' Society*, 79, 903–908.

Song, W. J., Kim, Y. H. & Kang, D. H. 2019. Effect of gamma irradiation on inactivation of Escherichia coli O157: H7, Salmonella Typhimurium and Listeria monocytogenes on pistachios. *Letters in Applied Microbiology*, 68, 96–102.

Srichamnong, W. & Srzednicki, G. 2015. Internal discoloration of various varieties of Macadamia nuts as influenced by enzymatic browning and Maillard reaction. *Scientia Horticulturae*, 192, 180–186.

Sudhakar, A., Chakraborty, S. K., Mahanti, N. K. & Varghese, C. 2021. Advanced techniques in edible oil authentication: A systematic review and critical analysis. *Critical Reviews in Food Science and Nutrition*, 1–29.

Suman, M., Cavanna, D., Sammarco, G., Lambertini, F. & Loffi, C. 2021. Fighting food frauds exploiting chromatography-mass spectrometry technologies: scenario comparison between solutions in scientific literature and real approaches in place in industrial facilities. *TrAC Trends in Analytical Chemistry*, 142, 116305.

Tapsell, L. C., Gillen, L. J., Patch, C. S., Batterham, M., Owen, A., Baré, M. & Kennedy, M. 2004. Including walnuts in a low-fat/modified-fat diet improves HDL cholesterol-to-total cholesterol ratios in patients with type 2 diabetes. *Diabetes Care*, 27, 2777–2783.

Taylan, O., Cebi, N., Yilmaz, M. T., Sagdic, O. & Bakhsh, A. A. 2020. Detection of lard in butter using Raman spectroscopy combined with chemometrics. *Food Chemistry*, 332, 127344.

Taylan, O., Cebi, N., Yilmaz, M. T., Sagdic, O., Ozdemir, D. & Balubaid, M. 2021. Rapid detection of green-pea adulteration in pistachio nuts using Raman spectroscopy and chemometrics. *Journal of the Science of Food and Agriculture*, 101, 1699–1708.

Teixeira, A. M. & Sousa, C. 2019. A review on the application of vibrational spectroscopy to the chemistry of nuts. *Food Chemistry*, 277, 713–724.

Tengku Mansor, T. S., Che Man, Y. B. & Rohman, A. 2011. Application of fast gas chromatography and Fourier transform infrared spectroscopy for analysis of lard adulteration in virgin coconut oil. *Food Analytical Methods*, 4, 365–372.

Timperley, C. M., Bird, M., Heard, S. C., Notman, S., Read, R. W., Tattersall, J. E. & Turner, S. R. 2005. Fluorinated pyridine derivatives: Part 1. The synthesis of some mono-and bis-quaternary pyridine salts of potential use in the treatment of nerve agent poisoning. *Journal of Fluorine Chemistry*, 126, 1160–1165.

Torres, I., Sánchez, M.-T., Vega-Castellote, M. & Pérez-Marín, D. 2021. Fraud detection in batches of sweet almonds by portable near-infrared spectral devices. *Foods*, 10, 1221.

Tsai, H.-C., Sheng, L. & Zhu, M. 2017. Antimicrobial efficacy of cinnamon oil against Salmonella in almond based matrices. *Food Control*, 80, 170–175.

Uesugi, A. R. & Harris, L. 2006. Growth of Salmonella Enteritidis phage type 30 in almond hull and shell slurries and survival in drying almond hulls. *Journal of Food Protection*, 69, 712–718.

Üner, B., Karaman, İ., Tanrıverdi, H. & Özdemir, D. 2011. Determination of lignin and extractive content of Turkish Pine (Pinus brutia Ten.) trees using near infrared spectroscopy and multivariate calibration. *Wood Science and Technology*, 45, 121–134.

Valasi, L. & Pappas, C. S. 2021. Chemometric differentiation of pistachios (pistacia vera, Greek 'Aegina variety) from two different harvest years using FTIR spectroscopy and DRIFTS and disk techniques. *AppliedChem*, 1, 62–74.

Valdés, A., Beltrán, A., Mellinas, C., Jiménez, A. & Garrigós, M. C. 2018. Analytical methods combined with multivariate analysis for authentication of animal and vegetable food products with high fat content. *Trends in Food Science & Technology*, 77, 120–130.

Valdés, A., Vidal, L., Beltran, A., Canals, A., Garrigós, M. C. 2015. Microwave-assisted extraction of phenolic compounds from almond skin byproducts (Prunus amygdalis): A multivariate analysis approach. *Journal of Agricultural and Food*, 63, 5395–5402.

Valletta, M., Ragucci, S., Landi, N., Di Maro, A., Pedone, P. V., Russo, R. & Chambery, A. 2021. Mass spectrometry-based protein and peptide profiling for food frauds, traceability and authenticity assessment. *Food Chemistry*, 365, 130456.

Venkitasamy, C., Brandl, M. T., Wang, B., Mchugh, T. H., Zhang, R. & Pan, Z. 2017. Drying and decontamination of raw pistachios with sequential infrared drying, tempering and hot air drying. *International Journal of Food Microbiology*, 246, 85–91.

Vivekanandan-Giri, A., Byun, J. & Pennathur, S. 2011. Quantitative analysis of amino Acid oxidation markers by tandem mass spectrometry. *Methods in enzymology*. Elsevier.

Von Wuthenau, K., Segelke, T., Müller, M.-S., Behlok, H. & Fischer, M. 2022. Food authentication of almonds (Prunus dulcis mill.). Origin analysis with inductively coupled plasma mass spectrometry (ICP-MS) and chemometrics. *Food Control*, 134, 108689.

Wadood, S. A., Boli, G., Xiaowen, Z., Hussain, I. & Yimin, W. 2020. Recent development in the application of analytical techniques for the traceability and authenticity of food of plant origin. *Microchemical Journal*, 152, 104295.

Wahl, H. G., Habel, S. Y., Schmieder, N. & Liebich, H. M. 1994. Identification of cis/trans isomers of methyl ester and oxazoline derivatives of unsaturated fatty acids using GC—FTIR—MS. *Journal of High Resolution Chromatography*, 17, 543–548.

Walter, E., Nascimento, M. & Kuaye, A. 2009. Efficacy of sodium hypochlorite and peracetic acid in sanitizing green coconuts. *Letters in Applied Microbiology*, 49, 366–371.

Walton, D. & Wallace, H. 2004. Dehusker effects on macadamia kernel quality. International Conference Postharvest Unlimited Downunder 2004 687, 417–418.

Walton, D. A., Wallace, H. M. 2009. Delayed harvest reduces quality of raw and roasted macadamia kernels. *Journal of the Science of Food and Agriculture*, 89, 221–226.

Weller, L. D., Daeschel, M. A., Durham, C. A. & Morrissey, M. T. 2013. Effects of water, sodium hypochlorite, peroxyacetic acid, and acidified sodium chlorite on in-shell hazelnuts inoculated with salmonella enterica serovar Panama. *Journal of Food Science*, 78, M1885–M1891.

Westerhuis, J. A., Van Velzen, E. J., Hoefsloot, H. C. & Smilde, A. K. 2010. Multivariate paired data analysis: multilevel PLSDA versus OPLSDA. *Metabolomics*, 6, 119–128.

Wihodo, M., Han, Y., Selby, T., Lorcheim, P., Czarneski, M., Huang, G. & Linton, R. July, 2005. Decontamination of raw almonds using chlorine dioxide gas. Institute of Food Technologist Annual Meeting, New Orleans, LA, 15–20.

Wijeratne, S. S., Abou-Zaid, M. M., Shahidi, 2006. Antioxidant polyphenols in almond and its coproducts. *Journal of Agricultural and Food*, 54, 312–318.

Wolff, R. L. & Bayard, C. C. 1995. Fatty acid composition of some pine seed oils. *Journal of the American Oil Chemists' Society*, 72, 1043–1046.

Wu, H., Wang, Q., Ma, T. & Ren, 2009. Comparative studies on the functional properties of various protein concentrate preparations of peanut protein. *Food Research International*, 42, 343–348.

Wuytack, E. Y., Diels, A. M., Meersseman, K. & Michiels, C. W. 2003. Decontamination of seeds for seed sprout production by high hydrostatic pressure. *Journal of Food Protection*, 66, 918–923.

Xu, Y., Zhong, P., Jiang, A., Shen, X., Li, X., Xu, Z., Shen, Y., Sun, Y. & Lei, H. 2020. Raman spectroscopy coupled with chemometrics for food authentication: A review. *TrAC Trends in Analytical Chemistry*, 131, 116017.

Yada, S., Huang, G., Lapsley, K. 2013. Natural variability in the nutrient composition of California-grown almonds. *Journal of Food Composition and Analysis*, 30, 80–85.

Yang, J., Bingol, G., Pan, Z., Brandl, M. T., Mchugh, T. H. & Wang, H. 2010. Infrared heating for dry-roasting and pasteurization of almonds. *Journal of Food Engineering*, 101, 273–280.

Yildirim, A. N., Akinci-Yildirim, F., Şan, B. & Sesli, Y. 2016. Total oil content and fatty acid profile of some almond (Amygdalus communis L.) cultivars. *Polish Journal of Food and Nutrition Sciences*, 66, 173–178.

Yu, J., Ren, S., Liu, C., Wei, B., Zhang, L., Younas, S. & Zheng, L. 2018a. Non-destructive detection and classification of in-shell insect-infested almonds based on multispectral imaging technology. *The Journal of Agricultural Science*, 156, 1103–1110.

Yu, P., Low, M. Y. & Zhou, W. 2018b. Design of experiments and regression modelling in food flavour and sensory analysis: A review. *Trends in Food Science & Technology*, 71, 202–215.

Zavahir, J. S., Nolvachai, Y. & Marriott, P. J. 2018. Molecular spectroscopy—Information rich detection for gas chromatography. *TrAC Trends in Analytical Chemistry*, 99, 47–65.

Zettl, D., Bandoniene, D., Meisel, T., Wegscheider, W. & Rantitsch, G. 2017. Chemometric techniques to protect the traditional Austrian pumpkin seed oil. *European Journal of Lipid Science and Technology*, 119, 1600468.

Zhang, J., Hong, Y., Cai, Z., Huang, B., Wang, J. & Ren, Y. 2019. Simultaneous determination of major peanut allergens Ara h1 and Ara h2 in baked foodstuffs based on their signature peptides using ultra-performance liquid chromatography coupled to tandem mass spectrometry. *Analytical Methods*, 11, 1689–1696.

Zhang, Y., Keller, S. E. & Grasso-Kelley, E. M. 2017. Fate of Salmonella throughout production and refrigerated storage of tahini. *Journal of Food Protection*, 80, 940–946.

Zhao, G., Liu, Y., Zhao, M., Ren, J. & Yang, B. 2011. Enzymatic hydrolysis and their effects on conformational and functional properties of peanut protein isolate. *Food Chemistry*, 127, 1438–1443.

Zhao, X., Zhu, H., Zhang, B., Chen, J., Ao, Q. & Wang, X. 2015. XRD, SEM, and XPS analysis of soybean protein powders obtained through extraction involving reverse micelles. *Journal of the American Oil Chemists' Society*, 92, 975–983.

Zhou, Y., Cui, Y., Wang, C., Yang, F., Yao, W., Yu, H., Guo, Y. & Xie, Y. 2021. Rapid and accurate monitoring and modeling analysis of eight kinds of nut oils during oil oxidation process based on Fourier transform infrared spectroscopy. *Food Control*, 130, 108294.

14 Authenticity and Chemometrics of Wine and Beer

Anastasios Koidis, Konstantina Pasvanka, Maria Tarapoulouzi, and Charalampos Proestos

CONTENTS

14.1 AUTHENTICITY AND CHEMOMETRICS OF WINE

14.1.1 Introduction

The authenticity of wines is a critical issue of great international interest. In many countries, specific quality control programs are mandatory to protect consumers from unintentional errors occurring during wine production, as well as to reduce counterfeiting. For this reason, advanced analytical techniques have been developed, for the measurement of trace elements or the simultaneous identification of a wide range of analyzers, for better data analysis and improved means for the traceability of wine, and for the distinction and classification of grapes and wines (mainly in terms of grape varieties, geographical origin and processing conditions; Versari et al., 2014).

The general term *fraud* refers to different types of fraud, such as diluting wines with water; adding alcohol, dyes and flavorings; mixing or substituting lower quality wines; affixing a false label (variety and geographical origin); and the addition of exogenous sugars to increase the alcoholic strength (Holmberg, 2010). However, this is a complex issue, as wine legislation varies by country.

Unfortunately, regulations, good winemaking practices and traceability procedures are not always sufficient to guarantee the authenticity of a given product, so one of the main challenges to wine authenticity is the need for accurate reference materials and standards that can be used to prove the identity of the wine (Takeoka & Ebeler, 2011). For this reason, it is necessary to develop appropriate analytical approaches for the analysis of the chemical composition and its changes related to adulteration in wines. The most common analytical methods are based on the creation of trace element profiles, phenolic compounds, volatile compounds and isotopes, using different chromatographic and spectroscopic techniques (Vaclavik et al., 2011). In particular, volatile compounds are used to characterize varieties, trace elements are used to distinguish wines geographically, and phenolic compounds are used for both variety and origin (Cuadros-Inostroza et al., 2010; Figure 14.1).

The successful classification of various wine products according to their variety, geographical origin, and certain aspects of the winemaking process is made through the combination of instrumental analyses with multivariate statistics. Multivariate statistical methods are a useful tool for the classification of grape and wines, based on complex sets of chemical or sensory data. Especially, principal component analysis (PCA) is very often used to establish the relationships and differentiation among the variables, cluster analysis (CA) is employed to discover natural groupings of samples, linear discriminant analysis (LDA) is applied to select the most important

DOI: 10.1201/9781003266167-14

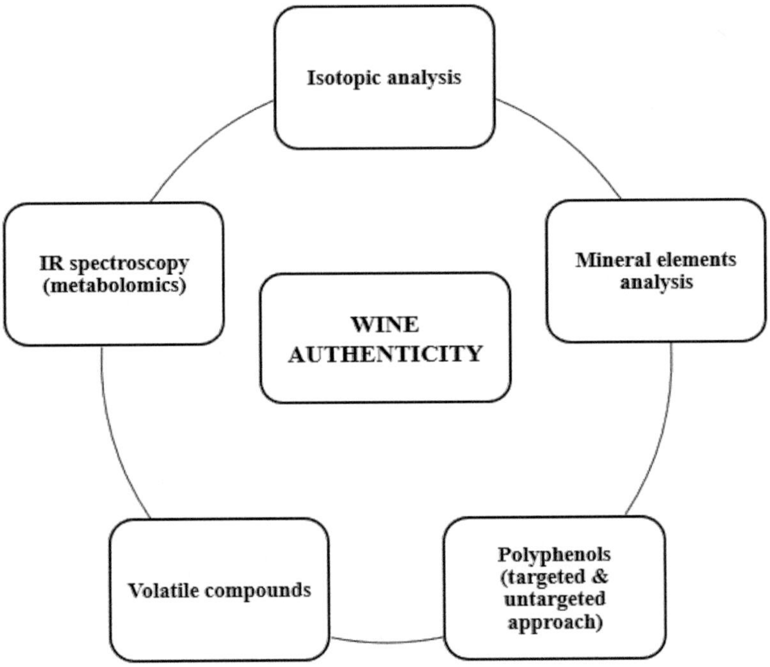

Figure 14.1 Analytical methods used to prove wine authenticity.

and significant variables, and canonical discriminant analysis (CDA) is applied to identify similarities between samples (Pasvanka et al., 2019).

14.1.2 Analytical Methods on Wine Authenticity

14.1.2.1 Mineral Elements

Recently, studies focused on food authenticity are in successive increase for preserving costumers against fraud and commercial disputes. For this purpose, the geographical origins of various food products have been verified developing quantitative methods with subsequent multivariate analysis (Pasvanka et al., 2019).

It has been observed that during the winemaking process, mineral elements are not transformed or metabolized, but they remain unaltered in the final product. For that reason, the classification of wines, as it concerns their geographical origin and variety, can rely on the element profile (Cozzolino et al., 2015). The analysis of inorganic compounds has been employed as one of the most promising methods, either on its own or in conjunction with other methods—referred to below—for detection of wine authenticity for both their variety and geographical origin.

The analyses of inorganic compounds in wines are usually carried out by atomic absorption spectroscopy (AAS), inductively coupled plasma atomic emission spectrometry (ICP-AES) or ICP-mass spectrometry (ICP-MS; Versari et al., 2014).

Wine contains 2–4 g/L of inorganic ingredients, which include mainly inorganic acids and some organic ones. The mineral elements are formed in the roots of the vine and transferred to the grapes, from where they finally pass to the must and then to the wine (Soufleros, 1997).

Mineral elements are used as indicators of the origin of wines, however as the composition of wines in inorganic components depends on many factors, the data should be studied carefully. Many of these factors, which affect the metal content of wines, are natural sources (e.g., the geochemistry of the vineyard soil), the external contamination of the vine during its development, the various contaminations during the vinification process, as well as grape variety, plant maturity and climatic conditions (Likar et al., 2015). Other possible sources that affect the mineral content of wines are the air pollution, pesticides and fertilizers that are used, irrigation water, aging processes, oenological practices and various additives during vinification (Vlope et al., 2009). For this reason, the classification of wines based on their mineral profile is influenced not only by geographical origin, but also by various exogenous factors.

Potassium (K) and calcium (Ca) are the natural ingredients of grapes and their concentration in wine reflects the condition of the vine in the final stages of grape ripening. However, K is also added to the wines as a metasulphite or carbonate and phosphorus (P) as a calcium phosphate or ammonia, thus increasing its final concentration in wines (Catarino et al., 2018).

The concentration of aluminum (Al) in both must and wine could be increased by the use of bentonite and, to a lesser extent, by contact with aluminum surfaces. However, winemaking equipment is now usually made of stainless steel, and these contaminants are rare. The elements manganese (Mn), magnesium (Mg), strontium (Sr), barium (Ba) and several rare earth metals are also reported as useful elements, even if Mn and Mg can be introduced through viticultural practices, such as the use of fertilizers and pesticides, while the concentrations of rare earth metals contained in wines may also increase due to the use of bentonites (Pasvanka, 2021).

Copper (Cu) can be transferred to the must by the use of various chemicals, however, the main source of copper in the final product comes from the addition of $CuSO_4$ in order to remove hydrogen sulfide. Residual copper can contribute to oxidative deterioration, which eventually leads to tanning of wine, especially the white ones. This phenomenon is one of the main oenological problems for winemakers and some metal ions (Cu, Fe, Mn) are the activators of this process. This affects the organoleptic characteristics of the wines, such as loss of aromatic freshness and the appearance of sediments in the wine. To minimize the incidence of these problems, it is generally recommended to keep the copper concentration below 0.3–0.5 mg/L. Moreover, some metal ions such as copper (Cu), iron (Fe), aluminum (Al), zinc (Zn) and nickel (Ni) contribute to the development of turbidity and alterations in aroma and taste (Pérez-Álvarez et al., 2019).

Elements in wine, the content of which depends on the geographical origin due to the direct relationship with the composition of the soil in which the vine is grown, are Al, Ba, Ca, Cu, Fe, K, Mg, Mn, Na, Sr and Zn. Other minerals, such as Fe, Zn, Cu, Cr, Se, Ca, Co and Ni, which are essential for the human body, are an integral part of one or more enzymes involved in metabolic or biochemical processes. These elements are also important for the successful completion of alcoholic fermentation and for the activation of yeast enzymes (Catarino et al., 2015). Alkaline elements, including rubidium (Rb), lithium (Li) and cesium (Cs), are good indicators of geographical origin, as they are not included in the group of elements resulting from contamination in the various stages of wine making, while boron (B) and strontium (Sr) are natural elements derived from their presence in the soil (Paneque et al., 2017). Knowledge of the metal content of wines is also essential for both producers and consumers, as some of them carry toxicological risks, such as cadmium (Cd) and lead (Pb) (Kruzlicova et al., 2013).

Concluding, the main factors influencing the mineral content in wines, are the soil in which the vine is grown, including its pollution, the vine's ability to absorb minerals and all the procedures during the winemaking processes. Finally, the division of the elements into groups according to their origin moves in three axes (Figure 14.2).

Many countries have imposed rules as a way to control element profile in wines. These rules must be followed by the producers to obtain the right to export their wine products to the markets.

In a study carried out with 44 red and white wines from different regions in Poland and different types of grapes, an attempt was made to reveal specific relationships between the wine samples or

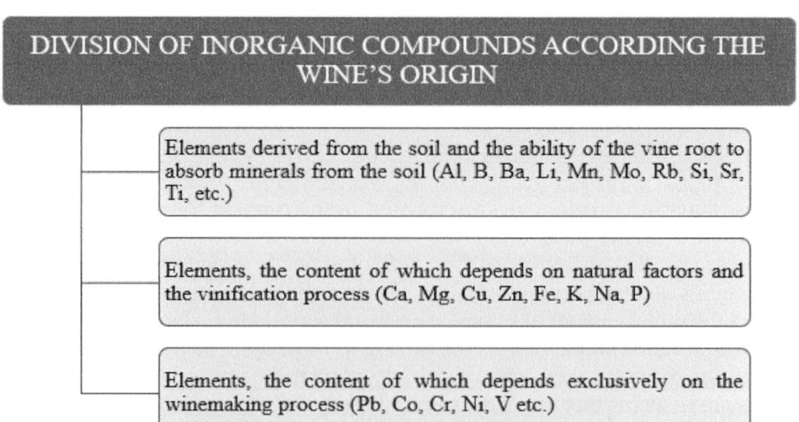

Figure 14.2 Classification of inorganic compounds in wines according their provenance.

between the chemical variables in order to classify the wines according to their metal content by application of chemometric analysis. For this discrimination, two techniques, inductively coupled plasma and mass spectrometry (ICP-MS) and inductively coupled plasma and optical emission spectrometry (ICP-OES), were applied. The chemometric methods used were cluster analysis (hierarchical and non-hierarchical with K-means algorithm) and principal components analysis (PCA), which revealed that two major factors are responsible for the wine quality and especially soil toxic metals content and soil natural composition. The wine samples could be readily separated by geographic objectives, and for each region some discriminating metal variables could be defined (Płotka-Wasylka et al., 2018).

In another study (Pérez-Álvarez et al., 2019), excellent classification results according to viticultural and oenological factors were obtained by application of inductively coupled plasma mass spectrometry (ICP-MS). The discriminant functions classify 100% of the 34 red and white wines, from different grape varieties, geographical zones, soil types, foliar N application, SO_2 addition and oak aging, with the exception of grape variety (97.0%) and oak aging (95.8%). A statistical analysis on mineral elements (major, minor and trace) of red, white and all wines together was performed using variance analysis (one-way ANOVA). Differences between averages were compared using the Duncan test at the 0.05 probability level. Discriminant analyses were carried out on data showing the concentration of elements found in the wines in order to classify them according to the aforementioned factors.

In another investigation, 639 Italian red wines from 18 different geographical regions were classified with partial least squares–discriminant analysis (PLS-DA). Thirty-one chemical elements—rare earth elements (REE), major and trace elements—all contributed to the discrimination of samples (Bronzi et al., 2020). In this study, the combination of ICP-MS with chemometrics was demonstrated to be useful for distinguish Chianti Italian red wines from samples of other Italian regions.

Rocha and collaborators investigated whether the wines from different Portuguese protected geographical indications (PGI) could be differentiated based on their elemental profile. For this purpose, 60 red and white wine samples from six different PGIs were analyzed for a wide range of inorganic compounds (53) by FAAS and ICP-MS. Discriminant analysis (DA) was used to find the elements with discriminant power. The results showed that the concentration of several elements (e.g., Ca, Mn, Co, Zn, Br, Sr, Cd, W Tl) was significantly different among wines and could reliably discriminate the wines from the different Portuguese PGI regions (Rocha et al., 2019).

A stepwise approach combined with Kruskal-Wallis non-parametric test to remove non-significant features, and linear discriminant analysis (LDA) to derive a feature importance index was used to classify wines samples according to place of origin. Aiming at improving categorization accuracy, different classification techniques were tested. When applied to a wine data set comprising 53 samples from four South America countries and 45 chemical elements concentrations reported by ICP-OES and ICP-MS, the proposed framework yielded 99.9% accurate classifications in the testing set (Soares et al., 2018).

The ICP-MS method also allowed to fully discriminate 180 samples of red and white wines from various regions in Greece, different vinification techniques and two contagious cultivation years (Pasvanka, 2021). The inorganic compounds were classified correctly up to 99% for both variety and region. In this study, the creation of the element "identity" of the wines and the creation of prediction models were achieved, in order to distinguish them in terms of varieties and winemaking regions. Through chemometrics (ANOVA, PCA, CA), it was confirmed that the elemental fingerprints, and especially the rare earth metals, are reliable indicators of authenticity, especially for the determination of the geographical origin.

14.1.2.2 *Polyphenol Imprint*

High-performance liquid chromatography (HPLC) is widely used in the determination of wine polyphenols. However, the development of HRMS or NMR techniques combined with the advanced chemometrics has been proven as a useful tool in order to not only identify but also recognize numerous wine and grape types (Diamantidou et al., 2018).

Metabolomic studies are divided into two main categories: the targeted and untargeted approaches (Figure 14.3). Targeted metabolomics is hypothesis testing approaches that aim to identify, detect and quantify a specific group of compounds, while untargeted are hypothesis-generating approaches that focus on the comprehensive detection of multiple endogenous and exogenous molecular with low-weight (<1 kDa) metabolites. Suspect screening metabolomics is a midway approach (Flamini et al., 2013), where specific mass-to-charge ratios are targeted without the use of commercial standards but only based on literature or mass spectral bases.

Several studies have been conducted worldwide contributing to wine authenticity and especially based on the polyphenol imprint to discriminate wine variety and geographical origin.

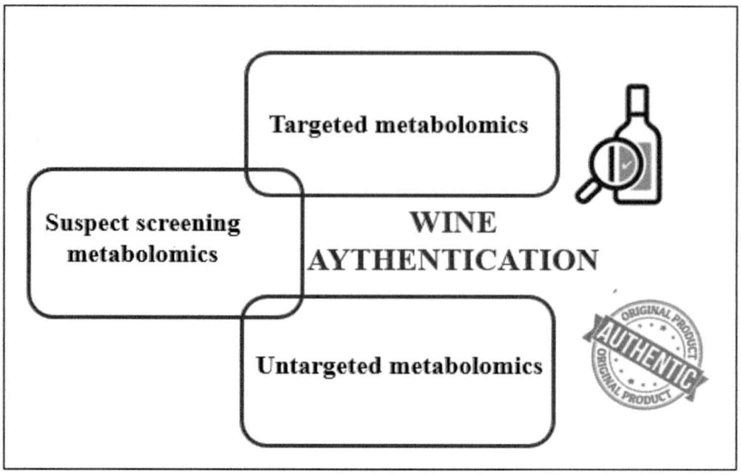

Figure 14.3 Categories of wine metabolomics.

A characteristic example is the resent study that Arapitsas et al. (2020) conducted, in which the characterization of the metabolome of 11 single-cultivar Italian red wines with the use of untargeted HRMS metabolomics was provided. In this study, quercetin was found more abundant in Sangiovese wines followed by Nebiolo and Nerello, isorhamnetin was more abundant in Nebiolo wines, and anthocyanin content was found higher in Teroldego wines.

A good example of the discriminating abilities of polyphenols was reported in 2020 (Stój et al., 2020) with wine samples from Poland. In this study were analyzed phenolic compounds in red monovarietal wines produced from Zweigelt (*Vitis vinifera*) and Rondo (non–*Vitis vinifera*) varieties while using the ultra-performance liquid chromatography (UPLC) with a PDA detector, coupled to mass spectrometry method. Fifty-five phenolic compounds belonging to five classes (anthocyanins, flavonols, flavan-3-ols, phenolic acids, stilbenes) were identified in Zweigelt and Rondo wines. The wines of the Zweigelt variety had lower concentrations of anthocyanin monoglucosides than those of the Rondo variety. Also, wines of the Zweigelt variety contained the highest concentrations of flavan-3-ols, and wines of the Rondo variety contained the highest concentrations of anthocyanin diglucosides. As it concerns data processing, hierarchical cluster analysis (HCA) revealed that Zweigelt wines and Rondo wines formed two separate groups.

In another study, determined the anthocyanin and proanthocyanidin profiles of 90 samples from Greek red grape varieties by HPLC-UV and MS detection. The authors reported significant differences in the anthocyanin and proanthocyanidin profiles of skin extracts of these varieties, while this was not observed in the seeds (Kyraleou et al., 2020). Data statistical analysis was conducted with variance analysis (ANOVA), while principal component analysis (PCA) was used to examine any possible grouping of samples.

Even though most reports on phenolic compounds are referred to red wines, there are also many in rosé or white wines. A specific statistical analysis involving ion ratios, discriminant analysis, and genetic algorithm (GA) was then applied to study the classification of rosé wines according to their geographical origin (Gil et al., 2020). They focused on polyphenols and distinguished that the compounds used for discrimination were vanillic acid, peonidin 3-O-acetyl-Glucoside-(epi) catechin, peonidin 3-O-Glucoside, and (epi)catechin-ethyl-(epi) catechin isomers.

As it concerns the statistical analysis of the data sets, tools such as the hierarchical cluster analysis (HCA) and the principal component analysis (PCA) have been used to depict differential or common trends among wine samples (Tzachristas et al., 2020). However, these approaches are less used when searching for discriminative markers among the groups. For that reason, various supervised methods including partial least squares (PLS) and orthogonal partial least squares (OPLS), artificial neural networks (ANN), canonical variate analysis (CVA) and support vector machine (SVM) have been previously employed (Alañón et al., 2015). A typical example is the study conducted from Portinale et al. (2017). In this survey was presented a data analysis approach, based on machine learning techniques, for the authenticity of 158 Italian samples. After analyzing, the data derived from spectrophotometric and chromatographic characterization of the wine phenolic fraction in both unsupervised and supervised approaches. The selected anthocyanins, phenolic acids, stilbenes and flavonoids were used for the authenticity assessment.

14.1.2.3 Volatile Compounds

Minor constituents in foods, especially volatile compounds, are inherently associated with the global quality and positive sensory attributes of various wines (Tahir et al., 2019). The role of flavor compounds in human health are very important due to their antioxidant, anticancer, anti-inflammatory and anti-obesity properties (Ayseli et al., 2016). The aroma of wines can be influenced by many factors, including grape variety, climate, fermentation condition, yeast strains, winemaking process, and aging and storage conditions (Versari et al., 2014; Weldegergis et al., 2011); therefore, there is a need of appropriate sample size to validate the authenticity of wines based on volatiles compounds. Several attempts have been made to establish highly sensitive and selective approaches for measuring of aroma components where comprehensive research have been performed on the various extraction and separation techniques such as SPE and SPME, as well as enhancing the chromatographics used, such as HPLC and GC-MS (Marín-San Román et al., 2020). However, these techniques have many drawbacks like time-consuming, expensive, and complex methods while extraction process that are used may affect the chemical composition of their aroma compounds (Boido et al., 2013). Therefore, scientists continue to make significant attempts to find alternative rapid, with no or minimal sample preparation, and cost-effective methods that might complement or replace conventional analytical tools in terms of measurement of aroma compounds. For these reasons, vibrational spectroscopies including infrared (e.g., FT-IR, NIR and MIR), Raman, and hyperspectral imaging (HSI) coupled with chemometric techniques could be used as alternatives for quantitative analysis of various aroma compounds in wines (Tahir et al., 2019).

Certain volatiles are characteristics for particular wine varieties (usually referred to as impact odorants; Figure 14.4). Especially, norisoprenoid compounds contribute to the varietal character

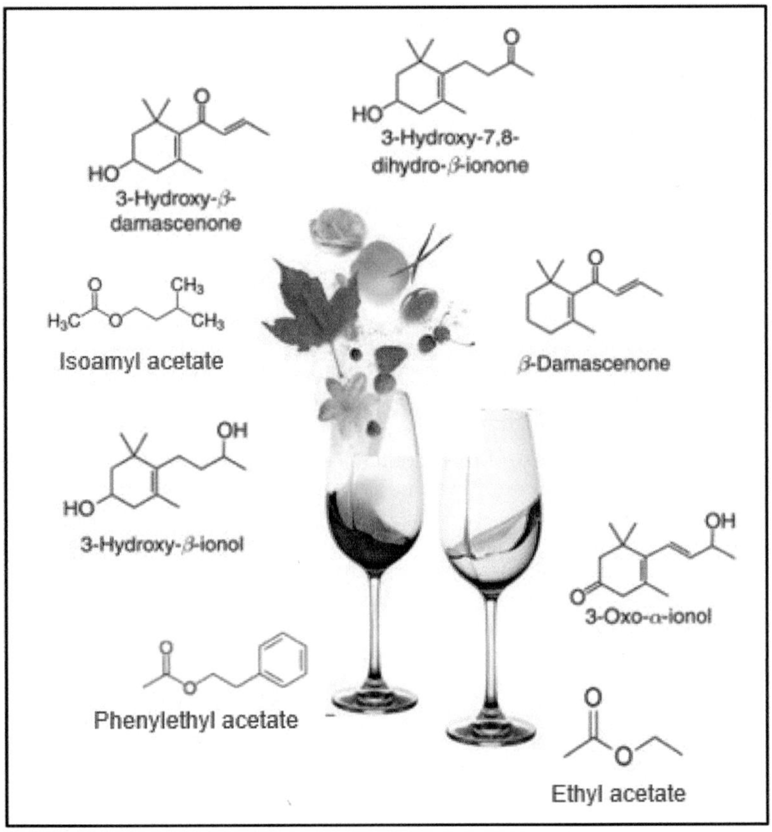

Figure 14.4 Indicative volatile compounds that give aroma to the wine.

of chardonnay wines, methoxypyrazines contribute to distinctive sauvignon blanc and cabernet sauvignon aroma, and isoamyl acetate was reported as an important aroma constituent of young pinotage wines. Also, furaneol (4-hydroxy-2,5-dimethylfuran-3(2H)- one) has been reported as a caramel odor contributor to merlot aroma, while a sesquiterpene, rotundone, has recently been shown to be responsible for the pepper aroma in shiraz wines. Aside from these impact odorants, the common and universal pattern of wine volatiles is dominated by the major fermentation products such as alcohols, esters and fatty acids (Weldegergis et al., 2011).

An overview of the main extraction techniques used for the analysis of volatile compounds in grape and wine samples were reported in the study of Marín-San Román et al. (2020). In their study, they notice that the sample preparation stage is necessary before doing the chromatographic analysis of the volatile compounds as the range of concentrations in which many of these compounds are found is quite low (ng/L). The main analytical techniques that presented in this review are LLE (liquid-liquid extraction), SPE (solid phase extraction), SPME (solid phase microextraction), SBSE (stir bar sorptive extraction) and TF-SPME (thin film solid phase microextraction), with the last two providing better results and advantages in comparison with the other techniques.

Another example of the discriminating abilities of volatile compounds was reported in 2011 (Weldegergis et al., 2011) in 334 wine samples of six different cultivars from South Africa. In this study, 39 volatile components partially responsible for the flavor of wine were quantified. In order to investigate possible correlation between volatile content and grape variety and geographical origin, chemometrics techniques were used. Factor analysis (FA), principal component analysis (PCA) and linear discriminant analysis (LDA) demonstrated significant differences in the levels of certain volatiles as a function of region and cultivar, with the latter factor proving to be more influential. A few volatile compounds were identified as potential predictors of the white wine cultivars while prediction for red wine cultivars was poor.

Isoamyl acetate and ethyl octanoate were seen as influential in the authentication of pinotage wines.

In another investigation, the headspace solid phase microextraction (HS-SPME) methodology combined with gas chromatography–ion trap/mass spectrometry (GC-IT-MS) and flame ionization detector (GC-FID) were applied to the extraction of volatile compounds in wines. The qualitative method allowed the extraction and identification of 60 volatile compounds in the sample studied, notably the classes of alcohols, esters, and fatty acids and the validation for the quantification of 55 volatile compounds of importance in wines and applied to 12 samples of merlot red wine from the south of Brazil. The calculation of the odor activity value (OAV) showed the most important components of the sample's aroma with ethyl isovalerate, ethyl hexanoate, octanoic acid, 1-hexanol and ethyl cinnamate had the greatest contribution to the aroma of the wines analyzed, which is predominantly fruity with the presence of fatty and herbal odors (Arcari et al., 2017).

The availability of chemical markers of wine aging is of great importance in winemaking as it detects frauds and also ensure the wine authenticity. In this view, the GC-GC-ToFMS has been used for the characterization of 23 monovarietal Madeira wines from five *Vitis vinifera* L. grape varieties (Marsala, Madeira, Porto and Jerez), aged from 3 to 20 years old and matured in oak casks. Independently of the variety and the type of wine, the chemical groups that could be used as potential age markers were predominantly acetals—1,1-diethoxyethane, diethoxymethane, 1,1-diethoxy-2-methylpropane, 1-(1-ethoxyethoxy)-pentane, 2-propyl-1,3-dioxolane and trans-dioxane—that are formed during the fermentation step, and their content increases remarkably during the aging process, most likely due to the oxidative condition that occurred during this step (Perestrelo et al., 2011).

14.1.2.4 Isotopic Analysis

In recent years, it has been observed that a powerful tool for geographical origin traceability and authenticity of food products is stable isotopes ratios in natural abundance (Francois et al., 2020). The isotopic ratio allows the identification of molecules that are present in food that have the same chemical structure but come from different raw materials and also allows the detection of molecules that are processed by different processes such as biological or industrial composition.

Isotopic methods have provided a powerful tool for detecting fraud, since it does not have to be done during the vinification of the product but can also be done in the final product, even after many years. The novelty of the isotopic analysis is the ability to identify molecules that have the same chemical structure but come from different raw materials and also to detect processed molecules. Furthermore, taking advantage of the different physicochemical properties of the isotopes of the same element can distinguish any mixtures, certify whether the alcohol in the wines comes from fermentation or from additives, investigate and identify whether sugars or other raw materials have been added, determine the geographical origin and certainly detect the isotopic imprint at all stages from the raw material to the final product, isotopic identity and quality, always with the aim of protecting consumers (Raco et al., 2015).

Table 14.1 depicts the verification of authenticity/adulteration of wines using fixed isotopes.

Stable isotope ratios are of great interest for verifying the authenticity of wines for several reasons. The non-homogeneous repartition of oxygen-18 ($\delta^{18}O$) and deuterium ($\delta^{2}H$) over the globe linked to the hydrological cycle provides information on plant growth localization and meteorological conditions and plant metabolism has a decisive impact on carbon-13 ($\delta^{13}C$) and deuterium (D/H) stable isotope ratios in the molecules metabolized by plants (Karalis et al., 2020).

A study carried out in Cyprus (Kokkinofta et al., 2017), a parallel implementation of SNIF-NMR, IRMS and ICP-AES, facilitated the investigation of the provenance, variety, vineyard and vintage year in relation to the classification of 76 red and white Cypriot wines. It has to be noted that the facilitation of both elements and stable isotope ratios and the combination with chemometrics (PCA) led to their classification and contributed to authenticity assurance of wines. Specifically, unsupervised method (principal component analysis) illustrated the importance of grape variety and origin and supervised analysis mentioned the vineyard effect and the contribution of the vintage year.

Wu et al. (2021) presented a classification of 240 red wine samples from France with an accuracy of 98.2% using multivariate analyses of the isotopic and elemental data. Inductively coupled plasma optical emission spectroscopy (ICPOES) was used to analyze five major elements (Ca, K, Mg, Na and Fe) and inductively coupled plasma mass spectrometry (ICP-MS) was employed to determine the trace elements. The $\delta^{13}C$ values of wine ethanol and glycerol and the $\delta^{18}O$ values of the wine water were analyzed by GC-C-IRMS. PLS-DA confirmed the wine origin and potentially detected the fake Bordeaux wines from other wine regions in France. Finally, ANN was used to carry out discriminant analysis of the above samples to investigate whether a different chemometric traceability model could improve on the accuracy found with PLS-DA.

Another example of light isotope observations for geographical origin characterization studied 51 authentic red wines from Argentina produced in two consecutive harvest years, 2007 and 2008 (Di Paola- Naranjo et al., 2011). The $\delta^{13}C$ showed the lowest values for the region with higher mean annual precipitation, but the effects were also connected to soil inputs (i.e., the content of organic matter from C3 and C4 plants). Analysis of variance (ANOVA) was performed with each sample, while discriminant analysis (DA) was used to evaluate whether wine and soil samples could be distinguished according to their geographical origin or grape variety. Furthermore, generalized Procrustes analysis (GPA) was applied in order to assess the relationship between wine and soil.

14.1.2.5 Spectroscopy

It has been observed that spectroscopic techniques are a more friendly and economical alternative to the traditional methodologies as they involve more complex instrumentation that may not readily be developable in an industry setting. Spectroscopic techniques (e.g., near-infrared, mid-infrared, NMR, UV-Vis) provide a great solution as they are simple and can afford speed of analysis,

Table 14.1 Presentation of the Isotopic Ratios Tested to Determine the Factors Affected and the Corresponding Types of Fraud in the Authenticity of Wines

Isotopic Ratio	Affected by	Type of Fraud
$^{13}C/^{12}C$	Photosynthetic cycle, water	It separates only the plants of different synthetic cycles
$^{18}O/^{16}O$, $^{2}H/^{1}H$	The origin of water	Add water (dilution)
$^{13}C/^{12}C$, D/H	Metabolism Photosynthetic cycle (C3, C4, CAM)	Illegal addition of alcohol from cane and/or sugar beet, cereals Illegal addition of sugar Mixtures of sugars and alcohols of different origins
$^{13}C/^{12}C$, D/H	The chemical composition	Add synthetic alcohol
$^{13}C/^{12}C$, $^{18}O/^{16}O$, $^{2}H/^{1}H$, D/H	All the above and the distillation time in relation to the temperature	Mixtures of different distillation fractions

simple sample preparation and environmental friendliness. It is not accidental that spectroscopic techniques have been utilized for different grape and wine studies, such as for targeted and non-targeted chemical analyses, prediction of sensory attributes, and wine traceability and authenticity. Chemometrics is frequently used in conjunction with spectroscopy for sample classification and discrimination (Ranaweera et al., 2021a).

UV-Vis spectroscopy is a fast, low-cost and reliable analytical technique that has been used in wine analysis (Aleixandre-Tudó et al., 2017). Spectra recorded at UV and visible wavelengths (typically 190–800 nm) provide information about compounds in wine containing a chromophore, such as hydroxybenzoic (280 nm) and hydroxycinnamic (320 nm) acids, flavan-3-ols (280 nm), flavonols (370 nm), and anthocyanin glucosides (520 nm; Ranaweera et al., 2021a). Azcarate et al. (2013) were able to correctly classify Argentinian sauvignon blanc wine samples with 100% accuracy according to their geographical origin by means of a non-targeted method combined with appropriate chemometric methods such as linear discriminant analysis (LDA) and partial least squares–discriminant analysis (PLS-DA). Another example with a high achievable percent (97.5% correct classification of grape variety) in four monovarietal Cretan wines from Greece showed that the latent variables resulting from orthogonal projections to latent structures–discriminant analysis (OPLS-DA) could be related to the absorption of aromatic compounds such as flavonols and phenolic acids (Philippidis et al., 2021). However, Tan et al. (2016) proposes the combination of UV-Vis spectroscopy with other spectroscopic methods (IR, fluorescence) to improve the classification models and provide suitable results for wine authentication.

IR spectroscopy is the most frequently applied spectroscopic method in comparison to other methods as it is a user-friendly and rapid technique that provides various information on many components in wines and can be used for the determination of parameters such as alcohol content, pH, volatile compounds, organic acids, sugars and polyphenols (Aleixandre-Tudó et al., 2020). Two main IR-based techniques are applied according to the range in the spectral region: near-infrared-NIR (14000–4000 cm^{-1}) and mid-infrared-MIR (4000 to 400 cm^{-1}). Near-infrared spectra contain less intense bands than mid-infrared, and it is not applicable to assign chemical groups specifically with NIR due to overlapping signals with ethanol and water around 1950 nm. On the other hand, in mid-infrared spectroscopy, there is a fingerprint region from 1500 to 400 cm^{-1} that includes unique absorption patterns of compounds in wines such as phenolics, and signals associated with various functional groups can be assigned, such as C=O related to organic acids at 1700 cm^{-1}, combinations of C–H vibrations and information related to ethanol and sugars at around 2300 to 2100 cm^{-1} (Geană et al., 2019). The introduction of techniques such as Fourier transform (FT) and attenuated total reflectance (ATR) increased the applicability of IR methods to wine authenticity. FT has improved data collection speed and reproducibility (Rodriguez-Saona et al., 2011), and ATR simplifies the sample handling process and is advantageous in routine analysis (Cozzolino et al., 2015). Roullier-Gall et al. (2019) combined ultra-high-resolution Fourier transform ion cyclotron resonance mass spectrometry (FT-ICR-MS) and electrochemical oxidation strategies to characterize the antioxidant property of chardonnay wines from France only differing in SO_2 added after prior pressing. They concluded that sulfur-containing compounds appeared to decrease with the electrochemical oxidation, whereas nitrogen-containing compounds were mostly formed. Another study (Hu et al., 2019) developed database models based on FT-MIR and FT-NIR coupled with chemometrics (PCA, SIMCA and DA) and proposed that it could be applied as a reference for geographical origin traceability of cabernet sauvignon wines (Australia, Chile and China) in a more comprehensive, effective and rapid manner. Especially, they achieved correct classification using SIMCA, 97%, 97%, and 92% (for Australian, Chilean and Chinese wines, respectively); and DA, 86%, 85%, and 77%, respectively.

Raman spectroscopy involves detecting the inelastic scattered light emitted from molecular vibrations of a sample, approximately in the range of 200 to 3600 nm (Ranaweera et al., 2021a). The Raman technique produces a weak signal, but the development of optimized detection capability provides the opportunity to obtain enough information regarding the chemical composition of the wine sample. There are two different regions in Raman spectroscopy, with Stokes Raman scattering having more dominant water, sucrose and ethanol peaks, and anti–Stokes Raman scattering from minor components such as aromatic compounds, including many phenolics (Magdas et al., 2018). Also, there are two types of Raman techniques that are applied in food analysis: FT-Raman spectroscopy and surface-enhanced Raman spectroscopy (SERS). The advantage of the latter over FT-Raman is the selectivity afforded by specific molecules being adsorbed to metal nanostructures, which enhances the intensity of Raman signals. A characteristic study that depicts the effectiveness of FT-Raman spectroscopy was Magdas's et al. (2019) survey. They manage to discriminate 126 white wines from Romania and France according to their variety (84%), provenance

(100%), and harvest year (95%) in conjunction with LDA chemometrics. Surface-enhanced Raman spectroscopy (SERS) was chosen from Zanuttin et al. (2019) in order to discriminate wines according to variety and producer combined with SIMCA chemometrics. Major metabolites were identified (purines, carboxylic acids, glutathione) with an overall correct classification of 87%. Although these methods show high rates of discrimination, there are some drawbacks, as it is necessary to prepare a colloidal dispersion of Ag nanoparticles to add to the sample and also requires spectral pre-processing such as multiplicative scatter correction to avoid the effect of fluorescence that can obscure Raman scattering, especially when analyzing wine (Xu et al., 2020).

Fluorescence spectroscopy is an analytical technique very useful due to its sensitivity, selectivity and rapidity (Airado-Rodríguez et al., 2011). It is based on the emission of longer wavelength light from a substance after absorption of energy in the UV or visible range. This analytical method can be well applied to wine analysis, with common fluorophores being a variety of phenolic compounds, vitamins and aromatic amino acids. According to the fluorophore's molecular and macromolecular constituents in the wine sample, a three-dimensional excitation-emission matrix (EEM) recorded over multiple excitations and emission wavelengths can be obtained and is considered to be the molecular fingerprint of the sample (Coelho et al., 2015). Therefore, this approach in combination with chemometrics can be utilized for authentication of wine and their discrimination according to geographical origin or variety. Indeed, in a survey conducted in 60 Slovakian white wines (Sádecká et al., 2020), synchronous fluorescence spectroscopy was applied to discriminate the samples according to their variety using LDA, achieving an overall rate of 100% correct classification in validation and 93% for prediction. In another newly published study (Ranaweera et al., 2021b), researchers used the absorbance–transmission and fluorescence excitation-emission matrix (A-TEEM) approach combined with extreme gradient boosting discriminant analysis (XGBDA) to classify 221 commercially produced Australian red wines by origin and variety with the remarkable percentages of 99.7% and 100% accuracy, respectively.

NMR spectroscopy can be used for quantitative analysis of analytes such as amino acids, sugars, alcohols, carboxylic acids and phenolic compounds, for qualitative analysis to determine molecular structures and for compositional profiling of a sample. NMR can be based on the acquisition of ^{1}H, ^{2}H, or ^{13}C spectra; however, for wine authentication, ^{1}H-NMR spectroscopy is the most effective analytical method as data acquisition is fast and highly reproducible when compared to the other spectroscopic techniques (Solovyev et al., 2021). A good example of wine authentication by the NMR method combined with classical multidisciplinary approaches was reported recently in 2019 by Gougeon et al. (2018). The analyzed samples were separated into three categories, the first was half-bottled wines topped up with wine of different origin, the second was suspect wine samples from foreign markets, and the third category was authentic samples. Moreover, Gougeon et al. (2018) developed a similarity score index in order to compare authentic with suspect wine samples and they determined that epicatechin, catechin, gallic acids and syringic acids were among the phenolic compounds evaluated as possible authenticity markers. In another recent study (Fan et al., 2018), ^{1}H-NMR was chosen as a nontargeted method for wine authentication. Indeed, researchers collected 99 red and 71 white wines from China and applied NMR analysis combined with PCA and LDA for separating red and white wine samples from different varieties. Furthermore, Mascellani et al. (2021) used NMR to classify over 900 wines from the Czech Republic according to type, based on color and residual sweetness, and variety using a random forest (RF) machine learning algorithm and LDA. They approve that metabolomics not only provides a better understanding of wine character but also an easy and rapid tool for assessing many aspects of wine quality. The application of the RF algorithm for the purpose and scale of wine authentication of this study represents a novel yet unexplored state-of-the-art approach with promising potential for food fraud.

Concluding, spectroscopic techniques have become a method of choice by many researchers for their routine analysis of wine and generally in food (Table 14.2). The advantages of easy to use in analysis, the availability of a wide range of instrumentation and sampling handling options allowed the preference of these methods by the wine and beer industries worldwide. However, some issues related to the poor understanding of the technology and also the high dependence on mathematics are still some of the few drawbacks of these techniques (Chapman et al., 2019).

14.2 AUTHENTICITY AND CHEMOMETRICS OF BEER

14.2.1 Introduction

14.2.1.1 Production and Types of Beer

Beer it the most consumed alcoholic beverage globally with evidence of use since the very first human civilizations. Beer is in increased demand globally fueled by emerging economies such as Brazil, India and China. It is produced from yeast fermentation of cereal malt, usually barley, in the presence of hops, which contribute to bitterness and act as preservative agents. Chemically, beer is a matrix of various volatile and non-volatile compounds contributing to flavor and aroma. There are many steps involved in the brewing process that determine the quality and properties of beer. The mashing step represents a fundamental part of the brewing process. This step generates fermentable sugars from the primary cereal carbohydrates from the cereal, commonly barley with the help of hydrolytic enzymes. These sugars are converted into alcohol during fermentation reaching a level between 3% and 12% depending on the specific origin and production techniques. They are different types of fermentation that can result in different types of beer. The top fermented beers involve the yeast *Saccharomyces cerevisiae*, which naturally floats to the top of the fermentation resulting in the production of the ale-type beers. The bottom fermented beers, which are the lager-type beers, the yeast *Saccharomyces pastorianus* typically sinks to the bottom of the tank during fermentation (Gresser, 2009).

The most popular type of beer worldwide is lager (90%) followed by ale and pale ale. Multinational multibillion companies such as Heineken, Carlsberg Budweiser and Guinness control the market for larger and other beers that are mainstream large-scale beers or industrial beers. American style lager, the most common type of lager, is a characteristically light beer with low alcohol content (3%–6%).

14.2.1.2 The Premium or Craft Beer and the Legislation Framework

In the last 30 years, mainly in Europe and the United States, the brewing market has witnessed an enormous development of artisanal microbreweries that produce the premium, craft beer or artisanal beer. The premium American-style lager is significantly more expensive than the standard lager. The major difference between these two types of American lager is the carbohydrate source; while premium lager uses only barley malt, the standard lager uses other starch sources such as wheat, exclusively or in conjunction with barley. In general, craft beer production is characterized partially by using of select raw materials, such as barley malt or wheat and hop, but also due to tapping into local and seasonal products such as fruit and spices, in combination with creative and traditional brewing technology. In some cases, such as the Trappist beers, the authenticity is connected to a specific place of production, the monasteries and brewers in specific places in France, Belgium and the Netherlands (Mannina et al., 2016; Engel et al., 2012).

In the EU space alone, the number of microbreweries increased exponentially in the last years (e.g., the UK has passed from 1032 units in 2011 to 1897 in 2020; Brewers of Europe Association (2022). With the increased consumption of premium and standard beers worldwide, increasing food fraud practices have emerged (Spink et al., 2019). In the beer sector, several cases of beer fraud have been reported, wherein the fraudsters have switched the labels and bottle caps of less expensive brands by the labels and caps of the more expensive market leader brands (Pereira et al. 2016). Although this illegal practice does not solely involve craft/premium beers, higher production cost and retail price, premium beers target counterfeit because there is a considerable economic gain to be made (da Silva et al., 2019).

Different legislation across countries does not help the situation as the term *craft* or *premium beer* is not often defined. According to German law, only malted barley, hops, yeast and water may be used to prepare bottom-fermented beer. This also applies to the preparation of top-fermented beer. However, among others, the use of different malt as well as pure cane, beet, or invert sugar are permitted for the preparation of top-fermented beer, which created an opening for craft beers to operate (Gresser, 2009). In Italy, craft breweries were defined for the first time as "small independent producers whose beer does not undergo to pasteurization and micro-filtration during the production process" (National Law n. 154 of 28/7/2016). In the United States, the market is less regulated. The American Brewers Association (www.brewersassociation.org) defines craft brewers as small, independent and traditional.

14.2.1.3 Research Gap and Need for Innovation

Craft beer samples are selected by consumers due to their unique taste, their origin and sometimes based on the information claimed on the label (e.g., "unfiltered product", "unpasteurized", "craft beer"). In fact, craft beers opened new opportunities for setting prices on the highly competitive beer market because of the willingness of consumers to pay for this type of premium beer (Kaufmann et al., 2021). Therefore, the authentication and traceability of beer products is an important issue for both beer consumers and producers. Conventional methods for beer analysis focus on their physicochemical characteristics (e.g., pH, special gravity, ethanol content), and they are not fit-for-purpose for the discrimination of different types of beers, whether these are beers from large scale breweries (industrial beers) and craft beers (both regular craft and specialty beers). An exception to this could be the organoleptic analysis but this is not implemented in the legislation in the same way that it is implemented for example for olive oil. Hence, all stakeholders agree that reliable analytical methods able to identify and discriminate products are needed to protect the beer market against fraud and counterfeit.

In the following few sections, we will review the recent scientistic literature for advances in the identification and discrimination of beer from a forensic point of view, as well as include advances in beer physicochemical profiling that would inform quality control (QC) and quality assurance (QC) in the factory. The common angle is that in both cases, untargeted analytical methods are used coupled with advanced multivariate methods, also known as chemometrics or machine learning algorithms, to gain insights in dealing with such large amount of information (big data).

14.2.2 Analytical Method for Authenticity and Traceability

14.2.2.1 Beer Authenticity and Traceability

As Figure 14.5 suggests, the classification of beers including distinguishing brands (brand identity), origins (geographical origin) and brewing processes (brewing type, fermentation type), as well as identifying authenticity of the label (craft vs. standard beer, the "beer style") has always attracted the attention of the research community and in some cases the industry. It is mainly achieved by several analytical instruments, such as liquid chromatography–mass spectrometry (LC-MS), gas chromatography–mass spectrometry (GC-MS), proton nuclear magnetic resonance (^{1}H-NMR) spectroscopy, near or standard infrared spectroscopy (NIR, FT-IR), ultraviolet–visible spectroscopy and fluorescence spectroscopy.

One must keep in mind that the chemical discrimination of beers is genuinely multivariate in nature. Thus, almost all the analytical approaches developed using the techniques mentioned above employed one or more chemometric tools to perform the multivariate analysis. Examples are LDA, PLS-DA and SIMCA and SVM, ANN for classification and PCA for exploration of the data, identification of clustering, outliers and so forth.

As in every food product, the chemical composition of beer strongly depends on the raw materials used, as well as the brewing production followed. Therefore, the chemical fingerprints and compositional profiles of the beer have a direct impact on the discrimination characteristics observed using the chemometric methods. Specific fundamental differences will be reflected in the chemical fingerprint and thus can aid the discrimination. For example, ale beers are produced at a warmer temperature than lager beers (typically 15–24°C), which allows yeasts to generate more volatiles such as esters, hence providing characteristic flavor notes. In parallel, the extraction of components from the raw materials is also favored by the elevated temperature, resulting in generally richer extracts in ale beers compared to lagers (Pérez-Ràfols & Saurina, 2015). The above would certainly help the discrimination in relation to the type or brewing process (ale vs. lager). Table 14.3 summarizes the research studies in the area by scope, analytes, type of beer used, number of samples in the study, production country when available and the analytical techniques and chemometric methods used.

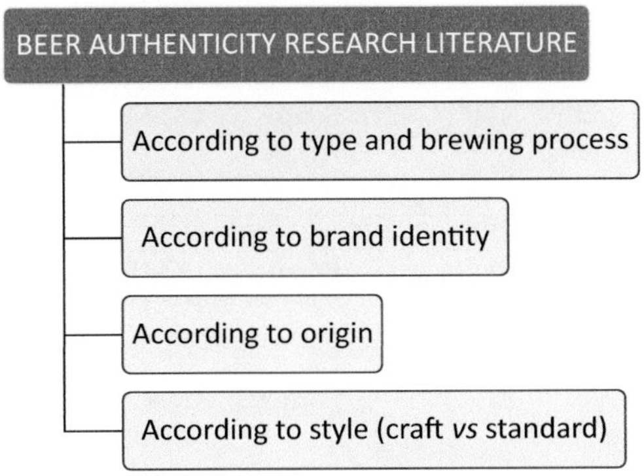

Figure 14.5 Classification of research studies according to their study aims in beer authenticity literature. Note that the size of the box roughly indicates the relative ratio of published papers in each area.

Table 14.2 Collection of Studies Referred to Wine Authenticity Assessment Combined with Chemometrics

Analyte(s)	Category of Sample	Number of Samples	Country	Analytical Technique(s)	Chemometric Method(s)	Reference
Metallic content of wines	Red and white wines	44	Poland	ICP-MS, ICP-OES	PCA, HCA	Płotka-Wasylka et al. (2018)
Categorization of wines in terms of variety, aging, region and the addition (or not) of sulfur	Red and white wines	34	Spain	ICP-MS	ANOVA	Pérez-Álvarez et al. (2019)
Geographic characterization of Italian red wines in terms of their metallic content	Red wines	639	Italy	ICP-MS	PLS-DA	Bronzi et al. (2020)
Multi-elemental analysis and differentiation of wines according to the origin	Red and white wines	60	Portugal	FAAS, ICP-MS	ANOVA, LDA	Rocha et al. (2019)
Authenticity of wines from different varieties and winemaking techniques	Red and white wines	180	Greece	ICP-MS	ANOVA, PCA, CA	Pasvanka (2021)
LC-MS metabolomic classification of 11 mono-cultivar Italian wines-metabolomic study	Red wines	110	Italy	UPLC-QTOF MS	PCA	Arapitsas et al. (2020)
Classification of wines produced from different grape varieties based on the analysis of phenolic compounds	Red wines	20	Poland	UPLC-PDA-MS/MS	PCA, HCA	Stój et al. (2020)
Discrimination of wines with a genetic algorithm and MS ion intensity ratios	Rosé wines	60	France	UPLC-QTOF MS	RF, LDA	Gil et al. (2020)
Discrimination of Greek grape varieties, according to the anthocyanin and proanthocyanidin profiles	Red wines	90	Greece	HPLC-MS	ANOVA, PCA	Kyraleou et al. (2020)
Chemometric investigation of the volatile content of South African wines	Red and white wines	334	South Africa	GC-MS	FA, PCA, LDA	Weldegergis et al. (2011)
Volatile composition of merlot red wine and its contribution to the aroma	Red wines	12	Brazil	GC-IT/MS, GC-FID	ANOVA	Arcari et al. (2017)
The establishment of potential age markers of Madeira wine	Madeira wines	23	Portugal	GC-GC-ToFMS	PCA	Perestrelo et al. (2011)
Elemental profile and classification of wine samples according to the country's origin	Red wines	53	South America	ICP-OES, ICP-MS	LDA	Soares et al. (2018)

Analyte(s)	Category of Sample	Number of Samples	Country	Analytical Technique(s)	Chemometric Method(s)	Reference
Isotopic and elemental profile of wine authenticity	Red and white wines	76	Cyprus	SNIF-NMR, IRMS, ICP-AES	PCA, OPLS-DA, PLS, CV-ANOVA	Kokkinofta et al. (2017)
Verification of wine's origin using isotope and elemental analyses coupled with chemometrics	Red wines	240	France	ICP-MS, ICP-OES, GC-C-IRMS	ANOVA, PLS-DA, ANN	Wu et al. (2021)
Fingerprints for main varieties of Argentinean red wines	Red wines	51	Argentina	Q-ICP-MS, TIMS	ANOVA, DA, GPA, CCA	Di Paola-Naranjo et al. (2011)
Classification and correlation between the sauvignon blanc wine samples by UV-Vis spectroscopy and chemometric methods	White wines	7	Argentina	UV-Vis	PCA, CA, LDA, PLS-DA	Azcarate et al. (2013)
Discrimination according to grape variety, aging process and barrel/container type	Red and white wines	4	Greece	UV-Vis	OPLS-DA	Philippidis et al. (2021)
Electrochemical triggering of the chardonnay wine metabolome	White wines	3	France	FT-ICR-MS	ANOVA	Roullier-Gall et al. (2019)
Geographical discrimination of cabernet sauvignon	Red wines	540	Australia, Chile, China	NIR, MIR	PCA, SIMCA, DA	Hu et al. (2019)
Discrimination of wines geographically, varietally, and by vintage	White wines	126	Romania, France	FT-RAMAN	LDA	Magdas et al. (2019)
Discrimination of wines according to variety and producer	White wines	180	Italy	SERS	PCA, SIMCA	Zanuttin et al. (2019)
Discrimination of wines according to variety	White wines	60	Slovakia	Fluorescence spectroscopy	PCA, LDA	Sádecká et al. (2020)
Discrimination of wines from different varieties and geographical origin	Red wines	221	Australia	Fluorescence spectroscopy	XGBDA	Ranaweera et al. (2021b)
Wine analysis and authenticity using ^{1}H-NMR metabolomics data	Red wines	19	China	NMR	ANOVA, PCA	Gougeon et al. (2018)
Varietal differentiation of wines produced in different regions	Red and white wines	170	China	NMR	PCA, LDA	Fan et al. (2018)
Metabolomic approach for wine varietal classification	Red and white wines	917	Czech Republic	NMR	RF, PCA	Mascellani et al. (2021)

Abbreviations: ANN, artificial neural network analysis; ANOVA, analysis of variance; CA, cluster analysis; CV-ANOVA, cross-validated analysis of variance; DA, discriminant analysis; FA, factor analysis; GPA, generalized Procrustean analysis; HCA, hierarchical cluster analysis; LDA, linear discriminant analysis; MANOVA, multivariate analysis of variance; OPLS-DA, orthogonal partial least squares–discriminant analysis; PCA, principal component analysis; PLS-DA, partial least squares–discriminant analysis; PLS, partial least squares; RF, random forest; SIMCA, soft independent modeling of class analogy; XGBDA, extreme gradient boosting discriminant analysis.

Table 14.3 Collection of Studies Referred to as Beer Authenticity Assessments Combined with Chemometrics

Scope	Analyte(s)	Category of Sample	Number of Samples	Country	Analytical Technique(s)	Chemometric Methods	Reference
Discrimination by type	Phenolic profile	Lager vs. ale	63	Spain	C18 HPLC/DAD	PLSDA, SIMCA	Perez-Rafalos & Saurina (2015)
Discrimination by type	fingerprints	Lager vs. ale	40	United States	^{1}H-NMR–800 MHz	OPLS-DA	Vasas et al. (2021)
Discrimination by type	metabolomic fingerprint	Lager vs. ale	24	Germany Belgium	UHPLC-HRMS	PLS-DA, VIP and p-value	Gallart-Ayala et al. (2016)
Discrimination by type	Polyphenol profile as markers	Lager vs. ale	63	Spain and other European	HPLC-DAD	PLS-DA	Perez-Rafols et al. (2015)
Discrimination by type	Spectral fingerprint	Beer types and styles	100	European	^{1}H-NMR	Peak-by-peak processing	Cavallini et al. (2021)
Classification per manufacturer	Volatile fingerprint	Trappist and Rochefort specialty beers	124	European	HS-SPME–GC-TOF-MS	PLS-DA, LDA and ANN-MLP	Cajka et al. (2010).
Classification per manufacturer	Multiple fingerprints	American-style lagers	135	China	Fluorescence, UV -Vis spectroscopy	LDA, data fusion	Tan et al. (2015)
Classification per manufacturer	Spectral fingerprint	Pale lagers	108	China	EEM fluorescence spectroscopy	PARAFAC, LDA, PLS-DA, k-NN	Fang et al. (2021)
Premium beer authenticity (brand identity)	Spectral Fingerprinting	Trappist and Rochefort specialty beers	233	Belgium, the Netherlands, Czech Republic	NMR–600 MHz	SIMCA, PLS-DA	Mannina et al. (2016)
Premium beer authenticity (brand identity)	Multi-method fingerprint	Different craft beer (pale ale)	60	Italy vs. other European	Thermogravimetry, MIR, NIR, UV-Vis	SIMCA, PLS-DA and data fusion	Biancolillo et al. (2014)
Premium beer authenticity (brand identity)	Spectral fingerprint	Trappist and specialty (Abbey)	267	Belgium, the Netherlands, Czech Republic	MIR (FT-IR)	ECVA and classifiers, LDA, k-NN, QDA	Engel et al. (2012)
Classification per production site	Spectral fingerprints	Lager	67	Spain	HS-MS e-nose	LDA	Vera et al. (2011)
Classification by origin	Multi-method fingerprint	Not specified	68	Germany, Portugal, Spain	Elemental analysis, UV-VIS, pH, etc.	SIMCA, ANN, SVM, LDA	Alcázar et al. (2012)

Discriminating by style and origin	Spectral Fingerprinting	Craft beers vs. industrial	31	Germany, Belgium, Italy, the Netherlands, Denmark	^{1}H-NMR–500 MHz (1D, 2D)	PCA, OPLS-DA	Palmioli et al. (2020)
Discriminating by style	Volatile fingerprint	Pilsner-style lager craft vs. industrial	79	Italy	HS-SPME/GC–MS	PLS-DA	Giannetti et al. (2019)
Discrimination by style	Spectral fingerprint	Different beer types and styles	61	Germany	^{1}H-NMR–600 MHz	PCA, PLS-R	Kaufmann et al. (2021)
Discrimination by style	Spectral fingerprint	Premium vs. standard lager	40	Brazil	^{1}H-NMR	SIMCA, PLS-DA	da Silva et al. (2019)
Discrimination by style	Electrochemical fingerprint	Craft beer of different types	60	China	Potentiometric measurements	LDA, LS-SVM	Yin et al. (2021)
Discrimination by style	Electrochemical fingerprint	Craft lager	113	United States	(Potentiostat/ galvanostat), e-nose	SIMCA, PLS-DA	Mutz et al. (2021)
Forensic discrimination	Spectral fingerprint	Lager beers (high/ low price)	141	United States	Direct mass spectrometry	PLSDA	Pereira et al. (2016)
Monitor craft beer manufacturing (PAT)	Spectral fingerprint	Pale ale	7	Belgium	NIR	PCA PLS, MSPC	Franca et al. (2021)
Other							
Predicting palate fullness	Multi-method fingerprint	Lager beer	41	Germany	Physicochemical and sensory descriptors	Correlation HCA and PLS-R	Krebs et al. (2021)
Classify beers according to their ethanol content and bitterness	Multi-method fingerprint	Lager and ales of different styles	29	Brazil and world beers	Physicochemical, phenolics, HPLC	PLS-R, PLS-DA	Moura-Nunes et al. (2016)
Assess aroma profiles using e-nose	E-nose multi-signal	Beer of different types and styles	20	Belgium, Czech, the Netherlands, Poland, Australia, United States, Mexico	MOS e-nose 9 sensors	ANN	Viejo et al. (2020)

Abbreviations: LS-SVM, least squares–support vector machine; MSPC, multivariate statistical process control.

14.2.2.1.1 According to the Type and Brewing Process

More studies have focused on differentiating beers according to type and brewing process and fewer on the origin and production style (craft vs. standard), which is arguably harder to do. To discriminate a unique target from other types of beers (e.g., Trappist from non-Trappist beers) is a much more defined analytical problem. In this example, UV spectroscopy (Weeranantanaphan & Downey, 2010) and FT-IR (Engel et al., 2012) were used to successfully confirm claimed identity for samples of a single beer brand based on their spectral profiles.

More sophisticated untargeted methods such as the high-resolution mass spectrometry (UHPLC-HRMS) provide more insights and sometimes can lead to identification of biomarkers. In the study of Gallart-Ayala et al. (2016), UHPLC-HRMS metabolomic fingerprint was used to distinguish between different beer types (a Pilsner lager and a Belgian ale). The analysis benefits from the fact that different types of yeast are involved in each type (top-fermenting ales vs. bottom-fermenting Pilsners) and fermentation temperatures (higher for ales), which in turn results in different molecular profiles produced. Using the MS-based biomarker discovery approach (variable-in-projection [VIP] and *p*-value from the analysis of variance statistical test), p-coumaric and ferulic acids, and 5-methylthioadenosine were tentatively identified as characteristic descriptors of Pilsner and Belgian ale beers, respectively. Other researchers have done something similar (e.g., Trappist vs. non-Trappist; Mattarucchi et al., 2010) and Rocheford versus non-Rocheford (Cajka et al., 2011). More recently, Giannetti et al. (2019) applied HS-SPME/GC-MS and chemometric modeling to characterize Pilsner-style lager craft beers by their volatile fraction, while Inui et al. (2013) explored the aroma profiling of different types of beers using a very advanced GC setup (GC × GC/time of flight–MS).

One of the most widely applied methods in classification of type or brewing processing is ^{1}H-NMR. The molecular fingerprint of NMR is also capable of discriminating beer types because it informs about the structural characteristics of the beer metabolome. Earlier studies such as the work of Duarte et al. (2002) paved the way to show that such discrimination differing in type (ale or lager) is possible even using 1D- and 2D-NMR and exploratory and non-classifying methods such as PCA. However, the authors admitted that high-resolution NMR alone is insufficient to enable the entire assignment of the beer spectra, which is however not needed for the chemometric approach. A more systematic study with robust multivariate analysis framework was followed by Cavallini et al. (2021). This study used 100 industrial production beer samples from diverse origins, brewery, beer style and fermentation type. Through innovative use of chemometrics that utilizes peak-by-peak processing, which systematically calculates interval-specific signals, they were able to show discrimination according to type and in some cases according to style (IPA vs. ale beer) within the same class of top-fermented beers. Other attempts for the discrimination of ale and lager beers were also demonstrated by Vasas et al. (2021). Although the sample size of some of the studies is generally small, the combination of a simpler analytical problem and a sophisticated technique (UHPLC-HRMS, GC-MS, or ^{1}H-NMR) indicates that discrimination of beer type is quite established at this point.

Many studies go as far as using the metabolomic approach to identify actual markers for the discrimination (Mannina et al., 2016; Cavallini et al., 2021). Flavonoids, such as certain phenolic acids (syringic and coumaric acids) can influence the flavor and the nutritional content of the beer. The phenolic profile can be easily resolved by liquid chromatography. Relatively new in the literature is their potential role as chemotaxonomical descriptors of foods and beverages. Perez-Rafalos & Saurina (2015) used a straightforward C18 HPLC/DAD method to quantify the polyphenol profile and applied simple chemometric techniques to successfully discriminate lager and ale beers, showcasing the polyphenol profile as another of the potential markers for beer type identity. Their research indicated that coumaric and ferulic acids were more abundant in lager while syringic and gentisic acids were more present in ale beer (Perez-Rafalos et al., 2015).

More experimental or niche methods have also been applied with generally good results. Early efforts in the chemometrics techniques can be traced back to the study of Bellido-Milla et al. (2000) who used flame atomic spectrometry and molecular absorption spectrometry coupled with PCA to showcase the differences. Through electro- and potentiometry, Yin et al. (2021) were able to generally discriminate between beers of different types and flavors (Weiss beer, draft beer, stout beer, craft 1903, and Pilsner beer). Ghasemi-Varnamkhasti et al. (2015) was able to discriminate between alcoholic and non-alcoholic beer using the signal of a metal-oxide semiconductor sensors (type MOS) e-nose and relatively standard chemometric techniques (SIMCA, PLS-DA, LDA, SVM). Recently, an interesting custom off-the-shelf MOS e-nose 9-sensor array was employed to successfully classify beers from different styles and types of fermentation with the help of advanced multivariate methods (ANN ML frameworks; Viejo et al., 2020).

14.2.2.1.2 According to Brand Identity

Although there meant to be some overlap between this and the category above (identification of types of beer) there are several very interesting studies explicitly dealing with identifying certain brands of beer even if they belong to the same type.

Whether it is a specific brand, or a beer with a designation of origin (like Trappist Beer) involved, consumers and the regulators feel very strongly toward potential fraud, which can have dire consequences for the industry at large. Using a very large data set from the EU TRACE project, Mannina et al. (2016) classified 238 beers in Trappist vs. non-Trappist, Rochefort vs. non-Rochefort categories with high accuracy (95% and 90%, respectively) using ^{1}H-NMR and relatively conventional chemometric techniques (PLS-DA and SIMCA). On the same sample set, a high-end approach (headspace–solid phase microextraction gas chromatography coupled to gas chromatography–time-of-flight–mass spectrometer [HS-SPME-GC-TOFMS]) could recognize beer brands based on their volatile fingerprint (Cajka et al., 2010). It is notable that the researchers used advanced multivariate calibration methods in the form of artificial neural networks with multilayer perceptrons (ANN-MLP) in parallel with the conventional methods (PLS-DA/SIMCA). In fact, the ANN-MLP classifier is one of the earliest uses in the literature of beer authenticity and paved the way for the follow-up studies.

More recently, excitation-emission matrix fluorescence spectroscopy coupled with multi-way chemometric techniques has been proposed for the non-invasive classification of Chinese pale lager beers produced by four different manufacturers (Fang et al., 2021).

Apart from single instrument techniques, the researchers in the area recognized the benefit of using signals from multiple analyses, techniques and instruments, which is also compatible with the concept of chemometrics.

The work of Biancolillo et al. (2014), who reported the fusion of responses from several instrumental techniques to enhance the discrimination power among two high-quality Italian beers, has been influential. The techniques involved were heterogenous: various spectroscopic (UV, visible, MIR and NIR) techniques and a thermogravimetric (TG) measurement; they were ingeniously selected to unambiguously characterize the samples. The different analytical signals and numerical parameters were fused to create revised PLS-DA and SIMCA calibration models and ultimately achieved better results than the models of the stand-alone techniques. Tan et al. (2015) also showed that chemometric classification of Chinese lager beers by manufacturer is possible using only a trio of spectroscopy techniques (fluorescence, UV and visible spectroscopies) coupled with a simple LDA.

The argument against the data fusion strategy using so many techniques is laborious and time-consuming. However, the truth is that the chemical information obtained is more comprehensive, and generally this facilitates classification or discrimination and not only an academic exercise.

14.2.2.1.3 According to Origin

Until now, research where beers were differentiated according to type and brewing was presented. The problem becomes exponentially more challenging when a beer of the same type can derive from different raw materials or production methods solely due to its geographical origin. Fewer examples of geographical discrimination of beers can be found in the literature.

In its simplest form, geographical identification could mean identifying a production site within the same country or within the same production company (brand). Such identification is usually linked with the presence of certain biomarker profile that is either deriving from the secondary analysis or is a key measuring parameter. Vera et al. (2011) demonstrated that headspace–mass spectroscopy (HS-MS) e-nose spectral fingerprints could differentiate the aroma of Spanish lager beers from different factories. Although the study used simple chemometrics (LDA), it was possible to discriminate between three or four production sites based on 111 discrete volatiles that were tentatively identified. Czech and non-Czech brands could be identified by their free amino acid content (Kabelová et al., 2008). Similarly, three Spanish production areas were identified according to their mineral profile, which links soil to food origin through the bioaccumulation of soil markers in the food from the plant nutrition pathway.

In the global market, however, beer origin detection is challenging. In a follow-up study, Alcazar et al. (2012) were able to identify the country of production (only among three countries: Germany, Portugal and Spain), where a plethora of chemical and physical descriptors were implored to aid with discrimination. These included multiple minerals as well as chemical indexes (total amino acids, total polyphenols, pH, real extract, and absorbance at 430 nm). Although a proper data fusion strategy was not followed, the level 1 SVM chemometric models showed very good accuracy (>99%).

14.2.2.1.4 According to Style (Craft versus Standard)

Style in beer refers to the production style, and it can vary usually between craft (both conventional and special beers) and industrial from well-known large breweries. The lines have been blurred because the large-scale manufacturers of industrial beers are now also producing craft beers to compete in price and in quality with the small-scale producers and the microbreweries that produce premium craft beers.

The identification of style is inherently a more difficult analytical problem. The distinction is generally impossible with the traditionally used beer analytical methods (apart from the organoleptic examination in special cases). In view of this, the question arises as to whether and to what extent it is possible to distinguish the different types of brewing based on untargeted analytical techniques and chemometrics. The beer style, however, is interconnected with origin and production methods. For example, the only thing that could differentiate a craft from an industrial beer could be the aroma, when the type (e.g., lager) and origin (e.g., Italy) are the same. In other words, the production method could result in different volatile and non-volatile chemical profile, without this necessary coming from the addition of adjuncts (fruits and different cereals) that are often used in craft beers. In this analytical problem, the research requires a higher level of innovation both in the techniques used and in the chemometric analysis to result in a tentative method. The most crucial element is the study design and how the samples are differentiated. Some researchers use the information exactly as supplied by the manufacturer on the label (e.g., premium vs. standard), whereas others use the retail price as the differentiator between the two classes.

With the application of NMR spectroscopy, it has been found that aromatic signals are more appropriate to discriminate among brewing styles. da Silva et al. (2019) employed ^{1}H-NMR spectroscopy combined with chemometrics to discriminate lager beer samples from two different classes, according to their style and information provided on the label. Using the same technique and different chemometric models (OPLS-DA), craft beers were well discriminated from industrials (Palmioli et al., 2020).

E-tongue is a new term for an emerging technique based on electrochemical fingerprints generated from potentiostat or galvanostat sensors in custom configurations. Mutz et al. (2021) proposed a testing procedure using commercially available screen-printed electrodes with promising results for the rapid differentiation of Brazilian lager beer (premium and standard American-style lager). *E-nose* is a similar technique based either on electrochemical fingerprints from their MOS sensors or, in its most sophisticated variant, mass fragment fingerprints from a mass spectrometer. In the implementation of Pereira et al. (2016), the e-nose system was based on paper spray–mass spectrometry (PS-MS) and applied in a forensic context for the differentiation of beers. In this study, the researchers let the market price define their class boundaries (premium vs. standard lager) and were still able to achieve almost 100% of reliability rate, without any false negative or false positive prediction, using partial least squares–discriminant analysis (PLS-DA) and ordered predictors selection (OPS) for variable selection.

14.2.2.2 Beer Rapid Analysis in Quality Control

The general composition of beer consists of water, carbon dioxide (CO_2), alcohol and extract mainly of organic acids, alcohols, amino acids and carbohydrates. Quality control (QC) is an essential element of beer production. For the industry, quality is defined in terms of maintaining batch-to-batch consistency and safety of the final products. In principle, QC can occur in any or all of the three major biochemical processes of beer production (malting, mashing and fermentation), and innovative ways to automate this process is always beneficial for the industry.

The concept of QC is first described in the influential work of Duarte et al. (2004), where FT-IR and the ^{1}H-NMR spectra of beers of different types (ale, lager, alcohol-free) were used to extract spectral information about factors affecting beer production. Although the chemometrics used are limited to PCA (which doesn't really offer classification, only exploration of data) and the methods were not rapid (30 mins for an NMR measurement), the researchers identified specific spectral regions that informed follow up research. Specifically, they identified that the 1200–800 cm^{-1} of the FT-IR and the aromatic region (6.0–10.0 ppm) of the NMR were able to give direct information about ethanol content (separating beers with more than 6.0% alcohol from the rest) and the carbohydrate source (different sugars indicating different cereals [FT-IR]) and the type of beers in general (ale, lager, and alcohol-free [NMR]). The seminal work of Lachenmeier et al. (2005) elevated the concept of QC by linking the ^{1}H-NMR signals to QC reference values that are used in the beer industry (original gravity, ethanol and lactic acid to MEBAK-guidelines) using PLS- regression. Rodrigues et al. (2011) offered insights into beer quality and beer aging, specifically by monitoring and modeling the storage with ^{1}H-NMR. Interestingly, when NMR spectra were classified as a function of their age, a clear aging trend was observed using PLS-DA, and a potential marker was identified (5-hydroxymethylfurfural).

In the framework of QA/QC, all previously mentioned works used a ^{1}H-NMR, a tool that is not possible to have industrial applications, at least in the production line (in-line) or in QC or new product development laboratory.

Lachenmeier (2007) explored the feasibility of FT-IR for rapid evaluation of quality and authenticity of beers and found that the majority (85%) of all samples tested were classified (PLS-R) as being in conformance with predetermined legal and quality requirements (relative density, ethanol, lactic acid and original gravity). About 10% of all samples exceeded these quality limits and were referred for confirmatory analysis. This is a big improvement because both the cost and time of FT-IR analysis are very low compared to NMR. Very recently, Franca et al. (2021) introduced advanced monitoring concepts for craft beer manufacturing using multivariate statistical process control as part of process analytical technology (PAT). This study demonstrated the potential of rapid non-invasive remote analytical sensors (here, the NIR) and the power of large data set chemometrics in a real-life scenario in the industry (7 batches of Belgian pale ale). NIR spectroscopy is certainly more applicable in an industrial setting than FT-IR on NMR and generally more suited for proper in-line QC, with many examples in other food and drink industries (REF).

Finally, as sensory acceptability is the important element of the beer quality for the industry, some work has been done toward linking sensory profiles with analytical fingerprints and subsequently predicting positive or negative sensory attributes of a beer sample from those fingerprints deriving from untargeted analysis and modeling with multivariate algorithms. An example is the palate fullness prediction using a battery of physicochemical parameters (original gravity, viscosity, indices of macromolecular distribution, total nitrogen and β-glucan) that are usually recorded for QC purposes (Krebs et al., 2021). A good correlation between palate fullness and other sensory characteristics (as assessed by a proper sensory panel) and the analytical parameters was established, and a regression model was built using PLS-R to predict the palate fullness intensity in beers showing decent performance ($R^2 = 0.7993$).

14.3 CONCLUSION

Wine authentication is of major concern to the whole food industry and also has great importance in quality control and consumer information. The capabilities of various analytical methods, used in combination with complicated chemometrics, on the discrimination and classification of wines have been widely proven in many surveys the last decades. For that reason, further developments should aim at improving the classification power of the proposed methods and design certification tools to allow the accurate classification of all the wine samples analyzed to develop a much larger database.

The literature agrees that beer authenticity is a multivariate problem that needs multivariate statistics (chemometrics in this case). However, the quality of the chemometrics methods, however, needs to be elevated as many of the research studies still perform linear classification using PLS-DA, LDA and SIMCA, and non-linear, next-generation classifiers such as the ANN or SNV are rarely used. Regarding the samples, it is noted that the researchers could not have access to the vast samples (and therefore untapped data) that the industry has. Naturally, the research studies show the potential and prove the concept so that these methods can be adopted, improved and scaled for industry applications. It is in this step that more work needs to be done, with academia and industry working closely in larger studies and big data. In terms of characterization of beer, resolution of the beer volatile and non-volatile profile, the research community has used the most modern and the most sophisticated equipment to analyze beers and identify chemical markers that could aid discrimination. Rapid methods have also been employed, including novel technologies such as the e-nose and the e-tongue. This literature review indicates that areas under investigation where robust solutions have still not been proposed are the rapid identification of beer style and geographical origin. But this is not an issue only found in beer; numerous other products, such as cheese, wine and milk, also face the same problem, and it is something that would require significantly more innovation in analytics and algorithms to achieve in the years to come.

REFERENCES

Airado-Rodríguez, D., Durán-Merás, I., Galeano-Díaz, T., & Wold, J.P. (2011). Front-face fluorescence spectroscopy: A new tool for control in the wine industry. *Journal of Food Composition and Analysis.*, 24, 257–264.

Alañón, M.E., Pérez-Coello, M.S., & Marina, M.L. (2015). Wine science in the metabolomics era. *TrAC-Trends Analytical Chemistry*, 74, 1–20.

Alcázar, Á., Jurado, J.M., Palacios-Morillo, A., de Pablos, F., & Martín, M.J. 2012. Recognition of the geographical origin of beer based on support vector machines applied to chemical descriptors. *Food Control*, 23, 258–262.

Aleixandre-Tudó, J.L., Buica, A., Nieuwoudt, H., Aleixandre, J.L., & du Toit, W. (2017). Spectrophotometric analysis of phenolic compounds in grapes and wines. *Journal of Agricultural and Food Chemistry*, 65, 4009–4026.

Aleixandre-Tudó, J.L., Castelló-Cogollos, L., Aleixandre, J.L., & Aleixandre-Benavent, R. (2020). Bibliometric insights into the spectroscopy research field: A food science and technology case study. *Applied Spectroscopy Reviews*, 55, 873–906.

Arapitsas, P., Ugliano, M., Marangon, M., Piombino, P., Rolle, L., Gerbi, V., Versari, A., & Mattivi, F. (2020). Use of untargeted LC-MS metabolome to discriminate Italian monovarietal red wines, produced in their different terroirs. *Journal of Agricultural and Food Chemistry*.

Arcari, S.G., Caliari, V., Sganzerla, M., & Godoy, H.T. (2017). Volatile composition of Merlot red wine and its contribution to the aroma: Optimization and validation of analytical method. *Talanta*, 174, 752–766.

Ayseli, M.T., & Ayseli, Y.İ. (2016). Flavors of the future: Health benefits of flavor precursors and volatile compounds in plant foods. *Trends in Food Science & Technology*, 48, 69–77.

Azcarate, S.M., Cantarelli, M.Á., Pellerano, R.G., Marchevsky, E.J., & Camiña, J.M. (2013). Classification of Argentinean Sauvignon blanc wines by UV spectroscopy and chemometric methods. *Journal of Food Science*, 78, 432–436.

Bellido-Milla, D., Moreno-Perez, J. M., & Hernández-Artiga, M. P. (2000). Differentiation and classification of beers with flame atomic spectrometry and molecular absorption spectrometry and sample preparation assisted by microwaves. *Spectrochimica Acta Part B: Atomic Spectroscopy*, 55(7), 855–864.

Biancolillo, A., Bucci, R., Magrì, A.L., Magrì, A.D., & Marini, F. 2014, "Data-fusion for multiplatform characterization of an Italian craft beer aimed at its authentication", *Analytica Chimica Acta*, vol. 820, 23–31.

Boido, E., Fariña, L, Carrau, F, Dellacassa, E, & Cozzolino, D. (2013). Characterization of glycosylated aroma compounds in tannat grapes and feasibility of the near infrared spectroscopy application for their prediction. *Food Analytical Methods*, 6(1), 100–111.

Brewers of Europe Association (accessed 24/2/22) https://brewersofeurope.org/site/countries/figures.php?doc_id=659

Bronzi, B., Brilli, C., Beone, G.M., Fontanella, M.C., Ballabio, D., Todeschini, R., & Buscema, M. (2020). Geographical identification of Chianti red wine based on ICP-MS element composition. *Food Chemistry*, 315, 126248.

Cajka, T., Riddellova, K., Tomaniova, M., & Hajslova, J. (2010). Recognition of beer brand based on multivariate analysis of volatile fingerprint. *Journal of Chromatography A*, 1217(25), 4195–4203.

Cajka, T., Riddellova, K., Tomaniova, M., & Hajslova, J. (2011). Ambient mass spectrometry employing a DART ion source for metabolomic fingerprinting/profiling: a powerful tool for beer origin recognition. *Metabolomics*, 7(4), 500–508.

Catarino, S., & Curvelo-Garcia, A.S. (2015). Composição mineral do vinho—Ocorrência de metais contaminantes. *Publindústria, Química Enológica métodos analíticos*, 1, 237–273.

Catarino, S., Madeira, M., Monteiro, F., Caldeira, I., Bruno de Sousa, R., & Curvelo-Garcia, A. (2018). Mineral composition through soil-wine system of Portuguese vineyards and its potential for wine traceability. *Beverages*, 4(4), 85.

Cavallini, N., Savorani, F., Bro, R., & Cocchi, M. (2021). A metabolomic approach to beer characterization. *Molecules*, 26(5), 1472.

Chapman, J., Gangadoo, S., Truong, V.K., & Cozzolino D. (2019). Spectroscopic approaches for rapid beer and wine analysis. *Current Opinion in Food Science*, 28, 67–73.

Coelho, C., Aron, A., Roullier-Gall, C., Gonsior, M., Schmitt-Kopplin, P., & Gougeon, R.D. (2015). Fluorescence fingerprinting of bottled white wines can reveal memories related to sulfur dioxide treatments of the must. *Analytical Chemistry*, 87, 8132–8137.

Cozzolino, D. (2015). The role of visible and infrared spectroscopy combined with chemometrics to measure phenolic compounds in grape and wine samples. *Molecules*, 20, 726–737.

Cuadros-Inostroza, A., Giavalisco, P., Hummel, J., Eckardt, A., Willmitzer, L., & Peña-Cortés, H. (2010). Discrimination of wine attributes by metabolome analysis, *Analytical Chemistry*, 82(9), 3573–3580.

da Silva, L.A., Flumignan, D.L., Tininis, A.G., Pezza, H.R., & Pezza, L. 2019, "Discrimination of Brazilian lager beer by ^{1}H NMR spectroscopy combined with chemometrics. *Food Chemistry*, 272, 488–493.

Di Paola-Naranjo, R. D., Baroni, M. V., Podio, N. S., Rubinstein, H. R., Fabani, M. P., Badini, R. G., & Wunderlin, D.A. (2011). Fingerprints for main varieties of Argentinean wines: terroir differentiation by inorganic, organic, and stable isotopic analyses coupled to chemometrics. *Journal of Agricultural and Food Chemistry*, 59(14), 7854–7865.

Diamantidou, D., Zotou, A., & Theodoridis, G. (2018). Wine and grape marc spirits metabolomics. *Metabolomics*, 14, 159.

Duarte, I. F., Barros, A., Almeida, C., Spraul, M., & Gil, A. M. (2004). Multivariate analysis of NMR and FTIR data as a potential tool for the quality control of beer. *Journal of Agricultural and Food Chemistry*, 52(5), 1031–1038.

Duarte, I. F., Barros, A., Belton, P. S., Righelato, R., Spraul, M., Humpfer, E., & Gil, A. M. (2002). High-resolution nuclear magnetic resonance spectroscopy and multivariate analysis for the characterization of beer. *Journal of Agricultural and Food Chemistry*, 50(9), 2475–2481.

Engel, J., Blanchet, L., Buydens, L.M., & Downey, G. (2012). Confirmation of brand identity of a Trappist beer by mid-infrared spectroscopy coupled with multivariate data analysis. *Talanta*, 99, 426–432.

Fang, H., Wu, H., Wang, T., Long, W., Chen, A., Ding, Y., & Yu, R. (2021). Excitation-emission matrix fluorescence spectroscopy coupled with multi-way chemometric techniques for characterization and classification of Chinese lager beers, *Food Chemistry*, 342, 128235.

Fan, S., Zhong, Q., Fauhl-Hassek, C., Pfister, M.K.H., Horn, B., & Huang, Z.B. (2018). Classification of Chinese wine varieties using ^{1}H NMR spectroscopy combined with multivariate statistical analysis. *Food Control*, 88, 113–122.

Flamini, R. (2013). Recent Applications of Mass Spectrometry in the Study of Grape and Wine Polyphenols. *ISRN Spectroscopy*, 813563.

França, L., Grassi, S., Pimentel, M.F., & Amigo, J.M. (2021). A single model to monitor multistep craft beer manufacturing using near infrared spectroscopy and chemometrics. *Food and Bioproducts Processing*, 126, 95–103.

Francois, G., Fabrice, V., & Didier, M. (2020). Traceability of fruits and vegetables. *Phytochemistry*, 173, 112291.

Gallart-Ayala, H., Kamleh, M. A., Hernández-Cassou, S., Saurina, J., & Checa, A. (2016). Ultra-high-performance liquid chromatography–high-resolution mass spectrometry based metabolomics as a strategy for beer characterization. *Journal of the Institute of Brewing*, 122(3), 430–436.

Geană, E.-I., Ciucure, C.T., Apetrei, C., & Artem, V. (2019). Application of spectroscopic UV—Vis and FT-IR screening techniques coupled with multivariate statistical analysis for red wine authentication: Varietal and vintage year discrimination. *Molecules*, 24, 4166.

Ghasemi-Varnamkhasti, M., Mohtasebi, S.S., Siadat, M., Ahmadi, H., & Razavi, S.H. (2015). From simple classification methods to machine learning for the binary discrimination of beers using electronic nose data. *Engineering in Agriculture, Environment and Food*, 8, 44–51.

Giannetti, V., Mariani, M.B., Torrelli, P., & Marini, F. (2019). Flavour component analysis by HS-SPME/GC—MS and chemometric modeling to characterize Pilsner-style Lager craft beers. *Microchemical Journal*, 149, 103991.

Gil, M., Reynes, C., Cazals, G., Enjalbal, C., Sabatier, R., & Saucier, C. (2020). Discrimination of rosé wines using shotgun metabolomics with a genetic algorithm and MS ion intensity ratios. *Scientific Reports*, 10, 1–7.

Gougeon, L., Da Costa, G., Le Mao, I., Ma, W., Teissedre, P.L., Guyon, F., & Richard, T. (2018). Wine analysis and authenticity using ^{1}H-NMR metabolomics data: Application to Chinese wines. *Food Analytical Methods*, 11, 3425–3434.

Gresser, A. (2009). *Properties and quality. Handbook of brewing.* Chapter 1. Wiley VCH Verlag GmbH & Co. KGaA.

Holmberg, L. (2010). Wine fraud. *International Journal of Wine Research*, 2, 105–113.

Hu, X.Z., Liu, S.Q., Li, X.H., Wang, C.X., Ni, X.L., Liu, X., Wang, Y., Liu, Y., & Xu, C.H. (2019). Geographical origin traceability of Cabernet Sauvignon wines based on Infrared fingerprint technology combined with chemometrics. *Scientific Reports*, 9, 8256.

Inui, T., Tsuchiya, F., Ishimaru, M., Oka, K., and Komura, H. (2013) Different beers with different hops. Relevant compounds for their aroma characteristics. *Journal of Agricultural and Food Chemistry*, 61, 4758–4764.

Kabelová, I., Dvořáková, M., Čížková, H., Dostálek, P., & Melzoch, K. (2008). Determination of free amino acids in beers: A comparison of Czech and foreign brands. *Journal of Food Composition and Analysis*, 21, 736–741.

Karalis, P., Poutouki, A.E., Nikou, T., Halabalaki, M., Proestos, C., Tsakalidou, E., Gougoura, S., Diamantopoulos, G., Tassi, M., & Dotsika, E. (2020). Isotopic Traceability (^{13}C and ^{18}O) of Greek Olive Oil. *Molecules*, 25, 5816.

Kaufmann, M., Schwarz, K.J., Dallmann, A., Kuballa, T., & Bergmann, M. (2021). ^{1}H NMR spectroscopic discrimination of different beer styles combined with a chemical shift-based quantification approach. *European Food Research and Technology*, 1–11.

Kokkinofta, R., Fotakis, C., Zervou, M., Zoumpoulakis, P., Savvidou, C., Poulli, K., & Loupasaki, S. (2017). Isotopic and elemental authenticity markers: A case study on Cypriot wines. *Food Analytical Methods*, 10(12), 3902–3913.

Krebs, G., Gastl, M., & Becker, T. (2021). Chemometric modeling of palate fullness in lager beers, *Food Chemistry*, 342, 128253.

Kruzlicova, D., Fiket, Ž., & Kniewald, G. (2013). Classification of Croatian wine varieties using multivariate analysis of data obtained by high resolution ICP-MS analysis. *Food Research International*, 54(1), 621–626.

Kyraleou, M., Kallithraka, S., Gkanidi, E., Koundouras, S., Mannion, D.T., & Kilcawley, K.N. (2020). Discrimination of five Greek red grape varieties according to the anthocyanin and proanthocyanidin profiles of their skins and seeds. *Journal of Food Composition and Analysis*, 92, 103547.

Lachenmeier, D. W. (2007). Rapid quality control of spirit drinks and beer using multivariate data analysis of Fourier transform infrared spectra. *Food Chemistry*, 101(2), 825–832.

Lachenmeier, D. W., Frank, W., Humpfer, E., Schäfer, H., Keller, S., Mörtter, M., & Spraul, M. (2005). Quality control of beer using high-resolution nuclear magnetic resonance spectroscopy and multivariate analysis. *European Food Research and Technology*, 220(2), 215–221.

Likar, M., Vogel-Mikuš, K., Potisek, M., Hančević, K., Radić, T., Nečemer, M., & Regvar, M. (2015). Importance of soil and vineyard management in the determination of grapevine mineral composition. *Science of the Total Environment*, 505, 724–731.

Magdas, D.A., Cozar, B.I., Feher, I., Guyon, F., Dehelean, A., Cinta Pinzaru, S. (2019). Testing the limits of FT-Raman spectroscopy for wine authentication: Cultivar, geographical origin, vintage and terroir effect influence. *Scientific Reports*, 9, 19954.

Magdas, D.A., Guyon, F., Feher, I., Pinzaru, S.C. (2018). Wine discrimination based on chemometric analysis of untargeted markers using FT-Raman spectroscopy. *Food Control*, 85, 385–391.

Mannina, L., Marini, F., Antiochia, R., Cesa, S., Magrì, A., Capitani, D., & Sobolev, A.P. (2016). Tracing the origin of beer samples by NMR and chemometrics: Trappist beers as a case study. *Electrophoresis*, 37, 2710–2719.

Marín-San Román, S., Rubio-Bretón, P., Pérez-Álvarez, E.P., & Garde-Cerdán, T. (2020). Advancement in analytical techniques for the extraction of grape and wine volatile compounds. *Food Research International*, 137, 109712.

Mascellani, A., Hoca, G., Babisz, M., Krska, P., Kloucek, P., & Havlik, J. (2021). ^{1}H NMR chemometric models for classification of Czech wine type and variety. *Food Chemistry*, 339, 127852.

Mattarucchi, E., Stocchero, M., Moreno-Rojas, J. M., Giordano, G., Reniero, F., Guillou, C. (2010) Authentication of Trappist beers by LC—MS fingerprints and multivariate data analysis, *Journal of Agricultural and Food Chemistry*, 58, 12089–12095.

Moura-Nunes, N., Brito, T.C., da Fonseca, N.D., de Aguiar, P.F., Monteiro, M., Perrone, D., & Torres, A.G. (2016). Phenolic compounds of Brazilian beers from different types and styles and application of chemometrics for modeling antioxidant capacity. *Food Chemistry*, 199, 105–113.

Mutz, Y.S., do Rosario, D., Silva, L.R., Santos, F.D., Santos, L.P., Janegitz, B.C., Filgueiras, P.R., Romão, W., de Q Ferreira, R., & Conte-Junior, C.A. (2021). Portable electronic tongue based on screen-printed electrodes coupled with chemometrics for rapid differentiation of Brazilian lager beer. *Food Control*, 127, 108163.

Palmioli, A., Alberici, D., Ciaramelli, C., & Airoldi, C. (2020). Metabolomic profiling of beers: Combining ^{1}H NMR spectroscopy and chemometric approaches to discriminate craft and industrial products. *Food Chemistry*, 327, 127025.

Paneque, P., Morales, M.L., Burgos, P., Ponce, L., & Callejón, R.M. (2017). Elemental characterization of Andalusian wine vinegars with protected designation of origin by ICP-OES and chemometric approach. *Food Control*, 75, 203–210.

Pasvanka, K. (2021). Wine authenticity from Greek grape varieties and different vinification techniques. *Doctoral thesis*, National and Kapodistrian University of Athens.

Pasvanka, K., Tzachristas, A., & Proestos, C. (2019). Quality tools in wine traceability and authenticity. In *Quality control in the beverage industry*. Academic Press, 289–334.

Pereira, H.V., Amador, V.S., Sena, M.M., Augusti, R., & Piccin, E. (2016). Paper spray mass spectrometry and PLS-DA improved by variable selection for the forensic discrimination of beers. *Analytica Chimica Acta*, 940, 104–112.

Perestrelo, R., Barros, A.S., Camara, J.S., & Rocha, S.M. (2011). In-depth search focused on furans, lactones, volatile phenols, and acetals as potential age markers of Madeira wines by comprehensive two-dimensional gas chromatography with time-of-flight mass spectrometry combined with solid phase microextraction. *Journal of Agricultural and Food Chemistry*, 59(7), 3186–3204.

Pérez-Álvarez, E.P, García, R., Barrulas, P., Dias, C., Cabrita, M.J., & Garde-Cerdán, T. (2019). Classification of wines according to several factors by ICP-MS multi-element analysis. *Food Chemistry*, 270, 273–280.

Pérez-Ràfols, C., & Saurina, J. (2015). Liquid chromatographic fingerprints and profiles of polyphenolic compounds applied to the chemometric characterization and classification of beers. *Analytical Methods*, 7(20), 8733–8739.

Pérez-Ràfols, C., Viñas, D., Hernández-Cassou, S., & Saurina, J. (2015). Experimental design for the determination of polyphenols by liquid chromatography: application to the chemometric characterization and classification of beers. *Analytical Methods*, 7, 3283–3290.

Philippidis, A., Poulakis, E., Kontzedaki, R., Orfanakis, E., Symianaki, A., Zoumi, A., & Velegrakis, M (2021). Application of ultraviolet-visible absorption spectroscopy with machine learning techniques for the classification of Cretan wines. *Foods*, 10(1), 9.

Płotka-Wasylka, J., Frankowski, M., Simeonov, V., Polkowska, Ż., & Namieśnik, J. (2018). Determination of metals content in wine samples by inductively coupled plasma-mass spectrometry. *Molecules*, 23(11), 2886.

Portinale, L., Leonardi, G., Arlorio, M., Coïsson, J.D., Travaglia, F., & Locatelli, M. (2017). Authenticity assessment and protection of high-quality Nebbiolo-based Italian wines through machine learning. *Chemometrics and Intelligent Laboratory Systems*, 171, 182–197.

Raco, B., Dotsika, E., Poutoukis, D., Battaglini, R., & Chantzi, P. (2015). O—H—C isotope ratio determination in wine in order to be used as a fingerprint of its regional origin. *Food Chemistry*, 168, 588–594.

Ranaweera, R.K., Capone, D.L., Bastian, S.E., Cozzolino, D., & Jeffery, D.W. (2021a). A review of wine authentication using spectroscopic approaches in combination with chemometrics. *Molecules*, 26(14), 4334.

Ranaweera, R.K.R., Gilmore, A.M., Capone, D.L. Bastian, S.E.P., & Jeffery, D.W. (2021b). Spectrofluorometric analysis combined with machine learning for geographical and varietal authentication, and prediction of phenolic compound concentrations in red wine. *Food Chemistry*, 130149.

Rocha, S., Pinto, E., Almeida, A., & Fernandes, E. (2019). Multi-elemental analysis as a tool for characterization and differentiation of Portuguese wines according to their Protected Geographical Indication. *Food Control*, 103, 27–35.

Rodrigues, J. A., Barros, A. S., Carvalho, B., Brandão, T., & Gil, A. M. (2011). Probing beer aging chemistry by nuclear magnetic resonance and multivariate analysis. *Analytica Chimica Acta*, 702, 178–187.

Rodriguez-Saona, L.E., & Allendorf, M.E. (2011). Use of FTIR for rapid authentication and detection of adulteration of food. *Annual Review of Food Science and Technology*, 2, 467–483.

Roullier-Gall, C., Kanawati, B., Hemmler, D., Druschel, G.K., Gougeon, R.D., & Schmitt-Kopplin, P. (2019). Electrochemical triggering of the Chardonnay wine metabolome. *Food Chemistry*, 286, 64–70.

Sádecká, J., & Jakubíková, M. (2020). Varietal classification of white wines by fluorescence spectroscopy. *Journal of Food Science & Technology*, 57, 2545–2553.

Soares, F., Anzanello, M.J., Fogliatto, F.S., Marcelo, M.C., Ferrão, M.F., Manfroi, V., & Pozebon, D. (2018). Element selection and concentration analysis for classifying South America wine samples according to the country of origin. *Computers and electronics in agriculture*, 150, 33–40.

Solovyev, P.A., Fauhl-Hassek, C., Riedl, J., Esslinger, S., Bontempo, L., & Camin, F. (2021). NMR spectroscopy in wine authentication: An official control perspective. Comprehensive Reviews in Food Science and Food Safety, 20, 2040–2062.

Soufleros, E. (1997). *Oenology. Science and knowledge*, Volume I. Thessaloniki.

Spink, J., Hegarty, P. V., Fortin, N. D., Elliott, C. T., & Moyer, D. C. (2019). The application of public policy theory to the emerging food fraud risk: Next steps. *Trends in Food Science & Technology*, 85, 116–128.

Stój, A., Kapusta, I., & Domagała, D. (2020). Classification of red wines produced from Zweigelt and rondo grape varieties based on the analysis of phenolic compounds by UPLC-PDA-MS/MS. *Molecules*, 25, 1342.

Tahir, H.E., Xiaobo, Z., Jianbo, X., Mahunu, G.K., Jiyong, S., Xu, J. L., & Sun, D.W. (2019). Recent progress in rapid analyses of vitamins, phenolic, and volatile compounds in foods using vibrational spectroscopy combined with chemometrics: A review. *Food Analytical Methods*, 12(10), 2361–2382.

Takeoka, G. R., & Ebeler, S. E. (2011). Progress in authentication of food and wine. In Progress in authentication of food and wine. *American Chemical Society*, 3–11.

Tan, J., Li, R., & Jiang, Z. (2015). Chemometric classification of Chinese lager beers according to manufacturer based on data fusion of fluorescence, UV and visible spectroscopies. *Food Chemistry*, 184, 30–36.

Tan, J., Li, R., Jiang, Z.T., Zhang, Y., Hou, Y.M., Wang, Y.R., Wu, X., & Gong, L. (2016). Geographical classification of Chinese Cabernet Sauvignon wines by data fusion of ultraviolet—visible and synchronous fluorescence spectroscopies: The combined use of multiple wavelength differences. *Australian Journal of Grape and Wine Research*, 22, 358–365.

Tzachristas, A., Pasvanka, K., Calokerinos, A., & Proestos, C. (2020). Polyphenols: Natural antioxidants to be used as a quality tool in wine authenticity. *Applied Sciences*, 10(17), 5908.

Vaclavik, L., Lacina, O., Hajslova, J., & Zweigenbaum, J. (2011). The use of high-performance liquid chromatography—quadrupole time-of-flight mass spectrometry coupled to advanced data mining and chemometric tools for discrimination and classification of red wines according to their variety. *Analytica Chimica Acta*, 685(1), 45–51.

Vasas, M., Tang, F., & Hatzakis, E. (2021). Application of NMR and chemometrics for the profiling and classification of ale and lager American craft beer. *Foods*, 10, 807.

Vera, L., Aceña, L., Guasch, J., Boqué, R., Mestres, M., & Busto, O. 2011. Characterization and classification of the aroma of beer samples by means of an MS e-nose and chemometric tools. *Analytical and Cioanalytical Chemistry*, 399, 6, 2073–2081.

Versari, A., Laurie, V.F., Ricci, A., Laghi, L., & Parpinello, G.P. (2014). Progress in authentication, typification and traceability of grapes and wines by chemometric approaches. *Food Research International*, 60, 2–18.

Viejo, C.G., Fuentes, S., Godbole, A., Widdicombe, B., & Unnithan, R.R. (2020). Development of a low-cost e-nose to assess aroma profiles: An artificial intelligence application to assess beer quality. *Sensors and Actuators B: Chemical*, 308, 127688.

Volpe, M.G., La Cara, F., Volpe, F., De Mattia, A., Serino, V., Petitto, F., Zavalloni, C., Limone, F., Pellechia, R., & De Prisco, P.P. (2009). Heavy metal uptake in the enological food chain. *Food Chemistry*, 117, 553–560.

Weeranantanaphan, J., & Downey, G. (2010). Identity confirmation of a branded, fermented cereal product by UV spectroscopy: A feasibility study involving a Trappist beer. *Journal of the Institute of Brewing*, 116(1), 56–61.

Weldegergis, B.T., de Villiers, A., & Crouch, A. M. (2011). Chemometric investigation of the volatile content of young South African wines. *Food Chemistry*, 128(4), 1100–1109.

Wu, H., Lin, G., Tian, L., Yan, Z., Yi, B., Bian, X., & Rogers, K.M. (2021). Origin verification of French red wines using isotope and elemental analyses coupled with chemometrics. *Food Chemistry*, 339, 127760.

Xu, Y., Zhong, P., Jiang, A., Shen, X., Li, X., Xu, Z., Shen, Y., Sun, Y., & Lei, H. (2020). Raman spectroscopy coupled with chemometrics for food authentication: A. review. *TrAC Trends Analytical Chemistry*, 131, 116017.

Yin, H., Hu, X., Huang, X., Zou, X., Xu, Y., Shi, J., & Yang, M. (2021). Rapid discrimination of beer flavors using ion-selective electrode array system combined with chemometrics. *Food Analytical Methods*, 1–7.

Zanuttin, F., Gurian, E., Ignat, I., Fornasaro, S., Calabretti, A., Bigot, G., & Bonifacio, A. (2019). Characterization of white wines from north-eastern Italy with surface-enhanced Raman spectroscopy. *Talanta*, 203, 99–105.

15 Legislation and Food Authenticity

Theodoros Varzakas

CONTENTS

15.1 INTRODUCTION

Food quality assurance and food quality control including authenticity are among the main pillars of food quality management. The extent of the importance of food quality detection, and more specifically food fraud, is the decisive factor on the effectiveness of the quality management strategy followed. Food quality control including food authenticity techniques is a subset of quality assurance/quality management activities.

An authentic food product is a food that has been produced or grown in such a way that it has a characteristic feature (e.g., nutritional value, sensory properties, specific composition, health-promoting properties) and/or has been produced under certain conditions (e.g., traditional, technologically advanced, or under intellectual property rights). Food authenticity is displayed on the original product packaging and labeled with a text or trademark. Food authenticity refers to product quality and characteristics that depend on food processing and environmental factors (e.g., geolocation, climate, soil type; Gvozdanovic et al., 2017).

Looking at the overall state of food, the most frequent adulterations are done in powders and liquids (e.g., olive oils, alcoholic beverages, fish, seasonings, sauces, cereals, flours) and much less in whole and fresh foods. In terms of food type, meat and meat products, fish and fish products, milk and dairy products, fruits, juices and honey are the most susceptible to fraud (Kyrova et al., 2017; Djekic et al., 2018; NSF, 2014).

15.1.1 Legislative Point of View

The world population is growing rapidly, and many food products repeatedly cross national boundaries to meet this rising demand for food. Especially nowadays, food prices are increasing due to the shortage of raw materials such as wheat. This means an increase in fraud and the need for international regulations, and legislation may contribute to effectively, safely and sustainably to facing this challenge by conduction of best practices in food production methods and testing across all stages of the food industry (from farm to fork), which in turn leads to superior final food product quality (ISO, 2017).

In more detail, food legislation should regulate the whole food production chain setting controls and provisions starting from animal feed, farm controls and pre-processing, and ending with food handling and commercial transactions, ensuring that the food product is not exposed to any fraudulent practices aimed at deceiving the consumer (FAO, 2020).

However, the biggest challenge for some of the food exporters located in developing countries is that some of the countries suffer from a lack of regulations and legislation on food quality on the one hand, and on the other hand, they are unable to enforce or validate such regulations. These exporters must comply with the food quality regulations and standards of importers (most of which are located in developed countries). It is wise that the developed countries should help developing countries to adapt to these regulations regarding the quality of food products in order

 DOI: 10.1201/9781003266167-15

to safeguard quality and safety and to avoid fraud (Hutt, 1984; International Trade Centre, 2010; Hyatt, 2016; Sagener, 2016; FSA, 2017).

Developing countries aiming to export to the lucrative markets of industrialized countries must have adequate information about buyers' requirements, as well as a solid system for collecting and presenting the necessary information. However, because the food supply chains in developing countries are typically fragmented—relying on many smallholder farmers—and are not necessarily organized into supportive producer associations that can provide the training the food producers would need to create traceability documentation, many of these smallholders will be excluded (Setboonsarng et al., 2009).

Related regulations in some developed countries are described in the following sections.

15.1.2 Regulations in the United Kingdom

Several key pieces of legislation in the United Kingdom tackle food quality issues, such as the amended Food Safety Act 1990, which provides the basis for all food regulations in England, Wales and Scotland. Under this law, all food companies subject to it (1) must not add or remove anything to food or carry out certain processing that would harm the health of consumers; (2) must ensure that the final food product is acceptable to consumers in terms of the expected quality; and (3) must ensure that food must be labeled and advertised and introduced in a manner that is not false or misleading (Walker, 2017).

The principle target of the Food Standards Act 1999 is to set up a Food Standards Agency. This Act provides the power to take action in the consumer's interest at any phase of the food production and supply chain.

15.1.3 Regulations in Canada

The food safety and nutritional legislation, policies and guidance in Canada were developed by Health Canada under the Food and Drugs Act. This act governs the food safety and quality sold in Canada. In Canada, food and drug regulations (C.R.C., c. 802) were last amended in 2020 (Consolidated Regulations of Canada, 2020). This legislation, where applicable, prescribes the standards of quality, composition, potency, purity or other property of the article of food or drug to which they refer. Regarding food, these regulations covering a diversity of products such as alcoholic beverages, spices, dairy products, fruits, vegetables, grains and bakery products, meat and its products. It is also noteworthy that this law tackles many other related topics, such as food packaging and food adulteration.

In June 2002, the Canadian federal government established an ambitious goal that, before 2008, the country would achieve tracing back 80% of agricultural products to its source, supporting the Brand Canada strategy (Lu Changhua et al., 2007), of which a mandatory identification system for cattle and beef on July 1, 2002, came into operation.

15.1.4 Regulations in South Africa

The importance of legislation and regulations in South Africa arose in order to protect the consumer from any potential risks that food may cause. It is important for South African consumers to have access to acceptable quality, safe and affordable food. Food carries potential risks that can cause harm to human health. Questions and answers will clarify this issue as reported by the Food Advisory Consumer Service (2020).

The main bodies responsible for food legislation in South Africa are the following departments:

- The Department of Agriculture, Forestry and Fisheries (DAFF);
- The National Department of Health;
- The Department of Trade and Industry.

DAFF regulates safety and quality of agriculture and animal products in terms of several acts of parliament:

- Agricultural Product Standards Act, 1990 (Act No. 119 of 1990);
- Animal Diseases Act, 1984 (Act No. 35 of 1984);
- Animal Identification Act, 2000 (Act No. 6 of 2002);
- Fertilisers, Farm Feeds, Agricultural Remedies and Stock Remedies Act, 1947 (Act no. 36 of 1984);

- Meat Safety Act, 2000 (Act No. 40 of 2000);
- Liquor Products Act, 1989.

15.2 THE CASE OF JAPAN

Japan's experience has the potential to provide lessons for other countries because Japan (1) has high standards and strict market requirements, (2) imports a large volume of food, (3) has a public that is responsive to food safety issues, (4) has a production system based on small farms and (5) uses a high-technology traceability system.

Japan's food market is notoriously hard to penetrate, due primarily to the high quality standards required by importers and consumers. Exporters who are able to meet the requirements of the Japanese market are well prepared to compete in other lucrative markets. Moreover, Japan is the biggest food importer in the world, importing more than 60% of its food, increasingly from developing countries. Compliance with Japan's food safety standards and traceability requirements opens the door to increased business opportunities.

The high rate of IT adoption in Japan provides a wide range of examples of information and communication technology (ICT) support in traceability systems.

Japanese regulations: Despite being a market known for strict food requirements, Japanese law requires a full traceability system only for domestic beef. For other foods, Article 3 of Japan's Food Sanitation Law (Food Marketing Research & Information Center, FMRIC, 2006; www.jetro.go.jp/en/market/regulations/pdf/food-e.pdf) requests that each operator keep records to identify all their suppliers and customers—a "one-step-back" and "one-step-forward" record. This request is similar to Article 18 of the European Union's EC Regulation 178/2002. However, in Japan this type of record-keeping is only recommended and is not compulsory. On the other hand, Japanese regulations do require labeling of the place of origin for fresh food and minimally processed food, not only at the retail level but also at the wholesale level. However, while origin labeling itself is required, a record-keeping system to verify origin area by providing documentation such as delivery slips and/or invoices is only recommended, not legally required, per Article 3 of the Food Sanitation Law.

Japanese Agricultural Standards (JAS): The JAS system was established in 1950 and is overseen by the Ministry of Agriculture, Forestry and Fisheries (MAFF, 2004, www.maff.go.jp/soshiki/syokuhin/hinshitu/e_label/).

One of the main components of the JAS system is its compliance certification systems. Products passing inspection in accordance with the JAS requirements are allowed to display the JAS logo. These standards are voluntary.

Originally, the JAS standards were intended to ensure general product quality and standardized labeling. However, in the 1990s, MAFF expanded the JAS system to cover specific methods of production. These expanded standards, called Specific JAS, cover (1) processed meat products (aged ham, sausage, and bacon); (2) free-range chicken (*Jidoriniku*); (3) organic foods; and (4) other products disclosing production history and methods (Setboonsarng et al., 2009).

Since 2001, MAFF had been subsidizing the development and introduction of traceability systems utilizing ICT that can be used throughout the food chain (FMRIC, 2006). From 2005 to 2007, MAFF spent 1–2 billion yen (about US$10–20 million) annually on various pilot projects and studies. Examples of these experiments include (1) *integrated circuit (IC) tags* to reduce the cost of reading and recording the unique ID code of food products at each stage of the food chain; (2) *handheld devices* to record electric data on farm inputs, processing and distribution without paper documentation; and (3) *web-based service technology* to keep and transfer data between servers through the internet (MAFF, 2006).

In Japan, many types of operators have implemented traceability systems *within* their enterprises. When an individual enterprise implements an internal traceability system, it is usually not a stand-alone system. Rather, operators commonly consider traceability as the basis of a specific or integrated management system. This integrated system may include a quality management system, safety management system, inventory management system or production history information disclosure system.

However, in order to establish a secure food chain traceability system, it is necessary to ensure consistent standards not only *within* individual organizations but also *between* food business operators along the supply chain, upstream to downstream. For traceability with wide-ranging application, it is also desirable to ensure consistency *across* food business operators who are at the same stage of the food supply chain, such as processors or packers. To do this, experience has shown that it is often most efficient for several food business operators to form an organization and work together to create a consistent traceability system.

Creators of traceability systems in Japan find themselves facing common challenges worldwide: (1) reaching a consensus on a traceability system that will fit the needs of all operators in the chain; (2) creating a system that is consistent *between* and *across* operators, and not only within an individual operator; and (3) creating a system that addresses the fragmented nature of supply chains with many small producers and operators as well as the ever-changing relationships within the supply chains (Setboonsarng et al., 2009).

15.3 THE UNITED STATES

In December 2003, the United States developed the statutes of tracking food safety, which required all enterprises involved in food transportation, distribution and import to record their trade information for tracking and tracing back (Cheng Hao., 2007). In addition, the US has included 70% of the cattle in the NAIS (National Animal Identity System) project at the end of 2009 (Xing Wenying., 2006).

In the United States, the Bioterrorism Act of 2002 declared that anyone who manufactures, processes, packs, transports, distributes, receives, holds or imports food has the responsibility to establish and maintain records. It also allows the Food and Drug Administration (FDA) to inspect those records if there is a reasonable belief that an article of food presents a serious health threat (Levinson, 2009).

The FDA's Food Safety Modernization Act (FSMA), which went into effect on January 4, 2011, requires registered food and feed facilities to evaluate the food safety hazards that could affect the food and feed they manufacture, process, pack or hold and to identify and implement preventive controls to address those hazards. This is to ensure the safety of both the imported and domestic food supplies by focusing on preventing contamination rather than responding to contamination (FDA, 2011; FSAI, 2004; FSA, 2002).

Regarding food fraud in the United States, the FDA has legally defined food adulteration and mislabeling (misbranding) as criminal offenses (Robson et al., 2021). The terms *food fraud* and *economically motivated adulteration* (*EMA*) are synonyms, according to the available literature (Reg. 178/2002; Everstine et al., 2013). Some US authors have subdivided EMAs into individual acts that include adulteration, tampering, theft, diversion, simulation, counterfeiting and misleading information. However, regardless of the definitions used, both sides point to the same problem, which is fraud against consumers to gain an economic advantage.

The FDA defines EMA as the false, intentional substitution or addition of compounds to increase apparent product value or reduce production costs and includes food, supplements, tobacco, cosmetics, medications, medical devices and equipment (FDA, 1992). EMA in the food industry can occur through manipulation within the original food composition, ingredient substitution or disapproved food processing, and mislabeling of food. Food fraud can include misrepresentation, theft, diversion, simulation, smuggling, counterfeiting and anything that can be classified as forgery, as described by the FDA, so it is not just the addition or substitution of raw materials (Spink and Moyer, 2011). Most commonly, criminal manipulation with food occurs in the form of adulteration and misbranding (mislabeling), as defined in sections 342 and 343 of the Federal Food, Drug, and Cosmetic Act (FDA, 1992).

According to the US literature, economically motivated food adulteration (e.g., adulteration of food or introduction of other materials and ingredients) is most similar to the EU concept of food fraud, with both sides emphasizing the intent to commit a food crime.

Moreover, economically motivated food fraud or adulteration is much more common than intentional contamination of food or water. For example, there were 200 reported cases of economically motivated food adulteration in the United States between 1980 and 2010 (Johnson, 2014). Such EMA incidents can cause serious harm to public health and are certainly a serious challenge to all of the parties involved, such as the food industry, inspection agencies and the legal system, as these acts are committed in a manner intended to avoid or minimize food fraud detection (Everstine et al., 2013).

15.4 THE EUROPEAN UNION

Traceability system was initially promoted by the EU to control the risks of mad cow disease in 1997. In July 2003, The EU published a White Paper on Food Safety proposing a new framework for food safety system. Most countries in the EU have implemented mandatory livestock and meat products traceability system (Zhu Haipeng., 2007). The EU also had a plan called TraceFish, the main objective of which is to study the full supply chain traceability of aquatic products (Liu Junrong., 2007).

EU Reg. 178/2002 (EU, 2002) narrows the definition to the food industry by defining traceability as the ability to trace and follow a food, feed, food-producing animal or substance intended to be or expected to be incorporated into a food or feed through all stages of production, processing and distribution.

The CAC (2006) offers a more concise definition of traceability as the ability to follow the movement of a food through specified stage(s) of production, processing and distribution.

Traceability is defined as a system able to maintain a credible custody of identification for animals or animal products through various steps within the food chain, from the farm to the retailer (McKean, 2001; Dalvit et al., 2007).

Food business operators (FBOs) have a legal obligation to comply with food hygiene requirements and to establish and maintain traceability and quality (of food for human and animal consumption, as well as ingredients used in food) at all stages of production, processing and distribution. They are also responsible for developing a reliable food supply system, with the ultimate goal of providing consumers with a safe food product. The implementation and certification of basic food safety standards (HACCP, ISO 22000), as well as extended standards (FSSC 22000, IFS, BRC) at FBOs, makes this possible (Jurica et al., 2021). Croatian national criminal legislation (there is no European general criminal law) defines certain general terms such as fraud, processing and marketing of food harmful to human health, or endangering life and property, by generally dangerous acts or media (OG Kazneni Zakon (Criminal Law, in Croatian), 2013).

For an incident to be classified as food fraud in the EU, it must meet the following four criteria (EC, 2017). First, there must be a breach of one or more regulations related to food and feed safety legislation. Second, there must be an intention to substitute high quality ingredients with low-quality ingredients (for large quantities of products). This must be verified against numerous factors that provide strong evidence that certain non-compatibilities are not incidental. In fact, some possible impurities in production, when ingredients are substituted for non-essential ingredients, could indicate an intent to deceive (i.e., imply adulteration). Third, the existence of significant economic profits that follow from the addition of non-compatibilities to the product must be established. Fourth, there must be an element of deception of the consumer, which is the final criterion linking all the other factors to food fraud. This implies that some form of deception, such as color or label changes, has concealed the true quality of the product (or, in worst-case scenarios, even the hazardous nature of the food). For instance, the element of food fraud can often imply concern for public health by hiding some actual harmful product characteristics (e.g., allergenicity, toxicity).

At the Member State level, food safety and food fraud were addressed by Reg. 852/2004/EC (EC, 2004) for food hygiene, while Reg. 178/2002/EC introduced the concepts of crisis management and the obligation to adopt a General Food Crisis Management Plan with the European Food Safety Authority (EFSA). Reg. 178/2002/EC, Article 14 prohibits the placing on the market of unsafe or harmful food, or anything that is unsuitable for human consumption (taking into account additives and pesticides at concentrations above maximum residue levels [MRLs]) according to Reg. 396/2005/EC (EC, 2005). The Reg. 510/2006/EC serves the protection of geographical origin and designations of origin of agricultural products and foodstuffs (EC Regulation, 2006). Many foods and beverages in the EU are protected by geographical origin schemes, such as protected geographical indications (PGI), where at least one of the stages of production, processing or preparation must take place in a specific region, and protected designations of origin (PDO), where all steps must take place in the same specific region (EU Quality Food and Drink, 2021).

The European Food Fraud Network (EU FFN) was established in response to the horsemeat scandal, and it helps EU countries to operate in accordance with Reg. 625/2017/EC regarding official control. The Directorate General for Health and Food Safety (DG SANTE)–Unit G5 is responsible for communication regarding food fraud incidents.

15.5 CHINA

China has also achieved important progress in food tracing. In April 2004, the State Food and Drug Administration and seven other departments chose the meat industry as a pilot industry, started meat and meat products traceability institution construction and system implementation (General Administration of Quality Supervision, 2002). The main tasks include developing suitable technical standards and management norms, publishing guidelines for implementing traceability system including "Meat products tracking and tracing Guide" and "Fresh product tracking and tracing Guide". In June 2004, administration of national barcode management promoting investigated on vegetable products traceability and started an application project on two vegetable production bases located in Shouguang and Luocheng, respectively, in Shandong province (Zhou

Yingheng et al., 2002). The project was successful in food quality control in the origin and enforcing standards of market access, product identification and recall. Integrated with electronic auctions and e-commerce, this project established a traceability system for pollution-free vegetables. The Shanghai Livestock Bureau legislated to build digital archives for pigs, cattle, sheep and other animals, and the residents can now get access to egg production information through the internet (Shanghai agricultural committee, 2001). In August 2008, Beijing enforced a food traceability system along the full supply chain for the food supplied for the Olympic Games to secure food quality and safety (Liu et al., 2009; Chang Xiang, 2007).

15.5.1 Health Hazards of Adulterated Foods

Many harmful effects can occur due to adulterated foods. Adulterated foods may cause chronic liver disease including cirrhosis, colon and other cancers, peptic ulcer disease, liver failure, electrolyte imbalance, and kidney failure. Adulterated food may also cause blood disorders, bone marrow disorders and heart disease, in addition to problems that may appear in the skin, including allergy symptoms. This is in addition to the fact that food adulteration may lead to food poisoning, paralysis or eventually death (Manasha and Janani, 2016; Bansal et al., 2017).

There are many examples highlighting the health risks of adulterated foods, including the addition of mineral oils to edible oils, which can lead to paralysis, cancer and so on (Bansal et al., 2017; Vyralakshmi and Jayasheela, 2017). Moreover, artificial colors added to foodstuffs can be a cause of liver damage, allergies and many more (Vyralakshmi and Jayasheela, 2017). In addition, food adulteration sometimes produces immediate side effects such as diarrhea, dysentery and vomiting. For example, coffee powder is adulterated by replacement with tamarind or tamarind seed powder, and these can cause diarrhea (Lakshmi et al., 2012). In addition, coffee can be adulterated by replacement or substitution with dandelion powder, and hence it can cause stomach upsets, dizziness and joint pain (Bagalkoti and Bagali, 2017). Moreover, adulteration of red chili powder with brick powder can cause vision loss and respiratory ailments (Manasha and Janani, 2016).

15.5.2 Traceability

Food quality is strongly linked to food traceability. Therefore, the importance of applying traceability technology is not only limited to the final food product but must also be applied to all stages of production, starting from raw materials and intermediate products. Only in this direction will safety and quality of the product be safeguarded. Traceability is forward and backward.

Depending on the direction in which information is recalled in the chain, backward traceability or tracing is the ability, at every point of the supply chain, to find the origin and characteristics of a product based on one or several given criteria. In contrast, forward traceability or tracking is the ability, at every point of the supply chain, to find the locality of products from one or several given criteria. It is important for an information system to support both types of traceability as the effectiveness for one type does not necessarily imply the effectiveness for the other (Varzakas, 2016).

For a specific definition of what we call traceability, we can refer to the definition given by the International Organization for Standardization (ISO) in 1994 in which traceability is defined as "the ability to track any food, feed, food-producing animal or substance that will be used for consumption through all stages of production, processing and distribution" (ISO, 1994).

Olsen and Borit (2013) redefined traceability based on the ISO definition as the ability to access any or all information relating to that which is under consideration throughout its entire life cycle by means of recorded identifications. Karlsen et al. (2010) highlighted that traceability is not the product and process information itself but a tool that makes it possible to find this information again at a later date. However, these definitions do not reflect the specific characteristics of food traceability.

The revised definition of Bosona and Gebresenbet (2013) is a very informative and comprehensive definition of food traceability. Food traceability is defined as a part of logistics management that capture, store and transmit adequate information about a food, feed, food-producing animal or substance at all stages in the food supply chain so that the product can be checked for safety and quality control, traced upward and tracked downward at any time.

Opara (2003) classified traceability into six important elements: product traceability, process traceability, genetic traceability, input traceability, disease and pest traceability, and measurement traceability, focusing on agricultural and the food supply chain.

On the other hand, consumers' desire and satisfaction focus on constant quality and safety along with a clear geographical identity for each product they buy. Hence, traceability systems have recently become one of the most important techniques used to ensure food safety as well as food quality (El Sheikha, 2015). The European Union has been among the pioneers in this field by enacting relevant laws and legislation, including encouraging participation in the Protected Food Name Systems (protected designation of origin [PDO], protected geographical indication [PGI], and traditional specialities guaranteed [TSG]). It is worth noting in this regard that tracking systems have the ability to be integrated into data systems where the consumer can receive information about any food product. Tracking systems such as barcode systems also allow for efficient food product recall and allow for fewer food products to be withdrawn, which helps in achieving significant cost savings, for the whole food chain (consumers, manufacturers and authorities; El Sheikha and Montet, 2016).

The advantages arising from the right implementation of traceability systems in food processing plants are:

- Timely and trustworthy recovery of required information for the tracking down of problematic points in the operations of the business, possibility for better organization, best use of raw materials and cost reduction;
- Possibility for better quality control of raw materials and support of the requirements of the business against its suppliers;
- Improvement of the relations with the direct clients of the business that might not necessarily be the final consumers;
- Increased protection of the company from hostile actions on its products and possibility of support of the firm against food safety, control and judicial authorities;
- Conformance to the current and forthcoming legislative actions;
- Capability for manufacturing and marketing of products targeted to certain consumer groups;
- These systems give the required information for a better process control (e.g., best control of raw materials, stock control, production plan, quality control), for clients and external agencies;
- They aid in crisis management (problems location, tracking down of defective lots, withdrawal of defective lots);
- They can verify at any moment the claims of the firm for product specifications (quality, origin, non-GM ingredients; Regattieri et al., 2007).

15.6 LOT NUMBER AND TRACEABILITY

In the traceability chain, the lot number plays a significant part. Lot number is determined by the food manufacturer who has the responsibility for its accuracy and thoroughness. In food packaging, the indication is always labeled and before that the letter L.

Traceability in a system depends on the definition of the size of a lot or traceable resource unit (TRU). The size of a lot with reference to quantities or number of suppliers is determined by the criteria set by the company. Product recall should be allowed and records regarding manufacturing lots should be planned in such a way so that raw materials could be delivered.

If a definition of a lot number is general, then the lot coding is less informational and elaborate, hence the larger the recall in case of an emergency.

The balance between the complexity and functionality of a traceability system should be achieved with the smallest feasible lot size. Receipt dates, identification of warehouse and quantities received are the only way to trace a product in a warehouse.

An example of traceability is radio frequency identification (RFID) technology. In the food sector in particular, there are three very important advantages (Sahin et al., 2002):

- Better management of perishable products;
- Improvements in monitoring, locating and solving product quality problems;
- Improvement in management of withdrawal product in cases of risk.

RFID is mandatory in the EU in the case of goats, sheep (EC, 2004) and cows (EC, 2000).

RFID technology enables objects or persons to be identified with a single identifier, the information being transmitted to a receiving device (reader) using radio frequency waves (Landt, 2005). One of the most promising applications for industry is in the field of logistics, being able to track a specific object. The most widely used system for identifying objects are bar codes that were mainly defined by two standards, the European Article Number (EAN) in Europe and another with a similar aim in the United States, the Universal Product Code (UPC). These two standards are today combined into a single standard known as GS1 (www.gs1.org/).

A basic RFID system consists of an identification device (tag) attached to the object or person to be identified, a reader/recorder capable of reading and/or writing the tags and a protocol defining both the information format and the reading/writing procedure.

15.7 CONCLUSIONS

Food fraud is an increasingly problematic phenomenon whose presence may be significantly reduced by FBOs, by implementing food quality management system and food defense principles within production (ISO 22000:2018; FSCC 22000, BRC Food Standard 8, PAS 96, IFS Food Version 6.1, SQF Quality Code), in order for manufacturers that have an interest to retain the quality of the final food product, as well as consumers' trust. Respectable food industries have already implemented such food quality systems, and the final products are made from certified ingredients. However, less considerate parties inside the food chain could have different aims, as their goal might be to use cheaper components, so they change products in their most important attributes. Hence, they might lie about the ingredients/composition on packaging or give false and misleading statements on labels.

Such violations of food law are criminal offenses that must be prosecuted by the police and the court.

The adoption of stringent quality control techniques along with authenticity methodologies along with effective audits and inspections will prevent such phenomena and will increase the trust and confidence of consumers to the food industry and the authorities.

Other possibilities include close cooperation with certified and official public health or veterinary institutes with accredited (ISO 17025:2017) analytical methods to detect food fraud.

Possible challenges are expensive and demanding analytical techniques, analyzing large data sets, and perhaps a lack of food-related regulations, which should be addressed in the near future. Intentional adulteration or unauthorized food manipulations (UFM) could cause serious consequences to public health or economy. Therefore, the general public could lose confidence in food safety at the global level, but also in the government's ability to protect them. The solution is in increasing the levels of risk-based inspections, monitoring and implementation of food defense systems throughout the entire food supply chain, and better control of ingredients and their certifications.

REFERENCES

Bagalkoti, A. K., & Bagali, C. S. (2017). Food adulteration and its health hazards. *Paryeshana Int. J. Ayurvedic Res*, *1*(6), 1–6. www.pijar.org/articles/Arch_Vol1_Issue6/1.Dr.Anil%20Bagalkoti.pdf (accessed March 19, 2022).

Bansal, S., Singh, A., Mangal, M., Mangal, A. K., & Kumar, S. (2017). Food adulteration: Sources, health risks, and detection methods. *Crit. Rev. Food Sci. Nutr*, *57*(6), 1174–1189. DOI: 10.1080/10408398.2014.967834.

Bosona, T., & Gebresenbet, G. (2013). Food traceability as an integral part of logistics management in food and agricultural supply chain. *Food Control*. 33: 32–48.

CAC. (2006). *Principles for traceability/product tracing as a tool within a food inspection and certification system*. 60/2006. www.codexalimentarius.net/. . ./CXG_060e.pdf.

Consolidated Regulations of Canada (C.R.C). *Food and drug regulations*. Food and Drug Act, C.R.C., c. 802, Regulations are current to August 25, 2020 and last amended on July 1, 2020. https://laws-lois.justice.gc.ca/PDF/C.R.C.,_c._870.pdf (accessed March 19, 2022).

Council Regulation (EC) No. 21/2004 of 17 December 2003 establishing a system for the identification and registration of ovine and caprine animals and amending Regulation (EC) No 1782/2003 and Directives 92/102/EEC and 64/432/EEC (2003) 7. Regulation (EC) No 1760/2000 of the European Parliament and of the Council of 17 July 2000, establishing a system for the identification and registration of bovine animals and regarding the labelling of beef and beef products and repealing Council Regulation (EC) No 820/97 (2000).

Dalvit, C., Marchi, M.D., & Cassandro, M. (2007). Genetic traceability of livestock products: A review. *Meat Sci.*, 77: 437–449.

Djekic, I., Režek Jambrak, A., Djugum, J., & Rajković, A. (2018). How the food industry experiences and perceives food fraud. *Qual. Assur. Saf. Crop. Foods*, 10, 325–333.

EC European Commission. (2017). *The EU food fraud network and the system for administrative assistance & food fraud*, Annual Report, EC European Commission: Luxembourg.

EC Regulation No 396/2005. *Maximum residue levels of pesticides in or on food and feed of plant and animal origin*. Available online: https://eur-lex.europa.eu/legal-content/EN/TXT/PDF/?uri=CELEX:32005R0396&from=EN (accessed on February 14, 2021).

EC Regulation No 510/2006. *The protection of geographical indications and designations of origin for agricultural products and foodstuffs*. Available online: https://eur-lex.europa.eu/legal-content/EN/TXT/PDF/?uri=CELEX:32006R0510&from=EN (accessed on February 14, 2021).

EC Regulation No 852/2004. *The hygiene of the foodstuffs*. Available online: https://eur-lex.europa.eu/legal-content/EN/TXT/PDF/?uri=CELEX:32004R0852&from=HR (accessed on February 14, 2021).

El Sheikha, A. F. (2015). New strategies for tracing foodstuffs: Biological barcodes utilizing PCR-DGGE. *Adv. Food Technol. Nutr. Sci. Open J.*, *1*(1), 1–7. DOI: 10.17140/AFTNSOJ-SI-1-101.

El Sheikha, A. F., & Montet, D. (2016). How to determine the geographical origin of seafood? *Crit. Rev. Food Sci. Nutr.*, *56*(2), 306–317. DOI: 10.1080/10408398.2012.745478.

EU Quality Food and Drink. *Discover European food and drinks protected by geographical indication schemes*. Available online: https://ec.europa.eu/info/food-farming-fisheries/food-safety-and-quality/certification/quality-labels/eu-qualityfood-and-drink_en (accessed on February 2, 2021).

Everstine, K., Spink, J., & Kennedy, S. (2013). Economically motivated adulteration (EMA) of food: Common characteristics of EMA incidents. *J. Food Prot.*, 76, 723–735.

FDA. (2011). *Food safety modernization act (FSMA), 111th Congress Public Law 353*. U.S. Government Printing Office. www.fda.gov/Food/FoodSafety/FSMA/ucm247548.htm (accessed May 2, 2021).

Food Advisory Consumer Service (FACS). (2020). *Regulation of food safety and quality in South Africa*. https://foodfacts.org.za/regulation-of-food-safety-and-quality-insouth-africa/ (accessed March 19, 2022).

Food and Agriculture Organization (FAO). (2020). *Food safety and quality, capacity development, food regulations*. www.fao.org/food/food-safetyquality/capacity-development/food-regulations/en/ (accessed March 19, 2022).

Food and Drug Administration (FDA). (1992). *Chapter 9—Federal Food, Drug, and Cosmetic Act (Sections 301–399d)*, FDA U.S.: Silver Spring, MD.

Food Marketing Research & Information Center (FMRIC). (2006). *Toresabiritei shisutemu kaihatsu jisshou jigyou jigo chousa houkokusho* (Report on the Pilot Demonstration of Developing a Traceability System). Tokyo: FMRIC. www.jetro.go.jp/en/market/regulations/pdf/food-e.pdf.

Food safety authority of Ireland. (2004). *Traceability from farm to fork*. https://www.fsai.ie/faq/traceability.html (accessed December 2021).

Food Standards Agency (FSA). (2017). *Exports of food and drink: Information for businesses on how to export food and drink products to other non-EU countries*. www.food.gov.uk/business-guidance/exports-of-food-and-drink (accessed March 19, 2022).

FSA. (2002). *Traceability in the food chain, a preliminary study*. https://www.adiveter.com/ftp_public/articulo361.pdf (accessed December 2021).

General administration of quality supervision, inspection and quarantine of the People's Republic of China. GB/T 15425–2002, 2002.

GS1, www.gs1.org/.

Gvozdanovic, K., Margeta, P., & Margeta, V. (2017). Application of analytical methods for food authentication. *Stočarstvo*, 71, 29–38.

Hao, Cheng. (2007). Animal product safety control and traceability technologies. *Mod. Agricult. Sci. Tech.*, 13, 169–170.

Hutt, P. B. (1984). A history of government regulation of adulteration and misbranding of food. *Food Drug Cosmet. Law J.*, 39(1), 2–73.

Hyatt, P. (2016). *4 of the biggest challenges export managers face today*. www.tradeready.ca/2016/topics/import-export-trade-management/4-biggestchallenges-export-managers-face-today/ (accessed March 19, 2022).

International Organization for Standardization (ISO). (2017). *ISO and food: Great things happen when the world agrees*. www.iso.org/files/live/sites/isoorg/files/store/en/PUB100297.pdf (accessed March 19, 2022).

International Trade Centre. (2010, 3). Challenges in agri-food exports: Building the quality infrastructure. *International Trade Forum Magazine*. www.cottonguide.org/challenges-in-agro-food-exports-building-the-qualityinfrastructure/ (accessed March 19, 2022).

ISO Standard 8402: 1994 International Organization for Standardization (ISO) (1994), www.iso.org.

Johnson, R. (2014). *Food Fraud and "Economically Motivated Adulteration" of Food and Food Ingredients.* Congressional Research Service: Washington, DC.
Jurica, K., Karaconji, Irena Brcic, Lasić, Dario, Bursać Kovacevic, Danijela & Putnik, Predrag. (2021). Unauthorized food manipulation as a criminal offense: Food authenticity, legal frameworks, analytical tools and cases. *Foods*, 10, 2570. https://doi.org/10.3390/foods10112570.
Karlsen, K.M., Olsen, P., & Donnnelly, K.A. (2010). Implementing traceability: Practical challenges at a mineral water bottling plant. *Br. Food J.*, 112(2): 187–197.
Kyrova, V., Surmanova, P., Ostry, V., Rehurkova, I., Ruprich, J., & Jechova, M. (2017). Sea fish fraud? A confirmation of Gadoid species food labelling. *Br. Food J.* 119, 122–130.
Lakshmi, V., Labs, R. V., Guntur, & Pradesh, A. (2012). Food adulteration. *Int. J. Sci. Inven. Today,* 1(2), 106–113.
Landt, J. (2005). The history of RFID. *IEEE Potentials*, 24(4), 8–11.
Levinson, D.R. (2009). *Traceability in the food supply chain.* http://oig.hhs.gov/oei/reports/oei-02-06-00210.pdf (accessed August 12, 2021).
Liu, Junrong. (2007). An new legal requirement on international fish market in the near future—EU tracefish project. *Fisheries Science*, 4, 42~43.
Liu, L., Zheng, H., Meng, H., Hu, H., Wu, J. & Li, C., (2009). IFIP international federation for information processing, Volume 295, *Computer and Computing Technologies in Agriculture II, Volume 3*, eds. D. Li, Z. Chunjiang, (Boston: Springer), pp. 2265–2273.
Lu, Changhua, Wang Changjiang, Hu Sinong etc. (2007). *Identification and traceability system for animals and animal products.* Chinese Agricultural Science and Technology Press, pp. 35–36.
Manasha, S., & Janani, M. (2016). Food adulteration and its problems (intentional, accidental and natural food adulteration). *Int. J. Res. Financial Market*, *6*(4), 131–140.
McKean, J.D. (2001). The importance of traceability for public health and consumer protection. *Sci. Techn. Rev.*, 20(2): 363–371.
Ministry of Agriculture, Forestry and Fisheries (MAFF). (2004). *Shitteokitai shokuhin no toresabiritei (What You Need to Know about Traceability Systems).* Tokyo: MAFF. www.maff.go.jp/soshiki/syokuhin/hinshitu/e_label/.
NSF. (2014). *Risk modelling of food fraud motivation: 'NSF fraud protection model' intelligent risk model scoping project FS 246004: Final Report*, The National Science Foundation (NSF): Alexandria, VA.
OG Kazneni Zakon (Criminal Law, in Croatian). (2013). *Official Gazette 56/2013.* www.zakon.hr/z/98/Kazneni-zakon (accessed on October 22, 2021).
Olsen, P., & Borit, M. (2013). How to define traceability. *Trends Food Sci. Tech.*, 29, 142–150.
Opara, L.U. (2003). Traceability in agriculture and food supply chain: A review of basic concepts, technological implications, and future prospects. *Food, Agricult. Environm.*, 1(1), 101–106.
Regattieri, A., Gamberi, M., & Manzini, R. (2007). Traceability of food products: General framework and experimental evidence. *J. Food Engineer.*, 81, 347–356.
Regulation (EC) No 178/2002 of the European Parliament and of the Council of 28 January 2002 laying down the general principles and requirements of food law, establishing the European Food Safety Authority and laying down procedures in matters of food safety L31/1–24.
Robson, K., Dean, M., Haughey, S., & Elliott, C. (2021). A comprehensive review of food fraud terminologies and food fraud mitigation guides. *Food Control*, 120, 107516.
Sagener, N. (2016). *Dependence on food imports threatens developing countries*, report finds. www.euractiv.com/section/agriculture-food/news/dependence-onfood-imports-threatens-developing-countries-report-finds/ (accessed March 19, 2022).
Sahin, E., Dallery, Y., Gershwin, S. (2002). Performance evaluation of a traceability system. *Proceedings of IEEE International Conference on Systems, Man and Cybernetics*, 3, 210–218, ISSN: 1062–922X.
Setboonsarng, S., Jun Sakai, & Lucia Vancura. (2009). Food safety and ICT traceability systems: Lessons from Japan for developing countries. Asian Development Bank Institute No. 139. Tokyo: Asian Development Bank Institute. www.adbi.org/working-paper/2009/05/28/3012.ict.food.traceability.system/ (accessed 19 March, 2022).
Shanghai agricultural committee. *Notification in carrying out animal identification in Shanghai. Shanghai agricultural committee* (2001) NO:126, 2001–08–24.
Spink, J., Moyer, D.C. (2011). Defining the public health threat of food fraud. *J. Food Sci*, 76, 1–7.
Varzakas, T. (2016). ISO 22000, HACCP, and other management tools for implementation of food safety-traceability case studies. In *Handbook of Food Processing: Food Safety, Quality, and Manufacturing Processes* (Varzakas, T., and Tzia, K. eds.). Taylor and Francis, CRC.

Vyralakshmi, G., & Jayasheela, G. (2017). Food adulteration and contamination-a catastrophe. *IOSR J. Environ. Sci. Toxicol. Food Technol., 11*(7), 62–70. DOI: 10.9790/2402-1107016270.

Walker, M. *Food and feed law: Compendium of UK food and feed legislation with associated context and changes during January–March 2017.* Government Chemist Programme Report. https://assets.publishing.service.gov.uk/government/uploads/system/uploads/attachment _data/file/617645/Foodfeedlaw_Jan-Mar_2017_v3.pdf (accessed March 19, 2022).

Wenying, Xing. (2006). The Policy of Traceability in Quality and Safety of Agricultural Products in USA. *World Agricult.*, 4.

Xiang, Chang. (2007.07.09) *Three food quality and safety traceability systems for Beijing Olympic games will be put into use in August.* www.chinanews.com.cn/

Zhou, Yingheng, Geng, Xianhui. (2002). Application of traceability in food safety. *Res. Agricult. Moderniz.*, 6.

Zhu, Haipeng. (2007) Research and implementation on traceability system of key cereal enterprise. *Chinese Aca. Agricult. Sci.*, 10–11.

16 Conclusions

Sofia Agriopoulou, Maria Tarapoulouzi, and Theodoros Varzakas

Nowadays, the human diet is shifting toward more healthy choices, including plant-based foods (e.g., "meats") where all the attention is focused. Hence, quality tools should be improved in order to safeguard food safety, quality, traceability and avoid fraud and in parallel, enhance authenticity. In this direction, chemometrics and authenticity will aid the efforts of food authorities to track down all fraudulent methods and will give the signal to companies to avoid food contamination and present consumers with authentic foods with nutritional value and not foods adulterated with microorganisms or cheaper ingredients. Consumers require and desire nutritious and authentic products and also need to read correct labeling that is not misleading.

Having examined these consumer needs and having reviewed other books of authenticity and chemometrics of foods of animal origin, we editors decided to focus on foods of plant origin due to their great interest to consumers. This work involved an exploration of the most important food products of plant origin, from cereals, nuts, legumes, table olives and olive oil to coffee, tea, fruits and vegetables, fruit juices, spices, mushrooms, beers and wines, and honey.

Different authentication techniques have been described either alone or in combination with chemometrics.

We believe this book will be the basis for further analysis and illustration of more state-of-the-art techniques coupled with new chemometric tools towards continuous improvement, quality, and food safety, which should not be jeopardized in the context of money or power. The new era of transparency has begun.

DOI: 10.1201/9781003266167-16

Index

Note: numbers in **bold** indicate a table; numbers in *italics* indicate a figure

For Product Safety Concerns and Information please contact our
EU representative GPSR@taylorandfrancis.com Taylor & Francis
Verlag GmbH, Kaufingerstraße 24, 80331 München, Germany